Mathematical Foundations of Deep Learning Models and Algorithms

GRADUATE STUDIES IN MATHEMATICS 252

Mathematical Foundations of Deep Learning Models and Algorithms

Konstantinos Spiliopoulos
Richard B. Sowers
Justin Sirignano

AMERICAN MATHEMATICAL SOCIETY
Providence, Rhode Island

2020 *Mathematics Subject Classification.* Primary 62-01, 65-01, 68-01, 68T07.

Library of Congress Cataloging-in-Publication Data

Names: Spiliopoulos, Konstantinos, 1980– author | Sowers, R. B. (Richard Bucher), 1965– author | Sirignano, Justin (Justin Anthony), 1988– author

Title: Mathematical foundations of deep learning models and algorithms / Konstantinos Spiliopoulos, Richard Sowers, Justin Sirignano.

Description: Providence, Rhode Island : American Mathematical Society, [2025] | Series: Graduate studies in mathematics, 1065-7339 ; volume 252 | Includes bibliographical references and index.

Identifiers: LCCN 2025030859 | ISBN 9781470481087 hardcover | ISBN 9781470483999 paperback | ISBN 9781470483982 ebook

Subjects: LCSH: Deep learning (Machine learning)–Mathematical models | Neural networks (Computer science)–Mathematical models | Regression analysis | Convergence | Kernel functions | AMS: Statistics – Instructional exposition (textbooks, tutorial papers, etc.) | Numerical analysis – Instructional exposition (textbooks, tutorial papers, etc.) | Computer science – Instructional exposition (textbooks, tutorial papers, etc.)

Classification: LCC Q325.73 .S65 2025

LC record available at https://lccn.loc.gov/2025030859

Graduate Studies in Mathematics ISSN: 1065-7339 (print); 2376-9203 (online)

DOI: https://doi.org/10.1090/gsm/252

∞ The paper used in this book is acid-free and falls within the guidelines
established to ensure permanence and durability.
Visit the AMS home page at `https://www.ams.org/`

10 9 8 7 6 5 4 3 2 1 30 29 28 27 26 25

Contents

Preface

This book is an outgrowth of a belief that the mathematics and, in general, the scientific community might be well served by an introduction to deep learning and neural networks in the language of mathematics. To borrow from Churchill, Shaw, and Wilde, mathematics and computer science are two disciplines separated by common notation. We believe that this book might help students, researchers, and practitioners more easily see and explore connections to this increasingly important collection of computational tools and ideas. Over several years of research and teaching in neural networks, the authors have come to the conclusion that

- There are many interesting and open mathematical research questions in the field of deep learning.
- Mathematical maturity gives students and researchers an advantage in thinking about machine learning.

Mathematical thinking innately has unique strengths in generalizing and abstracting ideas and also providing rigorous bounds on complex phenomena. We believe that a greater mathematical presence in the field of deep learning and neural networks can in turn contribute to the larger scientific community.

This book is aimed at advanced undergraduate students and graduate students as well as researchers and practitioners who want to understand the mathematics behind the different deep learning algorithms. The book is composed of two parts. Part 1 contains a mathematical introduction, while Part 2 discusses more advanced mathematical and computational topics, hinting at further research directions. This represents something of a "separation of scales" in our effort: the basics of deep learning are "microscopic", while large-scale structural analysis is more "macroscopic". Our hope is that the combination of both points of view will offer a better comprehension of the topic.

The first part of the book (Part 1) assumes knowledge of basic linear algebra, multivariate calculus, and some statistics and calculus-based probability. This part of the book should be accessible to an advanced mathematics, statistics, computer science, data science, or engineering undergraduate student. Part 1 starts with some classical topics in statistical learning theory (such as linear regression, logistic regression, and kernels), then gradually progresses to deep learning related topics (such as feed forward neural networks, backpropagation, stochastic gradient descent, dropout, batch normalization), and concludes with a broad spectrum of deep learning architectures and models (such as recurrent neural networks, transformers, convolution neural networks, variational inference, and generative models). We have also included sections and chapters on classical statistical topics as regularization, training, validation and testing, and feature importance. The purpose of the earlier chapters in Part 1 is to introduce the reader to the subject of deep learning in a gradual way through classical topics in statistics and machine learning. These early chapters allow us to illustrate known issues that come up in deep learning architectures through easier-to-present concrete settings that often allow explicit computations, the latter being rarely the case for general deep learning architectures.

The second part of the book (Part 2) contains material that is more advanced than Part 1, either mathematically, conceptually, or computationally. This part of the book should be accessible to advanced undergraduate students and graduate students aiming to go deeper in certain topics of deep learning. Certain aspects of the second part of the book (e.g., uniform approximation theorems, convergence theory for gradient and stochastic gradient descent, linear regime and the neural tangent kernel, feature learning regime and mean field field scaling, neural differential equations) would be easier to read given a basic understanding of real analysis and stochastic process. A self-contained appendix with more advanced required background material has been included to aid the reader. Other aspects of the second part of the book, e.g., distributed training and automatic differentiation, require less mathematical background but are more advanced either conceptually or computationally.

Part 1, potentially together with selected topics from Part 2, could serve as standalone material for an advanced undergraduate course or for a first year graduate introduction to the mathematics of deep learning (we have done so in related course offerings in our respective universities).

The topic of deep learning is already huge and is constantly growing. While we have attempted to provide a fairly broad overview, we do not claim to have covered all possible angles. Our aim has been to cover topics that we viewed as important, foundational, and reasonably well-developed at the time of writing. We have tried to establish a unified and consistent mathematical language, connecting those topics in a comprehensive way and keeping in mind that deep learning is both mathematically interesting and a tool in applied data analysis.

Our efforts have concentrated around the idea of presenting essential ideas as clearly as possible. As such, we may not have always presented the sharpest possible versions of the results, but we have pointed to research articles and other monographs where the interested reader can find more refined results. No attempt has been made to provide comprehensive historical attribution of ideas. We do however give appropriate references which will hopefully provide entry points into the literature.

The book is focused on developing a mathematical language for deep learning and unifying the presentation of concepts and ideas but also maintaining rigorous mathematical results. The goal is not only to understand the mathematical principles behind deep learning algorithms, but also to offer tools to quantify uncertainty in deep learning. Two of the questions that this book is trying to answer are

Why do things work the way they work?

and

How can we guarantee significance and robustness of our conclusions?

Website

Beyond reading the mathematical literature, the readers of this book will hopefully have the opportunity to experiment with the algorithms presented. Deep learning is a tool in data analysis. For the reader's convenience we have included Python code which will hopefully give the reader some appreciation for how deep learning might actually be used in practice. The datasets and Python codes referenced in various chapters of the book can be found and downloaded at the dedicated website for the book

https://mathdl.github.io/.

In addition, corrections and errata to the book will be updated there.

In many of the chapters of the book, exercises have been included to aid the reader in better comprehension of the material. Upon request, a solutions manual is available to the instructor of a class using this book.

We hope that this book will help open a door through which the mathematics and research community can pass in order to contribute even more to the ever-growing field of research and applications of deep learning. Whatever goals have motivated us to write these chapters, we admit to partial success only.

Acknowledgments

The authors of the book would like to thank family and friends who contributed in their own unique ways to the completion of the book.

- Konstantinos Spiliopoulos would like to thank first and foremost Anastasia for her constant support and patience, and his children Stavros and Evangelia for being a continuous driving force and source of inspiration!
- Richard Sowers would like to thank his students, who helped him learn deep learning. In particular, Rachneet Kaur, through a series of papers [**HBK$^+$20, KCM$^+$20, KKHS20, KLM$^+$23, KMZ$^+$19, KMSH22, KST$^+$20, KSZ$^+$19, SKZ$^+$19**], helped him bridge the gap from mathematical thinking to applications. Richard Sowers would also like to thank his wife Svetlana, and daughters Rebeka and Veronika. G. B. Shaw once opined that "A happy family is but an earlier heaven."

Finally, we would like to thank our friends, colleagues and students who looked at earlier versions of the manuscript and made many useful suggestions. In particular, we would like to thank in alphabetical order Nikos Georgoudios, Paul Glasserman, Samuel Isaacson, Markos Katsoulakis, Fotios Kokkotos, Eric Kolaczyk, Scott Robertson, Giorgos Zervas, Benjamin Zhang, and Zhuo Zhao. All of the remaining mistakes and imperfections are our own responsibility.

While working on this book, the authors were supported by the National Science Foundation (NSF) in the USA and the Engineering and Physical Sciences Research Council (EPSRC) in the UK (grant NSF-DMS 2311500).

KONSTANTINOS SPILIOPOULOS
Boston, MA, USA

RICHARD SOWERS
Urbana-Champaign, IL, USA

JUSTIN SIRIGNANO
Oxford, UK

May 2025

Notation

In this section we provide a reference list of the notation that is consistently used throughout the book. Most of the mathematical objects defined here are described in Chapter 1, Introduction.

Parameters

- θ: learnable parameter to be estimated from data
- η: learning rate in (stochastic) gradient descent methods
- M: number of datapoints in a given dataset
- N: number of hidden units in a given neural network layer

Data

- $(x_m)_{m=1}^M$: feature data
- $(y_m)_{m=1}^M$: label data
- $\mathcal{D} = \{(x_m, y_m)\}_{m=1}^M$: available dataset
- $\mathcal{D}_{\text{train}}$: train dataset
- $\mathcal{D}_{\text{test}}$: test dataset

Spaces

- $\mathbb{R}^D$: the Euclidean space of D-dimensional tuples of real numbers (typically arranged as column vectors)
- $\mathbb{R}^{D_1 \times D_2}$: the collection of $D_1 \times D_2$ matrices of real numbers

- $\mathcal{X}$: the space of features
- $\mathcal{Y}$: the space of labels
- Θ: the parameter space
- $L^q(\mathbb{R}^D)$ with $1 \le q < \infty$: the space of functions f on $\mathbb{R}^D$ such that

$$\|f\|_q = \left(\int_{\mathbb{R}^D} |f(x)|^q dx\right)^{1/q} < \infty$$

- $L^\infty(\mathbb{R}^D)$: the space of functions f on $\mathbb{R}^D$ such that

$$\|f\|_\infty = \sup_{x\in\mathbb{R}^D} |f(x)| < \infty$$

Norms

- $\|f\|_q = \left(\int_{\mathbb{R}^D} |f(x)|^q dx\right)^{1/q}$: the $L^q(\mathbb{R}^D)$ norm of a function f taking values in $\mathbb{R}^D$ for $1 \le q < \infty$
- $\|f\|_\infty = \sup_{x\in\mathbb{R}^D} |f(x)|$: the $L^\infty(\mathbb{R}^D)$ norm of a function f taking values in $\mathbb{R}^D$
- $\|x\|_q = \left(\sum_{i=1}^D x_i^q\right)^{1/q}$ for $q > 0$ and $x \in \mathbb{R}^D$
- $\|x\| = \|x\|_2$ for $x \in \mathbb{R}^D$
- $\|x\|_\infty = \max_{i=1,...,D} |x_i|$ for $x \in \mathbb{R}^D$

Functions

- $\sigma(x)$: activation function
- $S(x)$: logistic function
- $S_{\text{softmax}}(x)$: softmax function
- $\mathsf{ReLU}(x)$: rectified linear unit function
- $\bar{\mathfrak{m}}(x)$: function we want to estimate/learn
- $\mathfrak{m}(x;\theta)$: parametric model for $\bar{\mathfrak{m}}(x)$ with $\theta \in \Theta$ being the learnable parameter
- $\{\ell_y\}_{y\in\mathcal{Y}}$: collection of error functions
- $\lambda_{(x,y)}(\theta) \stackrel{\text{def}}{=} \ell_y(\mathfrak{m}(x;\theta))$: per datapoint loss function
- $\Lambda(\theta) \stackrel{\text{def}}{=} \frac{1}{|\mathcal{D}|}\sum_{(x,y)\in\mathcal{D}} \lambda_{(x,y)}(\theta)$: average (empirical) loss function
- $\Lambda_{\text{pop}}(\theta)$: population loss function

- $\mathbf{1}_A(z) \stackrel{\text{def}}{=} \begin{cases} 1 & \text{if } z \in A \\ 0 & \text{else} \end{cases}$: indicator function, where z is taken to be in some set S and A is a subset of S

Probabilities and Expectations

- $\mathbb{P}$: probability measure
- $\mathbb{E}$: expectation operator associated with a given probability measure
- $P_{\mathcal{D}}$: empirical probability measure
- $E_{\mathcal{D}}$: associated expectation operator

Special operations

- $(x \odot y)_d = x_d y_d$ for all $1 \leq d \leq D$: Hadamard multiplication for x and y in $\mathbb{R}^D$

Chapter 1

Introduction

1.1. Preliminaries

Deep Learning is a collection of ideas, algorithms, and computational implementations concerned with constructing functions, roughly, of the form

$$\begin{aligned}&(\text{Nonlinear function}) \circ (\text{linear transformation}) \circ (\text{Nonlinear function}) \circ \\ &\circ (\text{linear transformation}) \circ \cdots ; \end{aligned} \tag{1.1}$$

i.e., the composition of *layers* of linear transformations and nonlinear functions. The nonlinear functions in (1.1) are usually fairly simple, e.g., a vectorized logistic function or hyperbolic tangent. The *internal layers* may in fact be very high-dimensional, and there may be many layers. The iterative and large-scale combination of these simple operations yields a very powerful and flexible model for input-output relations.

Neural networks are not new; they have been used as far back as the 1980s. However, the real impact of deep learning started becoming apparent only in the late 2000s. Since then, deep learning has come to dominate some of the most important areas of machine learning, such as image, text, and speech recognition, and it is poised to have a significant impact on many other applications across science, engineering, medicine, and finance.

What precipitated the rapid rise of deep learning? It can be attributed to the fortuitous combination of several things:

- It turns out that the functional framework of (1.1) is broad enough (see Chapter 16, Universal Approximation Theorems) that it can approximate any "reasonable" function.

- The parameters in the linear transformations need to be *tuned*, usually by minimizing an appropriate error function on ground-truth data. More mathematically, formulas for sensitivities (derivatives) with respect to the parameters determining the linear transformation of (1.1) are used in gradient descent algorithms. Notwithstanding the number of parameters in large-scale modern networks (large networks of the form (1.1) with many internal layers may have millions to billions of parameters), these calculations can be written in a scalable and parallelizable way (Chapters 6, 7, 8, 23, and 24).
- High-dimensional neural networks and large training datasets require significant computational resources; modern widely available computational tools (viz., GPUs) can meet these needs.

Neural networks that are specifically of type (1.1) are called *feed forward*, and are the starting point for understanding deep learning (Chapter 5). More complicated networks, e.g., recurrent neural networks for sequential data, transformers (Chapter 13) and convolution neural networks (Chapter 14), take advantage of some unique structure in the problem. Although a given neural network is often too complex for rigorous statistical analysis, it typically is the combination of smaller classical parts (which on their own can be rigorously studied).

Deep learning is a departure from traditional statistics, which emphasizes hypothesis tests, confidence intervals, and other statistical properties. Despite the lack of such statistical guarantees, deep learning has had remarkable success. Although there are ad hoc aspects of deep learning, many of its advances are the result of careful and thoughtful design of network architectures and training methods. Examples include convolution networks for image recognition, long short-term memory (LSTM) networks for text recognition and time series data, transformers for machine translation and large language models, and more.

This book aims to cover the fundamental concepts underpinning deep learning and provide the computational methods to implement deep learning models. It focuses on mathematical principles but at the same time it aims to understand the practical components of training neural networks successfully as well as the possible factors that can cause the failure of training.

The focus of this book is mainly on the following topics of deep learning:

- *Approximation*: What types of functions and problems can be approximated and solved by neural networks?
- *Optimization and training*: How do we decide what types of architectures to use for a given problem? How do we train such models?

- *Generalization*: Assume that a finite dataset was used to obtain a good model approximating the unknown function of interest. Will this model perform well on data that have not been observed?

Deep learning has been very successful in many applications, but there are no guarantees that it will easily work in any given situation. Mathematical theory helps to provide rigorous guidelines for design of algorithms and also sheds light onto convergence properties of the algorithms.

1.2. Brief Historical Review of Deep Learning

Deep learning is part (a very significant part, in fact) of the much larger field of machine learning. In order to appreciate the scientific developments, let us briefly go over some of the history of deep learning.

It is largely accepted that the field of deep learning and the deployment of neural network models to describe functional relationships find their roots in neuroscience, starting with the work of McCulloch and Pitts in 1943 [**MP43**]. That work introduced the idea that simple components of our brain work together to perform complicated tasks. In such models, linear combinations of neurons become inputs to a nonlinear function, forming the basis of what is now called a neural network model.

The next big breakthrough was the work of Rosenblatt in 1958 on the perceptron model [**Ros58**]. In its basic form, the perceptron model is a single neuron for binary classification

$$\begin{aligned}\mathfrak{m}(x; w) &= \operatorname{sign}(w_1 x_1 + \cdots + w_D x_D + b) \\ &= \begin{cases} -1 & \text{if } w_1 x_1 + \cdots + w_D x_D + b < 0, \\ 1 & \text{if } w_1 x_1 + \cdots + w_D x_D + b \geq 0, \end{cases}\end{aligned}$$

where the weights w and the bias b are trainable parameters, and the x_d's are input signals. In this model the neuron fires if the output of the addition operation $w_1 x_1 + \cdots + w_D x_D + b$ is larger than a threshold (taken to be zero here). The perceptron model is a true milestone in deep learning, and we will study it in Chapter 4. In today's language the perceptron model is a single layer neural network.

When initially conceived, the perceptron model was very influential especially due to its ability to learn from data in a way similar to the human brain. However, soon it was realized that the perceptron model has limitations due to being a single-layer neural network with a single hidden unit, leading to difficulties when asked to distinguish between complicated data. In particular, the properties of perceptrons were analyzed in the book of Minsky and Rapert in 1969 [**MP17**] and their limitations were explored and pointed out. These limitations partially dampened interest in perceptrons and related models of artificial intelligence.

A second challenge with neural networks was in training; there was no effective and scalable way to find the weights and biases. The paper by Rumelhart, Hornik, and Williams in 1986 [**RHW86**] on backpropagation (see Chapter 6) resolved this impasse, iteratively using the chain rule of elementary calculus to obtain the gradient of the loss function with respect to model parameters. This led to a resurgence of interest in the field. Again, however, development slowed when applications to real problems increasingly required ad hoc processing and fine-tuning.

The story then again changed in the early 2000s with advances in computational power and growing amounts of available data. *Graphical processing units* (GPUs), initially developed for handling linear algebraic calculations needed for visualizations and video games in particular, turned out to be ideally suited to the calculations of training deep neural networks. Perhaps the first breakthrough was the work of Krizhevsky, Sutskever, and Hinton in 2012 [**KSH12**], where a convolutional neural network (CNN) was successfully trained to classify images from a large image dataset. At that point in time deep neural networks caught the attention of many people. Another breakthrough in 2017 was the paper [**VSP$^+$17**] where the authors proposed combining an *attention* mechanism and a feed forward neural network into what is now known as *transformer* architecture; this is the backbone of many very successful large language models.

Currently, very large models are trained on thousands of GPUs and can perform very complicated tasks. Deep learning is used in many practical applications ranging from machine translations to strategy games to image classification to solution of complex equations (such as high dimensional partial differential equations, optimization equations, etc.) to robotics, chemistry, biology, finance, and the list goes on and on. Typical applications have data that may be very high-dimensional but many have low-dimensional structure.

1.3. Overview and Notation

Mathematically, we are interested in maps from some space $\mathcal{X}$ to another space $\mathcal{Y}$. These maps are *parametrized* by elements of some other space Θ. We shall generically denote such a structure as $\mathfrak{m} : \mathcal{X} \times \Theta \to \mathcal{Y}$. We want to find a parameter $\theta^* \in \Theta$ such that the map $x \mapsto \mathfrak{m}(x; \theta^*)$ achieves some desired goal. We will often write $\mathfrak{m}(x; \theta)$ instead of $\mathfrak{m}(x, \theta)$ to emphasize the difference in the roles of x and θ; x refers to a point in the feature space, while θ refers to a tunable parameter.

In the case of supervised learning, $\mathcal{X}$ is usually the space of features and $\mathcal{Y}$ is the space of labels; we want to use $\mathfrak{m}(\cdot;\theta^*)$ to predict the label for future (out-of-sample) points in feature space. We want to find θ^* based on a collection of *ground-truth* feature-label pairs. These ground-truth points can typically be enumerated as a sequence $(x_m)_{m=1}^M$ of points in $\mathcal{X}$ and a corresponding sequence $(y_m)_{m=1}^M$ of points in $\mathcal{Y}$. When we are proving performance bounds, the pairs (x_m, y_m) for $m = 1, \dots, M$ are typically random samples originating from some *ground-truth distribution* on $\mathcal{X} \times \mathcal{Y}$. Since

- a pair (x_m, y_m) may be repeated (e.g., two cars with the same mileage (feature) may need the same repair (label)), and
- the ordering of the (x_m, y_m)'s is unimportant,

we can also think of the (x_m, y_m)'s as a *multiset* (an unordered collection allowing multiplicities) $\mathcal{D}$ of points in feature-label space $\mathcal{X}\times\mathcal{Y}$. Visualizing the ground-truth data as points in $\mathcal{X} \times \mathcal{Y}$ emphasizes that supervised learning is essentially a question of drawing a graph *near* an existing collection of existing points.

In the case of supervised learning, we want $\mathfrak{m}(x;\theta)$ to approximate y for as many ground-truth datapoints $(x, y) \in \mathcal{D}$ as possible. To quantify this, we will usually use a collection $\{\ell_y\}_{y\in\mathcal{Y}}$ of error functions. For each $y \in \mathcal{Y}$, $\ell_y : \mathcal{Y} \to [0, \infty)$ (preferably in some smooth way), with $\ell_y(y') = 0$ if and only if $y = y'$. For each ground-truth datapoint $(x, y) \in \mathcal{D}$ and $\theta \in \Theta$, the *per datapoint* loss function

$$\lambda_{(x,y)}(\theta) \stackrel{\text{def}}{=} \ell_y(\mathfrak{m}(x;\theta))$$

then quantifies how well $\mathfrak{m}(x;\theta)$ matches up with y.

The per datapoint loss functions are then aggregated into the *average* loss function

$$\Lambda(\theta) \stackrel{\text{def}}{=} \frac{1}{|\mathcal{D}|}\sum_{(x,y)\in\mathcal{D}} \lambda_{(x,y)}(\theta) = \frac{1}{M}\sum_{m=1}^{M} \lambda_{(x_m,y_m)}(\theta)$$

for each $\theta \in \Theta$. Sometimes, $\Lambda(\theta)$ as defined in the formula above, is also referred to as the empirical loss function. The parameter space Θ is usually Euclidean, so we can then try find the *best* parameter θ^* by some type of gradient descent on Λ.

Some of our notation and organization of thoughts differs from that of standard computer science literature; e.g., multisets, per datapoint, and average loss functions. We regularly force the ground-truth data into subscripts (e.g., ℓ_y and $\lambda_{(x,y)}$) to emphasize that we shouldn't differentiate with respect to x and y. Our notation reflects the privilege of trying to holistically present a developing field to a new audience. We believe that, once the ideas have been understood, any translation of notation should be fairly natural.

Much of our mathematical notation is standard. We use $\mathbb{R}^D$ to denote the Euclidean space of D-dimensional tuples of real numbers (typically arranged as column vectors) and $\mathbb{R}^{D_1 \times D_2}$ to denote the collection of $D_1 \times D_2$ matrices of real numbers. We use $\odot$ to denote Hadamard multiplication; for x and y in $\mathbb{R}^D$, $(x \odot y)_d = x_d y_d$ for all $1 \le d \le D$ (and similarly for matrices as needed). We also use

$$\mathbf{1}_A(z) \stackrel{\text{def}}{=} \begin{cases} 1 & \text{if } z \in A \\ 0 & \text{else} \end{cases}$$

to denote the indicator function, where z is taken to be in some set S and A is a subset of S.

For $1 \le q < \infty$, we say that a function $f \in L^q(\mathbb{R}^D)$, if f is such that $\|f\|_q = \left(\int_{\mathbb{R}^D} |f(x)|^q dx\right)^{1/q} < \infty$. If $q = \infty$, then $\|f\|_\infty = \sup_{x \in \mathbb{R}^D} |f(x)|$.

If $x \in \mathbb{R}^D$, then we define $\|x\|_q = \left(\sum_{i=1}^D x_i^q\right)^{1/q}$ for $q > 0$. When $q = 2$, $\|x\| = \|x\|_2$ is the standard Euclidean norm. In addition, we shall write $\|x\|_\infty = \max_{i=1,\ldots,D} |x_i|$.

Various algorithms rely on randomization, and a number of our advanced (rigorous) results are probabilistic. Generically, we can think of this randomness as coming from some call to numpy.rand on a computer.

For a given dataset $\mathcal{D} = \{(x_m, y_m)\}_{m=1}^M$, an element $(x_m, y_m) \in \mathcal{D}$ is typically considered to be coming from a given ground-truth distribution. We will use $\mathbb{P}$ to denote the background probability measure of this randomness and use $\mathbb{E}$ to denote the associated expectation operator. Very often, we will want to think of the empirical measure generated by the ground-truth data. We will use $P_\mathcal{D}$ and $E_\mathcal{D}$ to denote the associated empirical probability measure and the associated expectation operator, respectively. We use the distinct notation of $\mathbb{P}$ vs. $P_\mathcal{D}$ to reinforce the fact that applications of deep learning in practice are governed by a fixed dataset. This dataset may have been collected at great cost and, consequently, may be of limited size. Gold standard ground-truth datasets from academia are typically also collected under carefully thought-out protocols.

1.4. On the General Task of Machine Learning

Let us be a bit more explicit. Assume that we have observations $\{(x_m, y_m)\}_{m=1}^M$ of *ground-truth* points in $\mathcal{X} \times \mathcal{Y}$ and we want to model how the y_m's depend on the x_m's. We would like to do so by considering a collection $\{\mathfrak{m}(\cdot; \theta)\}_{\theta \in \Theta}$ of maps from $\mathcal{X}$ to $\mathcal{Y}$ and then learning the best parameter θ.

Supervised learning addresses the case where both input data (the x_m's) and output data (the y_m's) are available. Supervised learning is the main focus of this book and we briefly review the general structure in Section 1.5.

Unsupervised learning is concerned with the case where only the input data (the x_m's) are available, namely no labels are given to the learning. In contrast to supervised learning, unsupervised learning tries to find patterns in data (i.e., there are no designations as feature or label). *Generative Adversarial Networks* (GANs) in Chapter 15 are something of an exception; part of the GAN problem relies on algorithmically finding patterns in data.

Different machine learning algorithms make different choices for the model $\mathfrak{m}(x;\theta)$. They may also use different methods to estimate θ.

An important factor in the success of a particular choice of model for $\mathfrak{m}(x;\theta)$ depends on the suitability of the chosen class of models with respect to complexity of the data that we are dealing with and on the amount of data M that is available. For example, if $x \mapsto \mathfrak{m}(x;\theta)$ is a linear model, i.e.,

$$\mathfrak{m}(x;\theta) = \theta^\top x \tag{1.2}$$

and the data suggest quadratic behavior, there is little hope that $\mathfrak{m}(\cdot;\theta)$ will work. Complex, nonlinear relationships require general models $\mathfrak{m}(x;\theta)$ which are able to capture a rich enough collection of nonlinearities. Even if we do choose a correct general model, for example $\mathfrak{m}(x;\theta) = x^\top \theta x$, in the case of quadratically related data, it may be challenging to accurately estimate θ if the number of data samples M is very small. For example, suppose that the feature space $\mathcal{X}$ is d-dimensional. Even in this simple setting, the model has $d \times d$ degrees of freedom (θ is a $d \times d$ matrix) and a large number of data samples M is required to accurately estimate θ. If the number of data samples $M \ll d$, the model $\mathfrak{m}(x;\theta)$ will be inaccurate. Inaccuracy due to the dataset being much smaller than the model's degrees of freedom can oftentimes lead to *overfitting*; see Subsection 1.5.3.

The art in deep learning is trying to find the right collection of models. While computational methods and capacity are increasingly large, they are finite, as is the available amount of training data. In addition, in real-world applications, not only the exact functional form of the underlying relationship suggested by the data is unknown, but also the feature space $\mathcal{X}$ is typically very high-dimensional. It is impractical to consider *all* maps from feature space to label space. If the amount of training data is too small compared to the number of model parameters, we may *overfit*, potentially finding parameters which work well for the training data, but don't work for statistically similar new data (this is related to questions of *hyperparameter* selection, covered in Chapter 11).

There are broad epistemological properties which can guide in model selection. *Translation-invariance* is often appropriate for image classification; one wants to detect an object in an image regardless of its location in the image. Many systems are naturally *causal*; future events can't influence past or present events. *Convolutional neural networks* (Chapter 14) reflect translation invariance, while *recurrent neural networks* (Chapter 13) reflect causality. *Transformers* (Chapter 13) can also learn causality in data. In all of these cases, these architectures can be appropriately high-dimensional, but they are much lower-dimensional than the collection of *all* maps on, for example, raw pixel space or raw historical data or text in, say for instance, the English language.

Many widely used models can be iterated and can have many internal dimensions. One generally wants to consider model architectures that can capture intricate patterns in the data with corresponding model dimensions which are large enough to describe all reasonable maps from feature to labels but that at the same time fit the computational constraints.

Deep learning not only uses models that have a large number of degrees of freedom (multi-layer neural networks), but it has also been able to embed properties such as translation-invariance or causality into its model architectures. In addition, carefully designed optimization and statistical methods have been developed to control issues like overfitting. The combination of these aspects has been an important factor in the success of deep learning.

1.5. Quick Overview of Supervised Learning

Let us expand a bit on the task of supervised learning. Consider the problem of finding a model $\mathfrak{m}(x;\theta)$ for given data $\{(x_m, y_m)\}_{m=1}^M$. At a high level, we are *searching* for a parameter $\theta \in \Theta$ such that, on average across the data samples, $\mathfrak{m}(x_m;\theta)$ is *close* to y_m. Otherwise said, we want the graph of $\mathfrak{m}(\cdot;\theta)$ to optimally pass through the ground-truth data. That is,

$$\theta = \operatorname*{argmin}_{\theta' \in \Theta} \frac{1}{M} \sum_{m=1}^{M} \ell_{y_m}(\mathfrak{m}(x_m;\theta')),$$

where $\ell_y(y')$ is a measure of how close the prediction y' is to y or, stated otherwise, the error being made when y' is used to predict y (we will often normalize loss functions by the number of datapoints being considered; this gives scale invariance in the size of data). Recall that for a generically given function g with domain $\mathcal{Z}$, $\operatorname{argmin}_{z\in\mathcal{Z}} g(z)$ is the minimizer of g:

$$\operatorname*{argmin}_{z\in\mathcal{Z}} g(z) \stackrel{\text{def}}{=} \{z \in \mathcal{Z} : g(z) \leq g(z') \text{ for all } z' \in \mathcal{Z}\}.$$

Setting up this problem correctly involves several mathematical challenges. For example, for a given problem at hand:

- What is an appropriate loss function $\lambda_{(x,y)}(\theta) = \ell_y(\mathfrak{m}(x;\theta))$ to use?
- What is an appropriate class of models $\mathfrak{m}(x;\theta)$ to use?

- How should we solve the minimization problem to obtain the optimal value for the parameter θ?
- What if the dimension of the parameter space Θ and/or the number of datapoints M are very large? What if the number of available datapoints M is not large enough?
- How do we know whether our constructed model works well in new, unseen data from the same or a somewhat different class of problems?

The choices that the user makes for the error function, for the model and for the minimization algorithm to use can have a profound effect not only on how well the model works on in-sample or out-of-sample data, but also on the computational complexity of the resulting algorithm. In short, these choices (choices that the user needs to make) are of paramount importance as they impact the ability of the model to produce accurate predictions. Oftentimes this set of choices (or assumptions) that the algorithm uses to produce predictions goes by the name *inductive bias*.

In this book, we attempt to provide insights to these questions through a rigorous mathematical formulation. As a warm-up example let us briefly visit, in the next two subsections, two classical settings: the regression problem and the classification problem.

1.5.1. The regression problem. The choice of $\ell_y(y')$ depends upon the application. For example, suppose that $\mathcal{X} = \mathbb{R}^d$ and $\mathcal{Y} = \mathbb{R}$, and $\mathfrak{m}(x;\theta) : \mathbb{R}^d \to \mathbb{R}$. Then, a suitable choice for $\ell_y(y')$ would be the squared error $(y-y')^2$ and $\theta \in \Theta$ is the estimator from the familiar least-squares problem

$$\theta = \operatorname*{argmin}_{\theta' \in \Theta} \frac{1}{M} \sum_{m=1}^{M} (y_m - \mathfrak{m}(x_m;\theta'))^2 . \tag{1.3}$$

Defining a *loss* function

$$\Lambda(\theta) = \frac{1}{M} \sum_{m=1}^{M} \ell_y(\mathfrak{m}(x_m;\theta)), \tag{1.4}$$

the problem (1.3) leads to the more general problem of calibrating a model by minimizing a function. See Chapter 2 for a more comprehensive treatment of linear regression.

A *shallow* neural network adds a nonlinearity and is defined as follows:

$$\mathfrak{m}(x;\theta) = \sum_{n=1}^{N} c^n \sigma(w^n \cdot x + b^n), \tag{1.5}$$

where

- σ (typically nonlinear) is called the activation function.

- N is the number of hidden units.
- $\theta = (c^n, w^n, b^n)_{n=1,\dots,N}$ is the vector of parameters that need to be estimated from data.

If σ is the identity map, this reduces to a linear model $\mathfrak{m}(x;\theta) = w \cdot x + b$. We can again try to minimize (1.4), leading to parameter selection as in (1.3). Typically, we use *gradient descent* to iteratively reduce (1.4); we consider the recursion

$$\theta_{k+1} = \theta_k - \eta \nabla \Lambda(\theta_k), \tag{1.6}$$

where $\eta > 0$ is called the learning rate. We run this algorithm until we are convinced that we have converged.

Tracing through (1.4), differentiating Λ with respect to θ involves differentiating $\theta \mapsto (y_m - \mathfrak{m}(x_m;\theta))^2$ for each m. In more complicated (i.e., deeper) neural networks, each such computation may be computationally expensive (Chapter 6). *Stochastic gradient descent* (Chapters 7, 8, and 18) provides a theoretically justified way to randomly break this into *batches*, each one of which is more computationally viable.

1.5.2. The classification problem. Another common class of applications are *classification* problems where $\mathcal{Y}$ is a categorical (i.e., discrete-valued) variable. Suppose, for example, we are trying to decide if an image contains a car or not. Let's assign label 1 if the image contains a car, and label 0 if not, so that our label space is $\mathcal{Y} = \{0, 1\}$.

At a high level, we would like to calibrate a parametrized mapping from $\mathcal{X}$ to $\mathcal{Y}$. This is a bit problematic, however, now that the label space is discrete. The iterative calibration procedure (1.6) depended on the parametrized models being differentiable in the modeling parameter; this is impossible if the label space is discrete (a small change in the model can't change the label by only a *small* amount). For binary logistic regression, we instead take the label space to be $(0, 1)$, which we interpret as the probability of label 1. This is indeed better as $(0, 1)$ is a continuum. Informally, we would like to calibrate a parametrized probability $\mathfrak{m}(x;\theta) \in (0, 1)$ for the probability that the label is 1 if the feature value is x; more formally, we want

- $\mathfrak{m}(x;\theta)$ is the probability that the label is 1 if the feature is x.
- $1 - \mathfrak{m}(x;\theta)$ is the probability that the label is 0 if the feature is x.

Classical detection theory [**MV19**] then suggests model calibration by *maximum likelihood*; what is the parameter which gives observed data the highest probability? Namely, given ground-truth data $\{(x_m, y_m)\}_{m=1}^M$ of a collection of points in $\mathcal{X} \times \mathcal{Y} = \mathbb{R} \times \{0,1\}$, let's try to find

$$\text{argmax}_{\theta \in \Theta} \prod_{m=1}^{M} \mathfrak{m}^{y_m}(x_m;\theta)(1-\mathfrak{m}(x_m;\theta))^{1-y_m}. \tag{1.7}$$

This formula uses the convenience that

$$\mathfrak{m}^y(x;\theta)\,(1-\mathfrak{m}(x;\theta))^{1-y} = \begin{cases} \mathfrak{m}(x;\theta) & \text{if } y=1 \\ 1-\mathfrak{m}(x;\theta) & \text{if } y=0; \end{cases}$$

i.e., the left-hand side is the modeled probability that the label is y if the feature is x. In (1.7) we are then maximizing the probability that the sequence of labels is $(y_1, y_2 \cdots y_M)$ if the sequence of features is $(x_1, x_2 \cdots x_M)$ (implicit here is also an assumption that, conditioned on the feature sequence, the labels are independent).

With a few transformations we can rewrite (1.7) to look like the structure of Subsection 1.5.1. We can take logarithms (which preserve ordering) to get an additive structure like (1.4); (1.7) is equivalent to

$$\text{argmax}_{\theta \in \Theta} \sum_{m=1}^{M} \{y_m \ln \mathfrak{m}(x_m;\theta) + (1-y_m)\ln(1-\mathfrak{m}(x_m;\theta))\}.$$

We can switch also signs and normalize to get a minimization problem,

$$\underset{\theta \in \Theta}{\text{argmin}} \frac{1}{M} \sum_{m=1}^{M} \{-y_m \ln \mathfrak{m}(x_m;\theta) - (1-y_m)\ln(1-\mathfrak{m}(x_m;\theta)\}. \tag{1.8}$$

Writing

$$\ell_y(y') \stackrel{\text{def}}{=} -y \ln y' - (1-y)\ln(1-y'), \tag{1.9}$$

we can now rewrite (1.8) in terms of the loss function (negative log-likelihood)

$$\Lambda(\theta) \stackrel{\text{def}}{=} \frac{1}{M} \sum_{m=1}^{M} \ell_y(\mathfrak{m}(x;\theta)),$$

which is the same structure as (1.4). The error function (1.9) is in fact the *binary cross entropy* and more generally reflects a way to compare probability measures.

In standard logistic regression, the model $\mathfrak{m}(x;\theta)$ is given as a composition of a linear map and a logistic function; see Chapter 3.

As a final step, once we have found an optimal parameter value θ^*, we can decide upon a label in $\mathcal{Y}$ by *voting*; our final map from $\mathcal{X}$ to $\mathcal{Y}$ is

- Decide label 1 if $\mathfrak{m}(x;\theta^*) > 1-\mathfrak{m}(x;\theta^*)$. Equivalently, decide label 1 if $\mathfrak{m}(x;\theta^*) > 1/2$.

- Decide label 0 if $\mathfrak{m}(x;\theta^*) \leq 1 - \mathfrak{m}(x;\theta^*)$. Equivalently, decide label 0 if $\mathfrak{m}(x;\theta^*) \leq 1/2$.

In the example we just discussed, there were only two possible labels: an image could contain a car or not contain a car. One can easily imagine asking the same kind of question for a multiclass prediction problem. For instance, suppose one is trying to classify whether an image contains a car, an airplane, or neither; in this case we have three labels. We will study such questions in Section 3.7 through the lens of logistic regression. Then, in Chapters 7 and 8 we will study the problem for multiclass prediction problems when the model $\mathfrak{m}(x;\theta)$ is a softmax function of a neural network.

We remark here that in deep learning, in either the problem of regression or classification that we just described, two main issues would always come up: overfitting and generalization. We briefly introduce these notions in Subsections 1.5.3 and 1.5.4, respectively.

1.5.3. Overfitting. Overfitting occurs when a model is trained to closely match an observed dataset yet is inaccurate on new datapoints not in the training dataset. This is the result of a model with many more degrees of freedom d than the number of data samples M in the dataset.

In the regime where $d \gg M$, there are typically many different models which exactly fit the data samples. These models can vary considerably on new datapoints. For example, consider the model $\mathfrak{m}(x;\theta) = ax^2 + bx + c$, where $\theta = (a, b, c)$ and a dataset that is composed of a single datapoint (x^0, y^0). Then, there are an infinite number of models $\mathfrak{m}(x;\theta)$ which exactly fit the datapoint (x^0, y^0), i.e., choose $c = y^0$, $b = -ax^0$, and let a be any real number. The vast majority of these models will be inaccurate on new datapoints.

Complex models with large numbers of parameters, such as neural networks, are particularly susceptible to overfitting. The best way to reduce overfitting is to have a larger dataset, which will help fully resolve the degrees of freedom of the model. The deep learning field has also developed a number of techniques to control and reduce overfitting (dropout, data normalization, penalties, etc.), which will be discussed in later chapters; e.g., Chapters 9 and 10.

1.5.4. Generalization. Nature gives us data $\mathcal{X}$ and targets $\mathcal{Y}$, $\mathcal{X} \mapsto \mathcal{Y}$. However, nature does not necessarily tell us that a specific $x \in \mathcal{X}$ corresponds to a specific prediction $y \in \mathcal{Y}$.

Let's say that $\mathbb{P}$ is the probability distribution over our dataset $\mathcal{D}$. Assume that the training set is $\mathcal{D}_{\text{train}} = \{(x_i, y_i) \sim \mathbb{P}\}_{i=1}^k$. The goal of machine learning is to learn predictors that work well outside the training set $\mathcal{D}_{\text{train}}$. The training set $\mathcal{D}_{\text{train}}$ is only a source of information that nature gives us to find such a predictor.

Assume $\mathcal{D}_{\text{test}} = \{(x_i, y_i) \sim \mathbb{P}\}_{i=1}^{m}$ is a new collection of datapoints coming from the same distribution $\mathbb{P}$. Our model has not been constructed knowing $\mathcal{D}_{\text{test}}$ but needs to perform well on $\mathcal{D}_{\text{test}}$. To accomplish this, we need to search within a class of functions that is neither too big nor too small and that fits well both training and test data.

The error that our model makes on the training set $\mathcal{D}_{\text{train}}$ is called training error, whereas the error that it makes on the $\mathcal{D}_{\text{test}}$ set is called test error. We train the models on $\mathcal{D}_{\text{train}}$, but we want them to generalize well, i.e., to work well on $\mathcal{D}_{\text{test}}$. We discuss aspects of train error versus test error, that are of particular relevance to deep learning, in Section 1.6, and in further depth in Chapter 11.

1.6. Bias-Variance Tradeoff and Double Descent

Now that we have seen a bit, at a high-level, of the supervised learning paradigm, let us discuss the issues of under-parametrization, over-parametrization, and generalization for both classical statistics and deep learning.

Consider a feature-label dataset $\mathcal{D} = \{(x_m, y_m)\}_{m=1}^{M}$, and suppose we fit a parametric model $\mathfrak{m}(\cdot;\theta)$ to the data. We would of course like the best $\mathfrak{m}(\cdot;\theta^*)$ to both capture the feature-label structure of the training data, and also generalize well to unseen data. Let's understand *bias-variance* tradeoffs in trying to do so.

Assume that the data $\mathcal{D}$ comes from sampling an underlying distribution $\mathbb{P}$ on feature-label random variables (X, Y), with associated expectation operator $\mathbb{E}$. In the standard regression setting, we choose the parameter by minimizing the average square error between the true label and the predicted label; in terms of $\mathbb{P}$, this means we want to minimize

$$\mathbb{E}\left[(Y - \mathfrak{m}(X;\theta))^2\right]. \tag{1.10}$$

Theoretically, the *conditional expectation* $\mathfrak{m}_{\text{opt}}(X) \overset{\text{def}}{=} \mathbb{E}[Y|X]$ is the true minimum mean square error estimator of the label random variable given the feature random variable, the minimum being taken over all *measurable* maps from feature to label space. Our search over parametrized maps is thus an attempt to approximate the conditional expectation; generically, our parametric models $\{\mathfrak{m}(\cdot;\theta)\}_{\theta\in\Theta}$ are a proper subset of the collection of all such measurable maps.

Let θ^* be the parameter vector which minimizes (1.10), we can compare $\mathfrak{m}(x;\theta^*)$ to $\mathfrak{m}_{\mathrm{opt}}(x) = \mathbb{E}[Y|X = x]$ by writing

$$\begin{aligned}
\mathbb{E}\left[(Y - \mathfrak{m}(X;\theta^*))^2\right] &= \mathbb{E}\left[\{(Y - \mathfrak{m}_{\mathrm{opt}}(X)) + (\mathfrak{m}_{\mathrm{opt}}(X) - \mathfrak{m}(X;\theta^*))\}^2\right] \\
&= \mathbb{E}\left[(Y - \mathfrak{m}_{\mathrm{opt}}(X))^2\right] + \mathbb{E}\left[(\mathfrak{m}_{\mathrm{opt}}(X) - \mathfrak{m}(X;\theta^*))^2\right] \\
&\quad + 2\mathbb{E}\left[(Y - \mathfrak{m}_{\mathrm{opt}}(X))(\mathfrak{m}_{\mathrm{opt}}(X) - \mathfrak{m}(X;\theta^*))\right] \\
&= \underbrace{\mathbb{E}\left[(Y - \mathfrak{m}_{\mathrm{opt}}(X))^2\right]}_{\text{Bayes error}} + \underbrace{\mathbb{E}\left[(\mathfrak{m}_{\mathrm{opt}}(X) - \mathfrak{m}(X;\theta^*))^2\right]}_{\text{bias}^2+\text{variance}}.
\end{aligned}$$

The cross term disappears due to the projection property [**Bil95**] of conditional expectation,

$$\mathbb{E}\left[\{Y - \mathbb{E}[Y|X]\}\, G(X)\right] = 0$$

for all bounded measurable functions G; see [**Bil95**]. The first term is the *Bayes error*, which tells us how far $\mathbb{E}[Y|X]$ is from the true label (sometimes referred to as the *Nature's model*). However, it is important to note that the true model is not $\mathfrak{m}_{\mathrm{opt}}(x)$.

Let us now investigate the second term. The second term represents *bias* and *variance*, which tells us how far the best parametrized model is from the true minimum mean square error model. It is important to realize that the model $\mathfrak{m}(\cdot;\theta^*)$ has been constructed using a training dataset $\mathcal{D}$. This means that the model $\mathfrak{m}(\cdot;\theta^*)$ depends on the randomness of the dataset $\mathcal{D}$, which contains points (x, y) sampled from the true distribution $\mathbb{P}$.

At this point, we recall the definition of $E_{\mathcal{D}}$ denoting the expectation operator with respect to the dataset $\mathcal{D}$. For a *given* point x in the feature space, we add and subtract $E_{\mathcal{D}}\mathfrak{m}(x;\theta^*)$ within the expectation of the second term, and we get

$$\begin{aligned}
(\mathfrak{m}_{\mathrm{opt}}(x) - \mathfrak{m}(x;\theta^*))^2 &= \left(\mathfrak{m}_{\mathrm{opt}}(x) - E_{\mathcal{D}}\mathfrak{m}(x;\theta^*) + E_{\mathcal{D}}\mathfrak{m}(x;\theta^*) - \mathfrak{m}(x;\theta^*)\right)^2 \\
&= \left(\mathfrak{m}_{\mathrm{opt}}(x) - E_{\mathcal{D}}\mathfrak{m}(x;\theta^*)\right)^2 + (\mathfrak{m}(x;\theta^*) - E_{\mathcal{D}}\mathfrak{m}(x;\theta^*))^2 \\
&\quad - 2\left(\mathfrak{m}_{\mathrm{opt}}(x) - E_{\mathcal{D}}\mathfrak{m}(x;\theta^*)\right)(\mathfrak{m}(x;\theta^*) - E_{\mathcal{D}}\mathfrak{m}(x;\theta^*)).
\end{aligned}$$

Taking now the expectation over $\mathcal{D}$, the cross term will vanish, i.e.,

$$\begin{aligned}
&E_{\mathcal{D}}\left[\left(\mathfrak{m}_{\mathrm{opt}}(x) - E_{\mathcal{D}}\mathfrak{m}(x;\theta^*)\right)(\mathfrak{m}(x;\theta^*) - E_{\mathcal{D}}\mathfrak{m}(x;\theta^*))\right] \\
&\quad = \left(\mathfrak{m}_{\mathrm{opt}}(x) - E_{\mathcal{D}}\mathfrak{m}(x;\theta^*)\right) E_{\mathcal{D}}\left[(\mathfrak{m}(x;\theta^*) - E_{\mathcal{D}}\mathfrak{m}(x;\theta^*))\right] \\
&\quad = 0,
\end{aligned}$$

which then gives

$$E_{\mathcal{D}}\left(\mathfrak{m}_{\mathrm{opt}}(x)-\mathfrak{m}(x;\theta^*)\right)^2=\underbrace{\left(\mathfrak{m}_{\mathrm{opt}}(x)-E_{\mathcal{D}}\mathfrak{m}(x;\theta^*)\right)^2}_{\mathrm{bias}^2}$$
$$+\underbrace{E_{\mathcal{D}}\left(\mathfrak{m}(x;\theta^*)-E_{\mathcal{D}}\mathfrak{m}(x;\theta^*)\right)^2}_{\mathrm{variance}}.$$

In the expression above, the first term, i.e., the square of the bias, measures the difference of the optimal model $\mathfrak{m}_{\mathrm{opt}}(\cdot)$ and our constructed model $\mathfrak{m}(\cdot;\theta^*)$ across different experiments each one with a potentially different dataset $\mathcal{D}$. The second term (the variance) measures the sensitivity of our constructed model $\mathfrak{m}(\cdot;\theta^*)$ with respect to the dataset $\mathcal{D}$.

Hence, all in all, we can write

$$\mathbb{E}\left[(Y-\mathfrak{m}(X))^2\right]=\text{Bayes error}+\mathbb{E}\left[\mathrm{bias}^2+\mathrm{variance}\right]. \tag{1.11}$$

The formula (1.11) we derived is very insightful. In practice, we minimize the empirical loss function, creating a model using the available data $\mathcal{D}$. Even though not much can be done about the Bayes error component, the bias and variance components indicate that:

- If the dataset is too small and the model $\mathfrak{m}(x;\theta)$ is not trained on a sufficient number of datapoints, and then the model will have a large variance.
- If the model $\mathfrak{m}(x;\theta)$ is large or complex (e.g., many parameters) and training is done correctly, then the distance between the optimal model and the fitted model (on the dataset the model is calibrated to) will be small. In other words, large models typically lead to small bias.
- If the model is large (complex), then one typically needs a large dataset to properly calibrate the model. So, if one does not have a large enough dataset but has a large (complex) model, then this leads to small bias and large variance.

We can visualize this relationship with the graph in Figure 1.1.

Figure 1.1 shows the famous U-shaped curve for the squared-bias and variance terms. It shows that for a given dataset one should choose a model that minimizes both the squared-bias and the variance. We will revisit this topic in Chapter 11, where a proper decomposition of the dataset can be employed to to help address these issues.

However, in (very) deep learning Figure 1.1 does *not* always describe the full picture. As a matter of fact, it has been observed empirically in many instances that the loss keeps decreasing even as we fit larger and larger models in the same dataset. The typical picture is the so-called double-descent curve

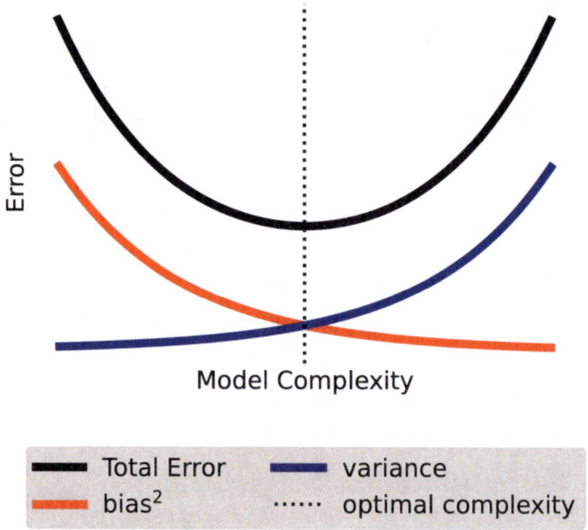

Figure 1.1. Bias-variance tradeoff.

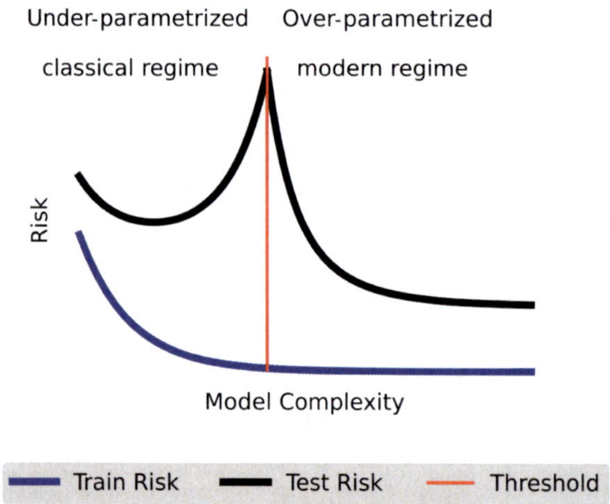

Figure 1.2. Double-descent curve in deep learning and the contrast to classical statistical wisdom.

of Figure 1.2. The papers [**HZRS16, BHMM19, NKB$^+$20**] contain ample empirical evidence of this phenomenon.

Figure 1.2 shows that after a certain threshold of model complexity, the loss function associated to deep neural networks keeps decreasing even if the dataset is not getting larger as the model gets larger. As discussed in [**BHMM19**], the transition from the classical regime to the modern deep neural network regime occurs when the model is large enough that it perfectly fits the training data. In conclusion, it is observed empirically that very large models trained with stochastic gradient descent lead to good generalization properties. As we shall see in Chapters 19 and 20, properly defined architectures also have an impact on generalization performance.

In the chapters that follow, we define, link, and understand the questions of approximation, optimization, training, and performance for different problems and for different neural network architectures. Section 1.8 presents the organization of the book.

1.7. Some Existing Related Books

This book aims to present deep learning from a mathematical lens, unifying the presentation of a number of concepts and ideas. We tried to do so always having in mind that deep learning is a tool in data analysis. The hope is that this book helps open a door through which the mathematical, broader scientific community, and practitioners can enter to better explore connections and advance our understanding and the state of the art in this increasingly important collection of computational tools and ideas.

There are many excellent books that have been written on the general topic of machine and statistical learning and a few that are more specialized for deep learning. Some of the existing books are more introductory than others, while others are more advanced mathematically. The books [**Bis06**, **HTF10**, **Mur22**, **Bac24**] cover various aspects of machine and statistical learning, while [**GBC16**, **Cal20**, **Pri23**, **BB24**] focus more on deep learning.

1.8. Organization of this Book

The big-picture idea of presenting deep learning that we have implemented in this book can be summarized in Figure 1.3.

Some of the elements of Figure 1.3 (even within each of its blocks) are more introductory whereas others are more advanced (either mathematically, computationally or conceptually). Therefore, we have organized this book in two parts. In Part 1 our goal is to introduce the different aspects of deep learning using an appropriate mathematical language. This part of the book should be accessible to an advanced mathematics, statistics, computer science, data science, or engineering undergraduate student. As we mentioned in the Preface, Part 1 together with select topics from Part 2 could serve as standalone teaching material for a course on a mathematical introduction to deep learning. Part 1 covers most of the aspects of Figure 1.3 with the exception of the optimization in the feature learning regime and some of the selected topics that require more advanced tools (both of which are covered in Part 2).

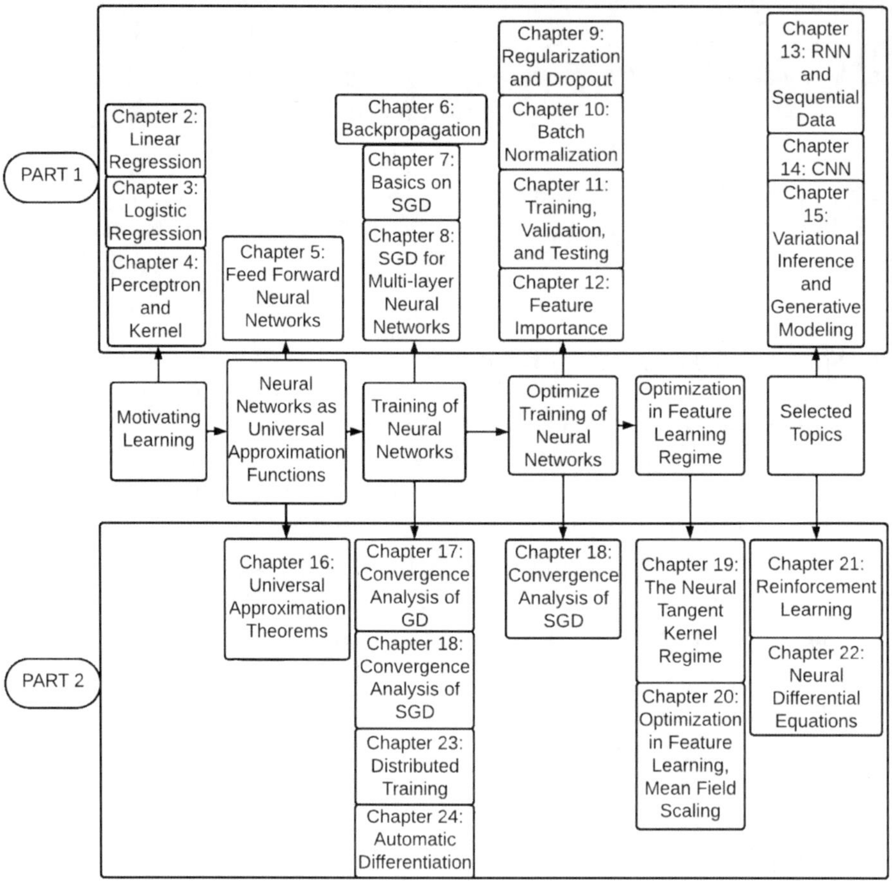

Figure 1.3. Conceptual organization of the book.

In Part 2 we go over deeper mathematical results regarding deep learning where we present the universal approximation theory, as well as convergence proofs for gradient descent, stochastic gradient descent, and different deep learning algorithms. In Part 2 we also discuss more involved computational aspects of deep learning including distributed training, message passing interface (MPI) and automatic differentiation. Part 2 should be accessible to graduate students and researchers aiming to go deeper into certain topics of deep learning, and it covers aspects of Figure 1.3 that were not covered in Part 1.

Certain chapters in Part 2 will require some prior exposure to analysis, probability theory, and stochastic processes. Some chapters of Part 2 are not necessarily more advanced mathematically than those in Part 1, but they are included in Part 2 because the topics are more advanced conceptually or computationally.

For the reader's easy reference we have added Appendixes A and B, where we have collected the main mathematical background in probability, stochastic processes, and analysis that is used in various parts of the book.

The idea of combining in one book both Part 1 and Part 2 is to give a more complete discussion of the topic and to offer to the reader, who wants to go deeper in certain topics, the ability to do so by jumping into Part 2. Alternatively, a reader may choose to go over the basic chapters of Part 1 and then selectively read other chapters. Our hope is that the combination of Parts 1 and 2 gives a good overview of the flow diagram 1.3. Figure 1.4 summarizes the prerequisites for each chapter and provides a useful guide for reading the book.

1.8.1. Part 1: Mathematical introduction to deep learning. We start in Chapters 2 and 3 with linear and logistic regression, respectively. These are classical statistical topics that one can find in many textbooks. However, here we introduce them in a way that is suitable for and motivates deep learning. Training of deep neural networks heavily relies on notions that are present in the simpler-to-present cases of linear and logistic regression. In Chapter 4 we motivate neural networks through the *perceptron* (one of the earliest neural network models) and its relation to kernels (another classical machine learning topic). The perceptron model also allows us to introduce the concept of gradient descent in an intuitive way. Then, Chapter 5 presents feed forward neural networks, probably the simplest form of a neural network. We define feed forward neural networks and connect them to truth tables. The fact that feed forward neural networks are universal approximators is discussed in some detail in Chapter 16 of Part 2 of the book.

The backbone of training of neural networks is backpropagation, which is explained next in Chapter 6. Essentially backpropagation is a smart way to apply chain rule and thus allows us to efficiently differentiate the loss function of complicated models and perform (stochastic) gradient descent. Stochastic gradient descent in turn is presented in Chapter 7 (for shallow neural networks) and in Chapter 8 (for deep neural networks). We also present some examples of Python code and discuss GPUs versus CPUs to aid the reader with coding aspects. We do emphasize however that the focus of the book is on the mathematical foundations of deep learning and not so much on its computational aspects, but as discussed in the Preface, accompanying code can be found in https://mathdl.github.io/. The well-known issue of vanishing gradient is also discussed.

Then, we move on to discussing regularization techniques that are widely used in deep learning in order to accelerate training and reduce overfitting. In particular, in Chapter 9 we discuss adding penalty terms to the loss function which is a well-developed method in statistics; see also classical texts in statistical learning such as [**HTF10**]. In Chapter 9 we also discuss dropout in some detail, which is a regularization method that is unique in deep learning and is very popular in practical applications. Another popular technique is batch normalization and is discussed in Chapter 10. The process of training, validation, and testing that is present in any application of deep learning is discussed in Chapter 11 while feature importance, a central topic in deep learning, is discussed in Chapter 12.

Next, we present three very popular deep learning models. In Chapter 13 we discuss recurrent neural networks that are designed to model time series and dependent data. In addition, we present and formulate mathematically the attention mechanism and the transformer architecture that are also used to model dependent data. We compare the attention mechanism to recurrent neural networks. In Chapter 14 we discuss convolution neural networks that have found great success in modeling imaging data.

In Chapter 15 we discuss variational inference and generative models. The goal here is to learn appropriate distributions so that we can generate data from them in order to match some empirical distribution. We achieve this via appropriately formulating a minimization problem with respect to an appropriate metric in the space of probability distributions. Variational inference is based on maximizing an appropriate lower bound stemming from a proper manipulation of the Kullback-Leibler divergence, called the evidence lower bound. Generative adversarial networks (GANs) are also discussed where we motivate them by revisiting the basic logistic classification problem leading to the discriminator-generator framework. Optimization in GANs and the Wasserstein GAN are also discussed.

Part 1 has few rigorous proofs; the goal is to help the reader understand what the major deep neural networks are and how to work with them. All chapters of Part 1 conclude with a "Brief Concluding Remarks" section summarizing what was covered in the specific chapter, what follows next, and oftentimes giving pointers to the literature for the interested reader.

1.8.2. Part 2: Advanced topics and convergence results in deep learning. In Part 2 of the book we dive into more advanced topics and theoretical aspects of deep learning. In Chapter 16 we present the main universal approximation theory going back to classical results from the 1980s, but we also present very recent theory developed for neural networks with ReLU nonlinearities. This part uses, to some extent, functional analytic notions and theorems that are reviewed in Appendix A.

In Chapter 17 we study gradient descent from a more theoretical perspective, reviewing convergence results and associated choices of learning rate schedules under both convexity and nonconvexity assumptions. We also discuss accelerated gradient descent methods, including second-order methods. Then, in Chapter 18 we turn our attention to theoretical properties of stochastic gradient descent. We study convergence rates and convergence properties of stochastic gradient descent under both convexity and nonconvexity assumptions, and we compare gradient descent with stochastic gradient descent. We also discuss accelerated methods as well as more advanced stochastic gradient descent methods (like RMSProp, ADAM, AdaMax) and their convergence properties.

In Chapters 19 and 20 we turn our attention to the asymptotic behavior of neural networks when trained with stochastic gradient descent. The question we want to answer is: will the algorithm recover the ground truth after training neural networks with stochastic gradient descent?

We answer this question via an investigation of the limit behavior of the neural network as the number of units per hidden layer and training steps grows to infinity. In order to obtain meaningful limiting behavior, we need to appropriately scale the neural network. The neural network limit training dynamics can then be analyzed to establish guarantees of convergence to the ground truth. An important practical byproduct of the mathematical analysis is that it offers insights into how to choose the learning rate hyperparameter. In particular, the learning rate hyperparameter needs to be chosen in specific ways with respect to the scalings in order for convergence to the ground truth to be possible. We discuss two of the main scalings: the neural tangent kernel (sometimes called the linear regime) in Chapter 19, and the mean field scaling (sometimes called the nonlinear regime) in Chapter 20. In Chapter 20 we also compare these different scalings and we comment on generalization performance.

In Chapter 21 we discuss reinforcement learning. Reinforcement learning could be a book by itself (probably multiple books), and there are many excellent textbooks on the topic. Our goal is to motivate deep reinforcement learning through a mathematical approach. We also present convergence results for neural Q-learning, which is reinforcement learning with neural network approximators.

In Chapter 22 we discuss neural ordinary differential equations (neural ODEs) and neural stochastic differential equations (neural SDEs). In particular, we model ODEs and SDEs using neural networks and then the goal is to learn the parameters of the neural network so that the resulting solution matches some predetermined profile in an appropriate metric (e.g., mean square sense). We show how adjoint equations can be used to efficiently optimize neural ODEs.

In Chapter 23 we discuss computational problems that arise in training of potentially very large deep learning models on potentially very large datasets. Due to the large model size and large amount of memory-per-data-sample, it may be challenging to evaluate and calculate the backpropagation step for a large minibatch on a GPU. Computational problems quickly arise due to issues including limited memory, parallelization problems, and more. Distributing training over multiple GPUs becomes advantageous. We discuss the topics of synchronous versus asynchronous gradient descent, parallel efficiency, and aspects of MPI communication. Illustrative computational examples in PyTorch MPI and Python MPI are presented.

In Chapter 24 we discuss automatic differentiation. In deep learning, we design and evaluate different model architectures that typically have a large number of hyperparameters. Deep learning would have faced a large obstacle if one had to derive from scratch the chain rule for implementing the backpropagation algorithm for each new model. Automatic differentiation addresses this challenge by automatically calculating the chain rule (and gradients with respect to the model parameters), facilitating model development and evaluation.

The chapters of Part 2 conclude with a "Brief Concluding Remarks" section summarizing what was covered in the specific chapter and giving pointers to the literature for the interested reader.

1.8.3. Appendixes. Appendix A has some background material on probability, stochastic processes and stochastic analysis. Appendix B has some background on basic inequalities used throughout the book and real and functional analysis. Notions that are discussed in these appendixes appear throughout the book and are particularly useful in Part 2 of the book.

1.8.4. Flow diagram of the book. Figure 1.4 shows a flow diagram of how the book is organized. Arrows demonstrate the background material needed to proceed to the next indicated chapter.

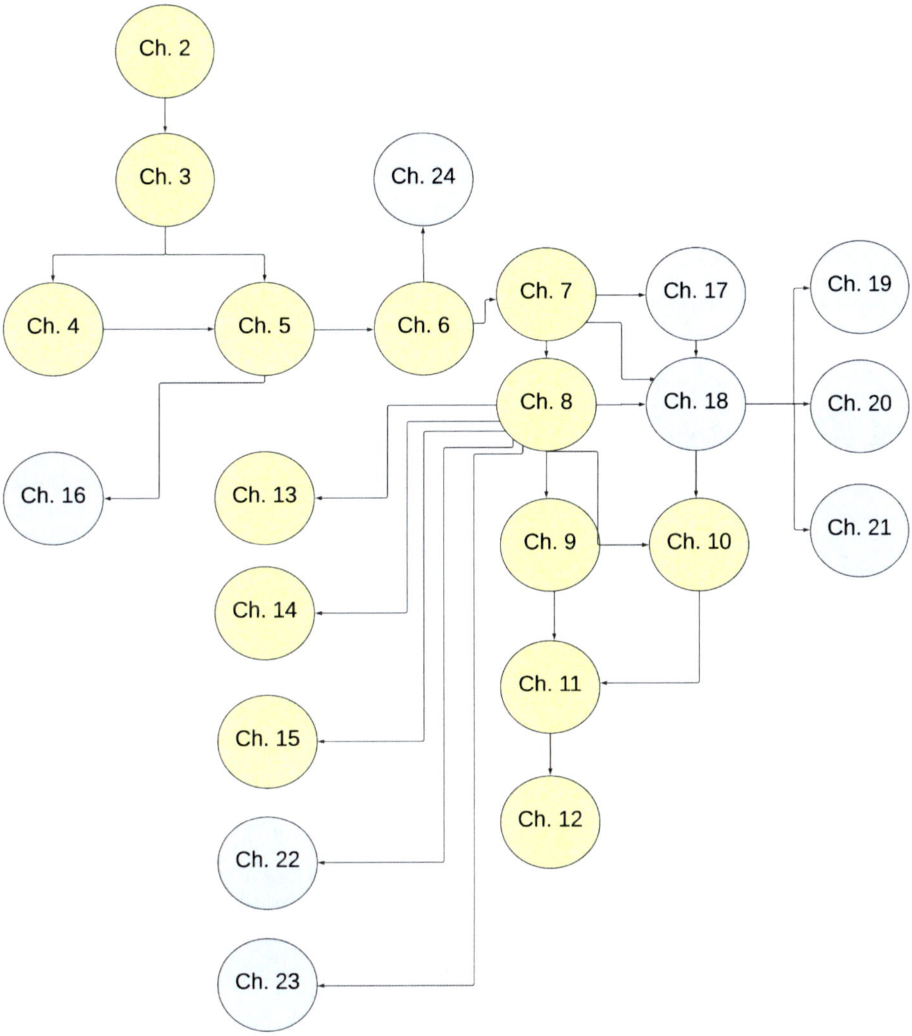

Figure 1.4. Chapter flow diagram of the book. Yellow color refers to chapters in Part 1, whereas light blue color refers to chapters in Part 2.

Part 1

Mathematical Introduction to Deep Learning

Chapter 2

Linear Regression

2.1. Introduction

Don't skip this chapter! We begin our journey with the obligatory sojourn in linear regression. This is a classical topic in statistics. Our goal is to set up a framework and notation in a familiar setting; we want to test some new ideas with an old friend.

Let's start with some sample ground-truth data (see the github repository), the first lines of which are in Table 2.1. The ground-truth data is a collection of (feature, label) pairs. Both feature and label are $\mathbb{R}$-values; i.e., they are both *numerical*.

Let's try to predict the label based on the feature. A scatter plot of this data is in Figure 2.1, and its data sort of clusters around a line. For $(w, b) \in \mathbb{R} \times \mathbb{R}$, define a linear *model*

$$\mathfrak{m}(x; \theta) \stackrel{\text{def}}{=} wx + b \qquad x \in \mathbb{R}, \tag{2.1}$$

Table 2.1. Sample data

x	y
5.8	19.0
1.5	3.5
2.7	10.4
9.4	19.2
6.5	6.6

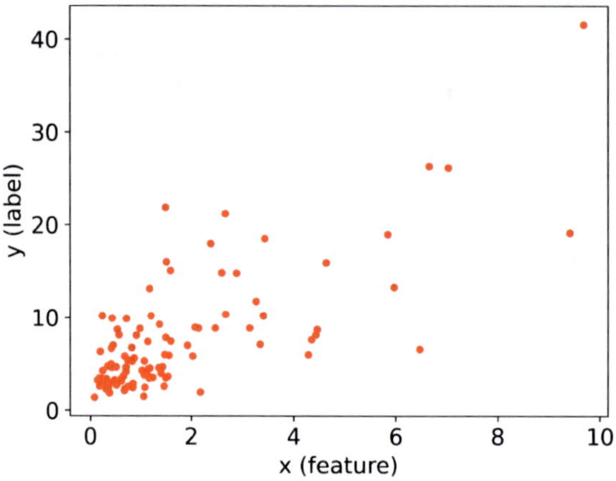

Figure 2.1. Scatter plot of ground-truth data for linear regression

where

$$\theta \stackrel{\text{def}}{=} \begin{pmatrix} w \\ b \end{pmatrix} \tag{2.2}$$

are the parameters of the model. We want to find the *best* parameter value θ^* in the parameter space $\Theta \stackrel{\text{def}}{=} \mathbb{R}^2$ such that the graph of $x \mapsto \mathfrak{m}(x; \theta^*)$ in some sense *optimally* passes through the ground-truth data. Namely, our predictor will be the map $x \mapsto \mathfrak{m}(x; \theta^*)$.

We have already introduced several important notions.

- The features lie in some Euclidean space $\mathbb{R}^F$ (here $F = 1$).
- The labels lie in some Euclidean space $\mathbb{R}^L$ (here $L = 1$).
- The parameters take value in some Euclidean space Θ (here $\Theta = \mathbb{R}^2$).
- We have selected a parametrized model $\mathfrak{m}(x; \theta)$ to predict the label based on the features.

This structure will underlie all of our efforts.

We want to *learn* (i.e., calibrate) our model to the ground-truth data of Table 3.1. so that the *graph* of $\mathfrak{m}(x; \theta)$ in some optimal way passes through a scatter plot of the ground-truth data. In Figure 2.2, we see two such possible graphs of $\mathfrak{m}(x; \theta)$. As we shall see later on in Figures 2.3 and 2.4, one can find a *better* curve when the choice of the best is done in a principled way.

The ground-truth data consists of a collection of points in feature $\times$ label space. Some of these may be repeated (for example, two different houses with the same square footage may also have the same price), so our ground-truth data should actually be a *multiset* $\mathcal{D}$ of points in feature $\times$ label space. Recall

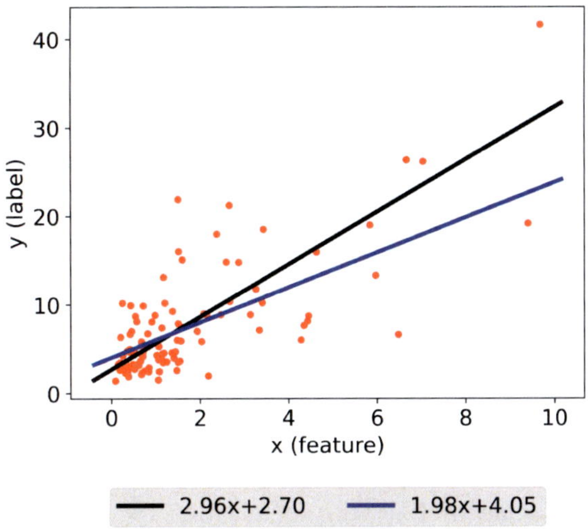

Figure 2.2. Variation of linear regression parameters

that a multiset consists of a collection of possibly repeated points (i.e., multiplicity larger than 1). The *cardinality* $|A|$ of a multiset A is the number of elements, with multiplicity included. In other words, if $A \stackrel{\text{def}}{=} \{1,1,2,4,5,5,5\}$, then $|A| = 7$. Directly thinking of multisets in feature-label space (as opposed to indexing by an enumeration) will allow us to retain notational access to features and labels. We believe that this will help with notational transparency when we consider stochastic gradient descent (Chapters 7 and 8) and training, validation, and testing (Chapter 11). We are also wary of having too many enumeration indices (gradient descent, dropout, epochs, and recurrent neural networks will all require their own indices).

2.2. Loss Function

We need to define the notion of best which we will use to define the optimal θ of (2.2). Let's first of all define an *error* function

$$\ell_y(y') \stackrel{\text{def}}{=} (y - y')^2 \tag{2.3}$$

for y and y' in $\mathbb{R}^L$, and then define a *per-datapoint* loss function

$$\lambda_{(x,y)}(\theta) \stackrel{\text{def}}{=} \ell_y\left(\mathfrak{m}(x;\theta)\right) \tag{2.4}$$

for $(x, y) \in \mathcal{D}$ and $\theta \in \mathbb{R}^P$, and then define an *average* loss

$$\Lambda(\theta) \stackrel{\text{def}}{=} \frac{1}{|\mathcal{D}|} \sum_{(x,y)\in\mathcal{D}} \lambda_{(x,y)}(\theta) \tag{2.5}$$

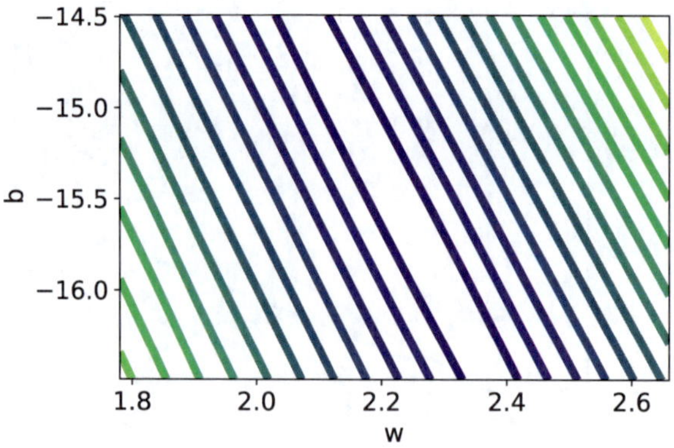

Figure 2.3. Contour plot of loss Λ for linear regression

for $\theta \in \mathbb{R}^P$. Our best parameter vector θ^* is given by

$$\theta^* = \operatorname{argmin}\left\{\Lambda(\theta) : \theta \in \mathbb{R}^P\right\}.$$

A few thoughts have guided our choice of notation:

- The per-datapoint loss of (2.4) separates the model from the error function; the same notation can be easily adapted to more complicated feed forward neural networks.
- The per-datapoint vs. average loss separates the effect of parameter variation from the effect of averaging; this notation can be easily adapted to stochastic gradient descent algorithms.

We have converted our search for a best line running through the points of $\mathcal{D}$ to a problem of minimizing a function on the parameter space $\mathbb{R}^P$. Since $P = 2$, we can construct a contour plot of Λ; see Figure 2.3. Computationally, it looks like Λ has a minimum (which we will more rigorously understand in a moment), and the minimizer looks like a good choice of the slope and intercept for a line passing through the data; see Figure 2.4.

2.3. Minimization

Our interest in linear regression is in setting up some generalizable ideas; let's see how we might carry out *gradient descent* on Λ (conveniently forgetting that the explicit solution of linear regression is well known). In Section 18.4.3 we will revisit the topic of (stochastic) gradient descent applied to linear regression. Gradient descent seeks to minimize a function by moving in the direction of largest descent, i.e., the negative gradient. Namely, we want to construct a sequence $(\theta_k)_{k=1}^\infty$ given by

$$\theta_{k+1} = \theta_k - \eta \nabla \Lambda(\theta_k), \tag{2.6}$$

where $\eta > 0$ is a *learning rate*.

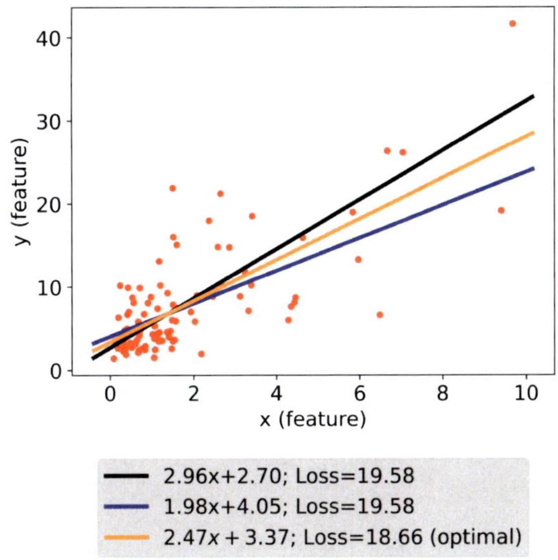

Figure 2.4. Best line through data

Explicitly,

$$\nabla\Lambda(\theta) = \frac{1}{|\mathcal{D}|} \sum_{(x,y)\in\mathcal{D}} \nabla\lambda_{(x,y)}(\theta)$$

and, by the chain rule,

(2.7)
$$\begin{aligned}\nabla\lambda_{(x,y)}(\theta) &= \ell_y'(\mathfrak{m}(x;\theta)) \begin{pmatrix} \partial\mathfrak{m}/\partial w(x;\theta) \\ \partial\mathfrak{m}/\partial b(x;\theta) \end{pmatrix} \\ &= -2\,(y - \mathfrak{m}(x;\theta)) \begin{pmatrix} x \\ 1 \end{pmatrix}.\end{aligned}$$

This allows us to explicitly write $\nabla\lambda_{(x,y)}$ as the sensitivity of the error function with respect to changes in the prediction model and in the sensitivity of the model with respect to the parameters.

Collecting things together, we have

(2.8)
$$\begin{aligned}\nabla\Lambda(\theta) &= -2\frac{1}{|\mathcal{D}|} \sum_{(x,y)\in\mathcal{D}} \begin{pmatrix} (y - \mathfrak{m}(x;\theta))\,x \\ y - \mathfrak{m}(x;\theta) \end{pmatrix} \\ &= -2\frac{1}{|\mathcal{D}|} \sum_{(x,y)\in\mathcal{D}} \begin{pmatrix} (y - wx - b)\,x \\ y - wx - b \end{pmatrix} \\ &= -2\begin{pmatrix} \overline{xy} - w\overline{x^2} - b\overline{x} \\ \overline{y} - w\overline{x} - b \end{pmatrix} \\ &= 2A\begin{pmatrix} w \\ b \end{pmatrix} - 2\begin{pmatrix} \overline{xy} \\ \overline{y} \end{pmatrix},\end{aligned}$$

where in turn

$$A \stackrel{\text{def}}{=} \begin{pmatrix} \overline{x^2} & \overline{x} \\ \overline{x} & 1 \end{pmatrix},$$

$$\overline{x} \stackrel{\text{def}}{=} \frac{1}{|\mathcal{D}|} \sum_{(x,y)\in\mathcal{D}} x,$$

$$\overline{y} \stackrel{\text{def}}{=} \frac{1}{|\mathcal{D}|} \sum_{(x,y)\in\mathcal{D}} y,$$

$$\overline{x^2} \stackrel{\text{def}}{=} \frac{1}{|\mathcal{D}|} \sum_{(x,y)\in\mathcal{D}} x^2,$$

$$\overline{xy} \stackrel{\text{def}}{=} \frac{1}{|\mathcal{D}|} \sum_{(x,y)\in\mathcal{D}} xy,$$

$$\overline{y^2} \stackrel{\text{def}}{=} \frac{1}{|\mathcal{D}|} \sum_{(x,y)\in\mathcal{D}} y^2.$$

Of course with an explicit (linear) formula for the $\nabla\Lambda$, the first-order conditions of optimality should help us explicitly find the minimizer. Explicit formulas are the exception rather than the rule, so we relegate these standard calculations to Section 2.6.

Gradient descent of (2.6) is thus (explicitly)

$$\begin{pmatrix} w_{k+1} \\ b_{k+1} \end{pmatrix} = \begin{pmatrix} w_k \\ b_k \end{pmatrix} - 2\eta A \begin{pmatrix} w_k \\ b_k \end{pmatrix} + 2\eta \begin{pmatrix} \overline{xy} \\ \overline{y} \end{pmatrix}.$$

Subtracting,

$$\begin{aligned} \begin{pmatrix} w_{k+1} - w_k \\ b_{k+1} - b_k \end{pmatrix} &= \begin{pmatrix} w_k - w_{k-1} \\ b_k - b_{k-1} \end{pmatrix} - 2\eta A \begin{pmatrix} w_k - w_{k-1} \\ b_k - b_{k-1} \end{pmatrix} \\ &= \{I_2 - 2\eta A\} \begin{pmatrix} w_k - w_{k-1} \\ b_k - b_{k-1} \end{pmatrix}, \end{aligned} \tag{2.9}$$

where

$$I_2 \stackrel{\text{def}}{=} \begin{pmatrix} 1 & 0 \\ 0 & 1 \end{pmatrix}.$$

Let's understand the eigenvalues of A. The characteristic equation of A is then

$$0 = \det \begin{pmatrix} \lambda - \overline{x^2} & -\overline{x} \\ -\overline{x} & \lambda - 1 \end{pmatrix} = \lambda^2 - \left(\overline{x^2} + 1\right)\lambda + \overline{x^2} - \overline{x}^2,$$

the two solutions of which are

$$\lambda_\pm = \frac{1}{2}\left\{\left(\overline{x^2} + 1\right) \pm \sqrt{\left(\overline{x^2} + 1\right)^2 - 4\left(\overline{x^2} - \overline{x}^2\right)}\right\}.$$

Since

$$
\begin{aligned}
(2.10)\qquad \left(\overline{x^2}+1\right)^2 - 4\left(\overline{x^2}-\overline{x}^2\right) &= \left(\overline{x^2}+1\right)^2 - 4\overline{x^2} + 4\overline{x}^2 \\
&= \left(\overline{x^2}-1\right)^2 + 4\overline{x}^2,
\end{aligned}
$$

which is positive, the eigenvalues of A are real (which of course also follows from the fact that A is symmetric). By the Cauchy-Schwarz inequality, $\overline{x}^2 \le \overline{x^2}$, so we can continue (2.10) as

$$
\left(\overline{x^2}+1\right)^2 - 4\left(\overline{x^2}-\overline{x}^2\right) \le \left(\overline{x^2}-1\right)^2 + 4\overline{x^2} = \left(\overline{x^2}+1\right)^2
$$

implying that

$$
(2.11)\qquad 0 \le \lambda_- \le \lambda_+ \le \overline{x^2}+1.
$$

The left inequality implies that A is nonnegative definite. Alternately, for any $(\alpha_1, \alpha_2) \in \mathbb{R}^2$,

$$
\begin{aligned}
\begin{pmatrix}\alpha_1 & \alpha_2\end{pmatrix} A \begin{pmatrix}\alpha_1 \\ \alpha_2\end{pmatrix} &= \alpha_1^2\overline{x^2} + 2\alpha_1\alpha_1\overline{x} + \alpha_2^2 \\
&= \frac{1}{|\mathcal{D}|}\sum_{(x,y)\in\mathcal{D}} \left\{\alpha_1^2 x^2 + 2\alpha_1\alpha_2 x + \alpha_2^2\right\} \\
&= \frac{1}{|\mathcal{D}|}\sum_{(x,y)\in\mathcal{D}} (\alpha_1 x + \alpha_2)^2 \ge 0.
\end{aligned}
$$

The eigenvalues of $I_2 - 2\eta A$ are then $1 - 2\eta\lambda_+$ and $1 - 2\eta\lambda_-$. The system (2.9) converges if it is a contraction, which occurs if

$$
|1 - 2\eta\lambda_\pm| < 1,
$$

i.e., if

$$
-1 < 1 - 2\eta\lambda_\pm < 1,
$$

i.e., if

$$
\eta < \min\left\{\frac{1}{\lambda_+}, \frac{1}{\lambda_-}\right\} = \frac{1}{\lambda_+}.
$$

Conversely, if $\eta > \frac{1}{\lambda_+}$, the loss will diverge under gradient descent for a generic initial condition. See Figures 2.6 and 2.5.

Following from (2.11), we are assured stability if

$$
\eta < \frac{1}{\overline{x^2}+1}.
$$

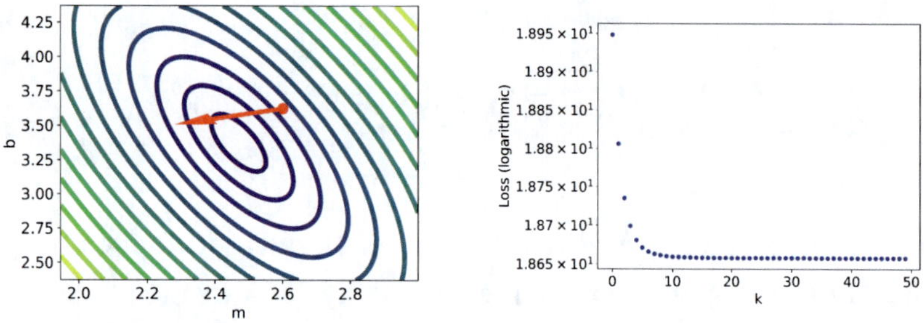

Figure 2.5. Converging gradient descent: $\theta_{k+1} = \theta_k - 0.119\nabla\Lambda(\theta_k)$, where $\eta_{\text{critical}} = 0.140$; trajectory of gradient descent (left) and loss values (right).

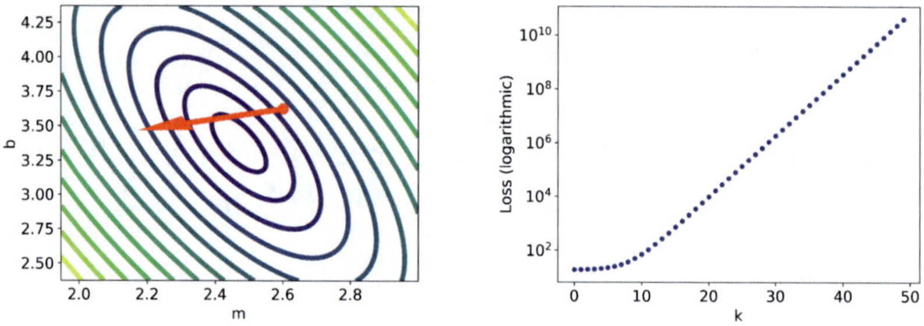

Figure 2.6. Nonconverging gradient descent: $\theta_{k+1} = \theta_k - 0.161\nabla\Lambda(\theta_k)$, where $\eta_{\text{critical}} = 0.140$; trajectory of gradient descent (left) and loss values (right).

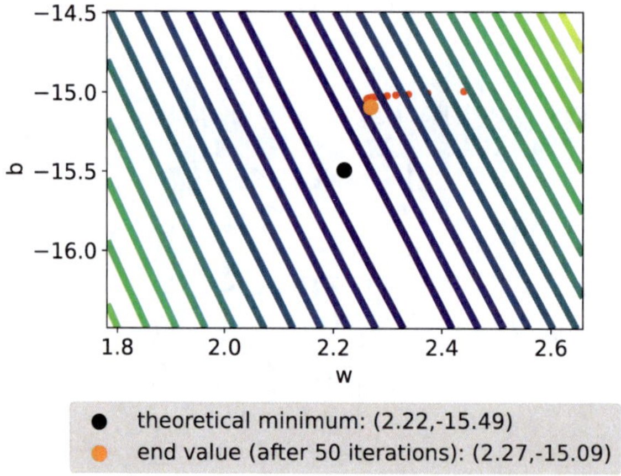

Figure 2.7. Iterations of gradient descent in loss landscape

2.4. Metric

The loss function Λ serves as an objective function to minimize and quantify best, but it may in fact be a mathematically convenient substitute for a *metric*, which we use to report the performance of our machine learning, and which may make more sense to stakeholders. If, for example, the labels are prices (and thus nonnegative), we might in fact be interested in the *relative* error of the prediction

$$\mu(\mathfrak{m}(\cdot;\theta^*)) \stackrel{\text{def}}{=} \frac{1}{|\mathcal{D}|} \sum_{(x,y)\in\mathcal{D}} \frac{|y - \mathfrak{m}(x;\theta^*)|}{y}.$$

In this case,

$$\text{loss} = 18.66 \qquad \text{and} \qquad \text{metric} = 0.52.$$

The absolute value function fails to be differentiable (at 0). Classification algorithms (see Chapter 3) in particular have a wealth of metrics reflecting relative combinations of different types of errors.

2.5. Computational Realization

If we look at how PyTorch implements linear regression, we will find that things are transposed. Suppose we have the matrix $A \in \mathbb{R}^{1\times 2}$

$$A \stackrel{\text{def}}{=} \begin{pmatrix} 3 & 2 \end{pmatrix},$$

and we want to use this as a linear map from $\mathbb{R}^2$ (feature space) to $\mathbb{R}$ (label space) via

$$A\begin{pmatrix} x_1 \\ x_2 \end{pmatrix} = 3x_1 + 2x_2,$$

and we want to apply it to a collection of $\mathbb{R}^2$ input (feature) vectors which would be computationally listed as

$$[[3,4],[-1,0],[4,2]]. \tag{2.12}$$

Mathematically, we want to compute

$$\begin{pmatrix} 3 & 2 \end{pmatrix}\begin{pmatrix} 3 \\ 4 \end{pmatrix} = 9 + 8 = 17,$$

$$\begin{pmatrix} 3 & 2 \end{pmatrix}\begin{pmatrix} -1 \\ 0 \end{pmatrix} = -3 + 0 = -3,$$

$$\begin{pmatrix} 3 & 2 \end{pmatrix}\begin{pmatrix} 4 \\ 2 \end{pmatrix} = 12 + 4 = 16.$$

The standard mathematical way to do this would be to *horizontally stack* all of these computations together as

$$\begin{pmatrix} 3 & 2 \end{pmatrix}\begin{bmatrix} 3 & -1 & 4 \\ 4 & 0 & 2 \end{bmatrix} = \begin{bmatrix} 17 & -3 & 16 \end{bmatrix}. \tag{2.13}$$

However, we usually think of the input (2.12) as

$$\begin{bmatrix} 3 & 4 \\ -1 & 0 \\ 4 & 2 \end{bmatrix}.$$

Transposing (2.13), we have the vertically stacked calculation

$$\begin{bmatrix} 3 & 4 \\ -1 & 0 \\ 4 & 2 \end{bmatrix} A^T = \begin{bmatrix} 17 \\ -3 \\ 16 \end{bmatrix}, \tag{2.14}$$

which corresponds to an output list

$$[[17], [-3], [16]]. \tag{2.15}$$

In practice, one rarely tries to access coefficients of linear transformations, so (2.14) is preferred as it directly maps (2.12) to (2.15).

2.6. Brief Concluding Remarks

Our review of linear regression was structured to introduce the framework we will use to understand deep neural networks. Deep neural networks will consist of parametrized models (generalizations of (2.1)) which we will try to fit to a ground-truth dataset. Error functions (generalizations of (2.3)) will allow us to construct per-datapoint losses (generalizations of (2.4)) which we can then aggregate to define an average loss (2.5). Gradient descent will help us minimize this average loss, once we use the chain rule (as in (2.7)) to calculate the gradient of the per-datapoint loss.

Of course with (2.8), the explicit solution of linear regression follows from the first-order conditions of optimality; $\nabla\Lambda = 0$ at the point

$$\begin{aligned} A^{-1}\begin{pmatrix} \overline{xy} \\ \overline{y} \end{pmatrix} &= \frac{1}{\overline{x^2} - \overline{x}^2}\begin{pmatrix} 1 & -\overline{x} \\ -\overline{x} & \overline{x^2} \end{pmatrix}\begin{pmatrix} \overline{xy} \\ \overline{y} \end{pmatrix} \\ &= \frac{1}{\overline{x^2} - \overline{x}^2}\begin{pmatrix} \overline{xy} - \overline{x}\,\overline{y} \\ -\overline{xy}\,\overline{x} + \overline{x^2}\,\overline{y} \end{pmatrix} \\ &= \frac{1}{\overline{x^2} - \overline{x}^2}\begin{pmatrix} \overline{xy} - \overline{x}\,\overline{y} \\ -\left(\overline{xy} - \overline{x}\,\overline{y}\right)\overline{x} + \left(\overline{x^2} - \overline{x}^2\right)\overline{y} \end{pmatrix}. \end{aligned}$$

This coincides with the standard formula: if we are given a new feature value x_{new}, our best guess of the label, under linear regression, is

$$\frac{\overline{xy} - \overline{x}\,\overline{y}}{\overline{x^2} - \overline{x}^2}x_{\text{new}} + \frac{-\left(\overline{xy} - \overline{x}\,\overline{y}\right)\overline{x} + \left(\overline{x^2} - \overline{x}^2\right)\overline{y}}{\overline{x^2} - \overline{x}^2} = \frac{\overline{xy} - \overline{x}\,\overline{y}}{\overline{x^2} - \overline{x}^2}\left(x_{\text{new}} - \overline{x}\right) + \overline{y}.$$

Explicit solutions of deep learning problems are the exception rather than the rule.

Linear regression and linear models in general are a classical topic in statistics and many excellent textbooks are available; see for example [**Agr15, RS07**] for an excellent comprehensive treatment of the topic. There are many other excellent sources on this as well ([**DS98, MPV21**]). The book [**HTF10**] also has good coverage with an eye towards statistical learning.

2.7. Exercises

Exercise 2.1. Consider the loss function $f(x) = 9x_1^2 + x_2^2$ for $x = (x_1, x_2) \in \mathbb{R}^2$.

(1) Write down the iteration $x(k+1) = x(k) - \eta \nabla f(x(k))$, where $x(k) = (x_1(k), x_2(k))$ and η is the learning rate.

(2) For what values of η will gradient descent converge?

Exercise 2.2. Consider the loss function $f(x) = 9x_1^2 + (x_2 - 3)^2$ for $x = (x_1, x_2) \in \mathbb{R}^2$.

(1) Identify the minimum point of $x^* = (x_1^*, x_2^*)$ of f.

(2) Write down the iteration $x(k+1) = x(k) - \eta \nabla f(x(k))$, where $x(k) = (x_1(k), x_2(k))$ and η is the learning rate.

(3) Define $y(k) = (y_1(k), y_2(k))$, with $y_1(k) = x_1(k) - x_1^*$ and $y_2(k) = x_2(k) - x_2^*$. Write down an iteration for y.

(4) For what values of η will the iteration for y converge?

Exercise 2.3. Consider the loss function $f(x) = 9x_1^2 + x_2^2$ for $x = (x_1, x_2) \in \mathbb{R}^2$.

(1) Fix $\eta > 0$ and consider the gradient descent $x(k+1) = x(k) - \eta \nabla f(x(k))$, where $x(k) = (x_1(k), x_2(k))$ and initial conditions $x(0) = (1, 2)$. Define $X_t^\eta = x(\lfloor t/\eta \rfloor)$ where $\lfloor \cdot \rfloor$ is the integer floor function. Find an ordinary differential equation (ODE) for $X_t = \lim_{\eta \to 0} X_t^\eta$.

(2) Explicitly solve the ODE for $X(t)$.

Exercise 2.4. Consider the loss function $f(x) = \lambda x^2$ for $x \in \mathbb{R}$.

(1) Write down the iteration $x(k+1) = x(k) - \eta f'(x(k))$, where η is the learning rate. Describe it explicitly.

(2) For what values of η will the iteration for x converge?

Exercise 2.5. What is the loss function for linear regression on the three points $(0, 1)$, $(2, 0)$, and $(1, 3)$?

Chapter 3

Logistic Regression

3.1. Introduction

Logistic regression, another classical topic in statistics, is a bit more complicated than linear regression because *i*) it involves a nonlinearity, and *ii*) it does not have an explicit solution. These two aspects will both be true for deep neural networks. Logistic regression is something of a bridge from linear regression to the full framework of deep neural networks. In thinking through logistic regression, we shall see that the way we framed linear regression is appropriate for the wider problems of deep neural networks.

Let's start with some sample ground-truth data (see the github repository), the first lines of which are in Table 3.1. The ground-truth data is a collection of (feature, label) pairs. While the feature is $\mathbb{R}$-valued (i.e., numerical), the label is now $\{0, 1\}$-valued; i.e., the label is *categorical* (and in fact binary).

Again, let's try to predict the label based on the feature. Let's construct a binned frequency plot of the labels as a function of the features; see Figure 3.1.

Table 3.1. Ground-truth data for binary classification

x	y
10.84	1
6.49	0
7.66	1
14.40	1
11.47	1

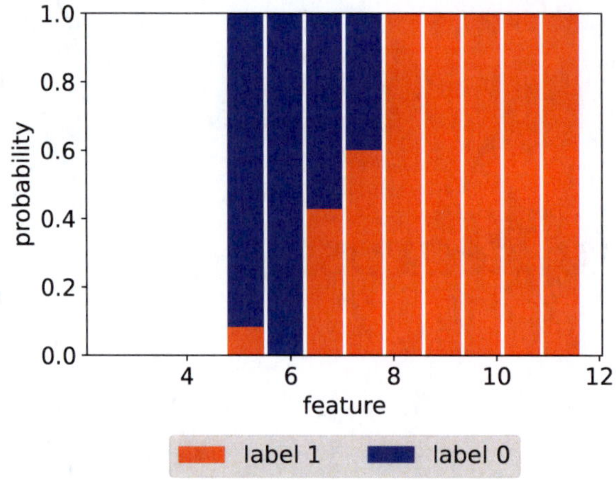

Figure 3.1. Binned frequency plot of ground-truth data for logistic regression

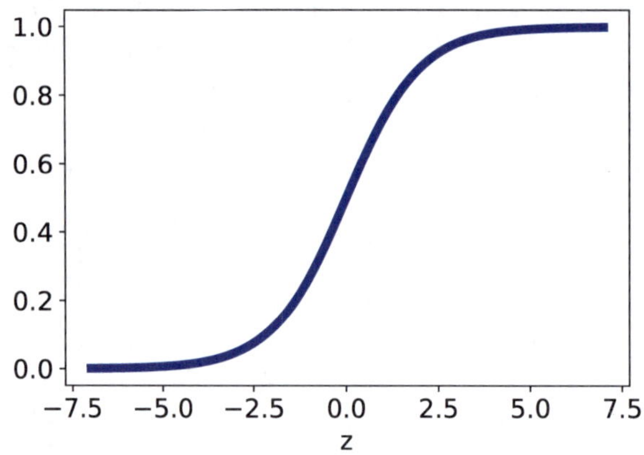

Figure 3.2. Logistic function

Roughly, we see that the normalized frequencies of the labels increase as the frequency increases.

Logistic regression tries to fit a shifted and scaled *logistic* function (plotted in Figure 3.2),

$$S(z) \stackrel{\text{def}}{=} \frac{e^z}{1+e^z} = \frac{1}{1+e^{-z}}, \qquad z \in \mathbb{R},$$

to Figure 3.1. In betting parlance, the $S(z)$ is the probability with odds of e^z to 1. Figure 3.3 gives a visual idea of the result. We want to do so in a robust way which bypasses the details of binning. Let's work through this, but do so in a way which mimics some of the notation of linear regression and to simultaneously introduce several tools which will be useful in our development of deep neural networks.

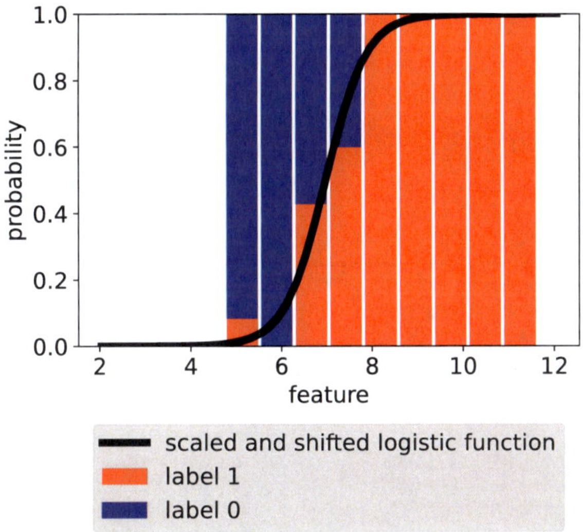

Figure 3.3. Binned frequency plot with fitted logistic function

3.2. Formalization of the Problem

Let's alter our perspective a bit. Instead of trying to directly predict the categorical label, let's combine the idea of binned frequency plots and the logistic function S to try to predict *probabilities*. Since probabilities are numerical (they take values in $[0,1]$) as opposed to categorical, we should be able to reuse some of the ideas of continuous optimization and gradient descent that we developed for linear regression in Chapter 2.[1] This becomes even more appealing when we realize that our labels in this case can easily be reinterpreted as probabilities that the label is 1:

$$\text{label} = \mathbb{P}\{\text{label is } 1\}.$$

Namely, if a ground-truth label is 1, it *must* (with probability 1) be of type 1. Conversely, if a ground-truth label is 0, it *can't* (i.e., it has probability 0) be of type 1. The binary labels 0 and 1 are thus extreme (i.e., the boundary of the collection $[0,1]$ of allowed probabilities), but we can nevertheless try to use S to approximate these probabilities.

Let's format logistic regression using some notation similar to that of Chapter 2. Let's try to predict the *probability* that a feature value x has label 1 with model

$$\mathfrak{m}(x;\theta) \overset{\text{def}}{=} S(wx+b) = \frac{e^{wx+b}}{1+e^{wx+b}} = \frac{1}{1+e^{-(wx+b)}}, \qquad x \in \mathbb{R}, \tag{3.1}$$

[1]The field of combinatorial optimization, on the other hand, is dedicated to optimization over finite sets, e.g., a finite collection of labels.

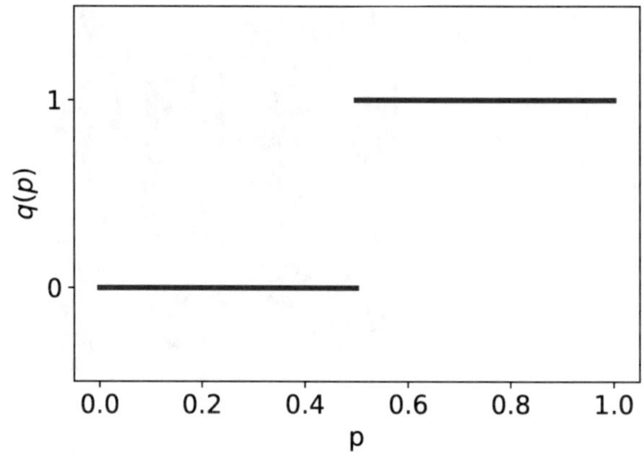

Figure 3.4. Voting quantization

where

$$\theta = \begin{pmatrix} w \\ b \end{pmatrix} \tag{3.2}$$

are the parameters of the model, taking values (again) in $\Theta = \mathbb{R}^2$. The logistic function S maps $\mathbb{R}$ in to $(0, 1)$ (a subset of the probability space $[0, 1]$), and $x \mapsto wx + b$ scales and shifts the features). In some way, we want to find the best parameter vector θ^* that captures the idea of Figure 3.3.

Once we have found an optimal θ^*, we can predict the probability that a feature value x has label 1 and then quantize by *voting*. Namely, for a feature value x, our prediction will be given by

$$\begin{aligned} &\begin{cases} \text{predict class 1 if } \mathfrak{m}(x; \theta^*) > 1/2 \\ \text{predict class 0 if } \mathfrak{m}(x; \theta^*) < 1/2 \end{cases} \\ &= q(\mathfrak{m}(x; \theta^*)) \\ &= \begin{cases} \text{predict class 1 if } w^* x + b^* > 0 \\ \text{predict class 0 if } w^* x + b^* < 0, \end{cases} \end{aligned} \tag{3.3}$$

where the voting quantizer q (see Figure 3.4) is

$$q(z) = \begin{cases} 1 & \text{if } z > 1/2 \\ 0 & \text{if } z < 1/2. \end{cases} \tag{3.4}$$

We'll have a bit more to say later about the discontinuity value of q at $z = 1/2$.

We want to find the optimal parameter vector θ^* by some optimization problem. Let's reuse some of the ideas of Chapter 2: for y and y' in our label space $[0, 1]$, let $\ell_y(y')$ be error function which compares y and y'. We want $\ell_y(y') \geq 0$ with equality if and only if $y = y'$. Given ℓ, we can construct a per-datapoint loss function

$$\lambda_{(x,y)}(\theta) \stackrel{\text{def}}{=} \ell_y(\mathfrak{m}(x;\theta)) \tag{3.5}$$

for each $(x,y) \in \mathbb{R} \times [0,1]$ and $\theta \in \mathbb{R}^P$, and then define the average loss

$$\Lambda(\theta) \stackrel{\text{def}}{=} \frac{1}{|\mathcal{D}|} \sum_{(x,y)\in\mathcal{D}} \lambda_{(x,y)}(\theta) \tag{3.6}$$

for each $\theta \in \mathbb{R}^P$. The argmin of Λ will be the best parameter vector θ^*.

For logistic regression, the error function $\ell_y(y')$ is given by the *binary cross entropy*,

$$\ell_y(y') \stackrel{\text{def}}{=} y \ln \frac{y}{y'} + (1-y) \ln \frac{1-y}{1-y'}. \tag{3.7}$$

More exactly $\ell_y(y')$ is the relative entropy of y with respect to y'. We define $0 \ln 0 \stackrel{\text{def}}{=} 0$, and since our ground-truth labels are in $\{0,1\}$, we have

$$\ell_y(y') = \begin{cases} -\ln y' & \text{if } y = 1 \\ -\ln(1-y') & \text{if } y = 0 \end{cases} = -y \ln y' - (1-y)\ln(1-y').$$

In Figure 3.5 we plot the entropy for labels 0 and 1. For $y' \in (0,1)$ (i.e., the range of S), $\ln y'$ and $\ln(1-y')$ are negative, so $\ell_y(y') > 0$. Asymptotically,

$$\ell_0(0+) \stackrel{\text{def}}{=} \lim_{y'\searrow 0} \ell_0(y') = 0$$

$$\ell_1(1-) \stackrel{\text{def}}{=} \lim_{y'\nearrow 0} \ell_1(y') = 0;$$

the error becomes small when y' is close to y. We also have that

$$\ell_0(1-) \stackrel{\text{def}}{=} \lim_{y'\nearrow 1} \ell_0(y') = \infty$$

$$\ell_1(0+) \stackrel{\text{def}}{=} \lim_{y'\searrow 0} \ell_1(y') = \infty,$$

so the error becomes infinite when y' gets close to the wrong binary label.

Let's write out $\lambda_{(x,y)}(\theta)$ of (3.5). We have that

$$1 - S(z) = \frac{1}{1+e^x}, \qquad x \in \mathbb{R}.$$

For $x \in \mathbb{R}$ and θ given by (3.2),

$$\lambda_{(x,0)}(\theta) = -\ln(1 - S(wx+b)) = \ln\bigl(1 + e^{wx+b}\bigr)$$
$$\lambda_{(x,1)}(\theta) = -\ln S(wx+b) = \ln\bigl(1 + e^{-(wx+b)}\bigr).$$

For a ground-truth datapoint $(x,0)$ (i.e., label 0), $\lambda_{(x,0)}(\theta)$ is small when $wx+b \approx -\infty$, while for a ground-truth datapoint $(x,1)$ (i.e., label 1), $\lambda_{(x,1)}(\theta)$ is

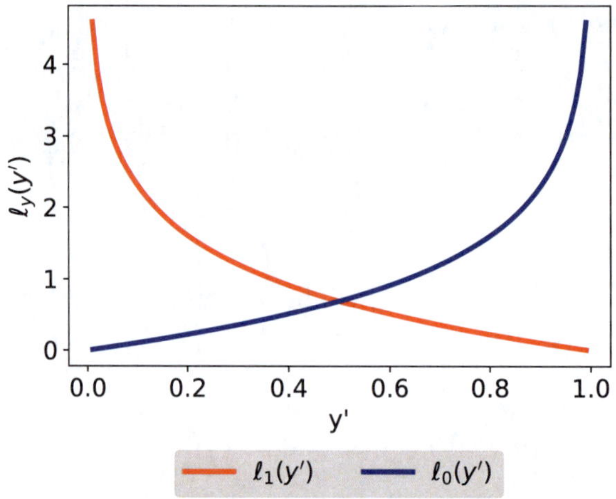

Figure 3.5. Entropy for labels 0 and 1

small when $wx+b \approx \infty$. This makes sense: for a ground-truth datapoint $(x, 0)$, minimizing $\lambda_{(x,0)}(\theta)$ should drive $wx + b \to -\infty$, implying that $\mathfrak{m}(x; \theta) \approx 0$ for all $x \in \mathbb{R}$, implying that all predictions end up being label 0 according to (3.3). On the other hand, for a ground-truth datapoint $(x, 1)$, minimizing $\lambda_{(x,1)}(\theta)$ should drive $wx + b \to \infty$, implying that $\mathfrak{m}(x; \theta) \approx 1$ for all $x \in \mathbb{R}$, implying that all predictions end up being label 1. Informally, minimizing Λ of (3.6) means that in each bin of Figure 3.1, we have a competition of minimizing $\lambda_{(x,0)}(\theta)$ and $\lambda_{(x,1)}(\theta)$ for x in that bin. See Figure 3.6.

As with linear regression, we want to carry out gradient descent. Namely, we want to construct a sequence $(\theta_k)_{k=1}^{\infty}$ given by

$$\theta_{k+1} = \theta_k - \eta \nabla \Lambda(\theta_k), \tag{3.8}$$

where $\eta > 0$ is a *learning rate.*

Once again,

$$\nabla \Lambda(\theta) = \frac{1}{|\mathcal{D}|} \sum_{(x,y) \in \mathcal{D}} \nabla \lambda_{(x,y)}(\theta).$$

Similarly to linear regression, the chain rule gives us

$$\nabla \lambda_{(x,y)}(\theta) = \ell_y'(\mathfrak{m}(x; \theta)) \begin{pmatrix} \partial \mathfrak{m}/\partial w(x; \theta) \\ \partial \mathfrak{m}/\partial b(x; \theta) \end{pmatrix}. \tag{3.9}$$

Here, however, the formula for ℓ_y' is a bit more complicated:

$$\ell_y'(y') = \begin{cases} -\frac{1}{y'} & \text{if } y = 1 \\ \frac{1}{1-y'} & \text{if } y = 0 \end{cases} = -\frac{y}{y'} + \frac{1-y}{1-y'}. \tag{3.10}$$

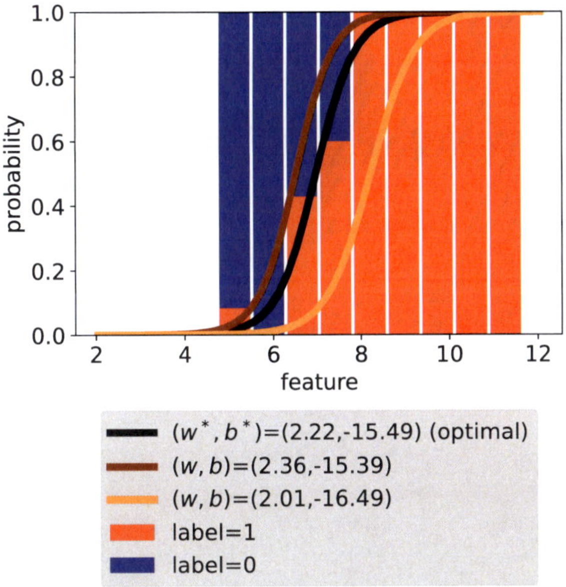

Figure 3.6. Binned frequency plot and variation of logistic regression parameters

We also have

$$S'(z) = \frac{e^{-z}}{(1+e^{-z})^2} = \frac{e^{-z}}{1+e^{-z}} \times \frac{1}{1+e^{-z}} = \frac{1}{(1+e^{z})(1+e^{-z})}, \qquad z \in \mathbb{R},$$

so

$$\begin{pmatrix} {}^{\partial m}/_{\partial w}(x;\theta) \\ {}^{\partial m}/_{\partial b}(x;\theta) \end{pmatrix} = \begin{pmatrix} S'(wx+b)x \\ S'(wx+b) \end{pmatrix} = S'(wx+b)\begin{pmatrix} x \\ 1 \end{pmatrix}.$$

For linear regression, the per-datapoint loss was a composition of the square error function and a linear map; here the per-datapoint loss is a composition of binary cross entropy, the logistic function, and a linear map. The chain rule is a bit more complex than for linear regression. We obtain the gradient of the per-datapoint loss (3.9) by combining (3.10) and (3.9). Combining everything in one place, we have

$$\nabla \lambda_{(x,y)}(\theta) = \ell_y'(S(wx+b))S'(wx+b))\begin{pmatrix} x \\ 1 \end{pmatrix}.$$

The results of gradient descent are in Figure 3.7. For *linear* regression, we had an explicit formula for the solution and could use that as a reference point for gradient descent. Here, we have used sklearn, i.e., another numerical algorithm, to give us a numerical approximation θ^* of the solution. The optimal values are

$$w^* = 2.22 \qquad \text{and} \qquad b^* = -15.49.$$

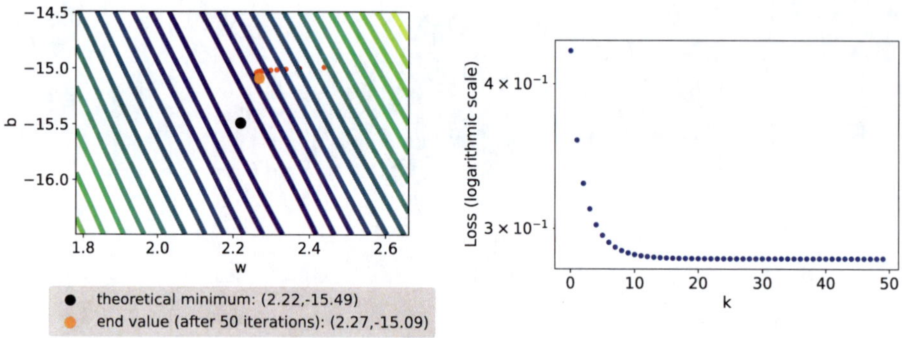

Figure 3.7. Converging gradient descent: $\theta_{k+1} = \theta_k - 0.01\nabla\Lambda(\theta_k)$; trajectory of gradient descent (left) and loss values (right).

3.3. Metric

How does our logistic regression algorithm perform? For the sake of specificity, let's take the optimal value θ^* of our parameters to be that given by sklearn;

$$w^* = 2.22 \qquad \text{and} \qquad b^* = -15.49 \tag{3.11}$$

We can then use (3.3). There are then a number of metrics we might use to assess our prediction algorithm. The simplest might be to compute an average classification error. Since a dataset might be imbalanced, we might alternately use a metric which combines classification error with the frequencies of the different labels.

3.4. Transitions and Scaling

Let's take a closer look at our optimal $x \mapsto \mathfrak{m}(x;\theta^*)$ for predicting the probability that a feature has label 1. With

$$\theta^* = \begin{pmatrix} w^* \\ b^* \end{pmatrix}$$

given by sklearn (3.11), we explicitly have

$$\mathfrak{m}(x;\theta^*) = S(w^*x + b^*) = S\left(\operatorname{sgn}(w^*)\frac{x - x_c}{s}\right),$$

where $\operatorname{sgn}(z) \stackrel{\text{def}}{=} z/|z|$ is the *sign* function for $z \neq 0$, and the *location* x_c and *scale* s are given by

$$\begin{aligned} x_c &= -b^*/w^* \\ s &= 1/|w^*|. \end{aligned} \tag{3.12}$$

For our data,

$$x_c = 6.9689 \qquad \text{and} \qquad s = 0.45. \tag{3.13}$$

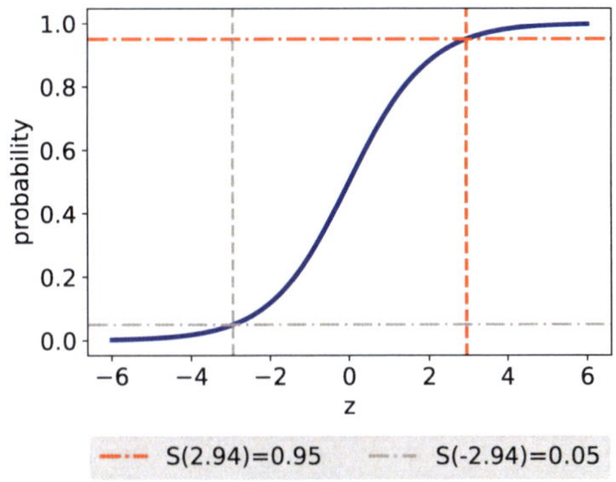

Figure 3.8. Transition of logistic function

We haven't defined sgn(0), but this is only relevant if $w^* = 0$, in which case our logistic regression would be $x \mapsto S(b^*)$.

The data of (3.13) also helps understand the natural mathematical question of what do we do if we are given a feature value x_{new} exactly such that $w^* x_{\text{new}} + b^* = 0$, i.e., $x_{\text{new}} = x_c$? Can you find a datapoint with feature *exactly* equal to x_c?

Returning to Figure 3.6, we see that the transition from 0 to 1 visually agrees with the value of x_c in (3.13). Analytically, we can invert the logistic function: if $S(z) = p$, then $e^z = p/1-p$, implying that

$$z = \ln \frac{p}{1-p}.$$

Thus 95% of the transition of the logistic function (i.e., from 0.05 to 0.95) occurs in the interval

$$\left(\ln \frac{0.05}{1-0.05}, \ln \frac{0.95}{1-0.95}\right) = \left(\ln \frac{0.05}{0.95}, \ln \frac{0.95}{0.05}\right) = (-2.95, 2.95).$$

See Figure 3.8. Solving

$$\frac{x - x_c}{s} = \pm 2.95,$$

we see that 95% of the transition in logistic regression should occur on the interval

$$(x_c - 2.95s, x_c + 2.95s).$$

See Figure 3.9.

Another nice feature of the logistic function is that it is symmetric around $(0, 1/2)$:

$$S(-z) = \frac{1}{1+e^z} = 1 - \frac{e^z}{1+e^z}.$$

In other words, given that the point

$$(z, S(z)) = \left(z, \tfrac{1}{2} + \left(S(z) - \tfrac{1}{2}\right)\right)$$

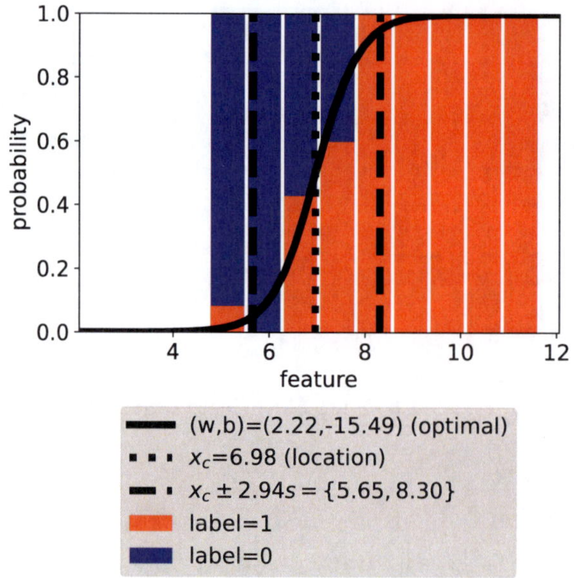

Figure 3.9. Empirical location and scale for fitted logistic function

is on the graph of S, the point

$$\left(-z, \frac{1}{2} - \left(S(z) - \frac{1}{2}\right)\right) = (-z, 1 - S(z))$$

is also on the graph of S. Consequently, the categorical predictions should be invariant under the choice of which category is given label 1 (as opposed to label 0).

3.5. Normalization

Returning to Figure 3.8, the transition from label 0 to label 1 occurred in the tail of S (compared to the fact that most of the transition in S is between -3 and 3. By *scoring* the feature data, we can use the intrinsic scale of the data, and make it more likely that the transition in the data matches the transition in S. In particular, gradient descent is likely to perform poorly in the tail, where the derivative of S is exponentially small. Define[2]

$$\mu_z \stackrel{\text{def}}{=} \frac{1}{|\mathcal{D}|} \sum_{(x,y)\in\mathcal{D}} x$$

$$\sigma_z \stackrel{\text{def}}{=} \left\{ \frac{1}{|\mathcal{D}| - 1} \sum_{(x,y)\in\mathcal{D}} (x - \mu_z)^2 \right\}^{1/2},$$

[2]Recall *Bessel's correction* https://en.wikipedia.org/wiki/Bessel%27s_correction for computing standard deviation.

where, we compute

$$\mu_z = 7.04 \qquad \text{and} \qquad \sigma_z = 2.30. \tag{3.14}$$

We then define a scored dataset

$$\mathcal{D}^z \stackrel{\text{def}}{=} \left\{ \left(\frac{x - \mu_z}{\sigma_z}, y \right) : (x, y) \in \mathcal{D} \right\}.$$

The feature data in this new dataset, by construction, has mean zero and variance 1; see Table 3.2.

Scoring allows us to consider a number of reference calculations.

- It allows us to easily identify *outlier* data, which is more than several standard deviations away from the mean.
- Relevant behavior in the graph of scored data ($\mathcal{D}^z$) will occur at feature values of order 1. This suggests that optimal values for models with scored data will also be of order 1. This suggests that gradient descent algorithms also be initialized at values of order 1.
- Scoring forces multidimensional data to all be at the same scale. This suggests that different elements of the optimal parameter values will also be of common order 1.

Table 3.3 gives the optimal parameter values for these different datasets, and Figure 3.11 gives the contour plot for the loss for the scored data. We see that the optimal parameter values for the scored and ground-truth data are of order 1. We also see that the level curves of the loss function are more regular for the scored data, suggesting that gradient descent algorithms will be more robust.

Of course the optimum parameters in the different coordinate systems are all related. If x is the feature, our model assigns label 1 with probability

$$S(w_o^* x + b_o^*) = S\left((1000 w_o^*) \frac{x}{1000} + b_o^* \right).$$

This shows the conversion between the optimum parameters (w_o^*, b_o^*) in the original coordinate system and the optimum parameters (w^*, b^*) in what we have called our ground-truth data:

$$\begin{aligned} w^* &= 1000 w_o^* \\ b^* &= b_o^*. \end{aligned}$$

Similarly, we can write

$$S(w_o^* x + b_o^*) = S\left((w_o^* \sigma) \left(\frac{x - \mu}{\sigma} \right) + (b_o^* + w_o^* \mu) \right),$$

Table 3.2. Scored ground-truth data

x	y
1.67	1
−0.24	0
0.27	1
3.24	1
1.95	1

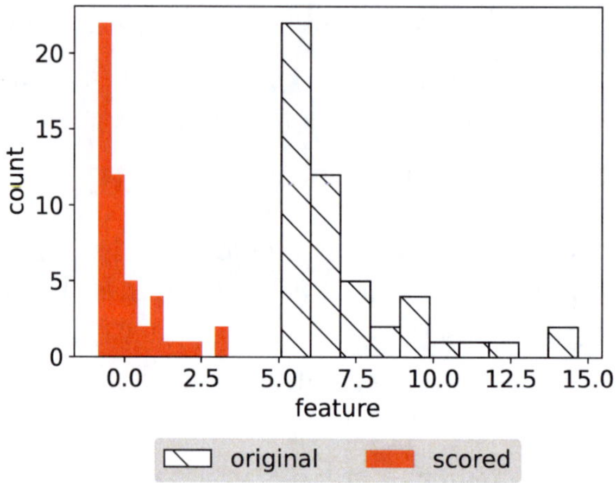

Figure 3.10. Scored data

Table 3.3. Optimal parameter values for original and scored ground-truth data

	original	scored
w:	2.22	5.06
b:	−15.49	0.17

which shows the conversion between (w_o^*, b_o^*) and the optimal parameters (w_z^*, b_z^*) in the scored coordinate system,

$$\begin{aligned} w_z^* &= w_o^* \sigma \\ b_z^* &= b_o^* + w_o^* \mu. \end{aligned}$$

The combination of (3.14) and Table 3.3 confirm this.

We will expand upon the notions of scoring in Chapter 10.

Of course the assumption of scoring is *statistical homogeneity*. If new data has different statistics, both the scoring and the parameters of logistic regression must be recalculated.

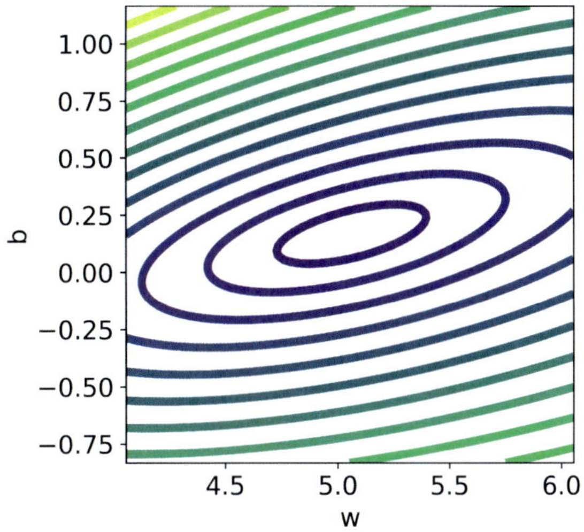

Figure 3.11. Contour plot of loss Λ for scored data

3.6. Perfect Data and Penalization

Perfect data for linear regression would mean that all feature-label points lie on a common line. Perfect data for logistic regression might be a collection of datapoints for all datapoints to the right of a threshold would all have label 1, and all datapoints to the left of the threshold would have label 0. What happens to logistic regression in this case?

Let's consider a collection of $N = 100$ feature values which are distributed normally with mean 0 and variance 1. Let's furthermore assume that all ground-truth points with positive feature values are assigned label 1, and all ground-truth datapoints with negative feature values are assigned label 0. See Table 3.4.

Informally, logistic regression should give us

$$S(\infty x)$$

and any positive feature value should always (i.e., probability 1) have label 1, so we should be evaluating S at a very large value. Conversely, any negative feature value should never (i.e., probability 0) have label 1. The effect of the bias b should be negligible.

Starting with $\{X_n\}_{n=1}^N$ normally chosen points, none of them are likely to be exactly 0, In fact, the transition from 0 to 1 in our perfect dataset will be given related to the minimum of $\{|X_n|\}_{n=1}^N$. This minimum will become smaller as N becomes larger, and (3.12) then implies that m should become larger as N

Table 3.4. Perfect ground-truth data

x	y
1.79	1
0.44	1
0.10	1
−1.86	0
−0.28	0

becomes larger. For any level $\ell > 0$,

$$\begin{aligned}\mathbb{P}\left\{\min_{1\le n\le N}|X_n| > \frac{\ell}{N}\right\} &= \prod_{n=1}^{N}\mathbb{P}\left\{|X_n| > \frac{\ell}{N}\right\}\\ &= \prod_{n=1}^{N}\left(1-\mathbb{P}\left\{|X_n|\le\frac{\ell}{N}\right\}\right)\\ &= \left(1-\int_{-\ell/N}^{\ell/N}\frac{e^{-z^2/2}}{\sqrt{2\pi}}dz\right)^N\\ &\approx \left(1-\frac{2\ell}{\sqrt{2\pi}}\frac{1}{N}\right)^N\\ &\approx \exp\left[-\frac{2}{\sqrt{2\pi}}\ell\right],\end{aligned}$$

where this approximation becomes more precise as N becomes larger (i.e., a limiting statement). In other words, $N\min_{1\le n\le N}|X_n|$ is approximately (as $N\nearrow\infty$) exponentially distributed with parameter $2/\sqrt{2\pi}$. Thus, the minimum $\min_{1\le N\le N}|X_n|$ should approximately have expectation

$$\mathbb{E}\left[\min_{1\le N\le N}|X_n|\right]\approx\frac{1}{N}\frac{2}{\sqrt{2\pi}},$$

which can also be interpreted as an asymptotic for the transition within our dataset. Taking m to be the reciprocal of this transition width (i.e., (3.12)), m should be of order N. In our case,

$$m_{\text{perfect}} = 252.95 \qquad \text{and} \qquad b_{\text{perfect}} = -0.71. \tag{3.15}$$

This is unfortunate. For perfect data, logistic regression diverges. We can *regularize* logistic regression by penalizing large values of m and b. Fixing a parameter $C > 0$, we can replace the per-datapoint loss of (3.5) with

$$\lambda^{(C)}_{(x,y)}(\theta) \stackrel{\text{def}}{=} \ell_y(\mathfrak{m}(x;\theta)) + C\|\theta\|^2. \tag{3.16}$$

Table 3.5. ground-truth data for multiclass classification

$\mathbf{x_1}$	$\mathbf{x_2}$	$\mathbf{y}$
3.73	1.30	2
2.16	6.82	2
3.94	−2.83	3
2.10	−0.35	2
−0.01	1.33	1

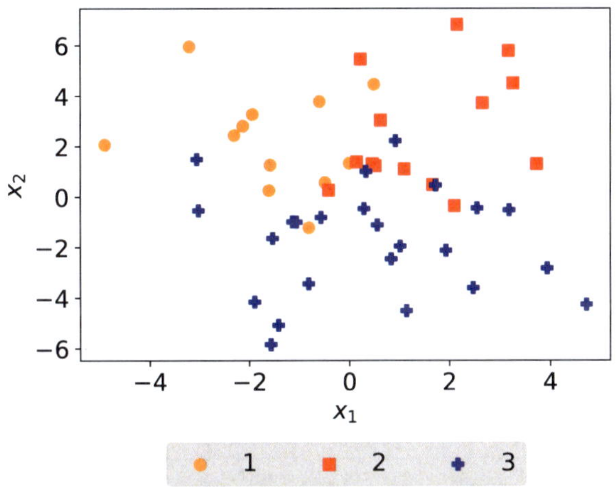

Figure 3.12. Scatter plot of ground-truth data for multiclass classification

This will be similar to the loss of (3.5) for reasonably small values of θ, but will penalize large values of θ. This makes more sense for scored data (Section 3.5), where we might expect the optimal parameters to be of order 1. In this case, the minimal loss for perfect data will significantly differ from the logistic regression loss when $|m| \geq 1/\sqrt{C}$, i.e., when the transition between labels is at scale less than $1/\sqrt{C}$. We will revisit this issue in Chapter 9 where we discuss regularization methods.

3.7. Multiclass prediction

How can we generalize this to multiple classes, as in Table 3.5? See also Figure 3.12. Here, the features are points $(x_1, x_2) \in \mathbb{R}^2$ and the labels are in $\{1, 2, 3\}$, corresponding to elements in $\mathbb{R}^3$, rather than in $\{0, 1\}$, which corresponded to probabilities.

In predicting binary labels, one-dimensional logistic regression compares the feature to a threshold; recall (3.3). The decision regions are thus half-lines.

With higher-dimensional features and multiclass labels, we might expect decision regions to be *polytopes*.

At a high level, binary one-dimensional logistic regression consisted of two parts; a calibrated probability model, viz. (3.1), and then a voting procedure, viz. (3.4). With a binary label, it was sufficient to optimize over a model for the probability of label 1; the probability of label 0 was complementary. This also meant that our loss function (3.7) could be written as a function of one argument, namely the probability of class 1. Similarly, with a binary label, the class with the higher probability was also the class with probability larger than 50%.

Logistic probabilities were essentially a function of one variable; the valued function

$$x \mapsto (1 - S(x), S(x)) = \left(\frac{1}{1+e^x}, \frac{e^x}{1+e^x}\right), \qquad x \in \mathbb{R}, \tag{3.17}$$

modeled the probability of class 0 and 1, respectively, as a function of $x \in \mathbb{R}$. The class of models we calibrated in (3.1) was the composition of logistic probabilities with a linear map $x \mapsto wx + b$. Let's now consider *softmax* probability map

$$S_{\text{softmax}}(x) \stackrel{\text{def}}{=} \left(\frac{e^{x_1}}{\sum_{i'=1}^{3} e^{x_{i'}}}, \frac{e^{x_2}}{\sum_{i'=1}^{3} e^{x_{i'}}}, \frac{e^{x_3}}{\sum_{i'=1}^{3} e^{x_{i'}}}\right), \qquad x = \begin{pmatrix} x_1 \\ x_2 \\ x_3 \end{pmatrix} \in \mathbb{R}^3.$$

Let's take our model of probabilities of the classes to be softmax probabilities composed with linear transformations from feature space into $\mathbb{R}^3$. In our example,

$$\mathfrak{m}(x;\theta) = S_{\text{softmax}}(wx + b),$$

where

$$\theta = (W, b)$$

is a parameter vector in $\Theta = \mathbb{R}^{3\times 2} \times \mathbb{R}^3$ consisting of weight and bias components of an affine transformation of $\mathbb{R}^2$. Once we have a model for probabilities, we can use *plurality* voting to predict the class. Writing

$$\mathfrak{m}(x;\theta) = \big(\mathfrak{m}_1(x;\theta), \mathfrak{m}_2(x;\theta), \mathfrak{m}_3(x;\theta)\big)$$

to denote the different components of $\mathfrak{m}(x;\theta)$, we will predict

$$\begin{cases} \text{predict class 1 if } \mathfrak{m}_1(x;\theta) > \max\{\mathfrak{m}_2(x;\theta), \mathfrak{m}_3(x;\theta)\} \\ \text{predict class 2 if } \mathfrak{m}_2(x;\theta) > \max\{\mathfrak{m}_1(x;\theta), \mathfrak{m}_3(x;\theta)\} \\ \text{predict class 3 if } \mathfrak{m}_3(x;\theta) > \max\{\mathfrak{m}_1(x;\theta), \mathfrak{m}_2(x;\theta)\}. \end{cases} \tag{3.18}$$

Extending (3.7), let's define

$$\ell_{p'}(p) \stackrel{\text{def}}{=} \sum_{i\in\{1,2,3\}} p_i \ln \frac{p_i}{p'_i} \tag{3.19}$$

for probability vectors (i.e., a vector of elements taking values in $[0,1]$ whose elements add up to 1) $p = \begin{pmatrix} p_1 & p_2 & p_3 \end{pmatrix}$ and $p' = \begin{pmatrix} p'_1 & p'_2 & p'_3 \end{pmatrix}$ such that $p' \in (0,1)^3$. We should write our ground-truth label data as vectors representing certainty of the different classes:

- assign label $\begin{pmatrix} 1 & 0 & 0 \end{pmatrix}$ if the label is 1,
- assign label $\begin{pmatrix} 0 & 1 & 0 \end{pmatrix}$ if the label is 2,
- assign label $\begin{pmatrix} 0 & 0 & 1 \end{pmatrix}$ if the label is 3.

These labels are *one-hot* vectors: one entry is 1, and the others are 0. One-hot vectors are a common way to encode categorical data. Our ground-truth data $\mathcal{D}$ is then a collection of points (x,y) in $\mathbb{R}^2 \times \mathbb{R}^3$, where the y's are one-hot probability vectors.

The per-datapoint loss is

$$\lambda_{(x,y)}(\theta) \stackrel{\text{def}}{=} \ell_y\left(\mathfrak{m}(x;\theta)\right), \qquad (x,y) \in \mathcal{D},$$

and the average loss is

$$\Lambda(\theta) \stackrel{\text{def}}{=} \frac{1}{|\mathcal{D}|} \sum_{(x,y)\in\mathcal{D}} \lambda_{(x,y)}(\theta).$$

Making (3.19) a bit more explicit, we have

$$\ell_{p'}(p) = -\ln p_i$$

if p' is the one-hot vector whose ith element is equal to 1 and the rest of the elements are equal to 0. We note that (3.19) thus naturally generalizes (3.7) to higher dimensions (this being one of the appeals of using entropy as a loss function).

To be specific, let's work through the quantization rule of (3.18) if the optimal parameter vector is

$$\theta^* = \begin{pmatrix} \begin{pmatrix} 2 & 1 \\ -1 & 0 \\ 1 & -3 \end{pmatrix} \\ \begin{pmatrix} 3 \\ 0 \\ -1 \end{pmatrix} \end{pmatrix},$$

i.e., our model is

$$\mathfrak{m}(x;\theta^*) = S_{\text{softmax}}\left(\begin{pmatrix} 2 & 1 \\ -1 & 0 \\ 1 & -3 \end{pmatrix}\begin{pmatrix} x_1 \\ x_2 \end{pmatrix} + \begin{pmatrix} 3 \\ 0 \\ -1 \end{pmatrix}\right)$$

$$= S_{\text{softmax}}\left(\begin{pmatrix} 2x_1 + x_2 + 3 \\ x_1 \\ x_1 - 3x_2 - 1 \end{pmatrix}\right) \qquad x = \begin{pmatrix} x_1 \\ x_2 \end{pmatrix} \in \mathbb{R}^2,$$

$$= \left(\frac{e^{2x_1+x_2+3}}{D}, \frac{e^{x_1}}{D}, \frac{e^{x_1-3x_2-1}}{D}\right)$$

where

$$D \stackrel{\text{def}}{=} e^{2x_1+x_2+3} + e^{x_1} + e^{x_1-3x_2-1}.$$

We can then rewrite (3.18) as

$$\begin{cases} \text{predict class 1 if } \frac{e^{2x_1+x_2+3}}{D} > \max\left\{\frac{e^{x_1}}{D}, \frac{e^{x_1-3x_2-1}}{D}\right\} \\ \text{predict class 2 if } \frac{e^{x_1}}{D} > \max\left\{\frac{e^{2x_1+x_2+3}}{D}, \frac{e^{x_1-3x_2-1}}{D}\right\} \\ \text{predict class 3 if } \frac{e^{x_1-3x_2-1}}{D} > \max\left\{\frac{e^{2x_1+x_2+3}}{D}, \frac{e^{x_1}}{D}\right\} \end{cases}$$

$$= \begin{cases} \text{predict class 1 if } e^{2x_1+x_2+3} > \max\{e^{x_1}, e^{x_1-3x_2-1}\} \\ \text{predict class 2 if } e^{x_1} > \max\{e^{2x_1+x_2+3}, e^{x_1-3x_2-1}\} \\ \text{predict class 3 if } e^{x_1-3x_2-1} > \max\{e^{2x_1+x_2+3}, e^{x_1}\} \end{cases}$$

$$= \begin{cases} \text{predict class 1 if } 2x_1 + x_2 + 3 > \max\{x_1, x_1 - 3x_2 - 1\} \\ \text{predict class 2 if } x_1 > \max\{2x_1 + x_2 + 3, x_1 - 3x_2 - 1\} \\ \text{predict class 3 if } x_1 - 3x_2 - 1 > \max\{2x_1 + x_2 + 3, x_1\}. \end{cases}$$

Looking a bit more closely at the requirements for predicting class 1, the inequality

$$2x_1 + x_2 + 3 > \max\{x_1, x_1 - 3x_2 - 1\}$$

is equivalent to

$$2x_1 + x_2 + 3 > x_1 \qquad \text{and} \qquad 2x_1 + x_2 + 3 > x_1 - 3x_2 - 1,$$

which is equivalent to

$$x_1 + x_2 > -3 \qquad \text{and} \qquad x_1 + 4x_2 > -4.$$

In other words, we decide class 1 if

$$\begin{pmatrix} 1 & 1 \end{pmatrix}\begin{pmatrix} x_1 \\ x_2 \end{pmatrix} > -3 \qquad \text{and} \qquad \begin{pmatrix} 1 & 4 \end{pmatrix}\begin{pmatrix} x_1 \\ x_2 \end{pmatrix} > -4.$$

This captures the decision rule for class 1 as the intersection of two hyperplanes (defined by their normals). Alternately, we could write the requirements for deciding class 0 as

$$x_2 > -x_1 - 3 \qquad \text{and} \qquad x_2 > -\tfrac{1}{2}x_2 - 1,$$

which captures the decision rule for class 0 as half-spaces defined by lines. As with univariate logistic regression, we are unlikely to ever face a feature variable (x_1, x_2) on the boundary lines or the triple-point where all inequalities are replaced by equalities. In summary, multiclass multivariate logistic regression leads to *polytopes*.

The framework of multiclass classification reduces to standard logistic regression for binary labels. If we would try to predict a $\{1, 2\}$-valued label on the basis of a scalar label x in the above way, multiclass prediction would search over probabilities of label 2 of the form

$$\begin{aligned}(1 - S(wx + b), S(wx + b)) &= \left(\frac{1}{1 + e^{wx+b}}, \frac{e^{wx+b}}{1 + e^{wx+b}}\right)\\ &= S_{\text{softmax}}(0, wx + b)\\ &= S_{\text{softmax}}\left(\begin{pmatrix}0\\w\end{pmatrix}x + \begin{pmatrix}0\\b\end{pmatrix}\right),\end{aligned}$$

so the collection of models considered by binary logistic regression is a subset of the collection of models considered by multiclass prediction. On the other hand, for general

$$W = \begin{pmatrix}w_1\\w_2\end{pmatrix} \qquad \text{and} \qquad B = \begin{pmatrix}b_1\\b_2\end{pmatrix},$$

we have

$$\begin{aligned}&S_{\text{softmax}}(Wx + B)\\ &= \left(\frac{e^{w_1x+b_1}}{e^{w_1x+b_1} + e^{w_2x+b_2}}, \frac{e^{w_2x+b_2}}{e^{w_1x+b_1} + e^{w_2x+b_2}}\right)\\ &= \left(\frac{1}{1 + e^{(w_2-w_1)x+(b_2-b_1)}}, \frac{e^{(w_2-w_1)x+(b_2-b_1)}}{1 + e^{(w_2-w_1)x+(b_2-b_1)}}\right)\\ &= (1 - S((w_1 - w_2)x + (b_1 - b_2)), S((w_1 - w_2)x + (b_1 - b_2))).\end{aligned}$$

Thus, the models of multiclass predictive probabilities collapse to the collection of models of logistic regression probabilities. Optimizing over models with logistic regression predictive probabilities is consequently the same as optimizing over models with multiclass predictive probabilities.

3.8. Brief Concluding Remarks

Logistic regression is another classical topic in statistics. Many excellent textbooks are available in the literature. The books [**Agr15**, **MSN08**] contain a nice exposition of logistic regression and more generally to generalized, linear, and mixed models, but there are many other excellent sources on this topic as well; see for example ([**DS98**, **MPV21**]). The book [**HTF10**] also has a good coverage with an eye towards statistical learning.

Our review of logistic regression was structured to introduce prediction of categorical variables. Deep neural networks heavily use the logistic function to model probabilities. Entropy, the softmax function, and one-hot encoding all scale well to higher-dimensional problems. We will extend our use of gradient descent to the more complex problems in deep neural networks.

3.9. Exercises

Exercise 3.1. Prove that the softmax function $S_{\text{softmax}}(x) : \mathbb{R}^D \to \mathbb{R}^D$ produces a probability distribution.

Exercise 3.2. Consider logistic regression with the two feature-label datapoints $(x_1, y_1) = (-2, 0)$ and $(x_2, y_2) = (1, 1)$. Compute the loss function at parameter values $\theta = (w, b) = (1/2, 1)$.

Exercise 3.3. Consider the relative entropy

$$H(p', p) = p' \ln \frac{p'}{p} + (1 - p') \ln \frac{1 - p'}{1 - p},$$

with $p, p' \in (0, 1)$. Show that the function $p' \mapsto H(p, p')$ is convex for each $p \in [0, 1]$.

Exercise 3.4. Consider the relative entropy

$$H(p', p) = p' \ln \frac{p'}{p} + (1 - p') \ln \frac{1 - p'}{1 - p},$$

with $p, p' \in (0, 1)$. Let us use entropy as a means to understand Euler's equations of optimality. Show that the Legendre-Fenchel transform of relative entropy, i.e., $L(\theta, p) = \max_{p' \in (0,1)} \{\theta p' - H(p', p)\}$, is the logarithm of the moment generating function of the Bernoulli random variable.

Exercise 3.5. Compute the Legendre-Fenchel transform

$$L(\theta, p) = \max_{p' \in (0,1)} \{\theta p' - \ln(p e^{\theta} + (1 - p))\},$$

and compare your answer to the setting of Exercise 3.4.

Chapter 4

From the Perceptron Model to Kernels to Neural Networks

4.1. Introduction

The goal of this chapter is to offer a different angle on how one can approach the classification problem with the ultimate goal of building towards the neural network formulation. We start with the simple perceptron model, show its connection to kernels, which then naturally leads to the conception of a neural network as a classifier and function approximator.

Kernel methods and kernel-based formulations are standard statistics and machine learning topics that one can find in many textbooks. In this chapter we use the framework of perceptron and kernels in order to introduce neural networks and motivate stochastic gradient descent. We describe the perceptron model in Section 4.2, which will motivate the stochastic gradient descent algorithm for neural networks. Then, in Section 4.3 we revisit the perceptron model and reformulate it through the lens of a kernel. In Section 4.4 we revisit linear regression and connect it to kernels. In Section 4.5, we motivate neural networks through the lens of kernels.

We introduce the idea of the neural network as a classifier and function approximator through the lens of the kernel formulation. As we will see, neural networks naturally come up as function approximators via the kernel perspective, building towards deep learning.

4.2. Perceptron Model and Stochastic Gradient Descent

Consider the problem of binary classification. In particular, assume that $y \in \{-1, 1\}$ and set (again with $\theta = (w, b)^\top$)

$$\mathfrak{m}(x; \theta) = \text{sign}(w \cdot x + b) = \begin{cases} -1 & \text{if } w \cdot x + b < 0, \\ 1 & \text{if } w \cdot x + b \geq 0. \end{cases}$$

This is the perceptron model introduced in [**Ros58**]. Consider, for example, the case where $w = (w_1, w_2)$. Here w_1, w_2 could for instance be length and weight, respectively, and we may, for example, be interested into classifying cats versus dogs.

An obvious loss function is the $0-1$ loss function

$$\Lambda_{0-1}(\theta) = \frac{1}{M} \sum_{m=1}^{M} 1_{\{y_m \neq \mathfrak{m}(x_m;\theta)\}}.$$

Note that the function $\lambda_{0-1}^m(\theta) = 1_{\{y_m \neq \mathfrak{m}(x_m;\theta)\}}$ is not a differentiable loss function. This lack of differentiability makes it hard to use optimization tools to minimize $\Lambda_{0-1}(\theta)$ in order to find the potential minimizer θ^*. Other (smoother) loss functions that can be used in place of $\lambda_{0-1}^m(\theta)$ are for example

$$\begin{aligned} \lambda_{\text{hinge}}^m(\theta) &= \mathsf{ReLU}(-y_m(w \cdot x_m + b)) = \max(0, -y_m(w \cdot x_m + b)), \\ \lambda_{\text{logistic}}^m(\theta) &= \log\big(1 + e^{-y_m(w \cdot x_m + b)}\big). \end{aligned}$$

Then, the optimization problem becomes

$$\text{Find } \theta^* = \text{argmin} \frac{1}{M} \sum_{m=1}^{M} \lambda^m(\theta) \tag{4.1}$$

for the chosen $\lambda^m(\theta)$, which, for example, could be any of the hinge loss, $\lambda_{\text{hinge}}^m(\theta)$, or the logistic loss, $\lambda_{\text{logistic}}^m(\theta)$.

A simple, but powerful, idea to solve this optimization problem is

Update the weights in the direction of the negative gradient.

Assume now that $b = 0$. Alternatively, note that b can be absorbed into w by extending the vector x to have its first element be defined to be equal to one. In either case, let the parameter θ be defined to be $\theta = w$. In the case of hinge loss for example, we then compute

$$\nabla \lambda_{\text{hinge}}^m(w) = \begin{cases} -y_m x_m & \text{if } y_m(w \cdot x_m) < 0 \text{ (i.e., } x_m \text{ incorrectly predicts)} \\ 0 & \text{otherwise.} \end{cases}$$

The *algorithm* becomes

- At time k, select a datapoint (x_k, y_k) sampled uniformly from $\mathcal{D}_{\text{train}}$.

- Update the weights

$$w_{k+1} = \begin{cases} w_k + y_k x_k & \text{if } y_k \neq \operatorname{sign}(w_k \cdot x_k) \\ w_k & \text{otherwise.} \end{cases} \tag{4.2}$$

Notice that the weight w_{k+1} becomes $w_k + y_k \cdot x_k$ only if a mistake happens. In fact, by direct substitution, using the fact that $y_k^2 = 1$ yields

$$y_k(w_{k+1} \cdot x_k) = y_k(w_k \cdot x_k) + \|x_k\|^2,$$

which says that an update leads to the quantity $y_k(w \cdot x_k)$ getting a push towards the *positive direction*, i.e., it increases. Thus, the algorithm inherently attempts to correct itself, eventually leading to a positive value for $y_k(w \cdot x_k)$.

This algorithm is essentially the *stochastic gradient descent* (SGD) algorithm that we study in detail in Chapter 7. In short, if we have an optimization problem like (4.1), SGD tries to approximate θ^* by iteratively computing

$$\theta_{k+1} = \theta_k - \eta \nabla \lambda^m(\theta_k)$$

for a randomly chosen $m \in \{1, \dots, M\}$. Here η is the *learning rate*. We study the SGD algorithm in detail in Chapters 7 and 8 and later its theoretical aspects in Chapter 18.

4.3. Perceptron Through the Lens of a Kernel

In this section, we present how one can go from a perceptron to a kernel which, as we will see in Section 4.5, then naturally leads to a neural network formulation.

Recall the update w_{k+1} via (4.2). Let $\alpha_m \in \{0, 1, 2, \dots\}$ denote how many times the perceptron sampled the datapoint (x_m, y_m) and led to an incorrect prediction. Then, we accumulate all those times in the linear combination

$$\hat{c} = \sum_{m=1}^{M} \alpha_m y_m x_m,$$

which naturally leads to the model, classifying the prediction to be -1 or 1,

$$\mathfrak{m}(x; \theta) = \operatorname{sign}(\hat{c} \cdot x) = \operatorname{sign}\left(\sum_{m=1}^{M} \alpha_m y_m x_m \cdot x \right),$$

where $\theta = \hat{c}$. Observing the last formula, we note that if for a given x we want to compute the classifier $\mathfrak{m}(x; \theta)$, we have two options. The first option is to keep track of $\hat{c}$ at the end of training. The second option is to keep track of $\{\alpha_1, \dots, \alpha_M\}$. As we shall see next, this second option can be thought of as a representation in which the kernel formulation arises in a natural way. In fact this second option carries the name of *dual representation* and the $\{\alpha_1, \dots, \alpha_M\}$ are called the dual variables. Another interpretation of the dual variables is that they constitute the weights that linearly combine with the training dataset $\{x_1, \dots, x_M\}$ to multiply the new input x to produce the corresponding prediction.

Let's assume now that we want to map x to a higher-dimensional space called the *feature space*, represented by, potentially, a high-dimensional vector $\psi(x)$, called the *feature vector*. To have a concrete example in mind, we may think of the situation of polynomial regression where a polynomial of higher order seems to fit the data the best instead of the straight line. For example, if $x = (a, b)$ is two-dimensional and the feature space is quadratic, we could set $\psi(x) = (a^2, \sqrt{2}ab, b^2)$ to obtain the quadratic feature space. We will see more such concrete examples in Chapter 11. Then, we can set

$$c_{k+1} = \begin{cases} c_k + y_k \psi(x_k) & \text{if } y_k \neq \operatorname{sign}(c_k \cdot \psi(x_k)) \\ c_k & \text{otherwise.} \end{cases}$$

Following the previous logic, we write for the accumulation of times that the perceptron predicted incorrectly

$$\hat{c} = \sum_{m=1}^{M} \alpha_m y_m \psi(x_m).$$

In this case, in order to do a prediction for a given input x, we use the model

$$\begin{aligned} \mathfrak{m}(x; \theta) = \operatorname{sign}(\hat{c} \cdot \psi(x)) &= \operatorname{sign}\left(\sum_{m=1}^{M} \alpha_m y_m \psi(x_m) \cdot \psi(x) \right) \\ &= \operatorname{sign}\left(\sum_{m=1}^{M} \alpha_m y_m \left(\psi(x_m) \cdot \psi(x) \right) \right). \end{aligned}$$

It is interesting to note that we no longer have a linear combination of the data $\{x_m\}_{m=1}^{M}$. Instead, we have a linear combination of the features $\{\psi(x_m)\}_{m=1}^{M}$. In addition, we also notice that to compute the prediction for a new datapoint x, we need to keep track of the product

$$K(x_m, x) = \psi^{\top}(x_m)\psi(x).$$

In the quadratic example above, where $x_m = (a_m, b_m) \in \mathbb{R}^2$ and $x = (a, b) \in \mathbb{R}^2$, we have $K(x_m, x) = (a_m a + b_m b)^2 = (x_m \cdot x)^2$.

It turns out that the perspective that led to the formulation above and the function $K(\cdot, \cdot)$ that we just defined are of broader interest. In fact, such functions K are called kernels.

Definition 4.1. If $K : \mathcal{X} \times \mathcal{X} \mapsto \mathbb{R}$ is symmetric and positive semidefinite then it is called a *kernel*.

Let us look at the perceptron model through the lens of a kernel based on the dual representation (see also [**STC04**]). This is sometimes referred to as the kernel perceptron update rule. Instead of adding $y \cdot x$ to w when x is misclassified (see (4.2)), we add 1 to the corresponding dual variable α:

- Initialize at $\alpha_m = 0$ for every $m \in \{1, \dots, M\}$.
- Sample at the mth iteration a datapoint $(x_{m^*}, y_{m^*}) \in \mathcal{D}_{\text{train}}$ uniformly at random.
- If there is an error, namely if we have that

$$y_{m^*} \neq \text{sign}\left(\sum_{m=1}^{M} \alpha_m y_m K(x_m, x_{m^*}) \right),$$

then update $\alpha_{m^*} = \alpha_{m^*} + 1$.

Note that, in this reformulation, we do not compute $\psi(x)$ anymore, which could be a very high-dimensional vector, and thus, is very expensive to compute. Instead, we compute the so-called Gram matrix of the data $G_{i,j} = K(x_i, x_j)$. We recall that we arrived at this formulation by setting $K(x_i, x_j) = \psi^\top(x_i)\psi(x_j)$.

Remark 4.2. Some comments on the kernel formulation versus the feature-space version are in order. While the kernel perceptron uses all available data samples at each iteration, the Gram matrix instead computes the kernel on all different pairs in the available dataset. Generally in machine learning, even though the Gram matrix formulation is preferable when the feature vector is large, the feature-space version is more attractive when the dataset is large. Low rank approximations methods can be used to make the calculation of the Gram matrix more tractable.

Let us now point out another interesting connection between Gram matrices and features. We have seen that given features, we can define the Gram matrix of the data G via the relation $G_{i,j} = K(x_i, x_j) = \psi^\top(x_i)\psi(x_j)$. It turns out that one can go in the other direction as well.

To see this, we first note that the Gram matrix of any dataset is a positive semidefinite matrix. Indeed, for all $\xi \in \mathbb{R}^M$, we have that

$$\xi^\top G \xi = \sum_{i,j=1}^{M} \xi_i \xi_j K(x_i, x_j) = \left(\sum_{m=1}^{M} \xi_m \psi(x_m) \right)^\top \left(\sum_{m=1}^{M} \xi_m \psi(x_m) \right) \geq 0,$$

showing that G is a positive semidefinite matrix.

Second, by Mercer's theorem, we have that for a function $K : \mathcal{X} \times \mathcal{X} \mapsto \mathbb{R}$, which is

- symmetric;
- square integrable, i.e.,

$$\int_{(x,x')\in\mathcal{X}\times\mathcal{X}} |K(x,x')|^2 dx dx' < \infty,$$

(which, in the finite-dimensional case $\mathcal{X} = \{x_1, \dots, x_M\}$, translates to $\sum_{i,j=1}^{M} |K(x_i, x_j)|^2 < \infty$); and such that,

- for all $h \in L^2(\mathcal{X})$

$$\int_{(x,x')\in\mathcal{X}\times\mathcal{X}} K(x,x')h(x)h(x')dx dx' \geq 0,$$

(which, in the finite dimensional case $\mathcal{X} = \{x_1, \dots, x_M\}$, translates to requiring that for any $\xi \in \mathbb{R}^M$, $\sum_{i,j=1}^{M} \xi_i \xi_j K(x_i, x_j) \geq 0$),

there exist $\psi_i : \mathcal{X} \mapsto \mathbb{R}$ and $\mu_i \geq 0$ such that

$$K(x,x') = \sum_{i\geq 1} \mu_i \psi_i^\top(x) \cdot \psi_i(x).$$

Mercer's theorem demonstrates that indeed we can also go from kernels to features. This is an important and powerful observation. In particular, kernels allow us to not have to worry much about feature spaces. Mercer's theorem essentially says that if the function K that we have chosen to work with is a kernel, then there will be a feature space to which K corresponds. We will further explore this key property in Section 4.5 to motivate deep learning.

4.4. Linear Regression and Kernels

Now that we have seen what a kernel is and how it applies in the perceptron model case, let us see how kernels naturally emerge in the setting of linear regression that we saw in Chapter 2.

Let us consider the case of linear regression with penalization. Similar to (3.16), we consider

$$\lambda_{(x,y)}^{(C)}(\theta) = \ell_y(\mathfrak{m}(x;\theta)) + C\|\theta\|^2, \tag{4.3}$$

where we set

$$\ell_y(\mathfrak{m}(x;\theta)) = \frac{1}{2}(y - \mathfrak{m}(x;\theta))^2, \text{ with } \mathfrak{m}(x;\theta) = w^\top \psi(x) + b,$$

$$\theta = w$$

(for the sake of exposition, we assume that b is known) and $\psi(x)$ is a potentially nonlinear feature space mapping. With $|\mathcal{D}| = M$, let the average loss be

$$\Lambda(\theta) = \frac{1}{M}\sum_{m=1}^{M} \lambda^{(C)}_{(x_m,y_m)}(\theta).$$

This can still be viewed as a linear regression problem, as even though the dependence on x may not be linear, the dependence on the unknown parameter $\theta = w$ is affine.

Taking the gradient $\nabla_w \Lambda(\theta) = \nabla_w \Lambda(w) = 0$, we obtain the equation for the corresponding w

$$w = -\frac{1}{2C}\sum_{m=1}^{M} (w^\top \psi(x_m) + b)\psi(x_m).$$

We deliberately do not solve this equation for w. Instead, we define $\alpha_m = -\frac{1}{2C}(w^\top \psi(x_m) + b)$. With this definition, we then write for the optimal w in terms of the α_m's,

$$w = \sum_{m=1}^{M} \alpha_m \psi(x_m). \tag{4.4}$$

The next step is to plug that into $\Lambda(w)$. For this purpose, we let $K(x_i, x_j) = \psi^\top(x_i)\psi(x_j)$ be a kernel, let $G = [G_{i,j}]_{i,j=1}^{M}$ be the Gram matrix with elements $G_{i,j} = K(x_i, x_j)$, $\tilde{b} = (b, \dots, b)^\top$, and let $\alpha = (\alpha_1, \dots, \alpha_M)^\top$. We can then write that

$$\Lambda(w) = \frac{1}{M}\left[\frac{1}{2}|\alpha^\top G|^2 + \alpha^\top G\tilde{b} + \frac{1}{2}\tilde{b}^\top\tilde{b} + C\alpha^\top G\alpha\right].$$

Now, we notice that instead of viewing Λ as a function of w, we can also view it as a function of α. Namely, abusing notation, we write

$$\Lambda(\alpha) = \frac{1}{M}\left[\frac{1}{2}|\alpha^\top G|^2 + \alpha^\top G\tilde{b} + \frac{1}{2}\tilde{b}^\top\tilde{b} + C\alpha^\top G\alpha\right].$$

The aforementioned expression is called the dual representation, and α are the dual variables.

Setting now the gradient $\nabla_\alpha \Lambda(\alpha) = 0$, we subsequently obtain for the solution with respect to the dual variables α

$$\alpha = -(G + 2CI)^{-1}\tilde{b}.$$

How can this be used for model prediction when a new datapoint x^* comes in? Going back to (4.4) and defining the design matrix Ψ to be the matrix whose kth column is given by $\psi(x_k)$ gives

$$w = \Psi\alpha,$$

giving for the corresponding model's prediction y^* (we assume b is known here for convenience)

$$y^* = w^\top \psi(x^*) + b = \alpha^\top \Psi^\top \psi(x^*) + b.$$

Note that in the last display, $\Psi^\top \psi(x^*)$ is simply the vector with elements $K(x_i, x^*)$.

Therefore, we have obtained that the linear regression problem can be formulated simply in terms of the Gram matrix G (through the dual variables α) and the kernel function K. Hence, as we show with the perceptron model in Section 4.3, in the linear regression setting we can work explicitly with the kernel function K. Then, thinking in the reverse direction, if we do indeed work with the kernel formulation, then Mercer's theorem guarantees that there would be a feature space to which the chosen kernel K corresponds.

4.5. From Kernels to Neural Networks

In the previous sections, we show that linear models such as linear regression and the perceptron can be written in the dual representation in terms of a Gram matrix G, or equivalently (in these cases) in terms of a kernel K. Mercer's theorem says that if K satisfies certain properties, then there is a feature space to which K corresponds.

A natural question arises. Is there a practical way to understand the underlying feature space? How could we choose ψ representing the feature space?

We take the perspective of learning the feature vectors ψ instead of fixing them a priori! This is one of the key ideas behind deep learning.

In the case of the perceptron we saw that

$$\mathfrak{m}(x;\theta) = \operatorname{sign}(c \cdot \psi(x)) = \operatorname{sign}\left(\sum_{n=1}^{N} c^n \psi^n(x)\right),$$

where $c = (c^1, \ldots, c^N)$ are weights and $\psi(x) = (\psi^1(x), \ldots, \psi^N(x))$ are features.

Let us set $\psi(x) = \sigma(w \cdot x)$ for a (potentially) unknown matrix w. Our model now has become

$$\mathfrak{m}(x; c, w) = \operatorname{sign}(c \cdot \sigma(w \cdot x)),$$

and we want to augment θ to include w as an unknown parameter. We define $\theta = (c^1, \ldots, c^N, w^1, \ldots, w^N)$, and we instead consider the model

$$\mathfrak{m}(x;\theta) = \operatorname{sign}\left(\sum_{n=1}^{N} c^n \sigma(w^n \cdot x)\right).$$

We can define the features to be $\psi^n(x) = \sigma(w^n \cdot x)$. We then find θ through the operation

$$\theta^* = \operatorname{argmin}_{\theta=(c,w)\in\Theta} \frac{1}{M} \sum_{m=1}^{M} \lambda\left(y_m, \sum_{n=1}^{N} c^n \sigma(w^n \cdot x_m)\right),$$

where λ could for example be the hinge loss that we saw earlier (or the logistic loss or any other loss function of our choice).

The latter step of modeling $\psi^n(x) = \sigma(w^n \cdot x)$ and considering $\theta = (c, w)$ as the parameter to be learned is very consequential. The problem suddenly becomes (a) nonlinear, (b) nonconvex, and (c) high dimensional. Indeed:

- **nonlinear**: the model is now a nonlinear function of (c, w).
- **nonconvex**: the optimization with respect to c was a convex problem, whereas now the optimization with respect to the augmented variable (c, w) is a nonconvex problem.
- **high dimensional**: previously we were only dealing with the vector c, whereas now we also need to learn the matrix w.

In fact, $\mathfrak{m}(x;\theta) = \operatorname{sign}\left(\sum_{n=1}^{N} c^n \sigma(w^n \cdot x)\right)$ is a neural network!

4.6. Brief Concluding Remarks

The conception of the perceptron model by Rosenblatt in 1958 [**Ros58**] was one of the biggest milestones in the development of neural network-based artificial intelligence. The book [**MP17**] goes deeper into the properties of perceptron learning.

Kernel methods for pattern analysis and statistical learning is a huge and very well-developed subject. Kernel functions enable us to be able to work with lower-dimensional algorithms and play a central role in the problem of linear approximation in the high-dimensional limit. Excellent texts that significantly expand on the topic of kernel methods include [**STC04, Bis06, HTF10, Bac24**]. There, kernel methods are discussed and analyzed for generic statistical and machine learning problems (not necessarily specific to deep learning).

The reader who is interested in how kernel methods compare to neural network-based methods is referred to [**GMMM20**] and the references therein. In particular, a number of empirical studies has shown that for many tasks, suitable kernel-based methods can replace neural network-based methods without much of a decline in performance. On the other hand, it has also been shown that neural networks suffer less from the curse of dimensionality and can approximate functions which cannot be learned through kernel-based methods.

Our goal in this chapter was more modest. Our exploration of kernels was centered around the objective of motivating the formulation of neural networks. The goal was to demonstrate in an intuitive way that the kernel formulation motivates the passage from linear problems to nonlinear problems, leading to the neural network formulation.

As we demonstrated in this chapter, kernels can motivate the formulation of neural networks; see also the lecture notes [**Cha22**] for a related discussion that partially also motivated aspects of the presentation in Sections 4.3 and 4.5, as well as [**Bis06**] for a related discussion regarding Section 4.4. In Chapter 5 we introduce one of the main classes of neural networks, that of feed forward neural networks, which is the building block for many of the deep learning algorithms.

Chapter 5

Feed Forward Neural Networks

5.1. Introduction

Feed forward networks can be thought of as compositions of logistic regression operators, which themselves are a composition of linear regression and the logistic function S.

To handle the higher-dimensional framework of feed forward networks, let's *overload* functions by vectorizing them (which numpy naturally does in Python). If $\phi : \mathbb{R} \to \mathbb{R}$, we define

$$\phi\left(\begin{pmatrix} x_1 \\ x_2 \\ \vdots \\ x_N \end{pmatrix}\right) \stackrel{\text{def}}{=} \begin{pmatrix} \phi(x_1) \\ \phi(x_2) \\ \vdots \\ \phi(x_N) \end{pmatrix},$$

i.e., ϕ of a vector is defined as ϕ applied to the elements of the vector. A feed forward neural network $\mathfrak{m}(x;\theta)$ chains together some parts of the models of multiclass prediction of Chapter 3.

Let's build what amounts to a feed forward network with two internal layers of dimensions 3 and 4 with internal activation functions of tanh. Let's also use a ReLU for the final layer, where

$$\mathsf{ReLU}(x) \stackrel{\text{def}}{=} \max\{x, 0\}, \qquad x \in \mathbb{R}.$$

Let's define

$$
\begin{aligned}
D_0 &= 2,\\
D_1 &= 3,\\
D_2 &= 4,\\
D_3 &= 1.
\end{aligned}
\tag{5.1}
$$

For $n \in \{1, 2, 3\}$ and $W_n \in \mathbb{R}^{D_n \times D_{n-1}}$ (i.e., a matrix of dimensions $D_n \times D_{n-1}$) and $B_n \in \mathbb{R}^{D_n}$, define the linear mapping $L^{(n)}_{W_n,B_n} : \mathbb{R}^{D_{n-1}} \to \mathbb{R}^{D_n}$ as

$$
L^{(n)}_{W_n,B_n}(x) \stackrel{\text{def}}{=} W_n x + B_n, \qquad x \in \mathbb{R}^{D_{n-1}}.
$$

For a parameter vector

$$
\theta \stackrel{\text{def}}{=} (W_1, B_1, W_2, B_2, W_3, B_3)
$$

in the parameter space $\Theta = \mathbb{R}^{D_1 \times D_0} \times \mathbb{R}^{D_1} \times \mathbb{R}^{D_2 \times D_1} \times \mathbb{R}^{D_2} \times \mathbb{R}^{D_3 \times D_2} \times \mathbb{R}^{D_3}$, let's define the feed forward neural network

$$
\mathfrak{m}(x;\theta) \stackrel{\text{def}}{=} \mathsf{ReLU}\left(L^{(3)}_{W_3,B_3}\left(\tanh\left(L^{(2)}_{W_2,B_2}\left(\tanh\left(L^{(1)}_{W_1,B_1}(x)\right)\right)\right)\right)\right), \quad x \in \mathbb{R}^F.
$$

To be very specific, if

$$
\theta = \left(\begin{pmatrix} 1 & -3 \\ 2 & 4 \\ -6 & 0 \end{pmatrix}, \begin{pmatrix} -2 \\ 0 \\ -1 \end{pmatrix}, \begin{pmatrix} -1 & -2 & 0 \\ 6 & 7 & -1 \\ 5 & 6 & -2 \\ 0 & 1 & -2 \end{pmatrix}, \begin{pmatrix} 4 \\ 1 \\ -2 \\ 3 \end{pmatrix}, \begin{pmatrix} 5 & -1 & 1 & 0 \end{pmatrix}, 6 \right), \tag{5.2}
$$

then

$$
\begin{aligned}
&\mathfrak{m}(x;\theta)\\
&= \mathsf{ReLU}\left(\begin{pmatrix} 5 & -1 & 1 & 0 \end{pmatrix} \tanh\left(\left(\begin{pmatrix} -1 & -2 & 0 \\ 6 & 7 & -1 \\ 5 & 6 & -2 \\ 0 & 1 & -2 \end{pmatrix} \tanh\left(\left(\begin{pmatrix} 1 & -3 \\ 2 & 4 \\ -6 & 0 \end{pmatrix} \begin{pmatrix} x_1 \\ x_2 \end{pmatrix} \right.\right.\right.\right.\\
&\qquad\qquad \left.\left.\left.\left. + \begin{pmatrix} -2 \\ 0 \\ -1 \end{pmatrix} \right) \right) + \begin{pmatrix} 4 \\ 1 \\ -2 \\ 3 \end{pmatrix} \right) \right) + 6 \right).
\end{aligned}
\tag{5.3}
$$

In practice we would rarely write out the parameter vector θ; we here hope that a fully expressed simple example will make things very concrete. We also remark that this example is a special case of the generic formula (1.1) for feed forward neural networks, which is the composition of *layers* of linear transformations and nonlinear functions. Oftentimes, it is customary to represent neural networks schematically, as in Figure 5.1.

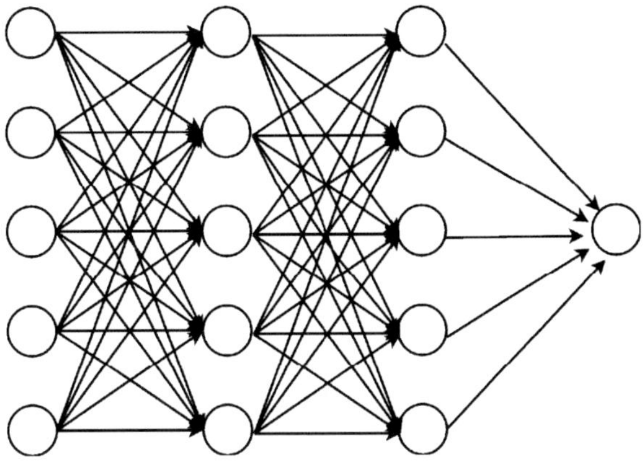

Figure 5.1. A schematic representation of a feed forward neural network with two hidden layers

In Figure 5.1 the vertices on the left represent the input data $x_1, \dots, x_5$. The two sets of vertices in the middle represent the two hidden layers, composed of hidden units. The arrows represent linear transformation of the inputs (either input feature data at the beginning or hidden units in what follows). The vertex on the far right represents the neural network output. Note that in a feed forward neural network any single input affects all the subsequent hidden layers.

Given our training data $\mathcal{D}$, which is a multiset in $\mathbb{R}^2 \times \mathbb{R}$, we might want to define an error function

$$\ell_y(y') \stackrel{\text{def}}{=} (y - y')^2, \qquad y' \in \mathbb{R},$$

for $y \in \mathbb{R}$ and then define the per-datapoint loss

$$\lambda_{(x,y)}(\theta) \stackrel{\text{def}}{=} \ell_y(\mathfrak{m}(x;\theta)), \tag{5.4}$$

for $(x, y) \in \mathcal{D}$ and $\theta \in \Theta$. Finally, we define the average loss function

$$\Lambda(\theta) \stackrel{\text{def}}{=} \frac{1}{|\mathcal{D}|} \sum_{(x,y)\in\mathcal{D}} \lambda_{(x,y)}(\theta), \qquad \theta \in \Theta.$$

The model (5.3) leads to several natural questions:

(1) How can we improve our choice (5.2) of parameters?

(2) Did we choose the number (2) of internal layers well?

(3) Did we choose the dimensions (3 and 4) of the internal layers well?

(4) Did we choose the internal *activation* functions (tanh) well?

We can use *gradient descent* to address the first question; this is typically called *training*. The second and third questions are typically thought of as *hyperparameter* selection, and it is typically addressed via *validation* procedures, see Chapter 11.

5.2. Truth Tables

Feed forward networks are one of the core architectures of deep neural networks. In practice, one selects the number of layers and their internal dimensions (i.e., (5.1)), the activation functions (i.e., tanh and ReLU), and the loss function (i.e., (5.4)), and then implements these in code.

In this section, we explore a nontrivial neural network without needing to explicitly compute the neural network. We also show using explicit basic constructions that linear combinations of indicators of rectangles can approximate generic functions. The latter can be viewed as a precursor to the well-developed universal approximation theory for neural networks that we explore in Chapter 16.

To make sure that we fully understand the basics, let's see how to implement two-dimensional *truth tables* as neural networks.

As an example, let's consider training data consisting of a Gaussian collection of points in the (x, y) plane. Let's assume that the statistics of the x and y coordinates have independent standard normal distribution. Let's assume that the points in the first quadrant (where x and y are both positive) have label 1 and the others have label 0. See Figure 5.2. We would like to train a neural network to recover the *rule* underlying these labeled data. Namely, we want to construct an analogue of (5.3) which approximates the labeling map

$$(x, y) \mapsto \begin{cases} 1 & \text{if } x \geq 0 \text{ and } y \geq 0 \\ 0 & \text{else} \end{cases} = \mathbf{1}_{\mathbb{R}_+ \times \mathbb{R}_+}(x, y). \tag{5.5}$$

Given that there are four quadrants (see Figure 5.3), there are $2^4 = 16$ possible labeling maps depending only on quadrant. Several benefits accrue from thinking through how to implement a number of these labeling maps in neural networks.

- It is a generic problem: any function can be approximated by a linear combination of indicators of rectangles (i.e., and maps, similar to (5.5)), see [**RF10**].
- We can use theoretically derived representations of neural networks as points of comparison for computationally derived representations.
- The labeling map (5.5) is an and map (it assigns label 1 when $x > 0$ and $y > 0$). More complicated truth tables can be decomposed into

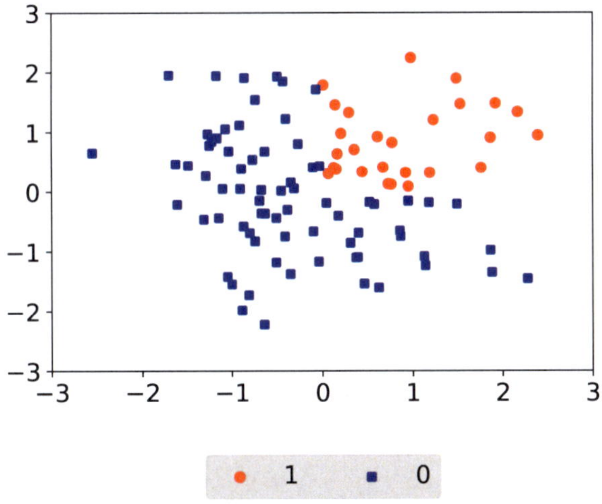

Figure 5.2. First quadrant

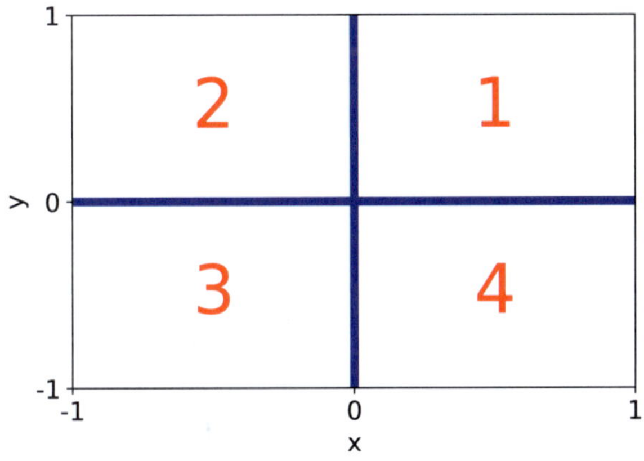

Figure 5.3. Quadrants

and, or, and not. More complicated logic *should* require deeper networks.

A basic building block will be an approximation of the step function $\mathbf{1}_{\mathbb{R}_+}$. For $\varepsilon > 0$, define

$$S_\varepsilon(x) \overset{\text{def}}{=} S\left(\frac{x}{\varepsilon}\right) = \frac{e^{x/\varepsilon}}{1 + e^{x/\varepsilon}}, \qquad x \in \mathbb{R}.$$

Then

$$\mathbf{1}_{\mathbb{R}_+}(x) \approx S_\varepsilon(x)$$

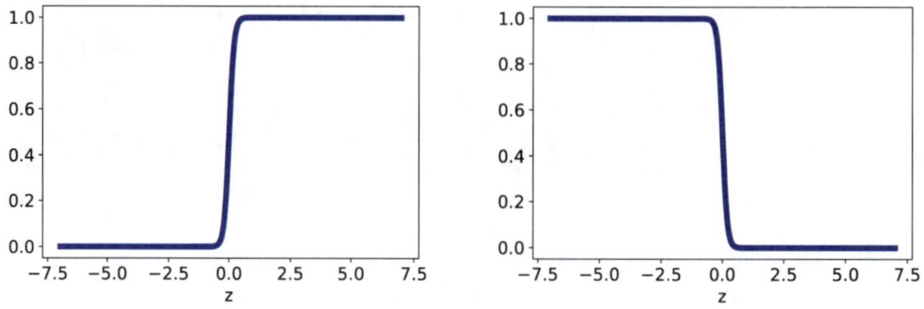

Figure 5.4. Scaled and reversed logistic functions

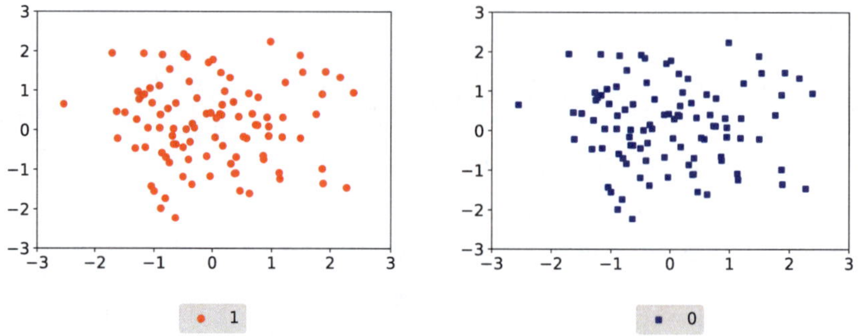

Figure 5.5. Degenerate cases

for (almost) $x \in \mathbb{R}$ as $\varepsilon \searrow 0$; see Figure 5.4. We can also write

$$\mathbf{1}_{\mathbb{R}_{-}}(x) \approx S_{\varepsilon}(-x)$$
$$\mathbf{1}_{\mathbb{R}_{-}}(x) \approx 1 - S_{\varepsilon}(x)$$

for (almost) all $x \in \mathbb{R}$ as $\varepsilon \searrow 0$. Thus, it should not surprise us that various representations will *not* be unique.

Let's start with some degenerate cases. Let's assume that all ground-truth datapoints are labeled 0; see Figure 5.5. We might write

$$\mathbf{1}_{\emptyset}(x, y) = S_{\varepsilon}(-1)$$

for all $(x, y) \in \mathbb{R}^2$ as $\varepsilon \searrow 0$; i.e., the underlying labeling map generating the data is approximated by $S_{\varepsilon}(-1)$. In fact, $\varepsilon \searrow 0$, $S_{\varepsilon}(z) \approx 0$ for all $z < 0$; for simplicity, let us take $z = -1$. Alternately, if all ground-truth points are labeled 1, we might write

$$\mathbf{1}_{\mathbb{R}^2}(x, y) \approx S_{\varepsilon}(1)$$

for all $(x, y) \in \mathbb{R}^2$ as $\varepsilon \searrow 0$.

We might next consider half-planes. Assume that all ground-truth datapoints $(x, y) \in \mathbb{R}^2$ with $x > 0$ (i.e., the right half-plane) are labeled 1, and all ground-truth points $(x, y) \in \mathbb{R}^2$ with $x < 0$ (the left half-plane); see Figure 5.6.

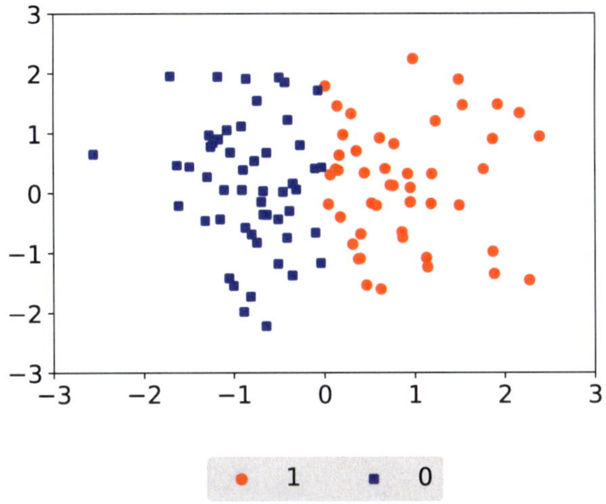

Figure 5.6. Half-planes

Let's here write

$$\mathbf{1}_{\mathbb{R}_+\times\mathbb{R}}(x,y) \approx S_\varepsilon(x)$$

as $\varepsilon \searrow 0$. Similarly, we can write

$$\begin{aligned}\mathbf{1}_{\mathbb{R}_-\times\mathbb{R}}(x,y) &\approx S_\varepsilon(x),\\ \mathbf{1}_{\mathbb{R}\times\mathbb{R}_+}(x,y) &\approx S_\varepsilon(y),\\ \mathbf{1}_{\mathbb{R}\times\mathbb{R}_-}(x,y) &\approx S_\varepsilon(-y).\end{aligned}$$

Let's now dive into a more interesting ground-truth set. Let's assume that all ground-truth datapoints (x,y) in the first quadrant $\mathbb{R}_+\times\mathbb{R}_+$ are labeled 1, and the rest are labeled 0 (see Figure 5.2). Ignoring points on the axes, and assuming that $\varepsilon \ll 1$, let's calculate that

(5.6)

$$S_\varepsilon(x)+S_\varepsilon(y) \approx \begin{cases} 2 & \text{if } (x,y)\in\mathbb{R}_+\times\mathbb{R}_+ \text{ (1st quadrant)}\\ 1 & \text{if } (x,y)\in(\mathbb{R}_+\times\mathbb{R}_-)\cup(\mathbb{R}_-\times\mathbb{R}_+) \text{ (2nd, 4th quadrants)}\\ 0 & \text{if } (x,y)\in\mathbb{R}_-\times\mathbb{R}_- \text{ (3rd quadrant).}\end{cases}$$

We can then single out the first quadrant as the collection of points (x,y) where $S_\varepsilon(x)+S_\varepsilon(y) > 3/2$ (and actually we could replace $3/2$ with any number in $(1,2)$). Let us implement applying a *shift* $z \mapsto S_\varepsilon(z-3/2)$ of S_ε. Namely,

$$\begin{aligned}S_\varepsilon\left(S_\varepsilon(x)+S_\varepsilon(y)-3/2\right) &\approx \begin{cases}1 & \text{if } (x,y)\in\mathbb{R}_+\times\mathbb{R}_+ \text{ (the first quadrant)}\\ 0 & \text{else}\end{cases}\\ &\approx \mathbf{1}_{\mathbb{R}_+\times\mathbb{R}_+}(x,y). \qquad (5.7)\end{aligned}$$

Returning to (5.6), we note that we can replace $3/2$ with any shift $z\in(1,2)$, as least in the regime where $\varepsilon \searrow 0$. Specific values of z and ε naturally come

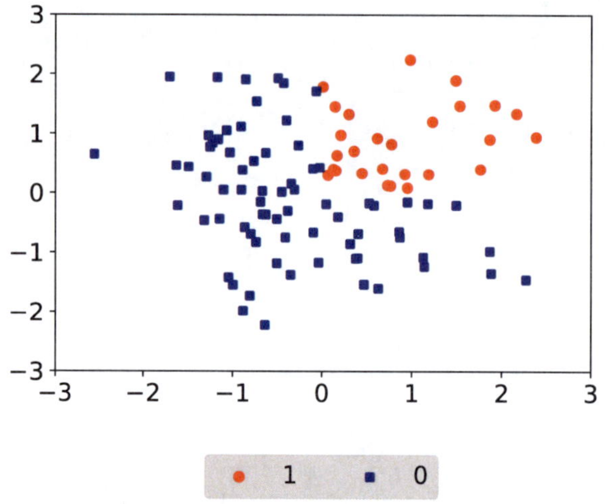

Figure 5.7. The first quadrant

from minimizing a loss function which matches the model to training data. In a certain regime, this loss function should thus be (asymptotically and locally) independent of z. Without delving into a more precise formulation, this nevertheless suggests that loss functions in deep neural networks may in practice be approximately constant along lower-dimensional submanifolds.

Let's next assume that the points (x, y) in our ground-truth dataset are labeled 1 only in the fourth quadrant $\mathbb{R}_+ \times \mathbb{R}_-$; see Figure 5.8. We can reflect the above calculations for the first quadrant across the x axis to get

$$\mathbf{1}_{\mathbb{R}_+\times\mathbb{R}_-}(x, y) \approx S_\varepsilon\left(S_\varepsilon(x) + S_\varepsilon(-y) - 3/2\right). \tag{5.8}$$

Using the symmetry of S, we of course also have that

$$\mathbf{1}_{\mathbb{R}_+\times\mathbb{R}_-}(x, y) \approx S_\varepsilon\left(S_\varepsilon(x) + \{1 - S_\varepsilon(y)\} - 3/2\right) = S_\varepsilon\left(S_\varepsilon(x) - S_\varepsilon(y) - 1/2\right). \tag{5.9}$$

This reflects the analogue of (5.6):

$$S_\varepsilon(x) - S_\varepsilon(y) \approx \begin{cases} 1-1 & \text{if } (x,y) \in \mathbb{R}_+ \times \mathbb{R}_+ \text{ (1st quadrant)} \\ 0-1 & \text{if } (x,y) \in \mathbb{R}_+ \times \mathbb{R}_+ \text{ (2nd quadrant)} \\ 0-0 & \text{if } (x,y) \in \mathbb{R}_- \times \mathbb{R}_- \text{ (3rd quadrant)} \\ 1-0 & \text{if } (x,y) \in \mathbb{R}_+ \times \mathbb{R}_- \text{ (4th quadrant)} \end{cases}$$

$$= \begin{cases} 0 & \text{if } (x,y) \in (\mathbb{R}_+ \times \mathbb{R}_+) \cup (\mathbb{R}_- \times \mathbb{R}_-) \text{ (1st and 3rd quadrants)} \\ -1 & \text{if } (x,y) \in \mathbb{R}_+ \times \mathbb{R}_+ \text{ (2nd quadrant)} \\ 1 & \text{if } (x,y) \in \mathbb{R}_+ \times \mathbb{R}_- \text{ (4th quadrant).} \end{cases}$$

Here, the shifted logistic map $z \mapsto S_\varepsilon(z - 1/2)$ allows us to single out the fourth quadrant.

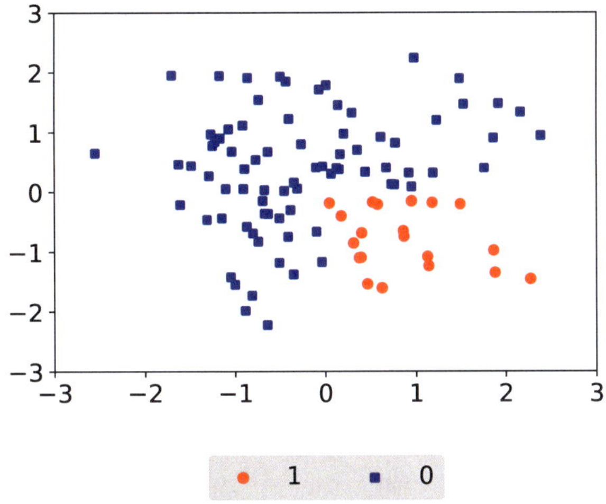

Figure 5.8. The fourth quadrant

Remark 5.1. A fairly profound consequence follows. Deep neural networks (as almost all other machine learning procedures) depend upon minimizing a loss function to best match coefficients to training data. We should *not* expect this loss function to have a unique minimum. In our example here of selecting a quadrant in $\mathbb{R}^2$ (a logical and), the loss function should have (at least) two minima, corresponding to (5.8) and (5.9).

We can of course easily generalize to the second and third quadrants:

$$\mathbf{1}_{\mathbb{R}_-\times\mathbb{R}_+}(x,y) \approx S_\varepsilon\left(S_\varepsilon(-x)+S_\varepsilon(y)-3/2\right) = S_\varepsilon\left(-S_\varepsilon(x)+S_\varepsilon(y)-1/2\right)$$
$$\mathbf{1}_{\mathbb{R}_-\times\mathbb{R}_-}(x,y) \approx S_\varepsilon\left(S_\varepsilon(-x)+S_\varepsilon(-y)-3/2\right) = S_\varepsilon\left(-S_\varepsilon(x)-S_\varepsilon(-y)+1/2\right).$$

We can of course take complements. Let's assume that all ground-truth datapoints (x, y) in the first quadrant $\mathbb{R}_+ \times \mathbb{R}_+$ are labeled 0, and the rest are labeled 1 (see Figure 5.9). Naturally,

$$\mathbf{1}_{\mathbb{R}^2\setminus(\mathbb{R}_+\times\mathbb{R}_+)}(x,y) \approx 1-\mathbf{1}_{\mathbb{R}_+\times\mathbb{R}_+}(x,y) = 1-S_\varepsilon\left(S_\varepsilon(x)+S_\varepsilon(y)-3/2\right).$$

To be parallel to our other representations (where the last *layer* is S_ε), let's again use the symmetry of S_ε to write

$$\mathbf{1}_{\mathbb{R}^2\setminus(\mathbb{R}_+\times\mathbb{R}_+)}(x,y) \approx S_\varepsilon\left(-S_\varepsilon(x)-S_\varepsilon(y)+3/2\right).$$

Similarly,

$$\mathbf{1}_{\mathbb{R}^2\setminus(\mathbb{R}_+\times\mathbb{R}_-)}(x,y) \approx S_\varepsilon\left(-S_\varepsilon(x)-S_\varepsilon(-y)+3/2\right) = S_\varepsilon\left(-S_\varepsilon(x)+S_\varepsilon(y)+1/2\right),$$
$$\mathbf{1}_{\mathbb{R}^2\setminus(\mathbb{R}_-\times\mathbb{R}_+)}(x,y) \approx S_\varepsilon\left(-S_\varepsilon(-x)-S_\varepsilon(y)+3/2\right) = S_\varepsilon\left(S_\varepsilon(x)-S_\varepsilon(y)+1/2\right),$$
$$\begin{aligned}\mathbf{1}_{\mathbb{R}^2\setminus(\mathbb{R}_-\times\mathbb{R}_-)}(x,y) &\approx S_\varepsilon\left(-S_\varepsilon(-x)-S_\varepsilon(-y)+3/2\right)\\ &= S_\varepsilon\left(S_\varepsilon(-x)+S_\varepsilon(-y)-1/2\right).\end{aligned}$$

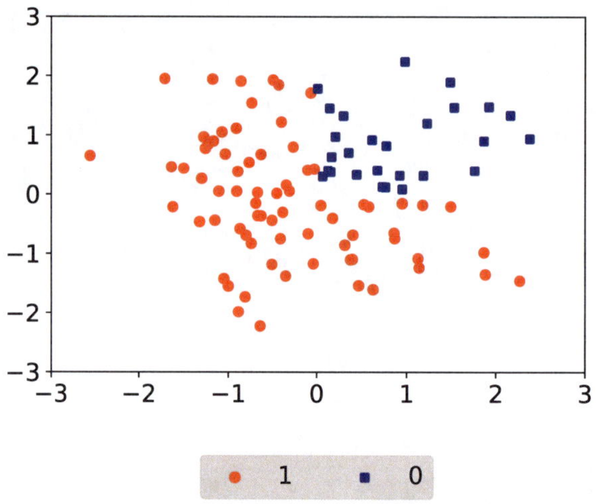

Figure 5.9. Complement of the first quadrant

Finally, let's consider opposing quadrants. Assume that all ground-truth datapoints (x, y) in either $\mathbb{R}_+ \times \mathbb{R}_+$ or $\mathbb{R}_- \times \mathbb{R}_-$ are labeled 1, and the remaining points (in $\mathbb{R}_+ \times \mathbb{R}_-$ and $\mathbb{R}_- \times \mathbb{R}_+$, i.e., the second and fourth quadrants) are labeled 0; see Figure 5.11. We can then write

$$\begin{aligned} \mathbf{1}_{(\mathbb{R}_+\times\mathbb{R}_+)\cup(\mathbb{R}_-\times\mathbb{R}_-)}(x,y) &\approx \mathbf{1}_{\mathbb{R}_+\times\mathbb{R}_+}(x,y) + \mathbf{1}_{\mathbb{R}_-\times\mathbb{R}_-}(x,y) \\ &\approx S_\varepsilon\left(S_\varepsilon(x) + S_\varepsilon(y) - 3/2\right) + S_\varepsilon\left(S_\varepsilon(-x) + S_\varepsilon(-y) - 3/2\right). \end{aligned} \tag{5.10}$$

We are approximating $\mathbf{1}_{(\mathbb{R}_+\times\mathbb{R}_+)\cup(\mathbb{R}_-\times\mathbb{R}_-)}$ as the sum of two functions which take values in $(0,1)$. We have no assurance that this sum will take values in $(0,1)$, particularly in the transition region near the origin $(0,0)$. We would also like the output of our approximate labeling map to be in $(0,1)$. Consider the map

$$z \mapsto S_\varepsilon(z - 1/2). \tag{5.11}$$

This function approximates the identity map on $\{0,1\}$, and takes values in $(0,1)$ (i.e., it *clips* values greater than 1). Let's replace (5.10) with

$$\begin{aligned} \mathbf{1}_{(\mathbb{R}_+\times\mathbb{R}_+)\cup(\mathbb{R}_-\times\mathbb{R}_-)}(x,y) \approx S_\varepsilon\big(&S_\varepsilon\left(S_\varepsilon(x) + S_\varepsilon(y) - 3/2\right) \\ &+ S_\varepsilon\left(S_\varepsilon(-x) + S_\varepsilon(-y) - 3/2\right) - 1/2\big). \end{aligned} \tag{5.12}$$

We can of course similarly write

$$\begin{aligned} \mathbf{1}_{(\mathbb{R}_+\times\mathbb{R}_-)\cup(\mathbb{R}_-\times\mathbb{R}_+)}(x,y) \approx S_\varepsilon\big(&S_\varepsilon\left(S_\varepsilon(x) + S_\varepsilon(-y) - 3/2\right) \\ &+ S_\varepsilon\left(S_\varepsilon(-x) + S_\varepsilon(y) - 3/2\right) - 1/2\big). \end{aligned} \tag{5.13}$$

The representations (5.12) and (5.13) have three layers of compositions of S_ε. Using (5.11), we can of course rewrite all of our representations with three

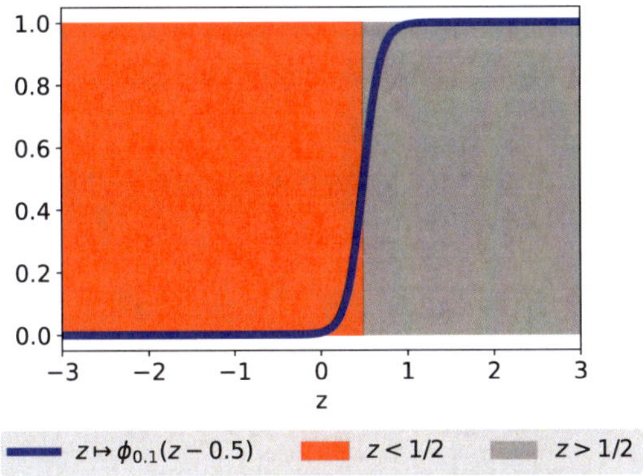

Figure 5.10. Clipping function

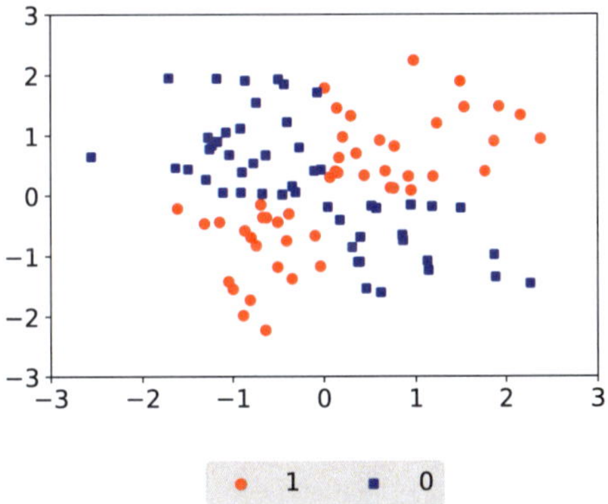

Figure 5.11. The first and third quadrants

layers. Collecting everything together, we have (for $\varepsilon \ll 1$)

$$
\begin{aligned}
\mathbf{1}_{\emptyset}(x,y) &= S_\varepsilon\left(S_\varepsilon\left(S_\varepsilon(-1) - 1/2\right) - 1/2\right),\\
\mathbf{1}_{\mathbb{R}^2}(x,y) &= S_\varepsilon\left(S_\varepsilon\left(S_\varepsilon(1) - 1/2\right) - 1/2\right),\\
\mathbf{1}_{\mathbb{R}_+\times\mathbb{R}}(x,y) &\approx S_\varepsilon\left(S_\varepsilon\left(S_\varepsilon(x) - 1/2\right) - 1/2\right),\\
\mathbf{1}_{\mathbb{R}_-\times\mathbb{R}}(x,y) &\approx S_\varepsilon\left(S_\varepsilon\left(S_\varepsilon(-x) - 1/2\right) - 1/2\right),\\
\mathbf{1}_{\mathbb{R}\times\mathbb{R}_+}(x,y) &\approx S_\varepsilon\left(S_\varepsilon\left(S_\varepsilon(y) - 1/2\right) - 1/2\right),\\
\mathbf{1}_{\mathbb{R}\times\mathbb{R}_-}(x,y) &\approx S_\varepsilon\left(S_\varepsilon\left(S_\varepsilon(-y) - 1/2\right) - 1/2\right),\\
\mathbf{1}_{\mathbb{R}_+\times\mathbb{R}_+}(x,y) &\approx S_\varepsilon\left(S_\varepsilon\left(S_\varepsilon(x) + S_\varepsilon(y) - 3/2\right) - 1/2\right),\\
\mathbf{1}_{\mathbb{R}_+\times\mathbb{R}_-}(x,y) &\approx S_\varepsilon\left(S_\varepsilon\left(S_\varepsilon(x) + S_\varepsilon(-y) - 3/2\right) - 1/2\right)\\
&= S_\varepsilon\left(S_\varepsilon\left(S_\varepsilon(x) - S_\varepsilon(y) - 1/2\right) - 1/2\right),\\
\mathbf{1}_{\mathbb{R}_-\times\mathbb{R}_+}(x,y) &\approx S_\varepsilon\left(S_\varepsilon\left(S_\varepsilon(-x) + S_\varepsilon(y) - 3/2\right) - 1/2\right)\\
&= S_\varepsilon\left(S_\varepsilon\left(-S_\varepsilon(x) + S_\varepsilon(y) - 1/2\right) - 1/2\right),\\
\mathbf{1}_{\mathbb{R}_-\times\mathbb{R}_-}(x,y) &\approx S_\varepsilon\left(S_\varepsilon\left(S_\varepsilon(-x) + S_\varepsilon(-y) - 3/2\right) - 1/2\right)\\
&= S_\varepsilon\left(S_\varepsilon\left(-S_\varepsilon(x) - S_\varepsilon(-y) + 1/2\right) - 1/2\right).
\end{aligned}
\tag{5.14}
$$

This allows us to represent two-dimensional truth tables as compositions of affine transformations and scaled logistic functions.

Define

$$\begin{aligned} D_0 &= 2, \\ D_1 &= 2, \\ D_2 &= 2, \\ D_3 &= 1. \end{aligned}$$

For $n \in \{1,2,3\}$ and $W_n \in \mathbb{R}^{D_n \times D_{n-1}}$ and $B_n \in \mathbb{R}^{D_n}$, define

$$L^{(n)}_{W_n,B_n}(X) \overset{\text{def}}{=} W_n X + B_n, \qquad X \in \mathbb{R}^{D_{n-1}}.$$

For a parameter vector

$$\theta \overset{\text{def}}{=} (W_1, B_1, W_2, B_2, W_3, B_3)$$

in the parameter space $\Theta = \mathbb{R}^{D_1 \times D_0} \times \mathbb{R}^{D_1} \times \mathbb{R}^{D_2 \times D_1} \times \mathbb{R}^{D_2} \times \mathbb{R}^{D_3 \times D_2} \times \mathbb{R}^{D_3}$, define for $(x, y) \in \mathbb{R}^2$

$$\mathfrak{m}\left(\begin{pmatrix} x \\ y \end{pmatrix}; \theta\right) \overset{\text{def}}{=} S_\varepsilon\left(L^{(3)}_{W_3,B_3}\left(S_\varepsilon\left(L^{(2)}_{W_2,B_2}\left(S_\varepsilon\left(L^{(1)}_{W_1,B_1}\left(\begin{pmatrix} x \\ y \end{pmatrix}\right)\right)\right)\right)\right)\right).$$

Let's start to go through the representations of (5.14), starting with $\mathbf{1}_\emptyset$ and $\mathbf{1}_{\mathbb{R}^2}$. Let's take

(5.15)
$$\begin{aligned} W^{(1)} &= \begin{pmatrix} 0 & 0 \\ 0 & 0 \end{pmatrix}, & B^{(1),\pm} &= \begin{pmatrix} \pm 1 \\ \pm 1 \end{pmatrix} \Big/ \varepsilon, \\ W^{(2)} &= \begin{pmatrix} 1 & 0 \\ 0 & 1 \end{pmatrix} \Big/ \varepsilon, & B^{(2)} &= -\begin{pmatrix} 1/2 \\ 1/2 \end{pmatrix} \Big/ \varepsilon, \\ W^{(3)} &= \begin{pmatrix} 1/2 & 1/2 \end{pmatrix} / \varepsilon, & B^{(3)} &= -1/2\varepsilon. \end{aligned}$$

Then

$$S\left(W^{(1)}\begin{pmatrix} x \\ y \end{pmatrix} + B^{(1),\pm}\right) = S\left(\begin{pmatrix} 0 & 0 \\ 0 & 0 \end{pmatrix}\begin{pmatrix} x \\ y \end{pmatrix} + \begin{pmatrix} \pm 1/\varepsilon \\ \pm 1/\varepsilon \end{pmatrix}\right) = \begin{pmatrix} S_\varepsilon(\pm 1) \\ S_\varepsilon(\pm 1) \end{pmatrix},$$

$$\begin{aligned} S\left(W^{(2)} S\left(W^{(1)}\begin{pmatrix} x \\ y \end{pmatrix} + B^{(1)}\right) + B^{(2)}\right) &= S\left(\begin{pmatrix} 1/\varepsilon & 0 \\ 0 & 1/\varepsilon \end{pmatrix}\begin{pmatrix} S_\varepsilon(\pm 1) \\ S_\varepsilon(\pm 1) \end{pmatrix} - \begin{pmatrix} 1/2\varepsilon \\ 1/2\varepsilon \end{pmatrix}\right) \\ &= S\left(\begin{pmatrix} S_\varepsilon(\pm 1) - 1/2 \\ S_\varepsilon(\pm 1) - 1/2 \end{pmatrix} \Big/ \varepsilon\right) \\ &= \begin{pmatrix} S_\varepsilon\left(S_\varepsilon(\pm 1) - 1/2\right) \\ S_\varepsilon\left(S_\varepsilon(\pm 1) - 1/2\right) \end{pmatrix}, \end{aligned}$$

and

$$
\begin{aligned}
&S\left(W^{(3)}S\left(W^{(2)}S\left(W^{(1)}\begin{pmatrix}x\\y\end{pmatrix}+B^{(1)}\right)+B^{(2)}\right)+B^{(3)}\right)\\
&= S\left(\begin{pmatrix}1/2\varepsilon & 1/2\varepsilon\end{pmatrix}\begin{pmatrix}S_\varepsilon\left(S_\varepsilon(\pm 1)-1/2\right)\\ S_\varepsilon\left(S_\varepsilon(\pm 1)-1/2\right)\end{pmatrix}-1/2\varepsilon\right)\\
&= S_\varepsilon\left(S_\varepsilon\left(S_\varepsilon(\pm 1)-1/2\right)-1/2\right)\\
&\approx \begin{cases}\mathbf{1}_{\mathbb{R}^2}(x,y) & \text{if input is } +1\\ \mathbf{1}_{\emptyset}(x,y) & \text{if input is } -1.\end{cases}
\end{aligned}
$$

The choice (5.15) of parameters directly implemented the first two lines of (5.14). We might also set

$$
W^{(3)} = \begin{pmatrix}0 & 0\end{pmatrix}/\varepsilon \qquad \text{and} \qquad B^{(3)} = -1/2\varepsilon, \tag{5.16}
$$

and we can then take $W^{(1)}$, $B^{(1)}$, $W^{(2)}$, and $B^{(2)}$ to be any elements of $\mathbb{R}^{2\times 1}$, $\mathbb{R}^2$, $\mathbb{R}^{2\times 2}$, and $\mathbb{R}^2$, respectively. This of course implies both a degenerate loss function (which does not depend on ($W^{(1)}$, $B^{(1)}$, $W^{(2)}$, or $B^{(2)}$ as long as (5.16) holds), and which has multiple global minima (at (5.15) and (5.16)).

Let's next consider the truth table which differentiates between the right- and left-hand planes. Let's consider $\mathbf{1}_{\mathbb{R}_+\times\mathbb{R}}$, and take

$$
\begin{aligned}
W^{(1)} &= \begin{pmatrix}1 & 0\\0 & 0\end{pmatrix}\Big/\varepsilon, & B^{(1)} &= \begin{pmatrix}0\\0\end{pmatrix}\Big/\varepsilon,\\
W^{(2)} &= \begin{pmatrix}1 & 0\\0 & 0\end{pmatrix}\Big/\varepsilon, & B^{(2)} &= -\begin{pmatrix}1/2\\0\end{pmatrix}\Big/\varepsilon,\\
W^{(3)} &= \begin{pmatrix}1 & 0\end{pmatrix}/\varepsilon, & B^{(3)} &= -1/2\varepsilon.
\end{aligned}
$$

Then

$$
\begin{aligned}
S\left(W^{(1)}\begin{pmatrix}x\\y\end{pmatrix}+B^{(1)}\right) &= S\left(\begin{pmatrix}1/\varepsilon & 0\\0 & 0\end{pmatrix}\begin{pmatrix}x\\y\end{pmatrix}+\begin{pmatrix}0\\0\end{pmatrix}\right)\\
&= S\left(\begin{pmatrix}x/\varepsilon\\0\end{pmatrix}\right) = \begin{pmatrix}S_\varepsilon(x)\\S(0)\end{pmatrix},\\
S\left(W^{(2)}S\left(W^{(1)}\begin{pmatrix}x\\y\end{pmatrix}+B^{(1)}\right)+B^{(2)}\right) &= S\left(\begin{pmatrix}1/\varepsilon & 0\\0 & 0\end{pmatrix}\begin{pmatrix}S_\varepsilon(x)\\S(0)\end{pmatrix}-\begin{pmatrix}1/2\varepsilon\\0\end{pmatrix}\right)\\
&= S\left(\begin{pmatrix}S_\varepsilon(x)-1/2\\0\end{pmatrix}\Big/\varepsilon\right)\\
&= \begin{pmatrix}S_\varepsilon\left(S_\varepsilon(x)-1/2\right)\\S(0)\end{pmatrix},
\end{aligned}
$$

and

$$\begin{aligned}
&S\left(W^{(3)}S\left(W^{(2)}S\left(W^{(1)}\begin{pmatrix}x\\y\end{pmatrix}+B^{(1)}\right)+B^{(2)}\right)+B^{(3)}\right)\\
&\quad= S\left(\begin{pmatrix}1/\varepsilon & 0\end{pmatrix}\begin{pmatrix}S_\varepsilon\left(S_\varepsilon(x)-1/2\right)\\ S(0)\end{pmatrix}-1/2\varepsilon\right)\\
&\quad= S_\varepsilon\left(S_\varepsilon\left(S_\varepsilon(x)-1/2\right)-1/2\right)\approx \mathbf{1}_{\mathbb{R}_+\times\mathbb{R}}(x,y).
\end{aligned}$$

Similar representations hold for the indicators of the upper, left, and lower half-planes.

Thirdly, we can represent the indicator of the first quadrant. Let's consider $\mathbf{1}_{\mathbb{R}_+\times\mathbb{R}_+}$ and take

$$\begin{aligned}
W^{(1)} &= \begin{pmatrix}1&0\\0&1\end{pmatrix}\Big/\varepsilon, & B^{(1)} &= \begin{pmatrix}0\\0\end{pmatrix},\\
W^{(2)} &= \begin{pmatrix}1&1\\0&0\end{pmatrix}\Big/\varepsilon, & B^{(2)} &= -\begin{pmatrix}3/2\\0\end{pmatrix}\Big/\varepsilon,\\
W^{(3)} &= \begin{pmatrix}1&0\end{pmatrix}/\varepsilon, & B^{(3)} &= -1/2\varepsilon.
\end{aligned}$$

Then

$$S\left(W^{(1)}\begin{pmatrix}x\\y\end{pmatrix}+B^{(1)}\right)=S\left(\begin{pmatrix}1/\varepsilon&0\\0&1/\varepsilon\end{pmatrix}\begin{pmatrix}x\\y\end{pmatrix}+\begin{pmatrix}0\\0\end{pmatrix}\right)=S\left(\begin{pmatrix}x/\varepsilon\\y/\varepsilon\end{pmatrix}\right)=\begin{pmatrix}S_\varepsilon(x)\\S_\varepsilon(y)\end{pmatrix},$$

$$\begin{aligned}
S\left(W^{(2)}S\left(W^{(1)}\begin{pmatrix}x\\y\end{pmatrix}+B^{(1)}\right)+B^{(2)}\right) &= S\left(\begin{pmatrix}1/\varepsilon&1/\varepsilon\\0&0\end{pmatrix}\begin{pmatrix}S_\varepsilon(x)\\S_\varepsilon(y)\end{pmatrix}-\begin{pmatrix}3/2\varepsilon\\0\end{pmatrix}\right)\\
&= S\left(\begin{pmatrix}S_\varepsilon(x)+S_\varepsilon(y)-3/2\\0\end{pmatrix}\Big/\varepsilon\right)\\
&= \begin{pmatrix}S_\varepsilon\left(S_\varepsilon(x)+S_\varepsilon(y)-3/2\right)\\S(0)\end{pmatrix},
\end{aligned}$$

and

$$\begin{aligned}
&S\left(W^{(3)}S\left(W^{(2)}S\left(W^{(1)}\begin{pmatrix}x\\y\end{pmatrix}+B^{(1)}\right)+B^{(2)}\right)+B^{(3)}\right)\\
&\quad= S\left(\begin{pmatrix}1/\varepsilon & 0\end{pmatrix}\begin{pmatrix}S_\varepsilon\left(S_\varepsilon(x)+S_\varepsilon(y)-3/2\right)\\ S(0)\end{pmatrix}-1/2\varepsilon\right)\\
&\quad= S_\varepsilon\left(S_\varepsilon\left(S_\varepsilon(x)+S_\varepsilon(y)-3/2\right)-1/2\right)\approx \mathbf{1}_{\mathbb{R}_+\times\mathbb{R}_+}(x,y).
\end{aligned}$$

Similar calculations hold for representations of indicators of other quadrants. Let's next represent the indicator of the complement of a single quadrant.

To approximate $\mathbf{1}_{\mathbb{R}^2\setminus(\mathbb{R}_-\times\mathbb{R}_+)}$, let's take

$$W^{(1)} = \begin{pmatrix} -1 & 0 \\ 0 & -1 \end{pmatrix} \Big/ \varepsilon, \qquad B^{(1)} = \begin{pmatrix} 0 \\ 0 \end{pmatrix},$$
$$W^{(2)} = \begin{pmatrix} 1 & 1 \\ 0 & 0 \end{pmatrix} \Big/ \varepsilon, \qquad B^{(2)} = -\begin{pmatrix} 1/2 \\ 0 \end{pmatrix} \Big/ \varepsilon,$$
$$W^{(3)} = \begin{pmatrix} 1 & 0 \end{pmatrix} / \varepsilon, \qquad B^{(3)} = -1/2\varepsilon.$$

Then

$$\begin{aligned} S\left(W^{(1)}\begin{pmatrix} x \\ y \end{pmatrix} + B^{(1)}\right) &= S\left(\begin{pmatrix} -1/\varepsilon & 0 \\ 0 & -1/\varepsilon \end{pmatrix}\begin{pmatrix} x \\ y \end{pmatrix} + \begin{pmatrix} 0 \\ 0 \end{pmatrix}\right) \\ &= S\left(\begin{pmatrix} -x/\varepsilon \\ -y/\varepsilon \end{pmatrix}\right) = \begin{pmatrix} S_\varepsilon(-x) \\ S_\varepsilon(-y) \end{pmatrix}, \\ S\left(W^{(2)}S\left(W^{(1)}\begin{pmatrix} x \\ y \end{pmatrix} + B^{(1)}\right) + B^{(2)}\right) &= S\left(\begin{pmatrix} 1/\varepsilon & 1/\varepsilon \\ 0 & 0 \end{pmatrix}\begin{pmatrix} S_\varepsilon(-x) \\ S_\varepsilon(-y) \end{pmatrix} - \begin{pmatrix} 1/2\varepsilon \\ 0 \end{pmatrix}\right) \\ &= S\left(\begin{pmatrix} S_\varepsilon(-x) + S_\varepsilon(-y) - 1/2 \\ 0 \end{pmatrix} \Big/ \varepsilon\right) \\ &= \begin{pmatrix} S_\varepsilon\left(S_\varepsilon(-x) + S_\varepsilon(-y) - 1/2\right) \\ S(0) \end{pmatrix}, \end{aligned}$$

and

$$\begin{aligned} &S\left(W^{(3)}S\left(W^{(2)}S\left(W^{(1)}\begin{pmatrix} x \\ y \end{pmatrix} + B^{(1)}\right) + B^{(2)}\right) + B^{(3)}\right) \\ &= S\left(\begin{pmatrix} 1/\varepsilon & 0 \end{pmatrix}\begin{pmatrix} S_\varepsilon\left(S_\varepsilon(-x) + S_\varepsilon(-y) - 1/2\right) \\ S(0) \end{pmatrix} - 1/2\varepsilon\right) \\ &= S_\varepsilon\left(S_\varepsilon\left(S_\varepsilon(-x) + S_\varepsilon(-y) - 1/2\right) - 1/2\right) \\ &\approx \mathbf{1}_{\mathbb{R}^2\setminus(\mathbb{R}_+\times\mathbb{R}_+)}(x, y). \end{aligned}$$

Finally, let's consider opposing quadrants. To approximate $\mathbf{1}_{(\mathbb{R}_+\times\mathbb{R}_+)\cup(\mathbb{R}_-\times\mathbb{R}_-)}$, let's take

$$W^{(1)} = \begin{pmatrix} 1 & 0 \\ 0 & 1 \end{pmatrix} \Big/ \varepsilon, \qquad B^{(1)} = \begin{pmatrix} 0 \\ 0 \end{pmatrix},$$
$$W^{(2)} = \begin{pmatrix} 1 & 1 \\ -1 & -1 \end{pmatrix} \Big/ \varepsilon, \qquad B^{(2)} = \begin{pmatrix} -3/2 \\ 1/2 \end{pmatrix} \Big/ \varepsilon,$$
$$W^{(3)} = \begin{pmatrix} 1 & 1 \end{pmatrix} / \varepsilon, \qquad B^{(3)} = -1/2\varepsilon.$$

Then, we have

$$S\left(W^{(1)}\begin{pmatrix}x\\y\end{pmatrix}+B^{(1)}\right)=S\left(\begin{pmatrix}1/\varepsilon & 0\\ 0 & 1/\varepsilon\end{pmatrix}\begin{pmatrix}x\\y\end{pmatrix}+\begin{pmatrix}0\\0\end{pmatrix}\right)=S\left(\begin{pmatrix}x/\varepsilon\\ y/\varepsilon\end{pmatrix}\right)=\begin{pmatrix}S_\varepsilon(x)\\ S_\varepsilon(y)\end{pmatrix},$$

$$\begin{aligned}S\left(W^{(2)}S\left(W^{(1)}\begin{pmatrix}x\\y\end{pmatrix}+B^{(1)}\right)+B^{(2)}\right)&=S\left(\begin{pmatrix}1/\varepsilon & 1/\varepsilon\\ -1/\varepsilon & -1/\varepsilon\end{pmatrix}\begin{pmatrix}S_\varepsilon(x)\\ S_\varepsilon(y)\end{pmatrix}+\begin{pmatrix}-3/2\varepsilon\\ 1/2\varepsilon\end{pmatrix}\right)\\ &=S\left(\begin{pmatrix}S_\varepsilon(x)+S_\varepsilon(y)-3/2\\ -S_\varepsilon(x)-S_\varepsilon(y)+1/2\end{pmatrix}\Big/\varepsilon\right)\\ &=\begin{pmatrix}S_\varepsilon\left(S_\varepsilon(x)+S_\varepsilon(y)-3/2\right)\\ S_\varepsilon\left(-S_\varepsilon(x)-S_\varepsilon(y)+1/2\right)\end{pmatrix},\end{aligned}$$

and

$$\begin{aligned}&S\left(W^{(3)}S\left(W^{(2)}S\left(W^{(1)}\begin{pmatrix}x\\y\end{pmatrix}+B^{(1)}\right)+B^{(2)}\right)+B^{(3)}\right)\\ &\quad=S\left(\begin{pmatrix}1/\varepsilon & 1/\varepsilon\end{pmatrix}\begin{pmatrix}S_\varepsilon\left(S_\varepsilon(x)+S_\varepsilon(y)-3/2\right)\\ -S_\varepsilon(x)-S_\varepsilon(y)-1/2\end{pmatrix}-1/2\varepsilon\right)\\ &\quad=S_\varepsilon\Big(S_\varepsilon\left(S_\varepsilon(x)+S_\varepsilon(y)-3/2\right)+S_\varepsilon\left(-S_\varepsilon(x)-S_\varepsilon(y)+1/2\right)-1/2\Big)\\ &\quad\approx \mathbf{1}_{(\mathbb{R}_+\times\mathbb{R}_+)\cup(\mathbb{R}_-\times\mathbb{R}_-)}(x,y).\end{aligned}$$

While we might have been a bit *too* thorough in writing out all of these cases, hopefully we have driven home the point that (sufficiently) deep neural networks can implement all logic.

5.3. Numerical Exploration

Some controlled computational experiments might help even more. The code for these examples can be found at https://mathdl.github.io/.

5.3.1. Half-space. We simulate some labeled data representing $\mathbf{1}_{\mathbb{R}_+\times\mathbb{R}}$ (see Figure 5.6) and then use PyTorch to train a two-layer neural network. We get

$$\begin{aligned}W^{(1)}&=\begin{pmatrix}7.380 & 0.174\\ 5.809 & 0.165\end{pmatrix}\approx\begin{pmatrix}6.6 & 0\\ 6.6 & 0\end{pmatrix},\\ B^{(1)}&=\begin{pmatrix}-0.283\\ -0.311\end{pmatrix}\approx\begin{pmatrix}0\\0\end{pmatrix},\\ W^{(2)}&=\begin{pmatrix}8.187 & 5.993\end{pmatrix},\\ B^{(2)}&=-6.287.\end{aligned}\tag{5.17}$$

Let's see how this compares to our theoretical representations. The first column of $W^{(1)}$ seems to be significantly larger than the second, and both elements of the first column are the same order of magnitude. Building upon this,

$B^{(1)}$ is also small. The first layer of the neural network corresponding to (5.17) is then about

$$S\left(W^{(1)}\begin{pmatrix}x\\y\end{pmatrix}+B^{(1)}\right)\approx S\left(\left(\begin{pmatrix}1&0\\1&0\end{pmatrix}\begin{pmatrix}x\\y\end{pmatrix}\Big/\varepsilon\right)+\begin{pmatrix}0\\0\end{pmatrix}\right)$$
$$=S\left(\begin{pmatrix}x\\x\end{pmatrix}\Big/\varepsilon\right)=\begin{pmatrix}S_\varepsilon(x)\\S_\varepsilon(x)\end{pmatrix}$$

with

$$\frac{1}{\varepsilon}=\frac{7.380+5.809}{2}\approx 6.595.$$

The second layer is approximately

$$S\,(8.187S_\varepsilon(x)+5.993S_\varepsilon(y)-6.287)\approx S_{\varepsilon'}\,(S_\varepsilon(x)-0.443)\,,$$

where

$$\frac{1}{\varepsilon'}=8.187+5.993\approx 14.180.$$

We thus approximately have that

$$S\left(W^{(2)}S\left(W^{(1)}\begin{pmatrix}x\\y\end{pmatrix}+B^{(1)}\right)+B^{(2)}\right)\approx S_{\varepsilon'}\,(S_\varepsilon(x)-0.5)\,,\tag{5.18}$$

which makes sense from our theoretical development.

Let's continue with this example, but start with a different initial condition. We here get

$$W^{(1)}=\begin{pmatrix}-4.730&-0.146\\7.795&0.169\end{pmatrix}\approx\begin{pmatrix}-6.3&0\\6.3&0\end{pmatrix},$$
$$B^{(1)}=\begin{pmatrix}0.099\\-0.136\end{pmatrix}\approx\begin{pmatrix}0\\0\end{pmatrix},$$
$$W^{(2)}=\begin{pmatrix}-5.676&9.608\end{pmatrix},$$
$$B^{(2)}=-1.754.$$

Then

$$S\left(W^{(1)}\begin{pmatrix}x\\y\end{pmatrix}B^{(1)}\right)\approx S_\varepsilon\left(\begin{pmatrix}-1&0\\1&0\end{pmatrix}\begin{pmatrix}x\\y\end{pmatrix}+\begin{pmatrix}0\\0\end{pmatrix}\right)=S_\varepsilon\left(\begin{pmatrix}-x\\x\end{pmatrix}\right)$$
$$=\begin{pmatrix}S_\varepsilon(-x)\\S_\varepsilon(x)\end{pmatrix},$$

where

$$\frac{1}{\varepsilon}=\frac{7.380+5.809}{2}\approx 6.595.$$

The second layer is approximately

$$\begin{aligned} &S(-5.676S_\varepsilon(-x) + 9.608S_\varepsilon(x) - 1.754) \\ &\quad\approx S(-5.676 + 5.676S_\varepsilon(x) + 9.608S_\varepsilon(x) - 1.754) \\ &\quad\approx S(15.284S_\varepsilon(x) - 7.43) \\ &\quad\approx S_{\varepsilon'}\left(S_\varepsilon(x) - 0.486\right), \end{aligned}$$

where

$$\frac{1}{\varepsilon'} = -5.676 + 9.608 \approx 15.284.$$

We again approximately have (5.18).

5.3.2. Quadrant. Let's do another example. We simulate some labeled data representing $\mathbf{1}_{\mathbb{R}_+\times\mathbb{R}_+}$ (see Figure 5.2), and we then use PyTorch to train a two-layer neural network. Here we get

$$\begin{aligned} W^{(1)} &= \begin{pmatrix} 0.096 & -7.474 \\ 7.514 & 0.063 \end{pmatrix} \approx \begin{pmatrix} 0 & -7.5 \\ 7.5 & 0 \end{pmatrix}, \\ B^{(1)} &= \begin{pmatrix} -0.237 \\ 0.054 \end{pmatrix} \approx \begin{pmatrix} 0 \\ 0 \end{pmatrix}, \\ W^{(2)} &= \begin{pmatrix} -9.638 & 9.194 \end{pmatrix}, \\ B^{(2)} &= -4.422. \end{aligned}$$

Here $W^{(1)}$ is approximately a large off-diagonal matrix, and $W^{(2)}$ is is approximately a scaled version of a vector of 1's. Then

$$S\left(W^{(1)}\begin{pmatrix} x \\ y \end{pmatrix} + B^{(1)}\right) \approx \begin{pmatrix} S_\varepsilon(-y) \\ S_\varepsilon(x) \end{pmatrix}$$

with

$$\frac{1}{\varepsilon} \approx 7.2.$$

The second layer is approximately

$$\begin{aligned} &S(-9.638S_\varepsilon(-y) + 9.194S_\varepsilon(x) - 4.422) \\ &\quad\approx S(-9.638 + 9.638S_\varepsilon(y) + 9.194S_\varepsilon(x) - 4.422) \\ &\quad\approx S(9.638S_\varepsilon(x) + 9.194S_\varepsilon(x) - 14.06) \\ &\quad\approx S_{\varepsilon'}\left(S_\varepsilon(x) + S_\varepsilon(y) - 1.493\right) \end{aligned}$$

with

$$\frac{1}{\varepsilon'} = \frac{-9.638 + 9.194}{2} \approx 9.416.$$

This exactly agrees with (5.7).

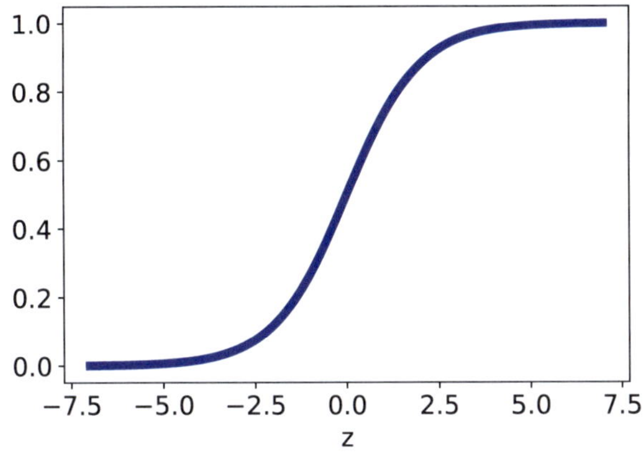

Figure 5.12. Logistic function

5.4. Activation Functions

Thus far, we have focused on logistic function nonlinearities. There are a number of other commonly used activation functions. Informally, activation functions have a *linear* regime, outside of which they *saturate* (see Figure 5.12). Composing activation functions with linear layers, we can represent local linear transformations; this is strongly reminiscent of a first-order Taylor expansion.

The *hyperbolic tangent* function

$$\tanh(x) = \frac{e^x - e^{-x}}{e^x + e^{-x}}, \qquad x \in \mathbb{R},$$

is often used in internal layers. It is linear near the origin and saturates at ± 1; see Figure 5.13.

Of course the tanh and logistic function S are related:

$$\begin{aligned} S(x) &= \frac{e^x}{1+e^x} = \frac{e^{x/2}e^{x/2}}{e^{x/2}\left(e^{x/2}+e^{-x/2}\right)} \\ &= \frac{e^{x/2}}{e^{x/2}+e^{-x/2}} = \frac{1}{2}\frac{\left(e^{x/2}+e^{-x/2}\right)+\left(e^{x/2}-e^{-x/2}\right)}{e^{x/2}+e^{-x/2}} \\ &= \frac{1}{2}\{1+\tanh(x/2)\}. \end{aligned}$$

This means that the set of compositions of linear layers and logistic functions is equivalent to the set of compositions of linear layers and tanh functions:

$$\begin{aligned} 5S(6x+8)+7 &= 10\left(\frac{1}{2}\left\{1+\tanh\left(\frac{6x+8}{2}\right)\right\}\right)+2 \\ &= 10\tanh(3x+4)+2. \end{aligned}$$

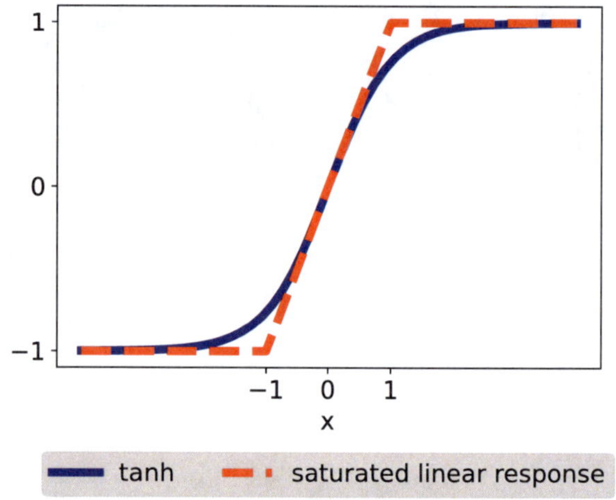

Figure 5.13. Tangent function

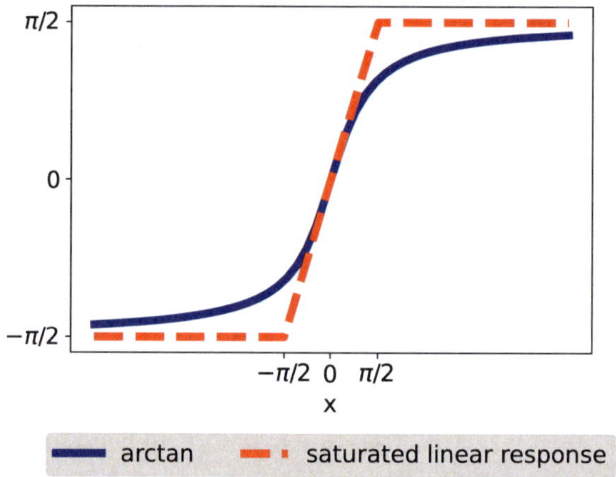

Figure 5.14. Arctangent function

The arctangent (inverse of tangent) function arctan can play a similar role as the tanh function, except that it saturates at $\pm\pi/2$. Since both arctan and tanh have derivative 1/4 at the origin, we can compare arctan to $x \mapsto \frac{\pi}{2}\tanh\left(\frac{2}{\pi}x\right)$. See Figure 5.15.

The ReLU (*rectified linear unit*) function

$$x \mapsto \max\{x, 0\}$$

is linear for positive argument, but is zero for negative values; see Figure 5.16. Note that the derivative of the ReLU function is the Heaviside function $\mathbf{1}_{\mathbb{R}_+}$. Gradient descent methods for finding the optimal scaling and shifting of models involving ReLUs can thus get stuck in dead zones where the derivative is zero.

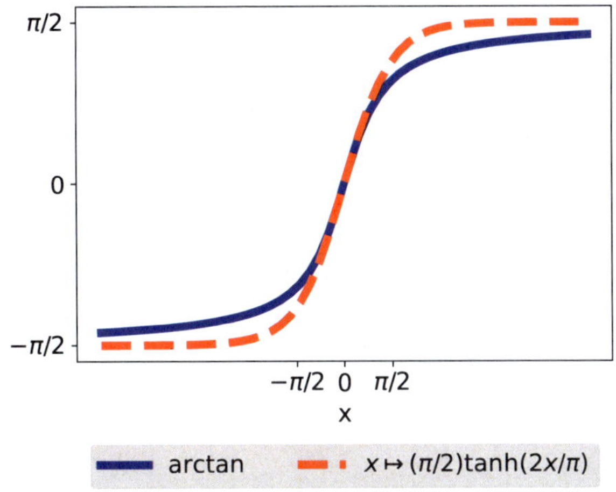

Figure 5.15. arctan and scaled tanh

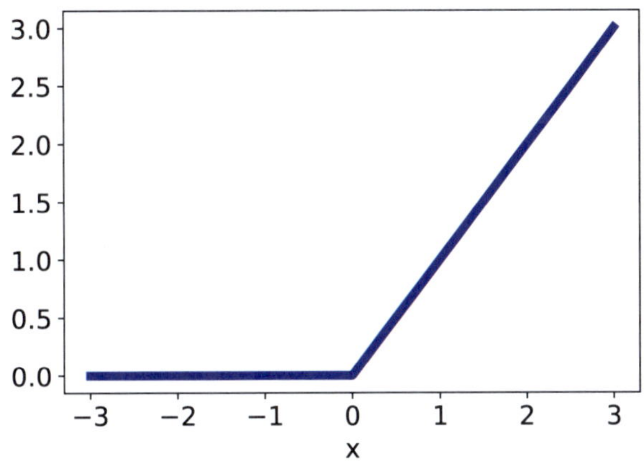

Figure 5.16. ReLU function

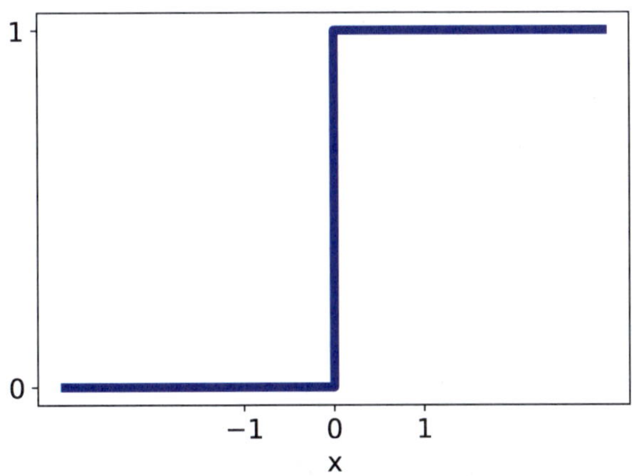

Figure 5.17. Heaviside function

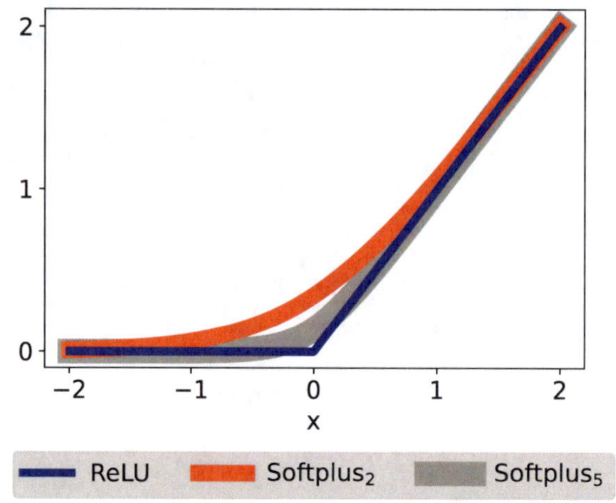

Figure 5.18. Softplus_β activation functions

The *Softplus* function Softplus is a (parametrized) approximation of the ReLU function; see also Exercise 5.1. For $\beta > 0$,

$$\text{Softplus}_\beta(x) \stackrel{\text{def}}{=} \frac{1}{\beta} \ln\left(1 + e^{\beta x}\right), \qquad x \in \mathbb{R}.$$

For $x \gg 1/\beta$ (typically $\beta \gg 1$), $\text{Softplus}_\beta(x) \approx x$, while for $x \ll -1/\beta$, $\text{Softplus}_\beta(x) \approx 0$. See Figure 5.18. The derivative of Softplus_β is

$$\text{Softplus}'_\beta(x) = \frac{e^{\beta x}}{1 + e^{\beta x}} = S(\beta x),$$

i.e., a scaled logistic function (which is strictly positive).

5.5. Brief Concluding Remarks

In this chapter we introduced, probably, the most basic (but not trivial) neural network architecture, i.e., feed forward neural networks. Part of the success of deep learning is due to the fact that neural networks are universal approximators. This means that they can approximate any reasonable function. In Chapter 16 of Part 2 of the book, we present the basic and fundamental results on universal approximation theory.

In Chapters 19 and 20 of Part 2 of the book, we study scaling limits of feed forward neural networks, namely the neural tangent kernel (oftentimes called the linear regime) and mean field scaling (oftentimes called the nonlinear regime). As we shall see there, such results bring more light into how, why, and when deep learning algorithms work in practice.

Now, the next step after defining feed forward neural networks is to train them so that we can estimate their parameter $\theta \in \Theta$ based on observed data.

To do so, we need to be able to quickly compute derivatives of the loss function $\Lambda(\theta)$ with respect to θ. This leads to the backpropagation algorithm, which is studied in Chapter 6, and to some extent is the backbone of all deep learning algorithms. The actual training is done via the various versions of the different gradient descent type of algorithms, studied in Chapters 7 and 8 of Part 1 from a practical perspective and in Chapters 17 and 18 of Part 2 of the book from a theoretical perspective.

5.6. Exercises

Exercise 5.1. Let $\beta > 0$ and consider the function $f^\beta : \mathbb{R}^d \mapsto \mathbb{R}$ such that for $x = (x_1, \dots, x_d)$,

$$f^\beta(x) = \beta \ln\left(e^{\frac{x_1}{\beta}} + \cdots + e^{\frac{x_d}{\beta}}\right).$$

Prove that for a given $x \in \mathbb{R}^d$,

$$\lim_{\beta \downarrow 0} f^\beta(x) = \max\{x_1, \dots, x_d\}.$$

What does this result imply for the approximation of ReLU activation functions by smooth functions?

Exercise 5.2. Consider the logistic activation function $S(x; c) = \frac{1}{1+e^{-cx}}$. Prove that for $c > 0$ and for all $x \in \mathbb{R}$ we have $0 \leq S'(x; c) \leq \frac{c}{4}$.

Exercise 5.3. Construct a two-layer neural network with a sigmoid activation function which gives

- Class 1, if $0 < y < 2x + 3$.
- Class 0, otherwise.

Exercise 5.4. Construct a two-layer neural network with a sigmoid activation function which gives

- Class 1, if $y < 2x + 3$ and $x > 0$.
- Class 0, otherwise.

Exercise 5.5. Consider the function

$$f(x) = \begin{cases} 0, & x < 0, \\ x, & 0 \leq x < 1, \\ 2x - 1, & x \geq 1. \end{cases}$$

Write the function $f(x)$ in the form

$$f(x) = W^{(2)} \mathrm{ReLU}(W^{(1)} x + B^{(1)}) + B^{(2)},$$

for some $W^{(1)}, B^{(1)} \in \mathbb{R}^{2\times 1}$, $W^{(2)} \in \mathbb{R}^{1\times 2}$, and $B^{(2)} \in \mathbb{R}$.

Exercise 5.6. Consider the function

$$f(x) = \begin{cases} -2x, & x < 0, \\ 0, & 0 \le x < 1, \\ 3(x-1), & x \ge 1. \end{cases}$$

Write the function $f(x)$ in the form

$$f(x) = W^{(2)}\mathsf{ReLU}(W^{(1)}x + B^{(1)}) + B^{(2)}$$

for some $W^{(1)}, B^{(1)} \in \mathbb{R}^{2\times 1}$, $W^{(2)} \in \mathbb{R}^{1\times 2}$, and $B^{(2)} \in \mathbb{R}$.

Chapter 6

Backpropagation

6.1. Introduction

Neural networks are trained by updating their weights using (stochastic) gradient descent on a desired loss function $\Lambda(\theta)$,

$$\theta_{k+1} = \theta_k - \eta \nabla \Lambda(\theta_k),$$

where we denote by η the learning rate (which oftentimes depends on the iteration k as well).

Backpropagation is an algorithm that allows us to compute $\nabla \Lambda(\theta_k)$ in an efficient way. It is essentially the chain rule done in an intelligent way. Backpropagation is a form of an automatic differentiation algorithm. In Chapter 24 we shall discuss automatic differentiation in more detail, while in this chapter we focus on backpropagation.

Let us assume that we want to compute the derivative $\frac{\partial F_{(L)}(\theta)}{\partial \theta}$ of a function $F_{(L)} : \mathbb{R}^{d_\theta} \mapsto \mathbb{R}^M$ that is the composition of L differentiable functions,

$$F_{(L)}(\theta) = \sigma_L\left(\sigma_{L-1}\left(\sigma_{L-2}\left(\cdots \sigma_1(\sigma_0(\theta)) \cdots\right)\right)\right),$$

where $\sigma_0(\theta) = \theta$, $\sigma_1 : \mathbb{R}^{d_\theta} \mapsto \mathbb{R}^K$, $\sigma_\ell : \mathbb{R}^K \mapsto \mathbb{R}^K$ for $\ell = 2, \ldots, L-1$ and $\sigma_L : \mathbb{R}^K \mapsto \mathbb{R}^M$. Let us denote the differential operator $D^\ell = \frac{\partial \sigma_\ell}{\partial \sigma_{\ell-1}}$. Then, using the chain rule, we can write informally

$$\frac{\partial F_{(L)}(\theta)}{\partial \theta} = D^L D^{L-1} D^{L-2} \ldots D^1.$$

Computing the chain rule from inside to outside (that is, first we compute D^1 followed by D^2, etc.) is usually referred to as the *forward mode of differentiation* and, in the case presented here, it has a cost of the order of $\mathcal{O}\left(K^2 d_\theta + (L-2)K^2 d_\theta + MKd_\theta\right)$.

Computing the chain rule from outside to inside (that is, first we compute D^L followed by D^{L-1}, etc.) to obtain $\frac{\partial F_{(L)}(\theta)}{\partial \theta}$ is usually referred to as the *reverse mode of differentiation* which, in the case presented here, has a cost of the order of $\mathcal{O}\left(K^2 d_\theta + (L-2)K^2 M + MKd_\theta\right)$.

By comparing the two costs, we see that

- Forward mode is better if $M > d_\theta$.
- Reverse mode is better if $M < d_\theta$.

In deep learning, the dimension of the parameter space is typically much larger than the dimension of the output space. Hence, reverse mode differentiation is preferable. As a matter of fact, backpropagation is an example of reverse mode differentiation.

In this chapter, we derive the backpropagation formula for calculating the gradient of a loss function with respect to the model parameters. We start with an example of a neural network with one hidden layer that has one neuron (the simplest possible case!) in Section 6.2. Section 6.3 has a slightly more general case of a feed forward neural network with two layers and a two-dimensional input. Backpropagation for general feed forward neural networks is presented in Section 6.4. Then, in Section 6.5 we present backpropagation in the context of learning. A common issue in training neural network models is the vanishing gradient problem that we describe in the setting of backpropagation in Section 6.6. We will visit backpropagation again in Chapters 7 and 8 when we formally discuss stochastic gradient descent for shallow and deep neural networks, respectively, as well as in Chapter 13 when we discuss recurrent neural networks.

6.2. Introductory Example

Let us consider first the simplest possible case: one hidden layer with one neuron. Let $\theta = (c, w)^\top$ and $\Lambda(\theta) = \frac{1}{2}(y - c\sigma(w \cdot x))^2$, where σ is some sufficiently smooth activation function. By the chain rule, the derivatives of the loss function with respect to the unknown parameters are

$$\frac{\partial \Lambda}{\partial c} = (y - c\sigma(w \cdot x))(-\sigma(w \cdot x))$$
$$\frac{\partial \Lambda}{\partial w} = (y - c\sigma(w \cdot x))(-c\sigma'(w \cdot x))x.$$

Let us define $Z = w \cdot x$ and $H = \sigma(Z)$. The idea of backpropagation (see [**RHW86**] for one of the first related applications to learning neural networks) is the following:

(1) The quantities Z and $(y - cH)$ are computed many times, so storing them and reusing them is a good idea. For instance, in the computation above, the same quantities appear in both derivatives. Hence we can compute them once, store them, and reuse them.

(2) Quantities like $(y - cH)$, $H = \sigma(Z)$, and Z can be viewed as outputs of consecutive layers. We need to be good with bookkeeping!

(3) There are two general steps: forward computation and backward computation. Activations are computed forward, sensitivities, i.e., derivatives, are computed backward.

Forward computation is where activations are computed from inside to outside.

$$\begin{aligned} Z &= w \cdot x, \\ H &= \sigma(Z), \\ \mathfrak{m} &= c \cdot H, \\ \Lambda &= \frac{1}{2}(y - \mathfrak{m})^2. \end{aligned}$$

Backward computation is where sensitivities are computed from outside to inside. We shall write $\frac{\partial \Lambda}{\partial Z} = \delta$ to denote the derivative (sensitivity) of the loss function with respect to a given layer. Sometimes in the literature, the δ sensitivities will be defined to be derivatives with respect to $H = \sigma(Z)$ (see Chapters 7 and 8 for related examples) instead of being with respect to Z. However, both formulations are essentially equivalent, given that one is a direct function of the other one.

Even though the concept of δ-sensitivities is not very important in this simple case, its importance will become much clearer in Section 6.3.

$$\begin{aligned} \frac{\partial \Lambda}{\partial \mathfrak{m}} &= \frac{\partial \Lambda}{\partial \Lambda}\frac{\partial \Lambda}{\partial \mathfrak{m}} = (-(y - \mathfrak{m})), \\ \frac{\partial \Lambda}{\partial c} &= \frac{\partial \Lambda}{\partial \mathfrak{m}}\frac{\partial \mathfrak{m}}{\partial c} = (-(y - \mathfrak{m})) \cdot H, \\ \frac{\partial \Lambda}{\partial H} &= \frac{\partial \Lambda}{\partial \mathfrak{m}}\frac{\partial \mathfrak{m}}{\partial H} = (-(y - \mathfrak{m})) \cdot c, \\ \frac{\partial \Lambda}{\partial Z} &= \frac{\partial \Lambda}{\partial h}\frac{\partial H}{\partial Z} = (-(y - \mathfrak{m}) \cdot c)\sigma'(Z) = \delta, \\ \frac{\partial \Lambda}{\partial w} &= \frac{\partial \Lambda}{\partial Z}\frac{\partial Z}{\partial w} = \delta \cdot x. \end{aligned}$$

It is interesting to note that in the above calculation each line uses the output of the previous line. Objects like $(y - \mathfrak{m})$, $\mathfrak{m} = c \cdot H$, $\sigma'(Z)$, and $Z = w \cdot x$ need only be computed once and are then simply reused.

6.3. Backpropagation in a More General Case

In the previous section we investigated the case of one hidden layer with one neuron. Let us now look into a slightly more general case. Let us consider a feed forward neural network with two layers and a two-dimensional input $x \in \mathbb{R}^2$,

$$\mathfrak{m}(x;\theta) = \sigma\big((W^2)^\top \sigma((W^1)^\top \cdot x + B^1) + B^2\big),$$

where we assume that $x \in \mathbb{R}^2$ and $\theta = (W^2, W^1, B^2, B^1)$. To make it a little bit more interesting, we will assume that $W^1 \in \mathbb{R}^{2\times 2}$ is a 2×2 matrix, $W^2, B^1 \in \mathbb{R}^2$ are two-dimensional vectors and $B^2 \in \mathbb{R}$. We have also made the convention that σ applied to a vector acts componentwise on its components.

We use the convention that superscripts correspond to the layer number, and subscripts correspond to the vector/matrix element. We emphasize that superscripts do not indicate powers; they indicate layer number.

Let us see now what backpropagation looks like in this case. As we shall see, just increasing the number of layers from one to two already makes things interesting and shows the importance of good bookkeeping. Let us define the output of the inner layer to be

$$\begin{aligned} Z^1 &= (W^1)^\top \cdot x^0 + B^1 \\ &= \begin{pmatrix} w^1_{11} & w^1_{21} \\ w^1_{12} & w^1_{22} \end{pmatrix} \begin{pmatrix} x^0_1 \\ x^0_2 \end{pmatrix} + \begin{pmatrix} b^1_1 \\ b^1_2 \end{pmatrix} \\ &= \begin{pmatrix} w^1_{11}x^0_1 + w^1_{21}x^0_2 + b^1_1 \\ w^1_{12}x^0_1 + w^1_{22}x^0_2 + b^1_2 \end{pmatrix} \\ &= \begin{pmatrix} Z^1_1 \\ Z^1_2 \end{pmatrix}. \end{aligned}$$

So, we have

$$x^1 = \begin{pmatrix} x^1_1 \\ x^1_2 \end{pmatrix} = \begin{pmatrix} \sigma(Z^1_1) \\ \sigma(Z^1_2) \end{pmatrix}.$$

Then, we have for the outer layer

$$\begin{aligned} Z^2_1 &= (W^2)^\top \cdot x^1 + B^2 \\ &= \begin{pmatrix} w^2_{11} & w^2_{21} \end{pmatrix} \begin{pmatrix} x^1_1 \\ x^1_2 \end{pmatrix} + B^2 \\ &= w^2_{11}x^2_1 + w^2_{21}x^2_2 + B^2. \end{aligned}$$

Let us now see what the derivatives of our loss function Λ with respect to the weights look like. We will use the δ-notation for sensitivities with respect to output of the different layers, namely we will set $\frac{\partial \Lambda}{\partial Z^j_i} = \delta^j_i$. As we shall now show, the δ's of the inner and the outer layer are related. We have, using the chain rule,

$$\begin{aligned}\delta_1^1 &= \frac{\partial\Lambda}{\partial Z_1^1} = \frac{\partial\Lambda}{\partial Z_1^2}\frac{\partial Z_1^2}{\partial Z_1^1} = \delta_1^2\frac{\partial Z_1^2}{\partial Z_1^1}\\ &= \delta_1^2 w_{11}^2\sigma'(Z_1^1).\end{aligned}$$

In a similar way we obtain

$$\delta_2^1 = \frac{\partial\Lambda}{\partial Z_2^1} = \frac{\partial\Lambda}{\partial Z_1^1}\frac{\partial Z_1^1}{\partial Z_2^1} = \delta_1^2 w_{21}^2\sigma'(Z_2^1).$$

Thus, we can write

$$\begin{pmatrix}\delta_1^1\\ \delta_2^1\end{pmatrix} = \delta_1^2\begin{pmatrix}w_{11}^2\sigma'(Z_1^1)\\ w_{21}^2\sigma'(Z_2^1)\end{pmatrix}.$$

What about δ_1^2? We compute

$$\delta_1^2 = \frac{\partial\Lambda}{\partial Z_1^2} = \frac{\partial\Lambda}{\partial \mathfrak{m}}\frac{\partial \mathfrak{m}}{\partial Z_1^2} = \frac{\partial\Lambda}{\partial \mathfrak{m}}\sigma'(Z_1^2).$$

If, for example, we choose the loss function $\Lambda(\theta) = \frac{1}{2}(y - \mathfrak{m}(x;\theta))^2$, then we shall have $\frac{\partial\Lambda}{\partial\mathfrak{m}} = (y - \mathfrak{m}(x;\theta))$.

The last displays demonstrate that the δ's of the inner layer are given in terms of the δ's of the outer layer.

Now that we demonstrated that the δ's of the different layers are related to each other, let us show that they can also determine the derivatives of the loss function Λ with respect to the parameters of interest $\theta = (W^2, W^1, B^2, B^1)$. In fact, by the chain rule again, we have

$$\begin{cases}\dfrac{\partial\Lambda}{\partial b_1^2} &= \dfrac{\partial\Lambda}{\partial Z_1^2}\dfrac{\partial Z_1^2}{\partial b_1^2} = \delta_1^2\\ \dfrac{\partial\Lambda}{\partial w_{11}^2} &= \dfrac{\partial\Lambda}{\partial Z_1^2}\dfrac{\partial Z_1^2}{\partial w_{11}^2} = \delta_1^2\cdot x_1^1\\ \dfrac{\partial\Lambda}{\partial w_{21}^2} &= \dfrac{\partial\Lambda}{\partial Z_1^2}\dfrac{\partial Z_1^2}{\partial w_{21}^2} = \delta_1^2\cdot x_2^1,\end{cases}$$

$$\begin{cases}\dfrac{\partial\Lambda}{\partial b_1^1} &= \dfrac{\partial\Lambda}{\partial Z_1^1}\dfrac{\partial Z_1^1}{\partial b_1^1} = \delta_1^1\\ \dfrac{\partial\Lambda}{\partial w_{11}^1} &= \dfrac{\partial\Lambda}{\partial Z_1^1}\dfrac{\partial Z_1^1}{\partial w_{11}^1} = \delta_1^1\cdot x_1^0\\ \dfrac{\partial\Lambda}{\partial w_{21}^1} &= \dfrac{\partial\Lambda}{\partial Z_1^1}\dfrac{\partial Z_1^1}{\partial w_{21}^1} = \delta_1^1\cdot x_2^0,\end{cases}$$

$$\begin{cases} \dfrac{\partial \Lambda}{\partial b_2^1} &= \dfrac{\partial \Lambda}{\partial Z_2^1}\dfrac{\partial Z_2^1}{\partial b_2^1} = \delta_2^1 \\ \dfrac{\partial \Lambda}{\partial w_{12}^1} &= \dfrac{\partial \Lambda}{\partial Z_2^1}\dfrac{\partial Z_2^1}{\partial w_{12}^1} = \delta_2^1 \cdot x_1^0 \\ \dfrac{\partial \Lambda}{\partial w_{22}^1} &= \dfrac{\partial \Lambda}{\partial Z_2^1}\dfrac{\partial Z_2^1}{\partial w_{22}^1} = \delta_2^1 \cdot x_2^0. \end{cases}$$

So, with $\ell = 1, 2$ we can summarize the previous calculations as

$$\begin{aligned} \frac{\partial \Lambda}{\partial b_j^\ell} &= \delta_j^\ell, \\ \frac{\partial \Lambda}{\partial w_{1j}^\ell} &= \delta_j^\ell \cdot x_j^{\ell-1}, \end{aligned}$$

which then allows us to write the stochastic gradient descent algorithm (more of this in Chapters 7 and 8) in terms of the sensitivities δ_j^ℓ, as follows

$$\begin{aligned} w_{ij,k+1}^\ell &= w_{ij,k}^\ell - \eta \delta_j^\ell x_{j,k}^{\ell-1} \\ b_{j,k+1}^\ell &= b_{j,k}^\ell - \eta \delta_j^\ell. \end{aligned}$$

6.4. Backpropagation for Multilayer Feed Forward Neural Networks

Let us consider now the case of a feed forward neural network of arbitrary depth $L < \infty$ of the form

$$H_j^\ell = \sigma\left(\sum_{i=1}^{N_\ell} w_{ij}^\ell H_i^{\ell-1} + b_j^\ell\right), \quad \ell = 1, \dots, L.$$

Here ℓ denotes the layer index and H_j^ℓ is the output of the corresponding jth neuron. The input variable is $H^0 = x$ and $H_i^0 = x_i$ is its ith component. The network's output, i.e., the model is $\mathfrak{m}(x; \theta) = H^L = (H_1^L, \dots, H_{N_L}^L)$, with $\theta = \left\{(w_{ij}^\ell, b_j^\ell), j = 1, \dots, N_\ell, i = 1, \dots, N_\ell, \ell = 1, \dots, L\right\}$. The loss function is

$$\Lambda(\theta) = \frac{1}{2}\sum_{j=1}^{N_L}(y_j - \mathfrak{m}_j(x; \theta))^2.$$

We use the same convention as before in that superscripts correspond to the layer number and subscripts to the vector/matrix element. We emphasize that, here, superscripts do not indicate powers, they indicate layer number.

Following the same process as in the previous section, we obtain that the sensitivities δ_j^ℓ among the different layers are related as follows

$$\delta_j^L = (H_j^L - y_j)\sigma'(Z_j^L), \text{ for } j = 1, \dots, N_L,$$

$$\delta_i^{\ell-1} = \sigma'(Z_i^{\ell-1}) \sum_j^{N_\ell} \delta_j^{(\ell)} w_{ij}^\ell,$$

while the derivatives of the loss function with respect to the parameters of the neural network can be written as

$$\frac{\partial \Lambda}{\partial b_j^\ell} = \delta_j^\ell,$$

$$\frac{\partial \Lambda}{\partial w_{ij}^\ell} = \delta_j^\ell H_i^{\ell-1}.$$

It is interesting to note that the formulas above are generalizations of the formulas for the two-layer neural network in Section 6.3.

In matrix notation, the neural network can be written as

$$H^\ell = \sigma\left((W^\ell)^\top H^{\ell-1} + B^\ell\right),$$

and the relation of the δ's among the different layers is compactly given by

$$\delta^L = (H^L - y) \odot \sigma'(Z^L),$$

$$\delta^{\ell-1} = (W^\ell \delta^\ell) \odot \sigma'(Z^{\ell-1}),$$

where $\odot$ is the elementwise product operation, and the derivatives of the loss function with respect to the network parameters take the form

$$\frac{\partial \Lambda}{\partial B^\ell} = \delta^\ell$$

$$\frac{\partial \Lambda}{\partial W^\ell} = H^{\ell-1}\left(\delta^\ell\right)^\top.$$

Remark 6.1. Note that one more use of the backpropagation formula is that if we change the objective function, the only thing that would change in the formulas above would be the formula for δ^L.

6.5. Backpropagation Applied to a Deep Learning Example

Let us consider in this section a simple example of learning with backpropagation. We will see more complex examples in Chapter 8. Suppose that the feature $\times$ label space is $\mathbb{R}^3 \times \mathbb{R}$ and that we want to train a three-layer neural network on the dataset

$$\mathcal{D} = \{((1,2,3),4),((5,6,7),8),\dots\}.$$

In particular, let the parameter be

$$\theta = (W^1, W^2, W^3, B^1, B^2, B^3),$$

and consider the model

$$\mathfrak{m}(x;\theta) = \sigma_3\left(W^3\sigma_2\left(W^2\sigma_1(W^1x + B^1) + B^2\right) + B^3\right),$$

where σ_1, σ_2, and σ_3 are given differentiable activation functions and the features $x \in \mathbb{R}^3$. We consider the loss function

$$\Lambda(\theta) = \frac{1}{|\mathcal{D}|}\sum_{(x,y)\in\mathcal{D}} \lambda_{(x,y)}(\theta),$$

where

$$\lambda_{(x,y)}(\theta) = \ell_y(\mathfrak{m}(x;\theta))$$

compares y and $\mathfrak{m}(x;\theta)$ in some differentiable way.

Gradient descent for the loss function gives us iterations for

$$\theta_k = (W_k^1, W_k^2, W_k^3, B_k^1, B_k^2, B_k^3),$$

where

$$\begin{aligned} W_{k+1}^j &= W_k^j - \eta\frac{\partial\Lambda}{\partial W^j}(\theta_k) \\ B_{k+1}^j &= B_k^j - \eta\frac{\partial\Lambda}{\partial B^j}(\theta_k), \end{aligned}$$

with $\eta > 0$ the learning rate. Backpropagation deals with the computation of the derivatives $\frac{\partial\Lambda}{\partial W^j}(\theta_k)$ and $\frac{\partial\Lambda}{\partial B^j}(\theta_k)$. The forward step computes

$$\begin{aligned} Z_k^1 &= W_k^1 x + B_k^1 \\ Z_k^2 &= W_k^2\sigma_1(Z_k^1) + B_k^2 \\ Z_k^3 &= W_k^3\sigma_2(Z_k^2) + B_k^3. \end{aligned}$$

The backward computation (sensitivities) is

$$\begin{aligned} \delta_k^3 &= \frac{\partial}{\partial\mathfrak{m}}\ell_y(\mathfrak{m}(x;\theta_k)) \odot \sigma_3'\left((Z_k^3)^\top\right) \\ \delta_k^2 &= \delta_k^3 W_k^3 \odot \sigma_2'\left((Z_k^2)^\top\right) \\ \delta_k^1 &= \delta_k^2 W_k^2 \odot \sigma_1'\left((Z_k^1)^\top\right). \end{aligned}$$

Therefore, we have the following explicit formulas for the derivatives involved in the gradient descent step:

$$\begin{aligned}
\frac{\partial \Lambda}{\partial W^3}(\theta_k) &= \frac{1}{|\mathcal{D}|} \sum_{(x,y)\in\mathcal{D}} \left(\delta_k^3\right)^\top \sigma_2\left((Z_k^2)^\top\right), \\
\frac{\partial \Lambda}{\partial B^3}(\theta_k) &= \frac{1}{|\mathcal{D}|} \sum_{(x,y)\in\mathcal{D}} \left(\delta_k^3\right)^\top, \\
\frac{\partial \Lambda}{\partial W^2}(\theta_k) &= \frac{1}{|\mathcal{D}|} \sum_{(x,y)\in\mathcal{D}} \left(\delta_k^2\right)^\top \sigma_1\left((Z_k^1)^\top\right), \\
\frac{\partial \Lambda}{\partial B^2}(\theta_k) &= \frac{1}{|\mathcal{D}|} \sum_{(x,y)\in\mathcal{D}} \left(\delta_k^2\right)^\top, \\
\frac{\partial \Lambda}{\partial W^1}(\theta_k) &= \frac{1}{|\mathcal{D}|} \sum_{(x,y)\in\mathcal{D}} \left(\delta_k^1\right)^\top x^\top, \\
\frac{\partial \Lambda}{\partial B^1}(\theta_k) &= \frac{1}{|\mathcal{D}|} \sum_{(x,y)\in\mathcal{D}} \left(\delta_k^1\right)^\top.
\end{aligned}$$

Note that there are many transposes to compute in the last expression. A more efficient implementation of this algorithm would first convert the dataset $\mathcal{D}$ into feature and label arrays

$$\begin{pmatrix} 1 & 2 & 3 \\ 5 & 6 & 7 \\ & \vdots & \end{pmatrix} \text{ and } \begin{pmatrix} 4 \\ 8 \\ \vdots \end{pmatrix},$$

and then work with transposes of the original calculations. In particular, we would have

$$\begin{aligned}
W_{k+1}^{j,\top} &= W_k^{j,\top} - \eta\left(\frac{\partial \Lambda}{\partial W^j}(\theta_k)\right)^\top \\
B_{k+1}^{j,\top} &= B_k^{j,\top} - \eta\left(\frac{\partial \Lambda}{\partial B^j}(\theta_k)\right)^\top,
\end{aligned}$$

with $\eta > 0$ the learning rate.

Then, we would define

$$\begin{aligned}
Z_k^{1,\top} &= x^\top W_k^{1,\top} + B_k^{1,\top} \\
Z_k^{2,\top} &= \sigma_1(Z_k^{1,\top}) W_k^{2,\top} + B_k^{2,\top} \\
Z_k^{3,\top} &= \sigma_2(Z_k^{2,\top}) W_k^{3,\top} + B_k^{3,\top}.
\end{aligned}$$

The backward computation (sensitivities) becomes

$$\delta_k^3 = \frac{\partial}{\partial \mathfrak{m}} \ell_y(\mathfrak{m}(x;\theta_k)) \odot \sigma_3'\left((Z_k^3)^\top\right),$$
$$\delta_k^2 = \delta_k^3 W_k^3 \odot \sigma_2'\left((Z_k^2)^\top\right),$$
$$\delta_k^1 = \delta_k^2 W_k^2 \odot \sigma_1'\left((Z_k^1)^\top\right).$$

Finally, we have the following explicit formulas for the derivatives involved in the gradient descent step:

$$\left(\frac{\partial \Lambda}{\partial W^3}(\theta_k)\right)^\top = \frac{1}{|\mathcal{D}|} \sum_{(x,y)\in\mathcal{D}} \sigma_2\left(Z_k^2\right)\delta_k^3,$$
$$\left(\frac{\partial \Lambda}{\partial B^3}(\theta_k)\right)^\top = \frac{1}{|\mathcal{D}|} \sum_{(x,y)\in\mathcal{D}} \delta_k^3,$$
$$\left(\frac{\partial \Lambda}{\partial W^2}(\theta_k)\right)^\top = \frac{1}{|\mathcal{D}|} \sum_{(x,y)\in\mathcal{D}} \sigma_1\left(Z_k^1\right)\delta_k^2,$$
$$\left(\frac{\partial \Lambda}{\partial B^2}(\theta_k)\right)^\top = \frac{1}{|\mathcal{D}|} \sum_{(x,y)\in\mathcal{D}} \delta_k^2,$$
$$\left(\frac{\partial \Lambda}{\partial W^1}(\theta_k)\right)^\top = \frac{1}{|\mathcal{D}|} \sum_{(x,y)\in\mathcal{D}} x\delta_k^1,$$
$$\left(\frac{\partial \Lambda}{\partial B^1}(\theta_k)\right)^\top = \frac{1}{|\mathcal{D}|} \sum_{(x,y)\in\mathcal{D}} \delta_k^1.$$

Notice that what makes the last computation more efficient is the fact that there are fewer transpose matrices to compute in the actual gradient descent.

6.6. Vanishing Gradient Problem

In this section we briefly describe an issue that is common across many deep neural network architectures, that of the vanishing gradient problem. We will revisit this issue in more detail in Section 8.4 as well as in Chapter 13, where we study recurrent neural networks.

Let us assume that the activation function $\sigma(z)$ is the logistic function $S(z)$. Then we will have that $S'(z) = S(z)(1-S(z))$. In that case the backpropagation algorithm yields

$$\delta_i^{\ell-1} = S(Z_i^{\ell-1})\left(1 - S(Z_i^{\ell-1})\right)\sum_j^{N_\ell} \delta_j^\ell w_{ij}^\ell.$$

Some remarks are now in order.

- If $Z_i^{\ell-1}$ is large in absolute value, then the logistic function is close to either zero or one. So $\delta_i^{\ell-1}$ would saturate to zero.
- By Exercise 5.2, we immediately see that $S'(z) = S(z)(1 - S(z)) \le \frac{1}{4}$. This suggests that we get a reduction of $\delta_i^{\ell-1}$ by a factor of 4.
- We also have

$$\frac{\partial \Lambda}{\partial w_{ij}^{\ell}} = \delta_j^{\ell} H_i^{\ell-1}$$
$$\frac{\partial \Lambda}{\partial b_j^{\ell}} = \delta_j^{\ell}.$$

Thus, if we propagate through several layers, the resulting gradient will eventually become very small. This phenomenon is called the *vanishing gradient problem.*

These suggest that $\theta \mapsto \Lambda(\theta)$ has flat regions, which of course is not good for stochastic gradient descent. In those cases momentum stochastic gradient descent studied in Chapters 17 and 18 may sometimes help. Using an activation function that would not saturate, like ReLU for instance, would also help; see also Chapter 13.

6.7. Brief Concluding Remarks

The backpropagation algorithm is the backbone of deep learning algorithms because it can lead to quick computation of the gradient of the loss function $\Lambda(\theta)$ with respect to θ. Backpropagation is part of the automatic differentiation algorithms that we discuss in Chapter 24. We do remark here though for completeness that automatic differentiation algorithms may not always be optimal, see for example [**Nau08**].

As we mentioned in the Introduction, the paper on backpropagation by Rumelhart, Hornik, and Williams in 1986 [**RHW86**] led to a considerable resurgence of interest in the field of neural network based artificial intelligence in a period where it was not clear how to efficiently train multilayer neural networks. An interesting mathematical framework for studying backpropagation from the lens of a Lagrangian formalism can be found in [**Lec88**].

Backpropagation is one of the main ingredients needed for the practical implementation and scaling to high-dimensional problems of the stochastic gradient descent algorithm. We explore the stochastic gradient descent algorithm in Chapters 7 and 8 for shallow and multi-layer neural networks, respectively. In Chapters 17 and 18 of Part 2 we discuss theoretical convergence properties of gradient descent and stochastic gradient descent.

6.8. Exercises

Exercise 6.1. Develop the backpropagation algorithm for the regular cross-entropy objective function when the activation function of the last layer is the logistic function.

Exercise 6.2. Assume that in a feed forward neural network the activation function in the ℓth layer is linear. Assume further that the weights and the inputs are independent random variables. Prove that δ_j^ℓ and $W_{i,j}^\ell$ are independent random variables.

Exercise 6.3. Consider the neural network

$$\mathfrak{m}(x;\theta) = h(\sigma_3(w_3\sigma_2(w_2\sigma_1(w_1x+b_1)+b_2)+b_3)),$$

where $\theta = (b_1, b_2, b_3, w_1, w_2, w_3) \in R^6$ and $\sigma_1, \sigma_2, \sigma_3, h$ are sufficiently smooth activation functions. We consider quadratic error loss, i.e.,

$$\Lambda(\theta) = \frac{1}{M}\sum_{m=1}^{M}(y_m - \mathfrak{m}(x_m;\theta))^2.$$

Then,

(1) Write down the forward propagation step in terms of the three different layers, Z^1, Z^2, Z^3.

(2) Compute the sensitivities $\delta^j = \dfrac{\partial\Lambda}{\partial Z^j}$ for $j = 1, 2, 3$, in terms of Z^j's.

(3) Compute the derivatives $\dfrac{\partial\Lambda}{\partial w_i}$ and $\dfrac{\partial\Lambda}{\partial b_i}$ for $i = 1, 2, 3$.

(4) Compare your results with Section 6.4.

Exercise 6.4. Define $\sigma_n(x) = \cos(2^n x)$ with $x \in \mathbb{R}$ and $n \in \{1, 2, 3\}$. Define the function $f(x, w_1, w_2, w_3) = e^{\pi\sigma_3(w_3\sigma_2(w_2\sigma_1(w_1x+b_1)+b_2)+b_3)}$. Write down the backpropagation algorithm to compute

(1) $\left(\dfrac{\partial f}{\partial w_1}, \dfrac{\partial f}{\partial w_2}, \dfrac{\partial f}{\partial w_3}\right)$,

(2) $\left(\dfrac{\partial f}{\partial b_1}, \dfrac{\partial f}{\partial b_2}, \dfrac{\partial f}{\partial b_3}\right)$.

Exercise 6.5. Define $\sigma_n(x) = \cos(2^n x)$ with $x \in \mathbb{R}$ and $n \in \{1, 2\}$. Define the function $f(x, w_1, w_2) = |\sigma_2(w_2\sigma_1(w_1x)) - 3|^2$. Write down the backpropagation algorithm to compute $\left(\dfrac{\partial f}{\partial w_1}, \dfrac{\partial f}{\partial w_2}\right)$.

Chapter 7

Basics of Stochastic Gradient Descent

7.1. Introduction

Stochastic gradient descent is the method of choice for training deep learning models. While the standard gradient descent algorithm uses all available data at each iteration of the algorithm, stochastic gradient descent uses only a randomly sampled subset (typically of small size) of the data each time. Even though the gradient descent algorithm can be thought to be more accurate, it becomes computationally prohibitive for large datasets. Stochastic gradient descent, despite not using all available information at each iteration, has been demonstrated to be both computationally efficient and accurate.

This chapter presents the mechanics of the *standard* stochastic gradient descent algorithm. In Chapter 17 we present convergence theory for the gradient descent algorithm, whereas in Chapter 18 we present the convergence theory for the stochastic gradient descent algorithm. As we shall see in Chapter 18 the extra randomness coming from the random choice of a subset of the data to be used at each iteration of the algorithm poses unique mathematical challenges.

In addition, besides the standard stochastic gradient descent algorithm presented in this chapter, there are many other variants (e.g., stochastic gradient descent with momentum, AdaGrad, RMSprop, ADAM, AdaMax) which we will cover in detail in Chapter 18.

In this chapter we focus on the application of the stochastic gradient algorithm to shallow neural networks. In Chapter 8 we apply the algorithm to multi-layer neural networks. Both chapters include implementation examples in Python.

7.2. The basic setup

Let us briefly review the basic setup. Machine learning estimates a statistical model for the relationship between an input X and an output Y. Formally, suppose there is data $(X, Y) \in \mathbb{R}^d \times \mathcal{Y}$ and a statistical model $\mathfrak{m}(x;\theta) : \mathbb{R}^d \to \mathcal{Y}$, where $\theta \in \Theta$ are the parameters in the model and must be estimated using the available data. The choice of the space $\mathcal{Y}$ depends on the problem at hand. It could for example be a Euclidean space $\mathbb{R}^J$ for some $J \in \mathbb{N}$ in a regression problem or a space of possible labels in a classification problem. We wish to find a model $\mathfrak{m}(x;\theta)$ such that $\mathfrak{m}(X;\theta)$ is an accurate prediction for Y.

To make this statement precise, for a given $y \in \mathcal{Y}$, recall that the function $\ell_y(z) : \mathcal{Y} \to \mathbb{R}$ measures how close a prediction $z \in \mathcal{Y}$ is to the actual observed outcome $y \in \mathcal{Y}$. Then, we recall the definition of the objective function

$$\Lambda(\theta) = \frac{1}{|\mathcal{D}|} \sum_{(x,y)\in\mathcal{D}} \lambda_{(x,y)}(\theta), \text{ where } \lambda_{(x,y)}(\theta) = \ell_y(\mathfrak{m}(x;\theta)). \tag{7.1}$$

The objective function (7.1) is a natural objective function for estimating the parameter θ. The quantity $\ell_y(\mathfrak{m}(x;\theta))$ measures the error being made when the model $\mathfrak{m}(x;\theta)$ is used to predict the value y. The error is then averaged over the distribution $P_{\mathcal{D}}$ of the data $\mathcal{D}$. The goal is to find a parameter θ such that the average error that the model $\mathfrak{m}(x;\theta)$ makes when predicting the outcome y is small.

The best model, within the class of models $\{\mathfrak{m}(x;\theta)\}_{\theta\in\Theta}$, is the model $\mathfrak{m}(x;\theta^*)$, where θ^* satisfies

$$\theta^* = \operatorname*{argmin}_{\theta\in\Theta} \Lambda(\theta). \tag{7.2}$$

For some simple models, (7.2) can be calculated exactly. However, for more complicated models such as neural networks, it cannot be exactly calculated. Instead, numerical methods are used to minimize the objective function (7.1). When (7.1) is convex, these numerical methods may converge to the exact solution (7.2). However, when (7.1) is nonconvex, the numerical methods are not guaranteed to converge to the exact solution of (7.2). Neural networks are nonconvex. In the nonconvex case, numerical methods are only guaranteed to converge to a point which satisfies certain optimization properties. We will discuss these important mathematical points later. The most widely used numerical method for minimizing (7.1) is stochastic gradient descent, which is the topic of this chapter.

Example 7.1. Consider a logistic regression model for classification where $\mathcal{Y} = \{l_1, l_2, \dots, l_J\}$, where l_j represents the jth label for $j = 1, 2, \dots, J$, and $\theta \in \Theta$, where $\Theta = \mathbb{R}^{J\times d}$. Note that we do not list the labels via enumeration, as in many cases it can mistakenly suggest that ordering of labels corresponds to ordering of feature values. Given an input $x \in \mathbb{R}^d$, the model $\mathfrak{m}(x;\theta)$ produces a probability of each possible outcome in $\mathcal{Y}$:

$$\mathfrak{m}(x;\theta) = S_{\text{softmax}}(\theta x),$$
$$S_{\text{softmax}}(z) = \frac{1}{\sum_{j=1}^{J} e^{z_j}}\left(e^{z_1}, e^{z_2}, \ldots, e^{z_J}\right),$$

where z_j is the jth element of the vector $z = \theta x \in \mathbb{R}^J$. The function $S_{\text{softmax}}(z)$: $\mathbb{R}^J \to \mathcal{P}(\mathcal{Y})$ is called the *softmax function* and is frequently used in deep learning. Here $\mathcal{P}(\mathcal{Y})$ is the set of probability measures on $\mathcal{Y}$. $S_{\text{softmax}}(z)$ takes a J-dimensional input and produces a probability distribution function on $\mathcal{Y}$. That is, the output of $S_{\text{softmax}}(z)$ is a vector of probabilities for the events $l_1, l_2, \ldots, l_J$. The softmax function can be thought of as a smooth approximation to the arg max function since it pushes its smallest inputs towards 0 and its largest input towards 1.

The objective function is the negative log-likelihood (commonly referred to in machine learning as the *cross-entropy error*):

$$\Lambda(\theta) = \frac{1}{|\mathcal{D}|}\sum_{(x,y)\in\mathcal{D}} \lambda_{(x,y)}(\theta),$$
$$\lambda_{(x,y)}(\theta) = -\sum_{j=1}^{J} \mathbf{1}_{y=l_j} \log \mathfrak{m}_j(x;\theta),$$

where $\mathfrak{m}_j(x;\theta)$ is the jth element of the vector $\mathfrak{m}(x;\theta)$ and $\mathbf{1}_{y=l_j}$ is the indicator function

$$\mathbf{1}_{y=l_j} = \begin{cases} 1, & y = l_j \\ 0, & y \neq l_j. \end{cases}$$

As we discussed in Section 3.7, the labels are typically encoded as one-hot vectors; one entry is 1 and the others are 0. Our ground-truth data $\mathcal{D}$ is then a collection of points (x, y) in $\mathbb{R}^d \times \mathbb{R}^J$, where the y's are one-hot probability vectors.

7.3. Stochastic gradient descent algorithm

The objective function (7.1) can be minimized via the well-known method of gradient descent:

$$\theta_{k+1} = \theta_k - \eta_k \nabla_\theta \Lambda(\theta_k). \tag{7.3}$$

Gradient descent repeatedly takes steps in the direction of *steepest descent*. The *negative* gradient of the objective function $\Lambda(\theta)$ is the direction of *steepest descent*. The negative gradient is the direction in which $\Lambda(\theta)$ is decreasing.

Gradient descent repeatedly takes small steps in the direction of the steepest descent. The magnitude of these steps is governed by the *learning rate* η_k, which is a positive scalar which may depend upon the iteration number k.

We can show that if the learning rate η_k is sufficiently small, the kth step of the gradient descent algorithm (7.3) is guaranteed to decrease the objective function. Assuming $\theta \in \mathbb{R}^{|\Theta|}$ and using a Taylor expansion,

$$\begin{aligned}\Lambda(\theta_{k+1}) - \Lambda(\theta_k) &= \nabla_\theta \Lambda(\theta_k)(\theta_{k+1} - \theta_k) + \frac{1}{2}(\theta_{k+1} - \theta_k)^\top \nabla_{\theta\theta}\Lambda(\bar{\theta})(\theta_{k+1} - \theta_k) \\ &= -\eta_k \Big(\nabla_\theta \Lambda(\theta_k) \Big)^\top \nabla_\theta \Lambda(\theta_k) + \frac{1}{2}\Big(\eta_k\Big)^2 \nabla_\theta \Lambda(\theta_k)^\top \nabla_{\theta\theta}\Lambda(\bar{\theta}_k)\nabla_\theta\Lambda(\theta_k),\end{aligned}$$

where $\bar{\theta}_k$ is a point on the line segment connecting θ_{k+1} and θ_k. As long as we are not already at a stationary point $\nabla_\theta \Lambda(\theta_k) = 0$, there is a choice of η_k such that the objective function will decrease, i.e., $\Lambda(\theta_{k+1}) - \Lambda(\theta_k) < 0$. It is also clear that if η_k is too large, the objective function may *increase* due to the second-order term. In practice, a careful choice of the learning rate is very important. The gradient descent algorithm uses only the first derivative $\nabla_\theta \Lambda(\theta)$ to update the parameter θ. If the algorithm takes too-large steps, the first derivative no longer accurately describes the change in the objective function.

Gradient descent requires computing the gradient $\nabla_\theta \Lambda(\theta_k)$, which can be computationally costly since it involves a summation over (potentially) many points (x, y):

$$\nabla_\theta \Lambda(\theta_k) = \nabla_\theta \frac{1}{|\mathcal{D}|} \sum_{(x,y)\in\mathcal{D}} \lambda_{(x,y)}(\theta_k) = \frac{1}{|\mathcal{D}|} \sum_{(x,y)\in\mathcal{D}} \nabla_\theta \lambda_{(x,y)}(\theta_k).$$

Stochastic gradient descent is a computationally efficient scheme for minimizing (7.1). It follows a *noisy* (but unbiased) descent direction:

$$\theta_{k+1} = \theta_k - \eta_k \nabla_\theta \lambda_{(x_k, y_k)}(\theta_k), \tag{7.4}$$

where (x_k, y_k) are independent and identically distributed (i.i.d.) samples from the distribution $\mathbb{P}_{(X,Y)}$. In particular, note that the *average* descent direction in (7.4) equals the descent direction in (7.3) since

$$E_{\mathcal{D}}\Big[\nabla_\theta \lambda_{(x_k,y_k)}(\theta_k) \Big| \theta_k \Big] = E_{\mathcal{D}}\Big[\nabla_\theta \lambda_{(X,Y)}(\theta_k) \Big| \theta_k \Big] = \nabla_\theta \Lambda(\theta_k).$$

Stochastic gradient descent is computationally efficient since it only requires the gradient of the loss from a *single data sample.* It can therefore perform many more iterations than gradient descent, given the same amount of time. In practice, stochastic gradient descent (7.4) typically converges much more rapidly than gradient descent (7.3).

Data samples $\{(x_m, y_m)_{m=1}^M\}$ are available from the distribution $\mathbb{P}_{(X,Y)}$. Then, (7.1) can be written as

$$\Lambda(\theta) = \frac{1}{M}\sum_{m=1}^{M} \lambda_{(x_m,y_m)}(\theta). \tag{7.5}$$

The gradient descent algorithm for (7.5) is

$$\theta_{k+1} = \theta_k - \eta_k \frac{1}{M}\sum_{m=1}^{M} \nabla_\theta \lambda_{(x_m,y_m)}(\theta_k). \tag{7.6}$$

The stochastic gradient descent algorithm for (7.5):

- Randomly initialize the parameter $\theta^{(0)}$.
- For $k = 0, 1, \ldots, K$:
  - Select a data sample (x_k, y_k) at random from the dataset $\{(x_m, y_m)\}_{m=1}^M$.
  - Calculate the gradient for the loss from the data sample (x_k, y_k),

$$G_k = \nabla_\theta \lambda_{(x_k,y_k)}(\theta_k).$$

  - Update the parameters

$$\theta_{k+1} = \theta_k - \eta_k G_k, \tag{7.7}$$

  where η_k is the learning rate.

The gradient descent algorithm (7.6) converges slowly since in order to take a single step, it must calculate the gradients for every data sample in the dataset. In contrast, the stochastic gradient descent algorithm (7.7) can rapidly take many steps since each step only requires calculating the gradient for a single data sample. For this reason, stochastic gradient descent is typically superior to gradient descent in practice. Stochastic gradient descent is especially advantageous when the size of the dataset M is large.

The gradient G_k in (7.7) determines the direction of the step. The learning rate η_k determines the *size* of the step. In order for (7.4) to converge, the learning rate must decay as $k \to \infty$. The decaying learning rate is required to average out the noise in the stochastic gradient descent step.

In fact, the learning rate must satisfy the following conditions in order for (7.4) to converge (see Theorem 7.3):

$$\sum_{k=0}^{\infty} \eta_k = \infty,$$

$$\sum_{k=0}^{\infty} (\eta_k)^2 < \infty. \tag{7.8}$$

In Chapter 18 we will justify the need for these choices and explain why the learning rate needs to decay. A learning rate which satisfies these conditions is

$$\eta_k = \frac{C_0}{C_1 + k},$$

where C_0 and C_1 are positive constants.

Example 7.2. We will derive the stochastic gradient descent algorithm for the logistic regression model stated in Example 7.1. The logistic regression model $\mathfrak{m}(x;\theta)$ is estimated from the dataset $(x_m, y_m)_{m=1}^{M}$ where $(x_m, y_m) \sim \mathbb{P}_{X,Y}$.

The gradient of the loss function for a generic data sample (x, y) is

$$\nabla_\theta \lambda_{(X,Y)}(\theta) = -\nabla_\theta \log S_{\text{softmax},y}(\theta x),$$

where $S_{\text{softmax},j}(z)$ is the jth element of the vector output of the function $S_{\text{softmax}}(z)$. Let $\theta_{j,:}$ be the jth row of the matrix θ. If for the jth label $y \neq l_j$,

$$\begin{aligned} \nabla_{\theta_{j,:}} \log S_{\text{softmax},y}(\theta x) &= -\frac{e^{\theta_{y,:}x}}{S_{\text{softmax},y}(\theta x)} \frac{e^{\theta_{j,:}x}}{\left(\sum_{j=1}^{J} e^{\theta_{j,:}x}\right)^2} x \\ &= -\frac{e^{\theta_{j,:}x}}{\sum_{j=1}^{J} e^{\theta_{j,:}x}} x \\ &= -S_{\text{softmax},j}(\theta x) x. \end{aligned} \tag{7.9}$$

If for the jth label $y = l_j$,

$$\begin{aligned} \nabla_{\theta_{j,:}} \log S_{\text{softmax},y}(\theta x) &= x - \frac{e^{\theta_{y,:}x}}{S_{\text{softmax},y}(\theta x)} \frac{e^{\theta_{j,:}x}}{\left(\sum_{j=1}^{J} e^{\theta_{j,:}x}\right)^2} x \\ &= x - S_{\text{softmax},j}(\theta x) x. \end{aligned} \tag{7.10}$$

Combining equations (7.9) and (7.10), we have that, for any j,

$$\nabla_{\theta_{j,:}} \log S_{\text{softmax},y}(\theta x) = \left(\mathbf{1}_{y=l_j} - S_{\text{softmax},j}(\theta x)\right) x.$$

Therefore,

$$\nabla_\theta \log S_{\text{softmax},y}(\theta x) = \Big(e(y) - S_{\text{softmax}}(\theta x)\Big) x^\top,$$

where

$$e(y) = \left(\mathbf{1}_{y=l_1}, \dots, \mathbf{1}_{y=l_J}\right), \tag{7.11}$$

represents the one-hot encoding vector. The stochastic gradient descent algorithm is:

- Select a data sample (x_k, y_k) at random from the dataset $\{(x_m, y_m)\}_{m=1}^{M}$.
- Calculate the gradient for the loss from the data sample (x_k, y_k):

$$G_k = -\Big(e(y_k) - S_{\text{softmax}}(\theta_k x_k)\Big)(x_k)^\top.$$

- Update the parameters:

$$\theta_{k+1} = \theta_k - \eta_k G_k.$$

7.3.1. Learning rates in practice. Although the conditions (7.8) are mathematically required for convergence, it is often sufficient in practice to simply use a piecewise learning rate schedule for $k = 0, 1, \dots, C_4$ such as

$$\eta_k = \begin{cases} C_0 & k \leq C_1 \\ 0.1C_0 & C_1 < k \leq C_2 \\ 0.01C_0 & C_2 < k \leq C_3 \\ 0.001C_0 & C_3 < k \leq C_4. \end{cases}$$

If the learning rate is too small, convergence may be very slow. However, if the learning rate is too large, the algorithm may oscillate and make no progress. Stochastic gradient descent takes unbiased but noisy steps. Therefore, too large of a learning rate may *amplify* this noise and cause oscillations. The larger the noise, the smaller the learning rate that is required. For this reason, gradient descent can use a larger learning rate than stochastic gradient descent. We will partially address this later by developing *minibatch* stochastic gradient descent in Section 7.3.4 which uses small batches of random samples to reduce the noise. In the end, the optimal learning rate heavily depends upon the specific problem and dataset.

7.3.2. Convergence. There is a large literature on the mathematical analysis of gradient descent as well as stochastic gradient descent. As a matter of fact we will go over the main convergence results in Chapters 17 and 18 for gradient descent and stochastic gradient descent, respectively. Nevertheless, this literature does not address many of the challenges of neural networks. To demonstrate, we present one of the strongest theorems regarding convergence for the stochastic gradient descent algorithm (7.4):

Theorem 7.3. *Suppose that* $\nabla_\theta \Lambda(\theta)$ *is globally Lipschitz and bounded. Furthermore, assume that the condition* (7.8) *holds and* $\Lambda(\theta)$ *is bounded. Then,*

$$\mathbb{P}\left[\lim_{k \to \infty} \|\nabla_\theta \Lambda(\theta_k)\|_2 = 0 \right] = 1. \tag{7.12}$$

The proof of Theorem 7.3 and of other related results for stochastic gradient descent are discussed in Chapter 18.

Theorem 7.3 states that, provided certain technical conditions are present, the parameter estimate θ_k will converge to a stationary point of the objective function $\Lambda(\theta)$. Theorem 7.3 is powerful since it covers *nonconvex* objective functions. The type of convergence in (7.12) is called *almost sure convergence* since with probability 1 the convergence occurs. For example, this is a stronger type of convergence than convergence in probability (see Appendix A for the definition of different modes of convergence).

Let us now examine the conditions necessary for Theorem 7.3 to hold. Recall that the function $\nabla_\theta \Lambda(\theta)$ is globally Lipschitz if

$$\|\nabla_\theta \Lambda(\theta) - \nabla_\theta \Lambda(\theta')\|_2 \leq L \|\theta - \theta'\|_2,$$

for any $\theta, \theta' \in \Theta$. Although there exist many models which satisfy this global Lipschitz condition, neural network models typically do not satisfy it. In fact, the gradient of a fully connected neural network with a single hidden layer will not be globally Lipschitz. It will also not be globally bounded. Therefore, Theorem 7.3 does not cover neural networks.

This discussion demonstrates some of the mathematical challenges of neural networks. The analysis of stochastic gradient descent algorithms for neural network models remains an interesting problem. We will visit related convergence results in Chapters 19 and 20. Stochastic gradient descent has proven very effective in practice and is the fundamental building block of nearly all approaches for training deep learning models.

7.3.3. Local Minima.

Stochastic gradient descent is not guaranteed to converge to the global minimum of the objective function $\Lambda(\theta)$ if the model $\mathfrak{m}(x;\theta)$ is a neural network. In fact, it is very unlikely to do so. A global minimum is a parameter θ^* such that

$$\Lambda(\theta^*) \leq \Lambda(\theta),$$

for any $\theta \in \Theta$. The global minimum for neural networks is typically not unique (i.e., there are multiple global minima).

Neural networks typically have many local minima. The point θ is a local minimum if there exists a $\delta > 0$ such that

$$\Lambda(\theta') \geq \Lambda(\theta) \quad \text{for every} \quad \|\theta' - \theta\|_2 < \delta.$$

Stochastic gradient descent may converge to a local minimum and not a global minimum. This is one of the challenges of nonconvex optimization. Neural networks are nonconvex and, as a consequence, the objective function $\Lambda(\theta)$ is also nonconvex. The issues of nonconvexity and existence of local minima for neural networks are presented in Exercises 7.9 and 7.6, respectively.

7.3.4. Minibatch Gradient Descent.

The stochastic gradient descent algorithm we presented earlier only uses a *single* data sample for computing the update direction. Although the update is unbiased, it may be very noisy (i.e., a large variance) since there is a large amount of randomness in the single data sample that is drawn. Very noisy updates can cause oscillations and slow down convergence.

The noise in stochastic gradient descent can be easily reduced by computing the gradient on a small *minibatch* of data samples instead of just a single data sample. More data samples reduce the variance in the update. The number of data samples M in the minibatch is still small compared to the size of the dataset N though, and therefore minibatch stochastic gradient descent still converges much more rapidly than gradient descent. Frequently, minibatch stochastic gradient descent and (one sample) stochastic gradient descent have the same computational speed since computations can be efficiently vectorized for moderate sized M ($\sim 100 - 1,000$) .

Since the noise in the updates is reduced, minibatch stochastic gradient descent can use a larger learning rate than stochastic gradient descent. Typically, the larger the minibatch size M, the larger the learning rate can be.

The minibatch stochastic gradient descent algorithm for (7.5):

- Randomly initialize the parameter θ_0.
- For $k = 0, 1, \ldots, K$:
  - Select $M_\circ$ data samples $\{(x_{(k,m)}, y_{(k,m)})_{m=1}^{M_\circ}\}$ at random from the dataset $\{(x_m, y_m)_{m=1}^{M}\}$, where $M_\circ \ll M$.
  - Calculate the gradient for the loss from the data samples:

$$G_k = \frac{1}{M_\circ} \sum_{m=1}^{M_\circ} \nabla_\theta \lambda_{(x_{(k,m)}, y_{(k,m)})}(\theta_k).$$

  - Update the parameters:

$$\theta_{k+1} = \theta_k - \eta_k G_k,$$

where η_k is the learning rate.

The minibatch update G_k is clearly still an unbiased estimate for the gradient $\nabla_\theta \Lambda(\theta_k)$. Furthermore, it is less noisy than the stochastic gradient descent update with a single sample, i.e.,

$$\begin{aligned}
\mathrm{Var}\left[G_k \middle| \theta_k\right] &= \mathrm{Var}\left[\frac{1}{M_\circ} \sum_{m=1}^{M_\circ} \nabla_\theta \lambda_{(x_{(k,m)}, y_{(k,m)})}(\theta_k) \middle| \theta_k\right] \\
&= \frac{1}{M_\circ} \mathrm{Var}\left[\nabla_\theta \lambda_{(x_k, y_k)}(\theta_k) \middle| \theta_k\right].
\end{aligned}$$

The conditional variance of a minibatch update is smaller by a factor of $\frac{1}{M_\circ}$ than stochastic gradient descent with a single sample, where $M_\circ$ is the minibatch size.

Although we have differentiated here between *minibatch stochastic gradient descent* and *stochastic gradient descent*, the former is also often referred to stochastic gradient descent. In practice, the term stochastic gradient descent is frequently used with the implicit assumption that it is in fact minibatch stochastic gradient descent. The term *batch* is also often used interchangeably with *minibatch*.

7.4. Applications to Shallow Neural Networks

A fully connected network with a single *hidden layer* (shallow neural network) can be written as follows

$$\begin{aligned} Z &= Wx + b^1, \\ H_i &= \sigma(Z_i), \quad i = 0, \ldots, d_H - 1, \\ \mathfrak{m}(x;\theta) &= CH + b^2. \end{aligned} \tag{7.13}$$

The neural network $\mathfrak{m}(x;\theta) : \mathbb{R}^d \to \mathcal{Y}$ takes an input x of size d and produces an output in $\mathcal{Y}$ and let's say $\mathcal{Y} = \mathbb{R}^J$. The parameters are $C \in \mathbb{R}^{J \times d_H}$, $b^1 \in \mathbb{R}^{d_H}$, $b^2 \in \mathbb{R}^J$, and $W \in \mathbb{R}^{d_H \times d}$. These parameters are collected in $\theta = \{C, b^1, b^2, W\}$.

Let us examine the architecture of the neural network (7.13). First, a linear transformation $Z = Wx + b^1$ of the input x is taken. Then, an elementwise nonlinearity $\sigma(\cdot) : \mathbb{R} \to \mathbb{R}$ is applied to each element of the vector $Z \in \mathbb{R}^{d_H}$. This elementwise transformation of Z produces the hidden layer $H \in \mathbb{R}^{d_H}$. The number of units in the hidden layer is d_H. The final output of the neural network is a linear transformation $CH + b^2$ of the hidden layer.

Typical choices for the nonlinearities $\sigma(z)$ are:

- $\tanh(z)$,
- Sigmoidal units: $\frac{e^z}{1+e^z}$,
- Rectified linear units (ReLU): $\max(z, 0)$.

In particular, ReLUs have proven very successful for multi-layer neural networks, and we will discuss them in more detail later.

The neural network model can be used to predict an outcome $Y \in \mathbb{R}^J$ given an input $X \in \mathbb{R}^d$. This is a *regression* problem, and the parameters θ must be chosen to minimize the error between the model prediction $\mathfrak{m}(X;\theta)$ and the actual outcome Y (e.g., the error here could be the squared Euclidean distance $\ell_y(z) = \|z - y\|_2^2$). Then, the goal is to select parameters θ that minimize the objective function

$$\Lambda(\theta) = \frac{1}{|\mathcal{D}|} \sum_{(x,y)\in\mathcal{D}} \|y - \mathfrak{m}(x;\theta)\|_2^2 . \tag{7.14}$$

A global minimum of the objective function (7.14) is

$$\theta^* \in \operatorname*{argmin}_{\theta \in \Theta} \Lambda(\theta).$$

The neural network (7.13) with a single hidden layer is the simplest neural network architecture. However, even in this basic setup, the objective function (7.14) is nonconvex. Therefore, it is not guaranteed that stochastic gradient descent will converge to the global minimum.

The neural network (7.13) is a nonlinear model due to the hidden layer H which involves the application of the elementwise nonlinearities $\sigma(\cdot)$. The *approximation power* of the neural network increases with the number of hidden units. First, we will make this statement mathematically precise. Then, we will discuss the practical implications.

Assume now that we would like to learn a model $\mathfrak{m}(x;\theta)$ for the relationship $y = \bar{\mathfrak{m}}(x)$ by observing data samples (X, Y). Under mild technical conditions (see Chapter 16 on the universal approximation theorems and [**HSW90**]) for any $\epsilon > 0$, there exists a neural network with d_H hidden units such that

$$\mathbb{E}_{X,Y}\big[\, \|Y - \mathfrak{m}(X;\theta^*)\|_2^2 \,\big] < \epsilon. \tag{7.15}$$

This result indicates that the neural network $\mathfrak{m}(x;\theta)$ can approximate the target function $\bar{\mathfrak{m}}(x)$ arbitrarily well if it has a sufficiently large number of hidden units.

It is important to understand that the result (7.15) does not necessarily mean that a neural network trained in practice will accurately approximate the target function $\bar{\mathfrak{m}}(x)$. Inequality (7.15) achieves the approximation error ϵ at a *global minimum*. However, numerically solving for the global minimum is intractable in practice. Instead, the objective function $\Lambda(\theta)$ is minimized using stochastic gradient descent, which may converge to a local minimum. Nonetheless, (7.15) implies that greater accuracy can be achieved by increasing the number of hidden units. In practice, increasing the number of hidden units will frequently increase the accuracy (as long as the neural network does not begin to overfit).

7.4.1. Classification with Neural Networks. The example (7.14) considers a regression problem where a model is trained to predict a *real-valued* output given an input. Neural networks can also be used for classification problems where a model is trained to predict a *categorical* outcome given an input. In this case, the outcome is one of a set of discrete values $\mathcal{Y} = \{l_1, l_2, \ldots, l_J\}$, where l_j represents the jth label for $j = 1, 2, \ldots, J$. The output of the model will be a vector of probabilities for these potential outcomes.

In order to perform classification, a softmax layer is added to the neural network. The neural network architecture becomes

$$\begin{aligned}
Z &= Wx + b^1, \\
H_i &= \sigma(Z_i), \quad i = 0, \dots, d_H - 1, \\
U &= CH + b^2, \\
\mathfrak{m}(x;\theta) &= S_{\text{softmax}}(U).
\end{aligned}$$

The dimensions of the W, C, b^1, and b^2 remain the same as in (7.13).

With the addition of the softmax layer, the neural network now maps the input x to a probability distribution on $\mathcal{Y}$, i.e., $\mathfrak{m}(x;\theta) : \mathbb{R}^d \to \mathcal{P}(\mathcal{Y})$. The objective function is the negative log-likelihood (commonly also called the cross-entropy error):

$$\begin{aligned}
\Lambda(\theta) &= \frac{1}{|\mathcal{D}|} \sum_{(x,y)\in\mathcal{D}} \lambda_{(x,y)}(\theta), \\
\lambda_{(x,y)}(\theta) &= -\sum_{j=1}^{J} \mathbf{1}_{y=l_j} \log \mathfrak{m}_j(x;\theta).
\end{aligned}$$

The neural network produces a vector of probabilities for all potential outcomes in $\mathcal{Y}$. In many typical applications, a single prediction is required for the *most likely* outcome. The most likely outcome is

$$\operatorname{argmax}_{j=1,2,\dots,J} \mathfrak{m}_j(x;\theta),$$

where $\mathfrak{m}_j(x;\theta)$ is the jth element of the output of the model $\mathfrak{m}(x;\theta)$.

7.4.2. Backpropagation Algorithm. In Chapter 6 we went over the basics of the backpropagation algorithm. We will now revisit this topic and derive the backpropagation formula for the classification problem in the previous subsection.

The stochastic gradient descent algorithm requires calculating the gradient of $\lambda := \lambda_{(X,Y)}(\theta)$ with respect to the parameters $\theta = \{C, b^1, b^2, W\}$. We will calculate this gradient using the chain rule.

First, similar to our calculations for logistic regression,

$$\begin{aligned}
\frac{\partial \lambda}{\partial U} &= -\Big(e(Y) - \mathfrak{m}(X;\theta)\Big), \\
e(y) &= \big(\mathbf{1}_{y=l_1}, \dots, \mathbf{1}_{y=l_J}\big).
\end{aligned}$$

Then, we immediately have the gradient with respect to b^2:

$$\frac{\partial \lambda}{\partial b^2} = \frac{\partial \lambda}{\partial U} \odot \frac{\partial U}{\partial b^2} = \frac{\partial \lambda}{\partial U},$$

Recall that $x \odot y$ denotes the elementwise multiplication of x and y. For example, if $x, y \in \mathbb{R}^d$ and $z = x \odot y$, then $z \in \mathbb{R}^d$ and $z_i = x_i y_i$. Similarly, if $x, y \in \mathbb{R}^{d_1 \times d_2}$, then $z \in \mathbb{R}^{d_1 \times d_2}$ and $z_{i,j} = x_{i,j} y_{i,j}$.

Next, we consider the gradient with respect to C.

$$\frac{\partial \lambda}{\partial C_{i,q}} = \sum_{j=1}^{J} \frac{\partial \lambda}{\partial U_j} \frac{\partial U_j}{\partial C_{i,q}} = \frac{\partial \lambda}{\partial U_i} H_q,$$

where $C_{i,q}$ is the element of the matrix C corresponding to the ith row and the qth column. In matrix notation,

$$\frac{\partial \lambda}{\partial C} = \frac{\partial \lambda}{\partial U} H^\top.$$

Define

$$\delta := \frac{\partial \lambda}{\partial Z}.$$

We have by the chain rule that

$$\begin{aligned}
\delta_i &= \sum_{j=1}^{J} \frac{\partial \lambda}{\partial U_j} \frac{\partial U_j}{\partial H_i} \frac{\partial H_i}{\partial Z_i} \\
&= \sum_{j=1}^{J} \frac{\partial \lambda}{\partial U_j} C_{j,i} \sigma'(Z_i) \\
&= \frac{\partial \lambda}{\partial U} \cdot C_{:,i} \sigma'(Z_i),
\end{aligned}$$

where $C_{:,i}$ is the ith column of the matrix C.

In matrix notation,

$$\delta = C^\top \frac{\partial \lambda}{\partial U} \odot \sigma'(Z),$$

where, with a slight abuse of notation, $\sigma'(Z)$ is understood as the elementwise application of $\sigma'(\cdot)$, i.e.,

$$\sigma'(Z) = \Big(\sigma'(Z_0), \sigma'(Z_1), \ldots, \sigma'(Z_{d_H - 1}) \Big).$$

Then, we immediately have the gradient with respect to b^1 in terms of δ:

$$\frac{\partial \lambda}{\partial b^1} = \delta.$$

Next, let's consider the gradient with respect to W, which can also be written in terms of δ:

$$\frac{\partial \lambda}{\partial W_{i,\ell}} = \delta_i X_\ell.$$

Therefore,

$$\frac{\partial \lambda}{\partial W} = \delta X^\top.$$

Collecting our results, the stochastic gradient descent algorithm for updating θ is:

- Randomly select a new data sample (X, Y).
- Compute the forward step $Z, H, U, \mathfrak{m}(X;\theta)$, and $\lambda_{(X,Y)}(\theta)$.
- Calculate the partial derivative

$$\frac{\partial \lambda}{\partial U} = -\Big(e(Y) - \mathfrak{m}(X;\theta)\Big).$$

- Calculate the partial derivatives

$$\begin{aligned}\frac{\partial \lambda}{\partial b^2} &= \frac{\partial \lambda}{\partial U},\\ \frac{\partial \lambda}{\partial C} &= \frac{\partial \lambda}{\partial U} H^\top,\\ \delta &= C^\top \frac{\partial \lambda}{\partial U} \odot \sigma'(Z).\end{aligned}$$

- Calculate the partial derivatives

$$\begin{aligned}\frac{\partial \lambda}{\partial b^1} &= \delta,\\ \frac{\partial \lambda}{\partial W} &= \delta X^\top.\end{aligned}$$

- Update the parameters $\theta = \{C, b^2, W, b^1\}$ with a stochastic gradient descent step

$$\begin{aligned}C_{k+1} &= C_k - \eta_k \frac{\partial \lambda}{\partial U} H^\top,\\ b^2_{k+1} &= b^2_k - \eta_k \frac{\partial \lambda}{\partial U},\\ b^1_{k+1} &= b^1_k - \eta_k \delta,\\ W_{k+1} &= W_k - \eta_k \delta X^\top,\end{aligned}$$

  where η_k is the learning rate.

Remark 7.4. We remark here that oftentimes an alternative definition is given in the literature for δ, namely $\delta := \frac{\partial \lambda}{\partial H}$. The two definitions are equivalent. In the latter case, one would instead arrive at the equations

$$\delta = \frac{\partial \lambda}{\partial H} = C^\top \frac{\partial \lambda}{\partial U},$$

$$\frac{\partial \lambda}{\partial b^1} = \delta \odot \sigma'(Z),$$
$$\frac{\partial \lambda}{\partial W} = \Big(\delta \odot \sigma'(Z)\Big) X^\top,$$

and the parameters $\theta = \{C, b^2, W, b^1\}$ are updated with a stochastic gradient descent step

$$\begin{aligned}
C_{k+1} &= C_k - \eta_k \frac{\partial \lambda}{\partial U} H^\top, \\
b^2_{k+1} &= b^2_k - \eta_k \frac{\partial \lambda}{\partial U}, \\
b^1_{k+1} &= b^1_k - \eta_k \delta \odot \sigma'(Z), \\
W_{k+1} &= W_k - \eta_k \Big(\delta \odot \sigma'(Z)\Big) X^\top.
\end{aligned}$$

Both formulations are equivalent. The difference in the case of a shallow neural network model is only whether $\sigma'(Z)$ appears in the formula for δ or in the subsequent formulas which are used to update the parameters via stochastic gradient descent. In the multilayer case that we study in Chapter 8 we will present this latter point of view instead, for completeness.

As we have already discussed, the stochastic gradient descent algorithm described above is frequently referred to as the *backpropagation algorithm*. It is composed of a forward step and a backward step. In the forward step, the output $\mathfrak{m}(X;\theta)$ and the intermediary network values (Z, H, and U) are calculated. In the backward step, the gradient of the loss function with respect to the parameter θ is calculated. The backward step relies upon the values calculated in the forward step.

The backward step is constructed in an efficient manner. For example, when calculating the gradient with respect to W, it reuses some of the calculations from the gradients for C and b^1. Essentially, a large number of the steps in the chain rule are shared across the different parameters, which avoids costly recalculations. In particular, the calculation for gradients of parameters in lower layers reuses portions of the chain rule which have already been evaluated for parameters in higher layers. We will discuss this again in more detail when multi-layer neural networks are presented.

A numerical implementation of the backpropagation algorithm can be verified by using finite differences. That is, for $\Delta > 0$ small enough, one can numerically estimate the gradient

$$\frac{\partial \mathfrak{m}}{\partial \theta_i}(x; \theta_i, \theta_{j\neq i}) \approx \frac{\mathfrak{m}(x; \theta_i + \Delta, \theta_{j\neq i}) - \mathfrak{m}(x; \theta_i - \Delta, \theta_{j\neq i})}{2\Delta},$$

and compare this against the result from the backpropagation algorithm.

The neural network architecture requires selecting a number of *hyperparameters* such as the number of hidden units, the type of activation function or the parameter initialization. Frequently, different choices of hyperparameters have to be tested in order to find the most optimal configuration (i.e., the network architecture which has the best performance); see Chapter 11.

The neural network becomes a more complex model as the number of hidden units is increased. That is, its approximation power will increase and it will be able to fit more complex relationships. However, the model will also become more likely to overfit as the number of hidden units increases.

7.5. Implementation Examples

In the previous section we went over the formulation of the stochastic gradient descent algorithm and the basics of neural networks and the backpropagation algorithm. In this section, we present a few coding examples in Python to demonstrate how the algorithm is implemented in practice.

7.5.1. PyTorch and TensorFlow. PyTorch and TensorFlow are software libraries which can perform automatic differentiation of deep learning models. In summary, these libraries will automatically calculate the backpropagation algorithm, even for complex models. This can significantly accelerate the development and testing of deep learning models.

PyTorch has a define-by-run framework while TensorFlow is a define-and-run framework. The define-by-run framework in PyTorch has certain modeling advantages and, in general, PyTorch is more seamlessly integrated with Python than TensorFlow. Both frameworks are widely used. Google developed TensorFlow, while Facebook is the main developer behind PyTorch.

TensorFlow specifies the model and the computational graph before training begins. (The computational graph is the forward and backward chain of relationships in the backpropagation algorithm.) Hence, it is called a *define-and-run* framework. The model is static and cannot be easily changed during training. By using a static model, in principle, TensorFlow can achieve certain computational efficiencies by a priori optimizing some of the procedures. Since the model does not change during training, the backpropagation algorithm also remains the same throughout training. In practice, certain *workarounds* can be used to modify the model during training; however, these are not necessarily straightforward to implement.

PyTorch allows the model and loss function to dynamically change during training and testing. PyTorch builds the computational graph on the fly. Hence, it is called a *define-by-run* framework. Thus, changes to the model which in turn cause changes to the backpropagation algorithm can be effortlessly handled, even when they occur during the training process. This produces a highly flexible computational framework for training and testing deep learning models. PyTorch's seamless integration with Python is also not to be underestimated. Typical applications will use Python for data preprocessing purposes, and better integration with the deep learning framework allows for a faster (and easier) development process.

Section 7.5.2 presents an example of PyTorch code for training a neural network on the MNIST dataset [**LBBH98**]. The MNIST dataset is a standard image recognition dataset of handwritten numbers (more details follow below) and is available from https://yann.lecun.com/exdb/mnist/. For the code examples in this chapter as well as in Chapters 8 and 23, the original dataset was downloaded and stored in an hdf5 file. The input data was normalized by the maximum value of a pixel (255). Section 7.5.3 provides an example to demonstrate the flexibility of PyTorch's define-by-run framework.

7.5.2. PyTorch Implementation for a Neural Network on the MNIST Dataset. PyTorch code is provided below for training a one-layer neural network on the MNIST dataset [**LBBH98**]. The MNIST dataset contains images of handwritten numbers $0, 1, \ldots, 9$. Each data sample is a pair (X, Y) where $X \in \mathbb{R}^{784}$ (the image is 28×28, and therefore there are 784 pixels which are inputs to the model) and $Y \in \mathcal{Y} = \{0, 1, \ldots, 9\}$ (typically modeled through a one-hot vector encoding). The goal is to train a model $\mathfrak{m}(x; \theta)$ to correctly classify an image given only the pixel data X.

Note that the training is divided into a sequence of *epochs*, where in each epoch the model is trained on the data from the entire training set. At the beginning of each epoch, the dataset is *randomly shuffled* so that the model is trained on a sequence of i.i.d. data samples.

Load the data:

```
import numpy as np
import torch
import torch.nn as nn
import torch.nn.functional as F
import torch.optim as optim
from torch.autograd import Variable

import h5py
import time

#load MNIST data
MNIST_data = h5py.File('MNISTdata.hdf5', 'r')
x_train = np.float32(MNIST_data['x_train'][:] )
y_train = np.int32(np.array(MNIST_data['y_train'][:,0]))
x_test = np.float32( MNIST_data['x_test'][:] )
y_test = np.int32( np.array( MNIST_data['y_test'][:,0]  ) )

MNIST_data.close()
```

Define the model:

```
#number of hidden units
N = 100

#Model architecture
class MnistModel(nn.Module):
    def __init__(self):
        super(MnistModel, self).__init__()
        # input is 28x28
        #These variables store the model parameters.

        self.fc1 = nn.Linear(28*28, N)
        self.fc2 = nn.Linear(N, 10)

    def forward(self, x):

        #Here is where the network is specified.

        x = F.tanh(self.fc1( x ))
        x = self.fc2( x )

        return F.log_softmax(x, dim=1)

model = MnistModel()
```

Define the optimization algorithm and training:

```
#Stochastic gradient descent optimizer
optimizer = optim.SGD(model.parameters(), lr=0.1)

batch_size = 100
num_epochs = 100
L_Y_train = len(y_train)
model.train()
train_loss = []

#Train Model
for epoch in range(num_epochs):

    #Randomly shuffle data every epoch
    I_permutation = np.random.permutation(L_Y_train)
    x_train = x_train[I_permutation,:]
    y_train = y_train[I_permutation]
    train_accu = []

    for i in range(0, L_Y_train, batch_size):
        x_train_batch = torch.FloatTensor( x_train[i:i+batch_size,:] )
        y_train_batch = torch.LongTensor( y_train[i:i+batch_size] )
        data, target = Variable(x_train_batch), Variable(y_train_batch)

        #PyTorch "accumulates gradients", so we need to set the stored
        #gradients to zero when there's a new batch of data.
```

```
        optimizer.zero_grad()

        #Forward propagation of the model, i.e. calculate the hidden
        #units and the output.

        output = model(data)

        #The objective function is the negative log-likelihood function.
        loss = F.nll_loss(output, target)

        #This calculates the gradients (via backpropagation)
        loss.backward()
        train_loss.append(loss.data)

        #The model parameters are updated using SGD.
        optimizer.step()

        #Calculate accuracy on the training set.
        prediction = output.data.max(1)[1]   # first column has actual
                                               prob.
        accuracy = ( float( prediction.eq(target.data).sum() )
        /float(batch_size)  )*100.0
        train_accu.append(accuracy)
    accuracy_epoch = np.mean(train_accu)
    print(epoch, accuracy_epoch)
```

Accuracy of the trained model:

```
#Calculate accuracy of trained model on the Test Set
model.eval()
test_accu = []
for i in range(0, len(y_test), batch_size):
    x_test_batch = torch.FloatTensor( x_test[i:i+batch_size,:] )
    y_test_batch = torch.LongTensor( y_test[i:i+batch_size] )
    data, target = Variable(x_test_batch), Variable(y_test_batch)
    optimizer.zero_grad()
    output = model(data)
    loss = F.nll_loss(output, target)
    prediction = output.data.max(1)[1]   # first column has actual prob.
    accuracy = (float( prediction.eq(target.data).sum()) /float(batch_size
                                                   ))*100.0
    test_accu.append(accuracy)
accuracy_test = np.mean(test_accu)
print(accuracy_test)
```

7.5.3. An Example Illustrating PyTorch's Define-by-Run Framework.

PyTorch's define-by-run framework allows significant flexibility when training models. The model architecture and data input can be *dynamically* changed during training. A simple example is provided below to illustrate this. The number of layers in the neural network is increased during training if certain criterion, which are only known during training, are satisfied.

Load the dataset:

```
import numpy as np
import torch
import torch.nn as nn
import torch.nn.functional as F
import torch.optim as optim
#from torchvision import datasets, transforms
from torch.autograd import Variable

import h5py
import time

#load MNIST data
MNIST_data = h5py.File('MNISTdata.hdf5', 'r')
x_train = np.float32(MNIST_data['x_train'][:] )
y_train = np.int32(np.array(MNIST_data['y_train'][:,0]))
x_test = np.float32( MNIST_data['x_test'][:] )
y_test = np.int32( np.array( MNIST_data['y_test'][:,0]  ) )

MNIST_data.close()
```

Define the model:

```
#number of hidden units
H = 50

batch_size = 100
num_epochs = 100
L_Y_train = len(y_train)
epoch_accuracy_list = []
epoch_accuracy_list.append(0.0)

#learning rate
LR0 = 0.1

W_list = []
W0 = torch.autograd.Variable(torch.randn((H,28*28)), requires_grad=True)
W_list.append(W0)

C = torch.autograd.Variable(torch.randn((10,H)), requires_grad=True)

Number_of_layers = 1
Max_Number_of_Layers = 3
```

Train the model:

```
#Train Model
for epoch in range(num_epochs):

    LR = LR0/float(Number_of_layers)
    #Set gradients to zero
    #W0.grad[:] = W0.grad[:]*0.0

    #Randomly shuffle data every epoch
    I_permutation = np.random.permutation(L_Y_train)
    x_train =  x_train[I_permutation,:]
```

```
y_train =  y_train[l_permutation]
train_accu = []

for i in range(0, L_Y_train, batch_size):
    x_train_batch = torch.FloatTensor( x_train[i:i+batch_size,:] )
    y_train_batch = torch.LongTensor( y_train[i:i+batch_size] )
    #data, target = Variable(x_train_batch).cuda(),
    Variable(y_train_batch).cuda()
    data, target = Variable(x_train_batch), Variable(y_train_batch)

    Z = torch.t( data )

    for i in range(len(W_list)):
        Z = torch.tanh( torch.mm(W_list[i], Z ) )

    V = torch.mm( C , Z)

    output = F.log_softmax( torch.t( V) , dim=1)
    loss = F.nll_loss(output, target)

    loss.backward()    # calculate gradients

    with torch.no_grad():
        for i in range(len(W_list)):
            W_list[i] -= LR * W_list[i].grad
        C -= LR * C.grad

    # Set the gradients to zero
    for i in range(len(W_list)):
        W_list[i].grad.zero_()
    C.grad.zero_()
```

Calculate accuracy:

```
    #calculate accuracy
    prediction = output.data.max(1)[1]   # first column has actual
                                           prob.
    accuracy = ( float( prediction.eq(target.data).sum() )
    /float(batch_size)  )*100.0
    train_accu.append(accuracy)
accuracy_epoch = np.mean(train_accu)

epoch_accuracy_list.append(accuracy_epoch)
```

Increase the number of layers if need be on the fly:

```
#Increase the number of layers in the neural network model
#if certain criteria are satisfied.

if ((epoch > 1) & (epoch_accuracy_list[-1]<epoch_accuracy_list[-2]+0.1
                                                                   )
& (Number_of_layers < Max_Number_of_Layers) & (epoch > 20) ):
    W_list.append(torch.autograd.Variable(torch.randn((H,H)),
                                          requires_grad=True))
Number_of_layers = len(W_list)

print(epoch, accuracy_epoch, Number_of_layers)
```

7.5.4. Accelerating Computations on Graphics Processing Units. The backpropagation algorithm for training neural networks is composed of a series of (large) matrix multiplications that can be efficiently parallelized on graphics processing units (GPUs). GPUs have thousands of cores which allows for highly parallelized computations. A drawback is that GPUs have significantly lower memory than CPUs. They can also be slower for sequential tasks. GPUs can provide up to a 10× speedup versus CPUs for deep learning, although performance can of course vary.

It is straightforward to train models on GPUs with PyTorch. It only requires a couple of modifications of the code from Section 7.5.3 as we observe below.

Load the data:

```
model = MnistModel()
@model.cuda()@

optimizer = optim.SGD(model.parameters(), lr=0.1)

batch_size = 100
num_epochs = 100
L_Y_train = len(y_train)
model.train()
train_loss = []
```

Train the model:

```
#Train Model
for epoch in range(num_epochs):

    I_permutation = np.random.permutation(L_Y_train)
    x_train = x_train[I_permutation,:]
    y_train = y_train[I_permutation]
    train_accu = []

    for i in range(0, L_Y_train, batch_size):
        x_train_batch = torch.FloatTensor( x_train[i:i+batch_size,:] )
        y_train_batch = torch.LongTensor( y_train[i:i+batch_size] )
       @data, target = Variable(x_train_batch).cuda(),
       Variable(y_train_batch).cuda()@

        optimizer.zero_grad()

        output = model(data)

        loss = F.nll_loss(output, target)

        loss.backward()
        train_loss.append(loss.data[0])

        optimizer.step()

        prediction = output.data.max(1)[1]
        accuracy = ( float( prediction.eq(target.data).sum() )/float(
                                                batch_size))*100.0
```

```
        train_accu.append(accuracy)

    accuracy_epoch = np.mean(train_accu)
    print(epoch, accuracy_epoch)
```

7.6. Brief Concluding Remarks

In the chapter we investigated aspects of the stochastic gradient descent algorithm and the associated backpropagation with an eye towards shallow neural networks. In Chapter 8 we generalize our investigations to cover multi-layer neural networks and we also discuss the vanishing gradient problem.

Condition (7.8) on the learning rate was introduced in the stochastic approximation algorithm of [**RM51**] in 1951 and oftentimes it goes by the name of the Robins-Monroe condition. Soon thereafter [**KW52**] proposed a similar stochastic approximation algorithm for estimating the maximum of a function. It was later proven by [**BT00**] in 2000 that under this condition and certain assumptions on the loss function, the SGD algorithm (7.4) converges to a critical point of the loss function. The paper [**BCN18**] contains related results as well. The book [**KY03**] has a detailed exposition on stochastic approximation theory.

Convergence theory results for gradient descent will be presented in Chapter 17 and for the stochastic gradient descent algorithm (7.4) in Chapter 18. In Chapter 18 we will also study some of the more advanced variants of the classical stochastic gradient descent algorithms, such as SGD with momentum, AdaGrad, RMSprop, ADAM, and AdaMax.

7.7. Exercises

Exercise 7.1. Consider a dataset $\{(x_m, y_m)_{m=1}^M\}$ where $x \in \mathbb{R}^d$ and $y \in \mathbb{R}$. Recall the least-squares objective function $\Lambda(\theta) = \frac{1}{M}\sum_{m=1}^M (y_m - \theta^\top x_m)^2$ for the linear model $\mathfrak{m}(x;\theta) = \theta^\top x$. Derive the stochastic gradient descent algorithm for this linear regression model.

Exercise 7.2. Consider a nonlinear model $\mathfrak{m}(x;\theta) = g(\theta^\top x)$ with an objective function

$$\Lambda(\theta) = \frac{1}{M}\sum_{m=1}^{M} |y_m - g(\theta^\top x_m)|,$$

where g is a nonlinear but sufficiently smooth function. Derive the stochastic gradient descent algorithm for this model.

Exercise 7.3. Derive the minibatch stochastic gradient descent algorithm for the logistic regression model.

Exercise 7.4. Use *minibatch* stochastic gradient descent to train a logistic regression model for classification of the MNIST dataset [**LBBH98**]. Analyze the effect of different learning rates and different minibatch sizes.

Exercise 7.5. Show that the global minimum of a neural network is not unique.

Exercise 7.6. Construct an example of a one-layer neural network which has many local minima that are not global minima.

Exercise 7.7. Consider the shallow neural network

$$\begin{aligned} Z &= Wx + b^1, \\ H_i &= \sigma(Z_i), \quad i = 0, \dots, d_H - 1, \\ \mathfrak{m}(x;\theta) &= CH + b^2, \end{aligned}$$

with the activation function $\sigma(z)$ being the clipped ReLU unit

$$\sigma(z) = \min\big(\max(z, 0), t\big),$$

where t is a hyperparameter. Consider the dataset $\mathcal{D} = \{(x_m, y_m)_{m=1}^M\}$ where $x_m \in \mathbb{R}^d$ and $y_m \in \mathbb{R}^K$. The loss function is

$$\Lambda(\theta) = \frac{1}{|\mathcal{D}|} \sum_{(x,y)\in\mathcal{D}} \lambda_{(x,y)}(\theta), \quad \text{where } \lambda_{(x,y)}(\theta) = |y - \mathfrak{m}(x;\theta)|.$$

Derive the backpropagation algorithm and the stochastic gradient descent algorithm to minimize this loss function for a neural network with clipped ReLU units.

Exercise 7.8. Show that the gradient (with respect to the parameters $\theta \in \Theta$) of the objective function for a one-layer neural network with $\ell_y(z) = (y - z)^2$ is not necessarily globally Lipschitz.

Exercise 7.9. Show that neural networks $\mathfrak{m}(x;\theta)$ are nonconvex functions of the unknown parameters $\theta \in \Theta$.

Exercise 7.10. Prove that there exists a constant learning rate $\eta > 0$ such that gradient descent always decreases the loss function $\Lambda(\theta) : \mathbb{R}^d \mapsto \mathbb{R}$ at every iteration if the second derivatives of $\Lambda(\theta)$ are uniformly bounded.

Chapter 8

Stochastic Gradient Descent for Multi-layer Networks

8.1. Introduction

In this chapter we consider a multi-layer feed forward neural network, see (8.1). As in Chapter 7, we develop the backpropagation stochastic gradient descent algorithm for this case. In addition, we revisit the vanishing gradient descent problem (which can become more profound in the case of multiple layers). We include an implementation example in PyTorch demonstrating the similarity to the shallow neural network case presented in Chapter 7.

8.2. Multi-layer Neural Networks

A fully connected, multi-layer neural network has multiple layers, where in each layer an elementwise nonlinearity is applied to the linear combination of the output from the previous layer.

$$\begin{aligned}
Z^1 &= W^1 x + b^1, \\
H^1 &= \sigma(Z^1), \\
Z^\ell &= W^\ell H^{\ell-1} + b^\ell, \quad \ell = 2, \dots, L, \\
H^\ell &= \sigma(Z^\ell), \quad \ell = 2, \dots, L, \\
U &= W^{L+1} H^L + b^{L+1}, \\
\mathfrak{m}(x; \theta) &= S_{\text{softmax}}(U).
\end{aligned} \tag{8.1}$$

The neural network has L hidden layers followed by a softmax function. Each layer of the neural network has d_H hidden units. The ℓth hidden layer is $H^\ell \in \mathbb{R}^{d_H}$. H^ℓ is produced by applying an elementwise nonlinearity to the input $Z^\ell \in \mathbb{R}^{d_H}$. Using a slight abuse of notation,

$$\sigma(Z^\ell) = \Big(\sigma(Z_0^\ell), \sigma(Z_1^\ell), \ldots, \sigma(Z_{d_H-1}^\ell)\Big).$$

The parameters are $\theta = \{W^1, \ldots, W^{L+1}, b^1, \ldots, b^{L+1}\}$. The input is $x \in \mathbb{R}^d$ and the input layer has parameters $W^1 \in \mathbb{R}^{d_H \times d}$ and $b^1 \in \mathbb{R}^{d_H}$. The parameters in the layers $\ell = 2, \ldots, L$ have dimensions $W^\ell \in \mathbb{R}^{d_H \times d_H}$ and $b^\ell \in \mathbb{R}^{d_H}$. The softmax layer has parameters $W^{L+1} \in \mathbb{R}^{K \times d_H}$ and $b^{L+1} \in \mathbb{R}^K$.

Similar to the situation studied in Chapter 7, the error (sometimes called the *loss*) for a data sample (x, y) is given by

$$\lambda_{(x,y)}(\theta) = -\sum_{j=1}^{J} \mathbf{1}_{y=l_j} \log\big(\mathfrak{m}(x;\theta)\big)_j.$$

Let $\lambda := \lambda_{(X,Y)}(\theta)$ and define

$$\delta^\ell := \frac{\partial \lambda}{\partial H^\ell}.$$

Note that here we have defined $\delta^\ell = \frac{\partial \lambda}{\partial H^\ell}$ instead of $\delta^\ell = \frac{\partial \lambda}{\partial Z^\ell}$ that we essentially did in Chapter 7; see also Remark 8.1.

By the chain rule, for $\ell = 1, \ldots, L-1$,

$$\begin{aligned}
\delta_i^\ell &= \sum_{j=1}^{d_H} \delta_j^{\ell+1} \frac{\partial H_j^{\ell+1}}{\partial H_i^\ell} \\
&= \sum_{j=1}^{d_H} \delta_j^{\ell+1} \sigma'(Z_j^{\ell+1}) W_{j,i}^{\ell+1} \\
&= \big(\delta^{\ell+1} \odot \sigma'(Z^{\ell+1})\big)^\top W_{:,i}^{\ell+1}.
\end{aligned}$$

Therefore, for $\ell = 1, \ldots, L-1$,

$$\delta^\ell = \big(W^{\ell+1}\big)^\top \big(\delta^{\ell+1} \odot \sigma'(Z^{\ell+1})\big).$$

Consequently, for $\ell = 1, \ldots, L-1$,

$$\begin{aligned}
\frac{\partial \lambda}{\partial b^\ell} &= \delta^\ell \odot \sigma'(Z^\ell), \\
\frac{\partial \lambda}{\partial W^\ell} &= (\delta^\ell \odot \sigma'(Z^\ell))\big(H^{\ell-1}\big)^\top,
\end{aligned}$$

where $H^0 := x$.

Finally, we have that

$$\delta^L = \left(W^{L+1}\right)^\top \frac{\partial \lambda}{\partial U},$$

and

$$\frac{\partial \lambda}{\partial b^{L+1}} = \frac{\partial \lambda}{\partial U},$$
$$\frac{\partial \lambda}{\partial W^{L+1}} = \frac{\partial \lambda}{\partial U}\left(H^L\right)^\top.$$

Collecting our results, the stochastic gradient descent algorithm for updating θ is:

- Randomly select a new data sample (X, Y).
- Compute the forward step $Z^1, H^1, \ldots, Z^L, H^L, U, \mathfrak{m}(X;\theta)$, and $\lambda := \lambda_{(X,Y)}(\theta)$.
- Calculate the partial derivative

$$\frac{\partial \lambda}{\partial U} = -\Big(e(Y) - \mathfrak{m}(X;\theta)\Big),$$

  where $e(Y)$ represents the one-hot encoding vector as in (7.11).
- Calculate the partial derivatives $\frac{\partial \lambda}{\partial b^{L+1}}$, $\frac{\partial \lambda}{\partial W^{L+1}}$, and δ^L.
- For $\ell = L-1, \ldots, 1$:
  - Calculate δ^ℓ via the formula

$$\delta^\ell = \left(W^{\ell+1}\right)^\top \left(\delta^{\ell+1} \odot \sigma'(Z^{\ell+1})\right).$$

  - Calculate the partial derivatives with respect to W^ℓ and b^ℓ.
- Update the parameters θ with a stochastic gradient descent step.

The backpropagation algorithm is computationally efficient since it does not recompute the chain rule for the parameters in different layers. Instead, layer ℓ reuses the gradient computed in the previous layer $\ell+1$ via the variable $\delta^{\ell+1}$. Furthermore, only δ^ℓ and $\delta^{\ell+1}$ need to be retained in memory in order to calculate the gradients for the parameters in layer ℓ.

Remark 8.1. Comparing to the backpropagation formulas that we derived in the shallow neural network case of Chapter 7, we note that we chose for completeness here to define $\delta^\ell := \frac{\partial \lambda}{\partial H^\ell}$, instead of $\delta^\ell := \frac{\partial \lambda}{\partial Z^\ell}$. As we remarked in Remark 7.4, both formulations are equivalent. We shall visit the formulation with the definition $\delta^\ell := \frac{\partial \lambda}{\partial Z^\ell}$ in Exercise 8.3.

8.3. Computational Cost

The computational cost of the backpropagation algorithm for multi-layer neural networks depends upon a number of factors, including the number of layers L, the number of units in each layer d_H, the size of the input d, and the number of classes J. The number of arithmetic operations required for the forward step (i.e., make a prediction) for the multi-layer neural network (8.1) is

$$2(L-1)(d_H^2 + d_H) + 2d_H(1 + d + J) + 2J.$$

The cost increases linearly in the number of layers L and quadratically in the number of hidden units d_H. The number of arithmetic operations required for the backward step (i.e., a stochastic gradient descent step on a single data sample) is

$$(L-1)(3d_H^2 + d_H) + d_H(2d_H + d + 3J + 1) + 2J.$$

The backpropagation step is more costly than the forward step. Note that in the cost estimate for the backpropagation step, we assume that we have stored all of the relevant values from the forward step. The number of arithmetic operations includes all addition, multiplication, and algebraic operations. If one is using minibatch stochastic gradient descent with a batchsize of $M_\circ$, each backpropagation step has

$$M_\circ\big[(L-1)(3d_H^2 + d_H) + d_H(2d_H + d + 3J + 1) + 2J\big]$$

arithmetic operations.

There is also a memory cost for the parameters θ. Large neural network models can require significant amounts of memory. The memory required to store the parameters for the multi-layer network (8.1) is

$$(L-1)(d_H^2 + d_H) + d_H(d + J) + J.$$

The backpropagation algorithm also requires δ^ℓ and $\delta^{\ell+1}$, which has size $2M_\circ d_H$ if the batchsize is $M_\circ$. Therefore, the total memory required for backpropagation is

$$(L-1)(d_H^2 + d_H) + d_H(d + J) + J + 2M_\circ d_H.$$

As an example, consider a neural network with five layers, 500 units per layer, 100 classes, an input vector of size 1,000, and a batchsize of 1,000. Each parameter is stored as a 32-bit floating point number. The memory required for such a neural network is approximately 0.08 GB, which is relatively small for neural networks. More sophisticated models (such as convolution networks) can require more memory. Since GPUs have smaller memory than CPUs, it can become a challenge to train large models with large batchsizes on the GPU.

The batchsize can be reduced to address this. Alternatively, there are also approaches for distributing the storage of the model across multiple GPUs or machines.

8.4. Vanishing Gradient Problem

The neural network model (8.1) becomes more complex as more layers are included. In principle, this means that the neural network can more accurately fit more complex nonlinear relationships. However, the numerical estimation of the neural network with stochastic gradient descent suffers a limitation called the *vanishing gradient problem* as the number of layers is increased.

As the number of layers L increases, the magnitude of the gradient with respect to the parameters in the lower layers becomes small (e.g., $\frac{\partial \lambda}{\partial W^{\ell}}$ for $\ell \ll L$). This leads to the (stochastic) gradient descent algorithm converging extremely slowly. Essentially, the lower layers take an impractically long amount of time to train.

For a fixed k, the magnitude of the gradient $\frac{\partial \lambda}{\partial W^{\ell}}$ will decrease as the total number of layers L increases. Thus, although increasing L leads to a more complex model, the numerical estimation of this model in practice becomes increasingly difficult. *Deep learning* is interested in models and methods with large numbers of layers L, i.e., very complex models, and has developed several approaches for overcoming the challenge of the vanishing gradient problem.

Example 8.2. Let us consider a simple case where we can analytically study the vanishing gradient problem. Consider the multi-layer network

$$
\begin{aligned}
Z^1 &= W^1 x + b^1, \\
H^1 &= \sigma(Z^1), \\
Z^{\ell} &= W^{\ell} H^{\ell-1} + b^{\ell}, \quad \ell = 2, \dots, L, \\
H^{\ell} &= \sigma(Z^{\ell}), \quad \ell = 2, \dots, L, \\
\mathfrak{m}(x;\theta) &= W^{L+1} H^L + b^{L+1},
\end{aligned}
$$

where each hidden layer has a single unit (i.e., $d_H = 1$) and $\sigma(\cdot)$ is a sigmoid function. Let's initialize $b^{\ell} = 0$ and $W^{\ell} = \frac{1}{2}$. The input dimension $d = 1$ and the output is also one dimensional. Assume $x = 1$ and let the loss function be $\ell_y(z) = (y - z)^2$.

$H^{\ell} = \sigma\left(\frac{1}{2} H^{\ell-1}\right)$ where we define $H^0 = x = 1$. Since $\sigma(\cdot)$ is a sigmoid function, $0 < \frac{1}{2} H^{\ell} \leq \frac{1}{2}$ for $\ell = 0, \dots, L-1$. Therefore, for $\ell = 1, \dots, L$,

$$
0 < H^{\ell} \leq \sigma\left(\frac{1}{2}\right) < 1,
$$

since $\sigma(\cdot)$ is a monotonically increasing function. Then, for $1 \leq \ell < L$,

$$\begin{aligned}\delta^\ell = \frac{\partial \lambda}{\partial H^\ell} &= \delta^{L+1}\sigma'(Z^{L+1})W^{L+1} \\ &= -2(y - \mathfrak{m}(x;\theta))W^{L+1} \prod_{j=k+1}^{L} \sigma'(Z^j)W^j.\end{aligned}$$

The derivative of a sigmoid function is $\sigma'(z) = \sigma(z)(1-\sigma(z))$, which implies that $|\sigma'(z)| \leq 1$. Therefore, we have that

$$|\delta^\ell| \leq |(y - \mathfrak{m}(x;\theta))| \times 2^{-(L-\ell)}.$$

The gradient with respect to the parameter W^ℓ is

$$\frac{\partial \lambda}{\partial W^\ell} = \frac{\partial \lambda}{\partial H^\ell}\frac{\partial H^\ell}{\partial W^\ell} = \delta^\ell \sigma'(Z^\ell)H^{\ell-1}.$$

Consequently, since $0 < H^\ell \leq 1$,

$$\left|\frac{\partial \lambda}{\partial W^\ell}\right| \leq |\delta^\ell| \leq C2^{-(L-\ell)}, \tag{8.2}$$

where C is a positive constant which may depend upon (x, y).

The bound (8.2) shows that the gradient with respect to the parameters in the ℓth layer decreases in magnitude as the total number of layers L increases. In fact, in this simple case, the magnitude decreases at an exponential rate in the total number of layers L. For large L, the gradient is so small that the lower layers in the network take an impractically long amount of time to train.

The vanishing gradient problem can also occur due to *saturation*. Saturation occurs when the inputs to the hidden units have very large magnitudes. For example, recall that if $\sigma(\cdot)$ is a sigmoidal function, then its derivative is

$$\sigma'(z) = \sigma(z)\big(1 - \sigma(z)\big).$$

Since $\lim_{\|z\| \to \infty} \sigma(z) \to 0$,

$$\lim_{\|z\| \to \infty} \sigma'(z) = 0.$$

Therefore, if the magnitudes of the inputs to the nonlinearities $\sigma(\cdot)$ are very large, the backpropagation rule will lead to very small gradients for parameters in the lower layers.

8.5. Implementation Example

Let's see now an implementation example for multi-layer neural networks. In fact, a multi-layer network is easily implemented in PyTorch. It only requires modifying the definition of the model in the code in Section 7.5.2.

We implement a multilayer neural network for the MNIST dataset; see [**LBBH98**]. In Chapter 7 we also worked with the MNIST dataset, but we implemented a shallow neural network instead.

```
#Multi-layer model architecture
#N is the number of units in each hidden layer
class MnistModel(nn.Module):
    def __init__(self):
        super(MnistModel, self).__init__()
        # input is 28x28
        #These variables store the model parameters.

        self.fc1 = nn.Linear(28*28, N)
        self.fc2 = nn.Linear(N, N)
        self.fc3 = nn.Linear(N, N)
        self.fc4 = nn.Linear(N, N)

        self.fc5 = nn.Linear(N, 10)

    def forward(self, x):

        #Here is where the network is specified.

        x = F.relu(self.fc1( x ))
        x = F.relu(self.fc2( x ))
        x = F.relu(self.fc3( x ))
        x = F.relu(self.fc4( x ))

        x = self.fc5( x )

        return F.log_softmax(x, dim=1)
```

The remainder of the code remains exactly the same as in the MNIST example of Chapter 7. That is, PyTorch is set up at a high level of abstraction where the user only needs to define (a) the network architecture and (b) the objective function. Once these are defined, PyTorch will automatically calculate the backpropagation rule and train the model.

8.6. Brief Concluding Remarks

In Chapters 7 and 8 we have explored the stochastic gradient descent algorithm for shallow and multi-layer neural networks, respectively. In Chapters 17 and 18 of Part 2 we will discuss theoretical convergence properties of gradient descent and stochastic gradient descent.

In Chapter 23 of Part 2, we elaborate on distributed training and on synchronous and asynchronous training, which allows us to scale stochastic gradient descent to high-dimensional problems and large datasets.

Overfitting is an important practical challenge that sometimes must be addressed when implementing deep learning algorithms. One method to reduce overfitting is to include regularization. We discuss two popular regularization methods, regularization by penalty terms and dropout, in Chapter 9.

8.7. Exercises

Exercise 8.1. Prove that the gradient (with respect to the unknown parameters $\theta \in \Theta$) of a multi-layer neural network is not globally Lipschitz and is not bounded.

Exercise 8.2. Implement the backpropagation algorithm for a multi-layer neural network on the MNIST dataset [**LBBH98**] from scratch in Python.

Exercise 8.3. Derive the SGD formulas of Section 8.2 using the definition $\delta^\ell := \frac{\partial \lambda}{\partial Z^\ell}$ instead of $\delta^\ell := \frac{\partial \lambda}{\partial H^\ell}$.

Exercise 8.4. Construct an example of a multi-layer neural network which has local minima that are not global minima.

Chapter 9

Regularization and Dropout

9.1. Introduction

In this chapter we present two of the main regularization techniques used in deep learning: regularization by penalty functions in Section 9.2 and dropout in Section 9.3. Regularization by penalty functions is a classical regularization method in statistics, and many classical textbooks cover this topic extensively, see for example [**HTF10**] and [**BD19**]. In Section 9.2 we will mainly provide definitions and go over some of the properties of regularization by penalty terms that are more closely related to deep learning. In Section 9.3 we describe in a greater detail dropout which is a regularization technique that is specific to deep neural networks. In Sections 9.4 and 9.5 we present details on the implementation of dropout in the case of shallow and of multi-layer neural networks, respectively. Dropout is very popular in deep learning due to its simplicity and general effectiveness.

Regularization is used in practice to reduce overfitting and model complexity. Both regularization methods that we will discuss (i.e., regularization by adding penalty terms to the error function and dropout) have been shown to generally lead to a reduction in overfitting.

9.2. Regularization by Penalty Terms

Let's think of the situation where we fit a complex model and we do not know if the variance is going to be large. As we discussed in the bias-variance tradeoff Section 1.6, one way to bring down the variance of a model is to collect more data. An alternative way is to potentially modify the loss function to encourage (said otherwise, *to gear*) training to simplify complexity. One way of doing so, is to redefine our loss function to be

$$\hat{\Lambda}(\theta) = \underbrace{\Lambda(\theta)}_{\text{Initial loss}} + \underbrace{\Omega(\theta)}_{\text{Regularization}} .$$

The simplest regularization to add is weight decay, which amounts to choosing

$$\Omega(\theta) = \frac{C}{2}\|\theta\|_2^2,$$

and is called the ℓ_2 regularization, where if $\theta = (p_1, \ldots, p_d) \in \mathbb{R}^d$, then $\|\theta\|_2^2 = \sum_{i=1}^d |p_i|^2$. Essentially, this regularized loss function penalizes for large values of $\|\theta\|_2$ but has small effect on Λ for not-too-large $\|\theta\|_2$. We note that we have already explored this point in some detail in the logistic regression case, Section 3.6. There we showed that for perfect data, logistic regression diverges but will be regularized if a penalty term is included. The penalty size C is a hyper-parameter that has to be tuned. The gradient of the penalized loss becomes

$$\nabla\hat{\Lambda}(\theta) = \nabla\Lambda(\theta) + C\theta,$$

and the stochastic gradient descent algorithm becomes

$$\begin{aligned}\theta_{k+1} &= \theta_k - \eta\left(\nabla\Lambda(\theta_k) + C\theta_k\right)\\ &= (1-\eta C)\theta_k - \eta\nabla\Lambda(\theta_k).\end{aligned}$$

Thus, the effect of the ℓ_2 regularization is to make the weights smaller in magnitude.

Example 9.1. As an example, consider the classical linear regression problem augmented by the regularization term. Let

$$\hat{\Lambda}(\theta) = \frac{1}{2}\|Y - X\cdot\theta\|^2 + \frac{C}{2}\|\theta\|_2^2.$$

This is called ridge regression and the estimator minimizing $\hat{\Lambda}(\theta)$ can be shown (see Exercise 9.1) to be

$$\theta^*(C) = \left(X^\top X + CI\right)^{-1} XY.$$

Thus, the presence of C makes the inverse smaller which subsequently makes $\theta^*(C)$ smaller when compared to the case of $\theta^*(0)$ which in turn corresponds to ordinary least squares.

Another popular regularization is the ℓ_1 regularization $\Omega(\theta) = C\|\theta\|_1$, where if $\theta = (p_1, \ldots, p_d) \in \mathbb{R}^d$, then $\|\theta\|_1 = \sum_{i=1}^d |p_i|$. The gradient of the penalized loss becomes

$$\nabla\hat{\Lambda}(\theta) = \nabla\Lambda(\theta) + C\,\mathrm{sign}(\theta),$$

where sign(θ) acts componentwise on the vector $\theta = (p_1, \dots, p_d) \in \mathbb{R}^d$, returning the sign of each component of the vector θ.

The effect of the ℓ_1 regularization is to create sparsity in the weights and as such this type of regularization is typical in signal processing for example. It has also been used for feature selection purposes, because it leads to some of the weights being zero or almost close to zero, suggesting that the corresponding features are not as vital. We do mention here that in regards to regularization by penalty terms, the ℓ_2 regularization is somewhat more common in deep learning than the ℓ_1 regularization.

9.2.1. Comparison Between the ℓ_2 and the ℓ_1 Regularization. As we mentioned earlier, the effect of the ℓ_2 regularization is to make the magnitude of the weights smaller, whereas the effect of the ℓ_1 regularization is to create sparsity. Let us now demonstrate why this is the case. The argument presented below is largely heuristic, but it is indicative of how one can think about the effect of regularization on the learned optimal parameter.

Let us assume that the objective function $\Lambda(\theta)$ is convex and smooth enough that we can apply a second-order Taylor expansion around the global minimum $\theta^\star$. Since $\nabla\Lambda(\theta^\star) = 0$, we shall have for θ sufficiently close to $\theta^\star$,

$$\Lambda(\theta) \approx \Lambda(\theta^\star) + \frac{1}{2}(\theta - \theta^\star)^\top H(\theta^\star)(\theta - \theta^\star),$$

where $\approx$ is there because we have not written the error term, and $H(\theta^\star)$ is the Hessian of $\Lambda(\theta)$, i.e., the matrix with the second-order partial derivatives of $\Lambda(\theta)$, evaluated at $\theta = \theta^\star$.

Differentiating this formula with respect to θ gives

$$\nabla\Lambda(\theta) \approx H(\theta^\star)(\theta - \theta^\star).$$

Let us now bring in the regularization effects. Let $\hat{\theta}$ be the optimal parameter for the regularized loss function (the ℓ_2 loss) $\hat{\Lambda}(\theta) = \Lambda(\theta) + \frac{C}{2}\|\theta\|_2^2$. Let us now further assume that $\hat{\theta}$ is in the range of $\theta^\star$ so that the approximate formula above still makes sense. Then, by manipulating the equation above (still ignoring the error terms from the Taylor expansion), we will have approximately

$$H(\theta^\star)(\hat{\theta} - \theta^\star) + C\hat{\theta} \approx 0,$$

yielding

$$\hat{\theta} \approx \left(H(\theta^\star) + CI\right)^{-1} H(\theta^\star)\theta^\star.$$

Hence, if $H(\theta^\star)$ is positive definite, then $\theta^\star \neq 0$ implies $\hat{\theta} \neq 0$. This heuristic argument then immediately suggests that ℓ_2 regularization does not induce sparsity in the parameters, but it does induce weight decay as measured by the coefficient $C > 0$.

On the other hand, the ℓ_1 regularization, similarly gives the condition

$$H(\theta^\star)(\theta - \theta^\star) + C\,\mathrm{sign}(\theta) \approx 0.$$

If, for simplification, we assume that $H(\theta^\star)$ is a diagonal matrix with positive diagonal elements (i.e., $H(\theta^\star) = \mathrm{diag}(H_{11}(\theta^\star), \dots, H_{dd}(\theta^\star))$ with $H_{ii}(\theta^\star) > 0$), then we can solve this equation yielding (recall the notation in this section $\theta = (p_1, \dots, p_d) \in \mathbb{R}^d$)

$$\hat{p}_i \approx p_i^\star - \frac{C}{H_{ii}(\theta^\star)}\,\mathrm{sign}(\hat{p}_i)$$

for the optimal parameter for the new regularized loss function $\hat{\Lambda}(\theta) = \Lambda(\theta) + \frac{C}{2}\|\theta\|_1$. Now letting $p_i^\star$ have the same sign as $\hat{p}_i$, we can rewrite this equation as

$$\hat{p}_i \approx p_i^\star - \frac{C}{H_{ii}(\theta^\star)}\,\mathrm{sign}(p_i^\star) = \mathrm{sign}(p_i^\star)\left(|p_i^\star| - \frac{C}{H_{ii}(\theta^\star)}\right).$$

Still, keeping in mind the hypothesis that $p_i^\star$ have the same sign as the $\hat{p}_i$'s, if we multiply both sides with $\mathrm{sign}(\hat{p}_i)$, we get

$$\left(|p_i^\star| - \frac{C}{H_{ii}(\theta^\star)}\right) \approx |\hat{p}_i| \geq 0.$$

This heuristic argument immediately shows that if C is large enough, then the ℓ_1 regularization can induce sparsity.

For completeness, we mention that one can also consider other type of regularizations. An example is affine additive combinations of the ℓ_1 and the ℓ_2 regularization, leading to what is called in the literature the elastic net regularization, and there are others too. We do not expand more on this topic here as there are many excellent resources in the literature discussing it in depth, such as [**HTF10**, **BD19**].

9.2.2. Effect of Overparametrization on Regularization. Let us now briefly discuss the effect of the dimensionality of the vector $\theta \in \Theta$, and let us set $d = \dim(\Theta)$. Let us denote by n the dimension of Y. In deep learning typically the dimension of θ is very large, oftentimes larger than the number of datapoints.

Let us focus for the moment on the linear regression setting, Example 9.1. In the setting of linear regression, X is a matrix of dimension $n \times d$. When there is no regularization, i.e., when $C = 0$, then the ordinary least squares take the form

$$\theta^*(0) = \left(X^\top X\right)^{-1} XY,$$

which is an unbiased estimator of θ and has small variance when $n \gg d$. However, when $d \gg n$, then $X^\top X$ becomes poorly conditioned which ultimately leads to overfitting. Introducing the ℓ_2 regularization or otherwise doing ridge regression ([**HK70, Tik63**]) leads to reduction of the error.

However, this does not necessarily mean that large models with little regularization do not generalize well, see [**KLS20**]. As a matter of fact as demonstrated in [**KLS20**], when $d \gg n$, the limit $C \to 0$ can have good generalization properties, and explicit ridge regression with $C > 0$ can fail to provide further improvements. This observation has also been pointed out (empirically) in the case of deep neural networks, see [**ZSBRV21**] for more details.

9.3. Dropout and its Relation to Regularization

Dropout is another regularization method that is very specific to deep learning. Dropout was initially developed in [**SHK$^+$14**] and has been extensively used since then. Deep neural networks (with many connections) can end up being unnecessarily complicated (sometimes leading to overfitting). Importantly, dropout takes the input features and zeros out a random subset of those. Dropout can be viewed as an implicit regularization method that manages to average among many approximate models without actually training many models separately.

Let us be more specific now on how dropout works. Let us set

$$\gamma = (\gamma_1, \dots, \gamma_d), \text{ with } \gamma_k = \begin{cases} 1, & \text{with probability } p \\ 0, & \text{with probability } 1-p, \end{cases}$$

where $p \in (0,1)$. Next, we define the dropout operator

$$\mathcal{D}(h) = \gamma \odot h, \tag{9.1}$$

where we recall that $\odot$ is an elementwise operation. Then for a given datapoint x in the dataset, the prediction using dropout for a shallow neural network looks as follows:

$$\hat{y} = \sum_{n=1}^{N} c^n \mathcal{D}(\sigma(w^n \cdot x + b^n)),$$

and analogously for a deep neural network. Dropout is a form of regularization as it reduces the model complexity; see Section 9.4 for a detailed example.

A schematic representation of dropout implemented on a feed forward neural network is shown in Figure 9.1. Notice that some of vertices (corresponding to input features or subsequent hidden units) and related connections have been dropped from further consideration and they do not affect the neural network's output anymore. At each implementation of the algorithm the vertices (input features or subsequent hidden units) are randomly selected to be dropped with probability p. Running the algorithm again would result in a different set of vertices being dropped from the network.

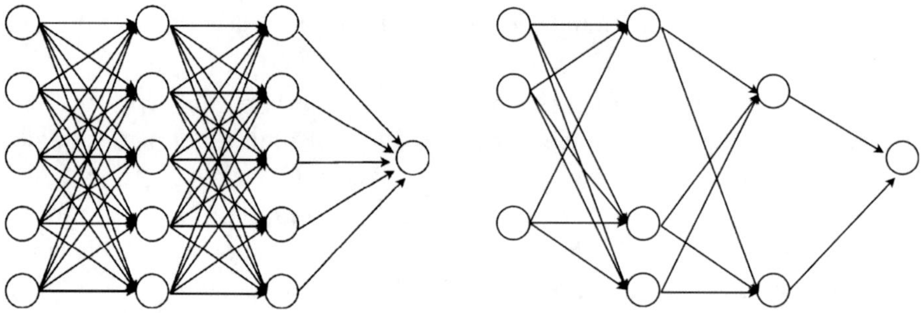

(a) Fully connected feed forward neural network

(b) An instance of the feed forward network on the left with dropout

Figure 9.1. Left: A fully connected feed forward neural network with two hidden layers. Right: An instance of the feed forward neural network on the left with dropout implemented. The missing vertices (corresponding to input features or hidden units) have been randomly selected to be dropped. Their connections have been automatically dropped as well.

Remark 9.2. In practice a typical value is $p = 0.5$, and as a matter of fact this is the default value for PyTorch. This means that on average for a given input point x, half of the activations are set to be zero. Of course, one typically experiments a little bit to figure out what the best value for p is in a given problem.

9.3.1. Dropout for Linear Regression. Let us now see how dropout works in the simple case of linear regression. In fact as we shall see below, in the case of linear regression, dropout effectively acts as ℓ_2 regularization that we visited in Section 9.2.

Define $\mathfrak{m}(x;\theta) = \mathcal{D}(x) \cdot \theta$. Then, denoting $\mathbb{E}_\gamma$ to be the expectation under the random variable γ, we consider the problem

$$\min_\theta \Lambda(\theta) = \min_\theta \mathbb{E}_\gamma \left|Y - (\gamma \odot X)\theta\right|^2 .$$

Since each entry of the vector $\gamma = (\gamma_1, \dots, \gamma_d)$ is a Bernoulli(p) random variable, direct calculations show that

$$\mathbb{E}_\gamma(\gamma \odot X) = pX,$$

$$\mathbb{E}_\gamma\left((\gamma \odot X)^\top(\gamma \odot X)\right)_{i,j} = \begin{cases} p^2\left(X^\top X\right)_{i,j}, & \text{for } i \neq j \\ p\left(X^\top X\right)_{i,j}, & \text{for } i = j. \end{cases}$$

Expanding the square and plugging in the formulas above, yield the following expression for the loss function:

$$\mathbb{E}_\gamma \left|Y - (\gamma \odot X)\theta\right|^2 = \left|Y - pX\theta\right|^2 + p(1-p)\theta^\top \operatorname{diag}(X^\top X)\theta.$$

If we now set $\hat{\theta} = p\theta$ and $\Omega(\hat{\theta}) = \frac{1-p}{p}\hat{\theta}^\top \operatorname{diag}(X^\top X)\hat{\theta}$, then we get

$$\mathbb{E}_\gamma \left|Y - (\gamma \odot X)\theta\right|^2 = \left|Y - X\hat{\theta}\right|^2 + \Omega(\hat{\theta}).$$

Thus indeed in the setting of linear regression, dropout acts as an ℓ_2 regularization and if a particular $(X^\top X)_{i,j}$ is large, then dropout forces its weight to be small.

9.4. A Neural Network Example with Dropout Implemented

Let us now implement by hand dropout on a simple neural network. As we mentioned in the beginning of Section 9.3 deep networks with many connections can end up being unnecessarily complicated (sometimes leading to overfitting). Suppose (perhaps in some intermediate layer) we are trying to write a map $(x_1, x_2) \mapsto y$ in terms of the features

$$x_1, \quad x_2, \quad x_1 + 0.1x_2.$$

It is clear that the third feature has redundant information. As a matter of fact linear maps of x_1, x_2, and $x_1 + 0.1x_2$ can be simplified:

$$W\begin{pmatrix} x_1 \\ x_2 \\ x_1 + 0.1x_2 \end{pmatrix} = W\begin{pmatrix} 1 & 0 \\ 0 & 1 \\ 1 & 0.1 \end{pmatrix}\begin{pmatrix} x_1 \\ x_2 \end{pmatrix}.$$

As we discussed earlier in Section 9.3 the main idea of dropout is to randomly remove (achieved by multiplying variables by zero) connections before doing gradient descent.

Suppose that we want to use a three-layer neural network to predict a binary label based on $\mathbb{R}^2$-valued features. Denoting by $S(\cdot)$ the usual logistic function, we have the model

$$\mathfrak{m}(x;\theta) = S\left(W^{(3)}S\left(W^{(2)}S\left(W^{(1)}X + B^{(1)}\right) + B^{(2)}\right) + B^{(3)}\right)$$

with the parameters being

$$\theta = \left(W^{(1)}, W^{(2)}, W^{(3)}, B^{(1)}, B^{(2)}, B^{(3)}\right) \in \mathbb{R}^{2\times 2} \times \mathbb{R}^{2\times 2} \times \mathbb{R}^{1\times 2} \times \mathbb{R}^2 \times \mathbb{R}^2 \times \mathbb{R}^1.$$

We note that the ideas below work equally well with any of the usual activation functions in place of the logistic function.

To train this model on a finite dataset $\mathcal{D} \subset \mathbb{R}^2 \times \{0,1\}$, we define

$$\lambda_{(x,y)}(\theta) = H(y, \mathfrak{m}(x;\theta)) \quad \text{with } (x,y) \in \mathbb{R}^2 \times \{0,1\},$$

where $H(y, m)$ is a per-datapoint error function of our choice. Then our goal is to minimize the loss function

$$\Lambda(\theta) = \frac{1}{|\mathcal{D}|}\sum_{(x,y)\in\mathcal{D}} \lambda_{(x,y)}(\theta).$$

Gradient descent with learning rate η for entire batch would be

$$\theta_{k+1} = \theta_k - \eta \frac{1}{|\mathcal{D}|} \sum_{(x,y)\in\mathcal{D}} \nabla \lambda_{(x,y)}(\theta_k).$$

Let us now add dropout to the second and third layers. Let us set

$$\gamma = \left(\left(\gamma_1^{(2)}, \gamma_2^{(2)}\right), \left(\gamma_1^{(3)}, \gamma_2^{(3)}\right)\right) \in [0,1]^2 \times [0,1]^2.$$

Let us define the matrices

$$D_2 = \begin{pmatrix} \gamma_1^{(2)} & 0 \\ 0 & \gamma_2^{(2)} \end{pmatrix} \quad \text{and} \quad D_3 = \begin{pmatrix} \gamma_1^{(3)} & 0 \\ 0 & \gamma_2^{(3)} \end{pmatrix},$$

and then define the model

$$\mathfrak{m}(x;\theta,\gamma) = S\left(W^{(3)} D_3 S\left(W^{(2)} D_2 S\left(W^{(1)} X + B^{(1)}\right) + B^{(2)}\right) + B^{(3)}\right),$$

with the parameters being as before, i.e,

$$\theta = \left(W^{(1)}, W^{(2)}, W^{(3)}, B^{(1)}, B^{(2)}, B^{(3)}\right) \in \mathbb{R}^{2\times 2} \times \mathbb{R}^{2\times 2} \times \mathbb{R}^{1\times 2} \times \mathbb{R}^2 \times \mathbb{R}^2 \times \mathbb{R}^1.$$

Going now into the training phase, let us fix some $p \in (0,1)$ (typical choice is $p = 0.5$) and choose the elements of γ as Bernoulli(p). For instance, if it turns out that $\gamma = ((1,1),(0,1))$, then we shall have that

$$D_2 = \begin{pmatrix} 1 & 0 \\ 0 & 1 \end{pmatrix} \quad \text{and} \quad D_3 = \begin{pmatrix} 0 & 0 \\ 0 & 1 \end{pmatrix}.$$

For $z \in \mathbb{R}^2$, we then have the dropout matrices

$$D_2 \begin{pmatrix} z_1 \\ z_2 \end{pmatrix} = \begin{pmatrix} z_1 \\ z_2 \end{pmatrix} \quad \text{and} \quad D_3 \begin{pmatrix} z_1 \\ z_2 \end{pmatrix} = \begin{pmatrix} 0 \\ z_2 \end{pmatrix}.$$

Note that D_2 keeps both inputs, whereas D_3 keeps only the second input. Depending on the choice of elements of γ, matrices D_2 and D_3 mask inputs to the second and third layers in the model $\mathfrak{m}(x;\theta,\gamma)$.

In the training phase, with per-datapoint loss

$$\lambda^{\text{dropout}}_{(x,y),\gamma}(\theta) = H(y, \mathfrak{m}(x;\theta,\gamma)),$$

we consider the minimization problem

$$\begin{aligned} \Lambda^{\text{dropout}}(\theta) &= \frac{1}{|\mathcal{D}|} \sum_{(x,y)\in\mathcal{D}} \mathbb{E}_{\gamma\sim\text{Bernoulli}(p)} \left[\lambda^{\text{dropout}}_{(x,y),\gamma}(\theta)\right] \\ &= \frac{1}{|\mathcal{D}|} \sum_{(x,y)\in\mathcal{D}} \mathbb{E}_{\gamma\sim\text{Bernoulli}(p)} \left[H(y, \mathfrak{m}(x;\theta,\gamma))\right] \\ &\approx \frac{1}{J|\mathcal{D}|} \sum_{(x,y)\in\mathcal{D}, 1\le j\le J} H(y, \mathfrak{m}(x;\theta,\gamma_j)), \end{aligned}$$

which is an average over all subneural networks to find the best parameter θ^*.

In the evaluation (testing) phase, we predict with

$$x \mapsto \mathfrak{m}(x; \theta^*, \mathbb{E}_{\gamma \sim \text{Bernoulli}(p)}[\gamma]).$$

In the case of the example above, we have

$$\mathbb{E}_{\gamma \sim \text{Bernoulli}(p)}[\gamma] = ((p, p), (p, p)),$$

which means that in the testing phase we replace the randomly chosen matrices D_2 and D_3 with their averages

$$\mathbb{E}_{\gamma \sim \text{Bernoulli}(p)}[D_2] = pI_2 \quad \text{and} \quad \mathbb{E}_{\gamma \sim \text{Bernoulli}(p)}[D_3] = pI_2,$$

where,

$$pI_2 = \begin{pmatrix} p & 0 \\ 0 & p \end{pmatrix} \quad \text{and} \quad pI_2 = \begin{pmatrix} p & 0 \\ 0 & p \end{pmatrix}.$$

This means that the model takes the form

$$\begin{aligned} &\mathfrak{m}(x; \theta^*, \mathbb{E}_{\gamma \sim \text{Bernoulli}(p)}[\gamma])) \\ &\quad = S\left(W^{(3),*} pI_2(S\left(W^{(2),*} pI_2 S\left(W^{(1),*} X + B^{(1),*}\right) + B^{(2),*}\right) + B^{(3),*}\right). \end{aligned}$$

To become even more detailed, let us investigate now the different possible outcomes. For this purpose, let us define

$$\begin{aligned} W^{(1)} &= \begin{pmatrix} w_{1,1}^{(1)} & w_{1,2}^{(1)} \\ w_{2,1}^{(1)} & w_{2,2}^{(1)} \end{pmatrix} \quad \text{and} \quad B^{(1)} = \begin{pmatrix} b_1^{(1)} \\ b_2^{(1)} \end{pmatrix}, \\ W^{(2)} &= \begin{pmatrix} w_{1,1}^{(2)} & w_{1,2}^{(2)} \\ w_{2,1}^{(2)} & w_{2,2}^{(2)} \end{pmatrix} \quad \text{and} \quad B^{(2)} = \begin{pmatrix} b_1^{(2)} \\ b_2^{(2)} \end{pmatrix}, \\ W^{(3)} &= \begin{pmatrix} w_1^{(3)} & w_2^{(3)} \end{pmatrix} \quad \text{and} \quad B^{(3)} = b^{(3)} \in \mathbb{R}, \\ x &= \begin{pmatrix} x_1 \\ x_2 \end{pmatrix}, \end{aligned}$$

with different matrices D_2 and D_3.

We shall have that

$$\lambda^{\text{dropout}}_{(x_1,x_2),\gamma}(\theta) = H(y, S\left(W^{(3)} D_3 S\left(W^{(2)} D_2 S\left(W^{(1)} X + B^{(1)}\right) + B^{(2)}\right) + B^{(3)}\right)).$$

In the gradient descent step, $-\nabla\lambda^{\text{dropout}}_{(x_1,x_2),\gamma}(\theta)$ is gradient descent in θ for minimizing

$$a^{p,(1)} = W^{(1)}x + B^{(1)} = \begin{pmatrix} w^{(1)}_{1,1}x_1 + w^{(1)}_{1,2}x_2 + b^{(1)}_1 \\ w^{(1)}_{2,1}x_1 + w^{(1)}_{2,2}x_2 + b^{(1)}_2 \end{pmatrix}, \tag{9.2}$$

$$\begin{aligned} a^{p,(2)} &= W^{(2)}D_2S(a^{p,(1)}) + B^{(2)} \\ &= \begin{pmatrix} w^{(2)}_{1,1}\gamma^{(2)}_1 S(a^{p,(1)}_1) + w^{(2)}_{1,2}\gamma^{(2)}_2 S(a^{p,(1)}_2) + b^{(2)}_1 \\ w^{(2)}_{2,1}\gamma^{(2)}_1 S(a^{p,(1)}_1) + w^{(1)}_{2,2}\gamma^{(2)}_2 S(a^{p,(1)}_2) + b^{(2)}_2 \end{pmatrix}, \end{aligned}$$

$$a^{p,(3)} = W^{(3)}D_3S(a^{p,(2)}) + B^{(3)} = w^{(3)}_1\gamma^{(3)}_1 S(a^{p,(2)}_1) + w^{(3)}_2\gamma^{(3)}_2 S(a^{p,(2)}_2) + b^{(3)},$$

$$\lambda^{\text{dropout}}_{(x_1,x_2),\gamma}(\theta) = H(y, S(a^{p,(3)})).$$

Then, depending on the choice for γ and the corresponding matrices D_2 and D_3, different scenarios emerge. In particular, here are some possibilities

- If $\gamma = ((1,1),(0,1))$, then
$$D_2 = \begin{pmatrix} 1 & 0 \\ 0 & 1 \end{pmatrix} \quad \text{and} \quad D_3 = \begin{pmatrix} 0 & 0 \\ 0 & 1 \end{pmatrix}.$$
Then in the model (9.2), $w^{(3)}_1, w^{(2)}_{1,1}, w^{(2)}_{1,2}, b^{(2)}_1$ are unused. Namely, the gradients of $\lambda^{\text{dropout}}_{(x_1,x_2),\gamma}(\theta)$ in those quantities vanish. The dimension of the network is effectively $\mathbb{R}^2 \to \mathbb{R}^2 \to \mathbb{R}^1 \to \mathbb{R}^1$.
- If $\gamma = ((1,1),(1,0))$, then
$$D_2 = \begin{pmatrix} 1 & 0 \\ 0 & 1 \end{pmatrix} \quad \text{and} \quad D_3 = \begin{pmatrix} 1 & 0 \\ 0 & 0 \end{pmatrix}.$$
Then in the model (9.2), $w^{(3)}_2, w^{(2)}_{2,1}, w^{(2)}_{2,2}, b^{(2)}_2$ are unused. Namely, the gradients of $\lambda^{\text{dropout}}_{(x_1,x_2),\gamma}(\theta)$ in those quantities vanish. The dimension of the network is effectively $\mathbb{R}^2 \to \mathbb{R}^2 \to \mathbb{R}^1 \to \mathbb{R}^1$.
- If $\gamma = ((1,1),(1,1))$, then
$$D_2 = \begin{pmatrix} 1 & 0 \\ 0 & 1 \end{pmatrix} \quad \text{and} \quad D_3 = \begin{pmatrix} 1 & 0 \\ 0 & 1 \end{pmatrix}.$$
Then in the model (9.2), all parameters are being used, i.e., the original λ is being used. The dimension of the network is effectively $\mathbb{R}^2 \to \mathbb{R}^2 \to \mathbb{R}^2 \to \mathbb{R}^1$.
- If $\gamma = ((1,1),(0,0))$, then
$$D_2 = \begin{pmatrix} 1 & 0 \\ 0 & 1 \end{pmatrix} \quad \text{and} \quad D_3 = \begin{pmatrix} 0 & 0 \\ 0 & 0 \end{pmatrix}.$$

Then in the model (9.2), the final feature-label map does not depend at all on any of the entries of $W^{(1)}$, $W^{(2)}$, $W^{(3)}$, $B^{(1)}$, or $B^{(2)}$. Namely, the gradients of $\lambda^{\text{dropout}}_{(x_1,x_2),\gamma}(\theta)$ in those quantities vanish. In this case we have a degenerate situation where $a^{p,(3)} = b^{(3)}$.

- If $\gamma = ((1,0),(1,1))$, then

$$D_2 = \begin{pmatrix} 1 & 0 \\ 0 & 0 \end{pmatrix} \quad \text{and} \quad D_3 = \begin{pmatrix} 1 & 0 \\ 0 & 1 \end{pmatrix}.$$

Then in the model (9.2), $w^{(1)}_{2,1}$, $w^{(1)}_{2,2}$, $b^{(1)}_2$ are unused. Namely, the gradients of $\lambda^{\text{dropout}}_{(x_1,x_2),\gamma}(\theta)$ in those quantities vanish. The dimension of the network is effectively $\mathbb{R}^2 \to \mathbb{R}^1 \to \mathbb{R}^2 \to \mathbb{R}^1$.

- If $\gamma = ((0,0),(1,1))$, then

$$D_2 = \begin{pmatrix} 0 & 0 \\ 0 & 0 \end{pmatrix} \quad \text{and} \quad D_3 = \begin{pmatrix} 1 & 0 \\ 0 & 1 \end{pmatrix}.$$

Then in the model (9.2), all the entries of $W^{(1)}$, $B^{(1)}$, and $W^{(2)}$ are unused. Namely, the gradients of $\lambda^{\text{dropout}}_{(x_1,x_2),\gamma}(\theta)$ in those quantities vanish. This means that $a^{p,(2)} = \begin{pmatrix} b^{(2)}_1 \\ b^{(2)}_2 \end{pmatrix}$ does not depend on the input x at all.

- If $\gamma = ((1,0),(0,1))$, then

$$D_2 = \begin{pmatrix} 1 & 0 \\ 0 & 0 \end{pmatrix} \quad \text{and} \quad D_3 = \begin{pmatrix} 0 & 0 \\ 0 & 1 \end{pmatrix}.$$

Then in the model (9.2), $w^{(1)}_{2,1}$, $w^{(1)}_{2,2}$, $b^{(1)}_2$, $w^{(2)}_{1,1}$, $w^{(2)}_{1,2}$, $b^{(2)}_1$, and $w^{(3)}_1$ are unused. Namely, the gradients of $\lambda^{\text{dropout}}_{(x_1,x_2),\gamma}(\theta)$ in those quantities vanish. The dimension of the network is effectively $\mathbb{R}^2 \to \mathbb{R}^1 \to \mathbb{R}^1 \to \mathbb{R}^1$. We essentially have

$$(x_1, x_2) \Rightarrow a^{p,(1)}_1 \Rightarrow a^{p,(2)}_2 \Rightarrow a^{p,(3)} \Rightarrow \text{label}.$$

9.5. Dropout on General Multi-layer Neural Networks

In this section, we redo Section 9.4 in a more general manner. A general multi-layer neural network with dropout can be realized as follows:

$$\begin{aligned} Z^1 &= W^1 x + b^1, \\ Z^\ell &= W^\ell H^{\ell-1} + b^\ell, \quad \ell = 1, \dots, L, \\ H^\ell &= \gamma^\ell \odot \sigma(Z^\ell), \quad \ell = 1, \dots, L, \\ U &= W^{L+1} H^L + b^{L+1}, \\ \mathfrak{m}(x;\theta) &= S_{\text{softmax}}(U). \end{aligned} \tag{9.3}$$

The neural network has L hidden layers followed by a softmax function. Each layer of the neural network has d_H hidden units. The ℓth hidden layer is $H^\ell \in \mathbb{R}^{d_H}$. H^ℓ is produced by applying an elementwise nonlinearity to the input $Z^\ell \in \mathbb{R}^{d_H}$. Here γ^ℓ is a vector of independent Bernoulli random variables with parameter p.

Let us denote $\Gamma = (\gamma^1, \dots, \gamma^L)$. Here Γ is typically called a *mask*. The role of Γ is to remove random subset of the hidden units in layer ℓ from the model. Hence, as we noted in Section 9.3, this is a form of regularization as it makes the model simpler.

At each update, say at step k, a random data sample (x_k, t_k) is drawn and a mask Γ is generated. Recalling the notation

$$\lambda^{\text{dropout}}_{(x_k,y_k),\Gamma}(\theta_k) = H(y_k, \mathfrak{m}(x_k; \theta_k, \Gamma)),$$

stochastic gradient descent with learning rate η_k for step k would be

$$\theta_{k+1} = \theta_k - \eta_k \nabla_\theta H(y_k, \mathfrak{m}(x_k; \theta_k, \Gamma_k)), \tag{9.4}$$

where Γ_k is a realization of Γ at the kth iteration.

The dropout algorithm seeks to minimize the objective function

$$\begin{aligned}\Lambda(\theta) &= \frac{1}{|\mathcal{D}|} \sum_{(x,y)\in\mathcal{D}} \mathbb{E}_{\Gamma\sim\text{Bernoulli}(p)} \left[\lambda^{\text{dropout}}_{(x,y),\Gamma}(\theta)\right] \\ &= \frac{1}{|\mathcal{D}|} \sum_{(x,y)\in\mathcal{D}} \mathbb{E}_{\Gamma\sim\text{Bernoulli}(p)} \left[H(y, \mathfrak{m}(x; \theta, \Gamma))\right],\end{aligned}$$

and similar to (x, y), the samples of Γ are i.i.d. as well. Therefore, under suitable conditions, one expects convergence as in Theorem 7.3.

However, it is important to keep in mind that dropout minimizes the *average loss* from a collection of models. Combined with the observation that the number of models grows exponentially with the total number of the hidden units $L \times \gamma_H$, we soon realize that optimizing

$$\frac{1}{|\mathcal{D}|} \sum_{(x,y)\in\mathcal{D}} \mathbb{E}_{\Gamma\sim\text{Bernoulli}(p)} \left[\lambda^{\text{dropout}}_{(x,y),\Gamma}(\theta)\right]$$

is not really feasible. Instead, in practice we apply stochastic gradient descent to the collection of models. Namely, we sample one specific model at each training step, hence the SGD (9.4) appears.

The next question we need to answer is what model to use for predictions, i.e., for the test dataset. In practice, the random variable γ^ℓ in (9.3) is replaced by $\mathbb{E}(\gamma^\ell) = (p, \ldots, p)$. Namely, we have

$$\begin{aligned} Z^1 &= W^1 x + b^1, \\ Z^{\ell+1} &= W^\ell H^\ell + b^\ell, \quad \ell = 1, \ldots, L-1, \\ H^\ell &= p \cdot \sigma(Z^\ell), \quad \ell = 2, \ldots, L, \\ U &= W^L H^L + b^L, \\ \mathfrak{m}(x; \theta) &= S_{\text{softmax}}(U). \end{aligned} \tag{9.5}$$

However, this is a heuristic, since it is equivalent to interchanging an expectation and a nonlinear function. Thus, the prediction rule (9.5) we are actually using corresponds to a different loss function:

$$\Lambda_{\text{prediction}}(\theta) = \frac{1}{|\mathcal{D}|} \sum_{(x,y) \in \mathcal{D}} H(y, \mathfrak{m}(x; \theta, \mathbb{E}[\Gamma])).$$

Of course, neural networks are nonlinear functions, so $\Lambda_{\text{prediction}}(\theta) \neq \Lambda(\theta)$. Nonetheless, the prediction network (9.5) has been proven effective in practice for many applications.

Remark 9.3. In our current definition of dropout, we have set $H^\ell = \gamma^\ell \odot \sigma(Z^\ell)$, $\ell = 1, \ldots, L$, during training in (9.3) and $H^\ell = p \cdot \sigma(Z^\ell)$, $\ell = 2, \ldots, L$, during the testing phase. An alternative and equivalent formulation (which is how PyTorch actually implements dropout) is to set $H^\ell = \frac{1}{p} \gamma^\ell \odot \sigma(Z^\ell)$, $\ell = 1, \ldots, L$, during training in (9.3) and $H^\ell = \sigma(Z^\ell)$, $\ell = 2, \ldots, L$, during the testing phase.

9.6. Brief Concluding Remarks

In this chapter we introduced the idea of regularization by penalty terms, covered in many other excellent textbooks such as [**HTF10, BD19**] for example. We also discussed dropout, which is a particularly successful regularization method in deep learning that was introduced in [**SHK⁺14**].

In many real data applications apart from including regularization, we also normalize by centering with the mean and scaling by the standard deviation. We discuss the essence of batch normalization in Chapter 10.

9.7. Exercises

Exercise 9.1. Prove that in the setting of ridge regression of Example 9.1, the minimiser of the loss function $\hat{\Lambda}(\theta)$ is indeed given by the formula

$$\theta^*(C) = \left(X^\top X + CI\right)^{-1} XY.$$

Exercise 9.2. Prove that in the setting of ridge regression of Example 9.1 and in the case $p \gg n$ that $\lim_{C \to 0} \theta^*(C) = \theta^*(0)$.

Exercise 9.3. In the context of the specific dropout example associated with (9.2) study which connections are being dropped and what is the final model in the cases of $\gamma = ((0,1),(1,1))$, $\gamma = ((0,1),(1,0))$, $\gamma = ((0,0),(1,0))$, $\gamma = ((0,0),(0,1))$, and $\gamma = ((0,0),(0,0))$.

Exercise 9.4. Consider a fully connected feed forward neural network mapping $\mathbb{R}^2 \mapsto (0,1)$ consisting of two internal three-dimensional layers. Introduce dropout matrices into the internal (second) and last (third) layer of the form

$$D_2 = \begin{pmatrix} 1 & 0 & 0 \\ 0 & 0 & 0 \\ 0 & 0 & 1 \end{pmatrix} \quad \text{and} \quad D_3 = \begin{pmatrix} 1 & 0 & 0 \\ 0 & 0 & 0 \\ 0 & 0 & 0 \end{pmatrix}.$$

Consider the dropout model (with logistic S activation function)

$$\mathfrak{m}(x;\theta,D_2,D_3) = S\left(W^{(3)}D_3 S\left(W^{(2)}D_2 S\left(W^{(1)}X + B^{(1)}\right) + B^{(2)}\right) + B^{(3)}\right),$$

and the corresponding per-datapoint loss

$$\lambda^{\text{dropout}}_{(x,y),D_2,D_3}(\theta) = H(y, \mathfrak{m}(x;\theta,D_2,D_3)),$$

with

$$\theta = (W^{(3)}, W^{(2)}, W^{(1)}, B^{(3)}, B^{(2)}, B^{(1)}) \in \mathbb{R}^{1\times 3} \times \mathbb{R}^{3\times 3} \times \mathbb{R}^{3\times 2} \times \mathbb{R} \times \mathbb{R}^3 \times \mathbb{R}^3.$$

For which elements of θ is the gradient of $\lambda_{(x,y),D_2,D_3}$ zero? Namely, which are the elements of $W^{(j)}$ and $B^{(j)}$ that are dropped out with this choice for D_2 and D_3?

Exercise 9.5. Derive the stochastic gradient descent algorithm for a single-layer fully connected neural network with dropout

$$\begin{aligned} Z^1 &= W^1 x + b^1, \\ H_i &= \gamma_i \odot \sigma(Z^1_i), \\ Z^2 &= W^2 H + b^2, \\ \mathbb{P}[Y = m] &= \frac{e^{Z_m}}{e^{Z_0} + e^{Z_1}}, \end{aligned}$$

where $x \in \mathbb{R}^d$, $W^1 \in \mathbb{R}^{L\times d}$, $b^1 \in \mathbb{R}^L$, $A \in \mathbb{R}^L$, $W^2 \in \mathbb{R}^{2\times L}$, $b^2 \in \mathbb{R}^2$, $Z^2 \in \mathbb{R}^2$, $Y \in \{0,1\}$, and $\sigma : \mathbb{R} \mapsto \mathbb{R}$. The loss function is cross-entropy loss (i.e., the negative log-likelihood) and γ_i is a Bernoulli random variable.

Chapter 10

Batch Normalization

10.1. Introduction

Let's return to some ideas from Chapters 2 and 3. In both of those chapters, we *normalized* the feature data by centering (by the mean) and scaling (by the standard deviation). This helped us in several ways:

- Generically, the inputs to each nonlinearity were likely to be where the nonlinearity has most effect (see Figure 10.1).
- It helped us to easily identify outlier training data.
- Generically, the optimal various weights and coefficients in the model $\mathfrak{m}$ were of order 1. This in particular suggested that we start our gradient descent algorithms to optimize coefficients in a neighborhood of the origin of size 1.
- Generically, the loss function was better behaved (see in particular Figures 3.7 vs. 3.11.

"Generically" here means (imprecisely) in a "typical" problem. We would like to adapt normalization to the inputs to each layer, hopefully allowing us to reap similar benefits in the internal layers [**IS15a**].

Batch normalization, initially proposed in [**IS15b**], adapts these ideas to inputs for the *internal* layers. Similar to dropout, studied in Chapter 9 the training and evaluation algorithms for batch normalization are distinct. The mean and standard deviation of the internal layers are computed in the training layer. The evaluation step uses these in predicting the output of the model $\mathfrak{m}$ for a new (out of sample) datapoint.

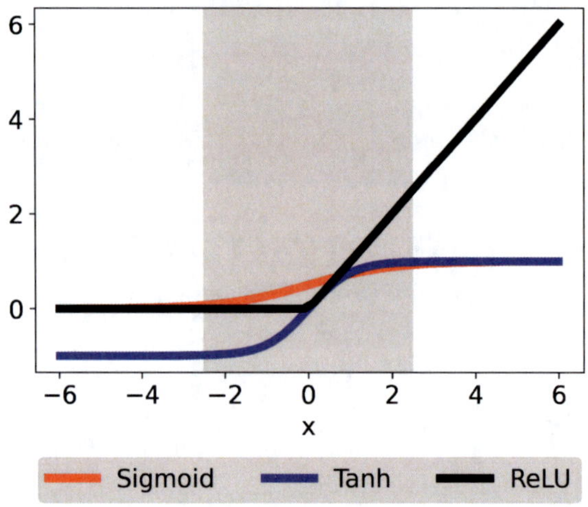

Figure 10.1. Graphical representation of typical nonlinearities

10.2. Batch Normalization Through an Example

To explain things, let's consider a simple example. Let's assume

- $\{\phi(\cdot, \theta^{(1)})\}_{\theta^{(1)}}$ is a collection of maps from from feature space $\mathbb{R}^F$ to $\mathbb{R}^2$,
- ψ is a fixed map from $\mathbb{R}^2$ to label space $\mathbb{R}$.

Our original model is the composition of ψ and $\phi(\cdot, \theta^{(1)})$'s which map $\mathbb{R}^F$ into $\mathbb{R}$; i.e.,

$$\mathbb{R}^F \xrightarrow{\phi(\cdot,\theta^{(1)})} \mathbb{R}^2 \xrightarrow{\psi} \mathbb{R},$$

or alternately

$$\mathfrak{m}(x;\theta^{(1)}) \stackrel{\text{def}}{=} \psi\left(\phi\left(x, \theta^{(1)}\right)\right).$$

We have a finite training dataset $\mathcal{D} \subset \mathbb{R}^F \times \mathbb{R}$ and an error function $\{\ell_y; y' \in \mathbb{R}\}$ which quantifies error between true and predicted labels. We want to select $\theta^{(1)}$ which minimizes

$$\Lambda(\theta) \stackrel{\text{def}}{=} \frac{1}{|\mathcal{D}|} \sum_{(x,y)\in\mathcal{D}} \ell_y\left(\mathfrak{m}(x;\theta^{(1)})\right).$$

Let's work through to batch normalize the input to the second layer, i.e., to ψ. This will help us understand how to normalize inputs to an activation function when these inputs are themselves outputs of a prior layer.

Let's (momentarily) fix $p = (\mu_1, \mu_2, \sigma_1, \sigma_2) \in \mathbb{R} \times \mathbb{R} \times (0, \infty) \times (0, \infty)$, and consider a collection $\{T_p(\cdot, \theta^{(2)})\}_{\theta^{(2)}}$ of parametrized maps from $\mathbb{R}^2 \to \mathbb{R}^2$

$$(10.1) \qquad T_p(\hat{x}, \theta^{(2)}) = \begin{pmatrix} w_1 \left(\frac{\hat{x}_1 - \mu_1}{\sigma_1}\right) + b_1 \\ w_2 \left(\frac{\hat{x}_2 - \mu_2}{\sigma_2}\right) + b_2 \end{pmatrix} \qquad \hat{x} = \begin{pmatrix} \hat{x}_1 \\ \hat{x}_2 \end{pmatrix},\ \theta^{(2)} = \begin{pmatrix} w_1 \\ w_2 \\ b_1 \\ b_2 \end{pmatrix},$$

and insert $T_p(\cdot, \theta^{(2)}) : \mathbb{R}^2 \to \mathbb{R}^2$ between the $\phi(\cdot, \theta^{(1)})$'s and ψ; i.e.,

$$\mathbb{R}^F \xrightarrow{\phi(\cdot,\theta^{(1)})} \mathbb{R}^2 \xrightarrow{T_p(\cdot,\theta^{(2)})} \mathbb{R}^2 \xrightarrow{\psi} \mathbb{R}.$$

Namely, consider the model

$$(10.2) \qquad \hat{\mathfrak{m}}_p(x; \theta) = \psi\left(T_p\left(\phi(x, \theta^{(1)}), \theta^{(2)}\right)\right), \qquad x \in \mathbb{R}^F,\ \theta = \begin{pmatrix} \theta^{(1)} \\ \theta^{(2)} \end{pmatrix},$$

with corresponding loss function

$$\begin{aligned} \hat{\Lambda}_p(\theta) &= \frac{1}{|\mathcal{D}|} \sum_{(x,y)\in\mathcal{D}} \ell_y\left(\hat{\mathfrak{m}}_p(x;\theta)\right) = \frac{1}{|\mathcal{D}|} \sum_{(x,y)\in\mathcal{D}} \left(\ell_y\left(\psi\left(T_p\left(\phi(x, \theta^{(1)}), \theta^{(2)}\right)\right)\right)\right) \\ &= \frac{1}{|\mathcal{D}|} \sum_{(x,y)\in\mathcal{D}} (\ell_y \circ \psi)\left(T_p\left(\phi(x, \theta^{(1)}), \theta^{(2)}\right)\right), \qquad \theta = \begin{pmatrix} \theta^{(1)} \\ \theta^{(2)} \end{pmatrix}. \end{aligned}$$

The parameter vector $\theta^{(2)}$ could in principle be subsumed into any linear transformation of the inputs of ψ. We also note that straightforward normalization of the input to ψ would be given by setting the w_i's to 1 and the b_i's to 0. The definition (10.1) of T_p gives extra degrees of freedom which contribute to numerical stability. In the literature the success of batch normalization has been explained from different angles, some examples of which are that (a) it often leads to reduced internal covariate shift (see [**IS15b**]) and (b) that it results to smoother gradients (see [**STIM18**]).

Let's understand gradient descent in θ. We will then combine this with choosing p to reflect actual means and variances.

Let's start by differentiating the second expression of (10.2) with respect to $\theta^{(1)}$. We get

$$\begin{aligned} \frac{\partial \hat{\Lambda}_p}{\partial \theta_i^{(1)}}(\theta) = \frac{1}{|\mathcal{D}|} \sum_{(x,y)\in\mathcal{D}} D(\ell_y \circ \psi)\left(T_p\left(\phi(x, \theta^{(1)}), \theta^{(2)}\right)\right) &\begin{pmatrix} w_1/\sigma_1 & 0 \\ 0 & w_2/\sigma_2 \end{pmatrix} \\ &\times \begin{pmatrix} \frac{\partial \phi_1}{\partial \theta_i^{(1)}}(x, \theta^{(1)}) \\ \frac{\partial \phi_2}{\partial \theta_i^{(1)}}(x, \theta^{(1)}) \end{pmatrix}. \end{aligned}$$

Varying the w_i's and then the b_i's, we get

$$\frac{\partial \hat{\Lambda}_p}{\partial w_1}(\theta) = \frac{1}{|\mathcal{D}|} \sum_{(x,y)\in\mathcal{D}} D\left(\ell_y \circ \psi\right)\left(T_p\left(\phi(x,\theta^{(1)}),\theta^{(2)}\right)\right)\begin{pmatrix} \frac{\phi_1(x,\theta^{(1)})-\mu_1}{\sigma_1} \\ 0 \end{pmatrix},$$

$$\frac{\partial \hat{\Lambda}_p}{\partial w_2}(\theta) = \frac{1}{|\mathcal{D}|} \sum_{(x,y)\in\mathcal{D}} D\left(\ell_y \circ \psi\right)\left(T_p\left(\phi(x,\theta^{(1)}),\theta^{(2)}\right)\right)\begin{pmatrix} ,0 \\ \frac{\phi_1(x,\theta^{(2)})-\mu_2}{\sigma_2} \end{pmatrix},$$

$$\frac{\partial \hat{\Lambda}_p}{\partial b_1}(\theta) = \frac{1}{|\mathcal{D}|} \sum_{(x,y)\in\mathcal{D}} D\left(\ell_y \circ \psi\right)\left(T_p\left(\phi(x,\theta^{(1)}),\theta^{(2)}\right)\right)\begin{pmatrix} 1 \\ 0 \end{pmatrix},$$

$$\frac{\partial \hat{\Lambda}_p}{\partial b_2}(\theta) = \frac{1}{|\mathcal{D}|} \sum_{(x,y)\in\mathcal{D}} D\left(\ell_y \circ \psi\right)\left(T_p\left(\phi(x,\theta^{(1)}),\theta^{(2)}\right)\right)\begin{pmatrix} 0 \\ 1 \end{pmatrix},.$$

Let's now let p be the desired means and variances of the image

$$\phi(\mathcal{D},\theta^{(1)}) \stackrel{\text{def}}{=} \left\{\phi(x,\theta^{(1)}); (x,y) \in \mathcal{D}\right\} \subset \mathbb{R}^2$$

of $\mathcal{D}$ through $\phi(\cdot,\theta^{(1)})$. For any $\mathcal{S} \subset \mathbb{R}^2$, define (with small regularization parameter ε (PyTorch sets $\varepsilon = 10^{-5}$) as its default value)

$$\mu_1(\mathcal{S}) \stackrel{\text{def}}{=} \frac{1}{|\mathcal{S}|} \sum_{(x_1,x_2)\in\mathcal{S}} x_1,$$

$$\mu_2(\mathcal{S}) \stackrel{\text{def}}{=} \frac{1}{|\mathcal{S}|} \sum_{(x_1,x_2)\in\mathcal{S}} x_2,$$

$$\sigma_{1,\varepsilon}(\mathcal{S}) \stackrel{\text{def}}{=} \left\{\frac{1}{|\mathcal{S}|} \sum_{(x_1,x_2)\in\mathcal{S}} (x_1 - \mu_1(\mathcal{S}))^2 + \varepsilon\right\}^{1/2},$$

$$\sigma_{2,\varepsilon}(\mathcal{S}) \stackrel{\text{def}}{=} \left\{\frac{1}{|\mathcal{S}|} \sum_{(x_1,x_2)\in\mathcal{S}} (x_2 - \mu_2(\mathcal{S}))^2 + \varepsilon\right\}^{1/2},$$

$$P_\varepsilon(\mathcal{S}) \stackrel{\text{def}}{=} \left(\mu_1(\mathcal{S}), \mu_2(\mathcal{S}), \sigma_{1,\varepsilon}(\mathcal{S}), \sigma_{2,\varepsilon}(\mathcal{S})\right).$$

Informally, we would like to optimize the model

$$\psi\left(T_{P_\varepsilon(\phi(\mathcal{D},\theta^{(1)}))}\left(\phi(x,\theta^{(1)}),\theta^{(2)}\right)\right), \tag{10.3}$$

where $\mathcal{D}$ is our training data. Let's now define a loss function. Set

$$\begin{aligned}\tilde{\Lambda}(\theta) &\stackrel{\text{def}}{=} \frac{1}{|\mathcal{D}|} \sum_{(x,y)\in\mathcal{D}} \ell_y\left(\psi\left(T_{P_\varepsilon(\phi(\mathcal{D},\theta^{(1)}))}\left(\phi(x,\theta^{(1)}),\theta^{(2)}\right)\right)\right) \\ &= \frac{1}{|\mathcal{D}|} \sum_{(x,y)\in\mathcal{D}} \left(\ell_y \circ \psi\right)\left(T_{P_\varepsilon(\phi(\mathcal{D},\theta^{(1)}))}\left(\phi(x,\theta^{(1)}),\theta^{(2)}\right)\right),\end{aligned}$$

which should correspond to (10.3). We note here that $\theta^{(1)}$ now appears in P_ε.

Let's compute the derivatives of $\tilde{\Lambda}$. Derivatives with respect to the w_i's and b_i's (i.e., the components of $\theta^{(2)}$) are similar to those of before. We have

$$\frac{\partial \Lambda}{\partial w_1}(\theta) = \frac{1}{|\mathcal{D}|} \sum_{(x,y)\in\mathcal{D}} (\ell_y \circ \psi)\left(T_{P_\varepsilon(\phi(\mathcal{D},\theta^{(1)}))}\left(\phi(x,\theta^{(1)}),\theta^{(2)}\right)\right) \times \begin{pmatrix} \frac{\phi_1(x,\theta^{(1)})-\mu_1(\phi(\mathcal{D},\theta^{(1)}))}{\sigma_{1,\varepsilon}(\phi(\mathcal{D},\theta^{(1)}))} \\ 0 \end{pmatrix},$$

$$\frac{\partial \Lambda}{\partial w_2}(\theta) = \frac{1}{|\mathcal{D}|} \sum_{(x,y)\in\mathcal{D}} (\ell_y \circ \psi)\left(T_{P_\varepsilon(\phi(\mathcal{D},\theta^{(1)}))}\left(\phi(x,\theta^{(1)}),\theta^{(2)}\right)\right) \times \begin{pmatrix} 0 \\ \frac{\phi_2(x,\theta^{(1)})-\mu_2(\phi(\mathcal{D},\theta^{(1)}))}{\sigma_{2,\varepsilon}(\phi(\mathcal{D},\theta^{(1)}))} \end{pmatrix},$$

$$\frac{\partial \Lambda}{\partial b_1}(\theta) = \frac{1}{|\mathcal{D}|} \sum_{(x,y)\in\mathcal{D}} (\ell_y \circ \psi)\left(T_{P_\varepsilon(\phi(\mathcal{D},\theta^{(1)}))}\left(\phi(x,\theta^{(1)}),\theta^{(2)}\right)\right)\begin{pmatrix}1\\0\end{pmatrix},$$

$$\frac{\partial \Lambda}{\partial b_2}(\theta) = \frac{1}{|\mathcal{D}|} \sum_{(x,y)\in\mathcal{D}} (\ell_y \circ \psi)\left(T_{P_\varepsilon(\phi(\mathcal{D},\theta^{(1)}))}\left(\phi(x,\theta^{(1)}),\theta^{(2)}\right)\right)\begin{pmatrix}0\\1\end{pmatrix}.$$

For $i \in \{1,2\}$,

$$\mu_i(\phi(\mathcal{D},\theta^{(1)})) \stackrel{\text{def}}{=} \frac{1}{|\mathcal{D}|} \sum_{(x,y)\in\mathcal{D}} \phi_i(x,\theta^{(1)}),$$

$$\sigma_{i,\varepsilon}(\phi(\mathcal{D},\theta^{(1)})) \stackrel{\text{def}}{=} \left\{\frac{1}{|\mathcal{D}|} \sum_{(x,y)\in\mathcal{D}} \left(\phi_i(x,\theta^{(1)}) - \mu_i(\phi(\mathcal{D},\theta^{(1)}))\right)^2 + \varepsilon\right\}^{1/2},$$

so

$$\frac{\partial \mu_i(\phi(\mathcal{D},\theta^{(1)}))}{\partial \theta_j^{(1)}} \stackrel{\text{def}}{=} \frac{1}{|\mathcal{D}|} \sum_{(x,y)\in\mathcal{D}} \frac{\partial \phi_i(x,\theta^{(1)})}{\partial \theta_j^{(1)}},$$

$$\frac{\partial \sigma_{i,\varepsilon}(\phi(\mathcal{D},\theta^{(1)}))}{\partial \theta_j^{(1)}} \stackrel{\text{def}}{=} \left\{\frac{1}{|\mathcal{D}|} \sum_{(x,y)\in\mathcal{D}} \left(\phi_i(x,\theta^{(1)}) - \mu_i(\phi(\mathcal{D},\theta^{(1)}))\right)^2 + \varepsilon\right\}^{-1/2} \times \left\{\frac{1}{|\mathcal{D}|} \sum_{(x,y)\in\mathcal{D}} \left(\frac{\partial \phi_i(x,\theta^{(1)})}{\partial \theta_j^{(1)}} - \frac{\partial \mu_i(\phi(\mathcal{D},\theta^{(1)}))}{\partial \theta_j^{(1)}}\right)\right\},$$

$$\frac{\partial \mu_i(\phi(\mathcal{D},\theta^{(1)}))}{\partial \theta_j^{(1)}} \stackrel{\text{def}}{=} \frac{1}{|\mathcal{D}|} \sum_{(x,y)\in\mathcal{D}} \frac{\partial \phi_i(x,\theta^{(1)})}{\partial \theta_j^{(1)}},$$

$$\frac{\partial \sigma_{i,\varepsilon}(\phi(\mathcal{D},\theta^{(1)}))}{\partial \theta_j^{(1)}} \stackrel{\text{def}}{=} \left\{\frac{1}{|\mathcal{D}|} \sum_{(x,y)\in\mathcal{D}} \left(\phi_i(x,\theta^{(1)}) - \mu_i(\phi(\mathcal{D},\theta^{(1)}))\right)^2 + \varepsilon\right\}^{-1/2} \times \left\{\frac{1}{|\mathcal{D}|} \sum_{(x,y)\in\mathcal{D}} \left(\frac{\partial \phi_i(x,\theta^{(1)})}{\partial \theta_j^{(1)}} - \frac{\partial \mu_i(\phi(\mathcal{D},\theta^{(1)}))}{\partial \theta_j^{(1)}}\right)\right\}.$$

Explicitly,

$$T_{P_\varepsilon(\phi(\mathcal{D},\theta^{(1)}))}\left(\phi(x,\theta^{(1)}),\theta^{(2)}\right) = \begin{pmatrix} w_1\left(\frac{\phi_1(x,\theta^{(1)})-\mu_1(\phi(\mathcal{D},\theta^{(1)}))}{\sigma_{1,\varepsilon}(\phi(\mathcal{D},\theta^{(1)}))}\right) + b_1 \\ w_2\left(\frac{\phi_2(x,\theta^{(1)})-\mu_2(\phi(\mathcal{D},\theta^{(1)}))}{\sigma_{2,\varepsilon}(\phi(\mathcal{D},\theta^{(1)}))}\right) + b_2 \end{pmatrix},$$

so

$$\frac{\partial T_{P_\varepsilon(\phi(\mathcal{D},\theta^{(1)}))}\left(\phi(x,\theta^{(1)}),\theta^{(2)}\right)}{\partial \theta_j^{(1)}} = \begin{pmatrix} w_1\left(\frac{\frac{\partial \phi_1(x,\theta^{(1)})}{\partial \theta_j^{(1)}} - \frac{\partial \mu_1(\phi(\mathcal{D},\theta^{(1)}))}{\partial \theta_j^{(1)}}}{\sigma_{1,\varepsilon}(\phi(\mathcal{D},\theta^{(1)}))}\right) \\ w_2\left(\frac{\frac{\partial \phi_2(x,\theta^{(1)})}{\partial \theta_j^{(1)}} - \frac{\partial \mu_2(\phi(\mathcal{D},\theta^{(1)}))}{\partial \theta_j^{(1)}}}{\sigma_{2,\varepsilon}(\phi(\mathcal{D},\theta^{(1)}))}\right) \end{pmatrix} - \begin{pmatrix} w_1\left(\frac{\phi_1(x,\theta^{(1)})-\mu_1(\phi(\mathcal{D},\theta^{(1)}))}{\sigma_{1,\varepsilon}^2(\phi(\mathcal{D},\theta^{(1)}))}\right)\frac{\partial \sigma_{1,\varepsilon}(\phi(\mathcal{D},\theta^{(1)}))}{\partial \theta_j^{(1)}} \\ w_2\left(\frac{\phi_2(x,\theta^{(1)})-\mu_2(\phi(\mathcal{D},\theta^{(1)}))}{\sigma_{2,\varepsilon}^2(\phi(\mathcal{D},\theta^{(1)}))}\right)\frac{\partial \sigma_{2,\varepsilon}(\phi(\mathcal{D},\theta^{(1)}))}{\partial \theta_j^{(1)}} \end{pmatrix}$$

and thus

$$\frac{\partial \tilde{\Lambda}(\theta)}{\partial \theta_j^{(1)}} = \frac{1}{|\mathcal{D}|} \sum_{(x,y)\in\mathcal{D}} D\left(\ell_y \circ \psi\right)\left(T_{P_\varepsilon(\phi(\mathcal{D},\theta^{(1)}))}\left(\phi(x,\theta^{(1)}),\theta^{(2)}\right)\right) \times \left\{\begin{pmatrix} w_1\left(\frac{\frac{\partial \phi_1(x,\theta^{(1)})}{\partial \theta_j^{(1)}} - \frac{\partial \mu_1(\phi(\mathcal{D},\theta^{(1)}))}{\partial \theta_j^{(1)}}}{\sigma_{1,\varepsilon}(\phi(\mathcal{D},\theta^{(1)}))}\right) \\ w_2\left(\frac{\frac{\partial \phi_2(x,\theta^{(1)})}{\partial \theta_j^{(1)}} - \frac{\partial \mu_2(\phi(\mathcal{D},\theta^{(1)}))}{\partial \theta_j^{(1)}}}{\sigma_{2,\varepsilon}(\phi(\mathcal{D},\theta^{(1)}))}\right) \end{pmatrix} - \begin{pmatrix} w_1\left(\frac{\phi_1(x,\theta^{(1)})-\mu_1(\phi(\mathcal{D},\theta^{(1)}))}{\sigma_{1,\varepsilon}^2(\phi(\mathcal{D},\theta^{(1)}))}\right)\frac{\partial \sigma_{1,\varepsilon}(\phi(\mathcal{D},\theta^{(1)}))}{\partial \theta_j^{(1)}} \\ w_2\left(\frac{\phi_2(x,\theta^{(1)})-\mu_2(\phi(\mathcal{D},\theta^{(1)}))}{\sigma_{2,\varepsilon}^2(\phi(\mathcal{D},\theta^{(1)}))}\right)\frac{\partial \sigma_{2,\varepsilon}(\phi(\mathcal{D},\theta^{(1)}))}{\partial \theta_j^{(1)}} \end{pmatrix}\right\}.$$

This gives us gradient descent for batch normalization.

10.3. Batch Normalization and Minibatches

Finally, let's connect our algorithm with minibatches. Directly, this would mean iterating on $P_\varepsilon\left(\phi\left(\mathcal{D}', \theta^{(1)}\right)\right)$ for a sampled subset $\mathcal{D}'$ of $\mathcal{D}$. Recall from the stochastic gradient Chapter 7 that we would in fact like to use sampled subsets of our training data to *update* various quantities, rather than recompute them entirely.

Let's fix a *momentum* parameter $\beta \in (0,1)$ (PyTorch takes $\beta = 0.1$ and Keras takes $\beta = 0.01$ (but with a reversed definition of momentum)). For each k, select a subset $\mathcal{D}_k$ of $\mathcal{D}$ and use stochastic gradient descent on $\mathcal{D}_n$ to update θ_k to θ_{k+1}, and define

$$p_{k+1} \stackrel{\text{def}}{=} \begin{cases} (1-\beta)p_k + \beta P_\varepsilon(\phi(\mathcal{D}_k, \theta_k^{(1)})) & \text{if } k \geq 1 \\ P_\varepsilon(\phi(\mathcal{D}_0, \theta_1^{(1)})) & \text{if } k = 0. \end{cases}$$

After N steps, let's predict using the model

$$\psi\left(T_{p_N}\left(\phi(x, \theta_N^{(1)}), \theta_N^{(2)}\right)\right).$$

- **Training** involves large (mini) batches, giving a sequence $\left((\theta_k^{(1)}, \theta_k^{(2)})\right)_{k=1}^{\infty}$ of values which decrease $\tilde{\Lambda}$ (perhaps minibatching is involved). $\theta_k^{(1)}$ and $\theta_k^{(2)}$ *should* tend to a limit, so $P_\varepsilon(\phi(\mathcal{D}, \theta_k^{(1)}))$ should tend to a limit (perhaps $\mathcal{D}$ is a minibatch).
- **Evaluation** involves using the model to predict a label for *several* (*one* or *few*) new datapoints. *Our goal is to estimate the limiting value of* $P_\varepsilon(\phi(\mathcal{D}, \theta_n^{(1)}))$.

10.4. Brief Concluding Remarks

Batch normalization has become a standard method to transform and normalize datasets in deep learning and was initially proposed in [**IS15b**]; see also [**STIM18**]. After we have formulated our dataset appropriately and defined our model, the next step is to train the model and validate it. This usually happens in three phases. In the training phase, we train the model. In the validation phase we validate the model and choose the best hyperparameters and in the test phase we see how well we are doing. We investigate these issues in Chapter 11.

Chapter 11

Training, Validation, and Testing

11.1. Introduction

At this point, we have a fairly good understanding of how to optimize parameters via (stochastic) gradient descent. Next, we need to understand how to handle the dataset we seek to analyze. Some motivation here comes from classical statistics. As we discussed in Section 1.6, the bias-variance tradeoff is an important phenomenon in statistics, but in order to quantify it, we would need to consider multiple datasets. However, in many situations, we only have a single dataset. The typical strategy is then to split the given dataset into three parts

$$\mathcal{D} = \text{training set} + \text{validation set} + \text{test set}.$$

- *Training:* minimizing *parameters* via (stochastic) gradient descent.
- *Validation:* optimizing over *hyperparameters* which characterize different architectures. The validation's set purpose is to compare different models.
- *Testing:* reporting the results (of the search for the best deep learning model). In order to avoid overfitting we test final accuracy on a test set.

Some practical insights of this decomposition are as follows:

- If we have low training error but high validation error, then we have high variance. In this case, we should use a simpler model and/or collect more data.

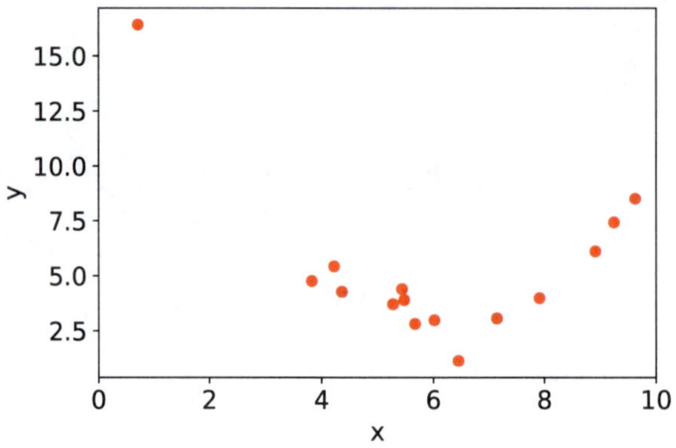

Figure 11.1. Sample (training) data

- If we have large train and test errors, then we have large bias. In this case, we can fit a more complex model.

Since the complexities of deep neural networks oftentimes do not yield theoretical guarantees of performance, well thought out training, testing, and validation steps are crucial to *defensible* claims about deep learning.

11.2. Polynomials

To understand the basic issues, let's consider a problem which is simple (in fact a linear regression problem) but which forces us to think through most of the relevant ideas. Let's *learn* (i.e., *train*) a *polynomial model*,

$$y = c_0 + c_1 x + c_2 x^2 + \cdots + c_D x^D$$

on a finite *training* set $\mathcal{D}^{\mathrm{tr}} \subset \mathbb{R} \times \mathbb{R}$ ($\mathbb{R} \times \mathbb{R}$ is the collection of (x, y)-pairs) of *training* data given in Figure 11.1. In this case, the *machine learning* part is easy; polynomials are particularly simple models for data, and we can get the c_d's by regressing the label on (engineered) feature set

$$\{1, x, x^2 \cdots x^D\}. \tag{11.1}$$

This allows us to focus on concepts of training, validation, and testing. We can think of the degree D as a model complexity *hyperparameter*. Furthermore, the data in Figure 11.1 looks quadratic (and was so constructed); the optimum D should be 2. This allows us a simple check on a number of conclusions.

11.3. Training

For each non-negative integer D, let

$$\mathcal{P}_D \stackrel{\mathrm{def}}{=} \left\{ x \mapsto \sum_{d=0}^{D} c_d x^d \ : \ (c_0, c_2 \cdots c_D) \in \mathbb{R}^{D+1} \right\}$$

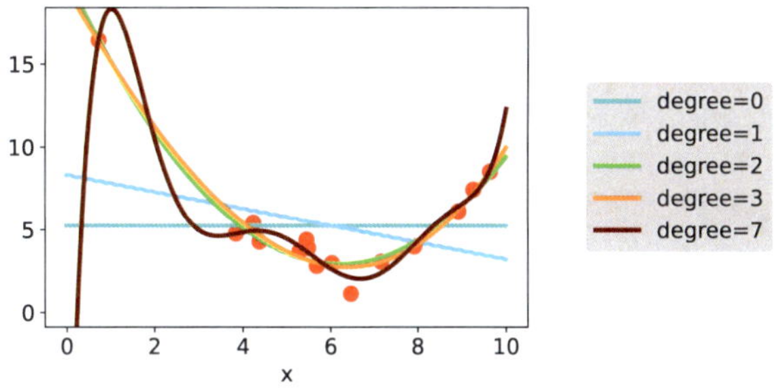

Figure 11.2. Polynomial approximations of training data. Low degrees (blue) underfit and high degrees (red) overfit. Intermediate degrees (green) show good approximation

be the collection of D-dimensional polynomials in x. We can use *regression* to find coefficients; for each $D \in \{0, 1 \dots\}$, let's compute the polynomial

$$P_D^*(x) \stackrel{\text{def}}{=} \operatorname*{argmin}_{P \in \mathcal{P}_D} \left\{ \frac{1}{|\mathcal{D}^{\text{tr}}|} \sum_{(x,y) \in \mathcal{D}^{\text{tr}}} (y - P(x))^2 \right\}$$

which best fits the data $\mathcal{D}^{\text{tr}}$. Rewriting this in more standard notation, we want to equivalently compute the coefficients

$$(c_d^{*,(D)})_{d=0}^D = \operatorname{argmin} \left\{ \frac{1}{|\mathcal{D}^{\text{tr}}|} \sum_{(x,y) \in \mathcal{D}^{\text{tr}}} \left(y - \sum_{d=0}^{D} c_d x^d \right)^2 \right\}, \tag{11.2}$$

minimizing the mean square error. We note that the minimization problem (11.2) is *linear* in the c_d's. Implicitly, we are trying to minimize a quadratic *loss* function. We also note that

- If D is too small (i.e., 0 or 1), it *underfits* the data and doesn't capture variation.
- If D is too large (larger than 2), it *overfits* the known data, but may not fit new data.

See Figure 11.2.

11.4. Validation

Let's assume that we have a new *validation* dataset $\mathcal{D}^{\text{va}}$ (which is statistically similar to the training dataset $\mathcal{D}^{\text{tr}}$) of feature-label examples. This should be able to help us find the best hyperparameter (degree) D. See Figure 11.3. In our case (which reflects what might generically happen), the validation data $\mathcal{D}^{\text{va}}$ fills in *gaps* in the training data $\mathcal{D}^{\text{tr}}$. Comparing our trained models (for different hyperparameters) on these trained models allows us to optimize over hyperparmeters.

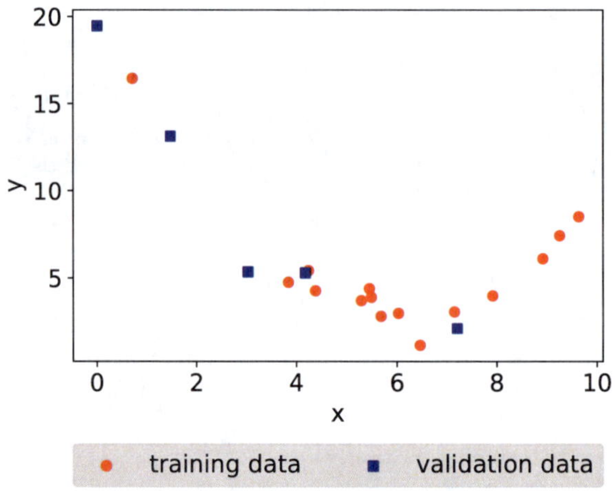

Figure 11.3. Scatter plot of training and validation data.

In carrying out validation, let's use the absolute value (not the square error) as a *metric* to compare predicted and ground-truth labels. Namely, let's minimize

$$\frac{1}{|\mathcal{D}^{\text{va}}|} \sum_{(x,y)\in\mathcal{D}^{\text{va}}} |y - P_D^*(x)|$$

over D; our best hyperparameter is

$$D^* \stackrel{\text{def}}{=} \operatorname{argmin}\left\{ \frac{1}{|\mathcal{D}^{\text{va}}|} \sum_{(x,y)\in\mathcal{D}^{\text{va}}} |y - P_D^*(x)| \ : \ D \in \mathbb{N} \right\}.$$

In our sample data, $D^* = 3$; the best model is in fact cubic, although quadratic and quartic curves also show low validation errors.

A validation curve as in Figure 11.4 captures *bias-variance* tradeoffs. If our model is too simple (in our case, D is too small), we can't capture the important structure (in this case, quadratic dependence) of the training data; i.e., we underfit. If our model is too complex (in our case, D is too large), we can fit the training data, but the structure of our model (in this case, polynomials) forces our trained models to do poorly on new data. This is mathematically natural in our case; given N (x, y) points, where no two x's are the same, we can find an at most $N-1$ degree polynomial [**HJ20**] which passes through these points. This high-degree polynomial may, however, suffer significant oscillations elsewhere.

Since the validation is over a discrete set of hyperparameters (degree D of a polynomial), we use discrete minimization, allowing us to use more meaningful and non-differential metrics (i.e., the absolute value function $|\cdot|$ as opposed to the square error $(\cdot)^2$) (see also the discussion of metrics in linear and logistic

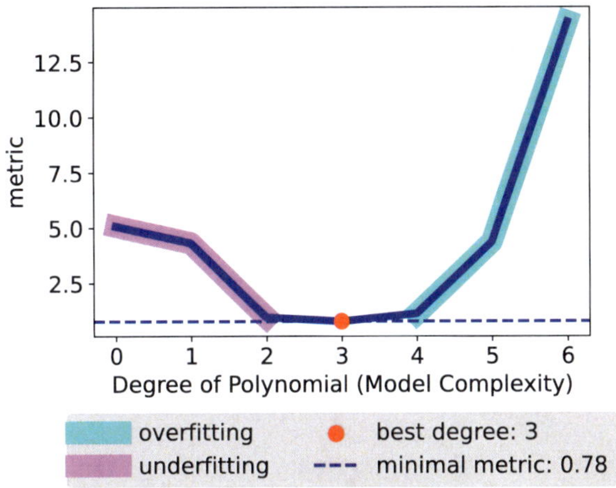

Figure 11.4. Validation curve

regression). By comparison, the *loss* function is used in training to minimize over parameters.

Once we have found the optimum D^*, we can retrain our model

$$P^* \operatorname*{argmin}_{P \in \mathcal{P}_{D^*}} \left\{ \frac{1}{|\mathcal{D}^{\mathrm{tr}} \cup \mathcal{D}^{\mathrm{va}}|} \sum_{(x,y) \in \mathcal{D}^{\mathrm{tr}} \cup \mathcal{D}^{\mathrm{va}}} |y - P(x)| \right\}$$

with the entire training and validation dataset $\mathcal{D}^{\mathrm{tr}} \cup \mathcal{D}^{\mathrm{va}}$; see Figure 11.5. For our example data, we get

$$P^*(x) = 0.01x^3 + 0.23x^2 - 4.67x + 19.54. \tag{11.3}$$

Our validation step suggested that we use a cubic polynomial. The result is a polynomial with in fact a small coefficient in the cubic terms. Roughly, this agrees with our intuition that our data is quadratic. Recall that *our goal is a good approximation of the dependence of label on feature*; the *exact* degree is relatively unimportant.

How do we report the performance of our algorithm? Suppose that we have a third (finite) *test* (holdout) dataset $\mathcal{D}^{\mathrm{te}} \subset \mathbb{R} \times \mathbb{R}$; see Figure 11.6. Let's again use the metric to compare predicted and ground-truth labels. The performance of our polynomial approximation is then

$$\frac{1}{|\mathcal{D}^{\mathrm{te}}|} \sum_{(x,y) \in \mathcal{D}^{\mathrm{te}}} |y - P^*(x)| .$$

The performance in this case is

$$\text{performance} = 1.69. \tag{11.4}$$

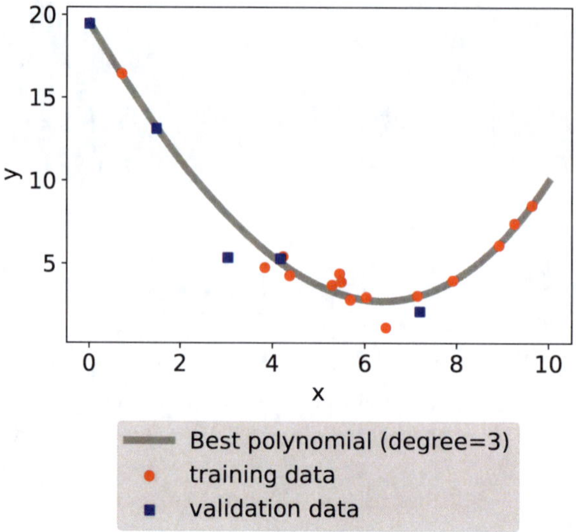

Figure 11.5. Retrained model using optimal hyperparameters and training and validation data. Training data gave us coefficients of polynomials. Validation data gave us degree of polynomial.

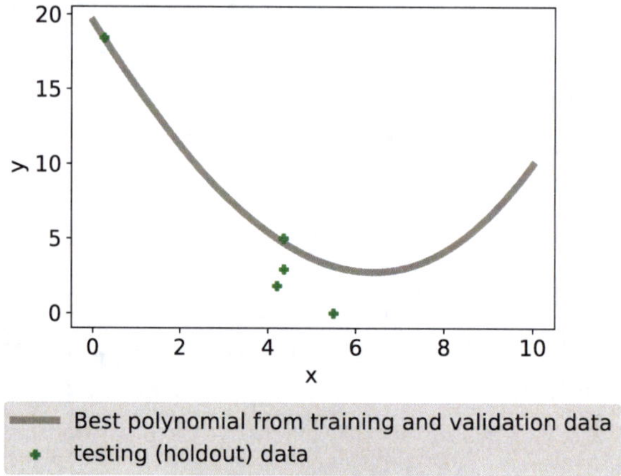

Figure 11.6. Test data

11.5. Cross-Validation

We were admittedly a bit careful in constructing our above datasets. Our x values (the features) were sampled from a standard Gaussian distribution with mean 0 and standard deviation 10. The y values (labels) were constructed by additive perturbing

$$P_{\text{true}}(x) = 0.5x^2 - 6x + 20 \tag{11.5}$$

with points sample from a standard Gaussian distribution. The fitted polynomial (11.3) is reasonably close to (11.5). Polynomial regression is in fact statistically more complicated than linear regression. We tried to adapt ideas from linear regression to the engineered features of (11.1), but we should have added some corrections to account for biases in higher powers of Gaussian noise. However, there is similarly no firm statistical ground for multilevel feed forward networks as in Chapter 5. The complex feature-label relations which deep neural networks try to model are often beyond what pure statistical theory can address.

In constructing our training set, we were also a bit careful in selecting a seed (for the randomization algorithm) which led to a gap in feature space near $x = 2$; that gap highlighted the need for a new (validation) dataset. Similarly, our training dataset consisted of only 15 points; a larger sample size would have started to fill in the gap near $x = 2$.

In fact, our above example was built upon a remarkably small collection of ground-truth datapoints; see Table 11.2. Our test dataset (Figure 11.6) was in particular *very* small; it consisted of only five points. A commonly accepted ratio of sizes of training, validation, and testing sets is 70-15-15; see Table 11.1. The breakdown in our example is in Table 11.2.

Table 11.1. Training, validation, and testing breakdown

Training	70%
Validation	15%
Testing	15%

Table 11.2. Dataset sizes

Role	Count	Percent
Training	15	60
Validation	5	20
Testing	5	20

Suppose that we are given a ground-truth dataset $\mathcal{D}$. Building upon our understanding of training, validation, and testing, let's reverse our above discussion. Let's first randomly select the test (holdout) dataset $\mathcal{D}^{\text{te}}$ used to report our final performance. We want to use the remaining data,

$$\mathcal{D}^{\text{tr\&va}} \stackrel{\text{def}}{=} \mathcal{D} \setminus \mathcal{D}^{\text{te}},$$

for selecting parameters (training) and hyperparameters (validation); i.e, for optimal *model building*. To make our training and validation steps as robust

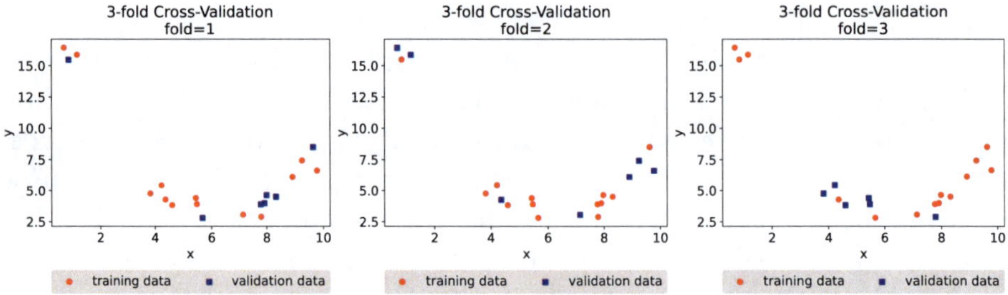

Figure 11.7. Three-fold cross validation curves, training, and validation data

as possible, let's *average* over ways to subselect training and validation sets. In *K-fold cross validation*, let's partition the non-testing data into K *folds* (subsets) $\{\mathcal{F}_k\}_{k=1}^K$. Let's take $K = 3$ in our example problem; see Figure 11.7. Let's then write

$$\mathcal{D}^{\text{tr\&va}} = \bigcup_{k=1}^{K} \mathcal{F}_k.$$

For each k, we can then train on

$$\mathcal{D}^{\text{tr\&va}} \setminus \mathcal{F}_k = \bigcup_{\substack{1 \le k' \le K \\ k' \neq k}} \mathcal{F}_{k'}$$

and find

$$P_D^{(k),*} \stackrel{\text{def}}{=} \underset{P \in \mathcal{P}_D}{\operatorname{argmin}} \left\{ \frac{1}{|\mathcal{D}^{\text{tr\&va}} \setminus \mathcal{F}_k|} \sum_{(x,y) \in \mathcal{D}^{\text{tr\&va}} \setminus \mathcal{F}_k} (y - P(x))^2 \right\}.$$

Then let's construct a validation function

$$m^{(k)}(D) \stackrel{\text{def}}{=} \frac{1}{|\mathcal{F}_k|} \sum_{(x,y) \in \mathcal{F}_k} \left| y - P_D^{(k),*}(x) \right| \qquad D \in \mathbb{N},$$

for each fold, and then minimize the average metric

$$D^*_{K\text{-fold}} \stackrel{\text{def}}{=} \operatorname{argmin} \left\{ \frac{1}{K} \sum_{k=1}^{K} m^{(k)}(D) \, : \, D \in \mathbb{N} \right\},$$

over all hyperparameters D. Again, we get underfitting for low D and overfitting for high D (see Figure 11.8), and in our example, we get $D^*_{K\text{-fold}} = 2$ with best polynomial

(11.6) $$P^*(x) = 0.43x^2 - 5.51x + 20.5.$$

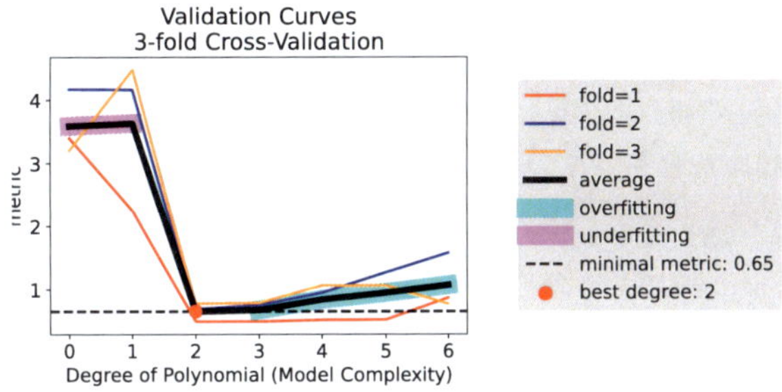

Figure 11.8. Three-fold cross validation curves

As before, once we have the best hyperparameter $D^*_{K\text{-fold}}$, we can retrain our model using all of the available data $\mathcal{D}^{\text{tr\&va}}$;

$$P^* \stackrel{\text{def}}{=} \operatorname*{argmin}_{P \in \mathcal{P}_{D^*_{K\text{-fold}}}} \left\{ \frac{1}{|\mathcal{D}^{\text{tr\&va}}|} \sum_{(x,y)\in\mathcal{D}^{\text{tr\&va}}} (y - P(x))^2 \right\}.$$

Again, we can measure performance using the metric on the test dataset $\mathcal{D}^{\text{te}}$,

$$\frac{1}{|\mathcal{D}^{\text{te}}|} \sum_{(x,y)\in\mathcal{D}^{\text{te}}} |y - P^*(x)|.$$

In our example, we get

$$\text{performance} = 0.72. \tag{11.7}$$

In our example, we get better performance (11.7) with three-fold validation than we did in our original analysis (11.4). Similarly, the estimated polynomial (11.6) from using four-fold validation is closer to the true polynomial (11.5) than the original estimated polynomial (11.3). Averaging over the folds lessens the effect of idiosyncrasies in training and validating over only one selection of $\mathcal{D}^{\text{tr}}$ and $\mathcal{D}^{\text{va}}$.

11.6. Brief Concluding Remarks

In this chapter we saw that well thought out training, testing, and validation steps are crucial to the development of robust and meaningful deep learning algorithms.

In the next chapter we study feature importance. The usefulness of deep learning is that it allows us to represent high-dimensional data. On the other hand, the high-dimensionality of the data makes clearer the need to rank the importance of the various features of the model. Hence, feature importance becomes practically relevant. This is the content of Chapter 12.

Chapter 12

Feature Importance

12.1. Introduction

The appeal of deep neural networks is that they can easily represent high-dimensional datasets. Given a deep neural network, we can explore *feature importance* to rank how important various features are.

Let's work through a problem we can easily understand and visualize. Assume that we have a collection of (x, y) points in feature space $\mathbb{R}^2$ with labels $\ell \in \{0, 1\}$ which, roughly, are

- labeled 1 above the x-axis,
- labeled 0 below the x-axis.

In a real problem,

- the boundary will not be exactly horizontal.
- there may be some noisy observations near the boundary.

Table 12.1 and Figure 12.1 shows a simulation of such points.

Table 12.1. Sample points

x	y	label
1.76	0.40	0
0.98	2.24	1
1.87	−0.98	0
0.95	−0.15	0
−0.10	0.41	0

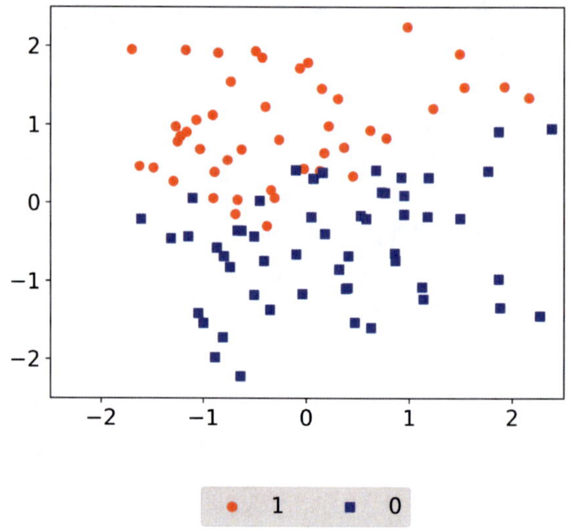

Figure 12.1. Scatter plot of labeled points

Nevertheless, since the boundary is horizontal-*ish*,

the y-feature is more important than the x-feature.

How can we quantify this?

We usually look at feature importance within the framework of some algorithm that we have trained on some dataset $\mathcal{D} \subset \mathbb{R}^2 \times \{0,1\}$. Let's use logistic regression in (x, y) to construct a model $\mathfrak{m}$ from feature space $\mathbb{R}^2$ to label space $\{0,1\}$. For our dataset,

$$\begin{cases} \text{choose 1} & \text{if } y > 0.37x + 0.28 \\ \text{choose 0} & \text{else.} \end{cases}$$

Dividing by the coefficient of y (and retaining the direction of the inequalities, since this coefficient is positive), we can simplify this as

$$\mathfrak{m}^*(x,y) \overset{\text{def}}{=} \begin{cases} 1 & \text{if } y > 0.37x + 0.28 \\ 0 & \text{else.} \end{cases}$$

(In fact, we constructed the points of Figure 12.1 by sampling $N = 100$ points $\{(x_n, y_n)\}_{n=1}^N$ from a standard two-dimensional Gaussian distribution, and then labeling the points 1 if and only if $y_n > 0.3x_n + 0.2 + \frac{1}{2}\eta_n$, where the η_n are samples of a standard Gaussian distribution independent of the (x_n, y_n)'s.)

The starting point of analyses of feature importance is typically some assessment of performance of an algorithm. Namely if $\mathfrak{m} : \mathbb{R}^2 \to \{0,1\}$, let's

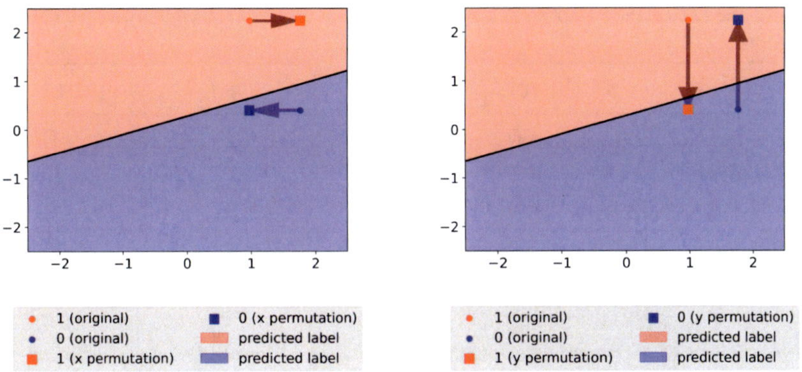

Figure 12.2. Permutation of x (left) and y (right) data

define the *accuracy* of the model $\mathfrak{m}$ applied to the dataset $\mathcal{D}$ as

$$\mathcal{A}_{\mathcal{D}}(\mathfrak{m}) \stackrel{\text{def}}{=} 1 - \frac{1}{|\mathcal{D}|} \sum_{((x,y),\ell)\in\mathcal{D}} |\mathfrak{m}(x,y) - \ell|, \tag{12.1}$$

which, in this case, is the relative number of correct classifications. Computing the accuracy of our logistic regression model $\mathfrak{m}^*$, we have

$$\mathcal{A}_{\mathcal{D}}(\mathfrak{m}^*) = 92.0\%. \tag{12.2}$$

We want to think of (12.2) as *baseline* accuracy, and it reflects the feature-label relationship for some fixed model (e.g., $\mathfrak{m}^*$). Informally, we want to quantify importance of x vs. y in our data by perturbing x and y in our model, and seeing how much $\mathcal{A}$ deteriorates. From Figure 12.1, we expect that perturbing y will lead to a greater deterioration in performance than perturbing x; correct classification depends more on y than x, so y is more important than x.

12.2. Feature Permutation

If x is less important than y, rearrangements of x should have less importance than rearrangements of y. Namely, if we rearrange x, the accuracy should decrease *less* than if we rearrange y. Permutations in y are more likely to move points across the decision boundary, causing classification errors. See Figure 12.2.

Let's define an *x-shuffle* of $\mathcal{D}$ as a copy of $\mathcal{D}$ where we have shuffled the x values. Similarly, a *y-shuffle* of $\mathcal{D}$ is a copy of $\mathcal{D}$ where we have shuffled the y-values. Fix an x-shuffle $\tilde{\mathcal{D}}_x$ of $\mathcal{D}$ and a y-shuffle $\tilde{\mathcal{D}}_y$ of $\mathcal{D}$; see Tables 12.2 and 12.3. Let's then calculate the performance (accuracy) of $\mathfrak{m}$ (i.e., $\mathcal{A}_{\tilde{\mathcal{D}}_x}(\mathfrak{m})$ and $\mathcal{A}_{\tilde{\mathcal{D}}_y}(\mathfrak{m})$) on each of these test sets. See Table 12.4. As expected, the accuracy has deteriorated much more on $\mathcal{D}_y$ than on $\mathcal{D}_x$; information about y is more important than information about x.

Table 12.2. Permutation of x

x	y	label
−0.51	0.40	0
0.86	2.24	1
1.87	−0.98	0
1.87	−0.15	0
−0.07	0.41	0

Table 12.3. Permutation of y

x	y	label
1.76	−1.18	0
0.98	−0.65	1
1.87	−0.98	0
0.95	0.91	0
−0.10	1.71	0

Table 12.4. Accuracy of $\mathfrak{m}^*$ for $\mathcal{D}$ and for an x-shuffle and a y-shuffle.

$\mathcal{A}_{\mathcal{D}}(\mathfrak{m}^*)$	92.0%
$\mathcal{A}_{\mathcal{D}_x}(\mathfrak{m}^*)$	80.0%
$\mathcal{A}_{\mathcal{D}_y}(\mathfrak{m}^*)$	55.0%

Table 12.5. Accuracy of $\mathfrak{m}^*$ for $\mathcal{D}$ and average accuracy for its x-shuffles and y-shuffles.

$\mathcal{A}_{\mathcal{D}}(\mathfrak{m}^*)$	92.0%
$\bar{\mathcal{A}}_{y\text{-shuffles}}(\mathfrak{m}^*)$	82.9%
$\bar{\mathcal{A}}_{y\text{-shuffles}}(\mathfrak{m}^*)$	54.3%

Of course Table 12.2 and 12.3 are only one way of shuffling x and y. There are in fact $N!$ ways to shuffle x and y. We should actually *average* over these ways to shuffle x and y and compute

$$\bar{\mathcal{A}}_{x\text{-shuffles}}(\mathfrak{m}^*) \stackrel{\text{def}}{=} \frac{N!}{\text{number of } x\text{-shuffles } \tilde{\mathcal{D}}_x} \mathcal{A}_{\tilde{\mathcal{D}}_x}(\mathfrak{m}^*),$$

$$\bar{\mathcal{A}}_{y\text{-shuffles}}(\mathfrak{m}^*) \stackrel{\text{def}}{=} \frac{N!}{\text{number of } y\text{-shuffles } \tilde{\mathcal{D}}_y} \mathcal{A}_{\tilde{\mathcal{D}}_y}(\mathfrak{m}^*),$$

and compare them to $\mathcal{A}_{\mathcal{D}}(\mathfrak{m}^*)$. As with Table 12.4, the larger the decrease from $\mathcal{A}(\mathfrak{m}^*)$ to $\bar{\mathcal{A}}_{x\text{-shuffles}}(\mathfrak{m}^*)$ or $\bar{\mathcal{A}}_{y\text{-shuffles}}(\mathfrak{m}^*)$, the more important the feature. By Stirling's formula,

$$N! \stackrel{N\nearrow\infty}{\approx} \sqrt{2\pi N}\left(\frac{N}{e}\right)^N$$

is typically very large, so $\bar{\mathcal{A}}_{y\text{-shuffles}}(\mathfrak{m}^*)$ and $\bar{\mathcal{A}}_{y\text{-shuffles}}(\mathfrak{m}^*)$ are approximated by sampling. Table 12.5 shows the results in our case (approximating $\bar{\mathcal{A}}_{y\text{-shuffles}}(\mathfrak{m}^*)$ and $\bar{\mathcal{A}}_{y\text{-shuffles}}(\mathfrak{m}^*)$ with 100 samples). As with Table 12.5, y is more important.

Our specific formula (12.1) $\mathcal{A}_{\mathcal{D}}(\mathfrak{m})$ of accuracy would have given an accuracy of 1 if $\mathfrak{m}^*$ would have been able to perfectly classify $\mathcal{D}$. More generally,

$$\text{accuracy of model } \mathfrak{m} \text{ on dataset } \mathcal{D} = \text{constant} - \text{error of model } \mathfrak{m} \text{ on dataset } \mathcal{D}.$$

Feature importance can thus be quantified as increases in error as various features are shuffled.

12.3. Shapley Value

Our second way of understanding feature importance stems from [**LL17**], and it quantifies how much a feature adds to all possible predictions.

Again, we start with our fixed trained model, $\mathfrak{m}^*$, which maps a pair $(x, y) \in \mathbb{R}^2$ into a label $\ell \in \{0, 1\}$, and our ground-truth dataset $\mathcal{D} \subset \mathbb{R}^2 \times \{0, 1\}$. Let's build off of ideas of *conditional expectation* and *project* $\mathfrak{m}^*$ onto maps from lower-dimensional sets of features (using $\mathcal{D}$). Namely, let's define

$$\begin{aligned}
\mathfrak{m}^*_{1,1}(x, y) &\stackrel{\text{def}}{=} \mathfrak{m}^*(x, y), \\
\mathfrak{m}^*_{1,0}(x) &\stackrel{\text{def}}{=} \frac{1}{|\mathcal{D}|} \sum_{((x',y'),\ell)\in\mathcal{D}} \mathfrak{m}^*(x, y'), \\
\mathfrak{m}^*_{0,1}(y) &\stackrel{\text{def}}{=} \frac{1}{|\mathcal{D}|} \sum_{((x',y'),\ell)\in\mathcal{D}} \mathfrak{m}^*(x', y), \\
\mathfrak{m}^*_{0,0} &\stackrel{\text{def}}{=} \frac{1}{|\mathcal{D}|} \sum_{((x',y'),\ell)\in\mathcal{D}} \mathfrak{m}^*(x', y').
\end{aligned} \tag{12.3}$$

The function $\mathfrak{m}^*_{1,1}$ is simply $\mathfrak{m}^*$, and predicts a label based on both features. At the other extreme, $\mathfrak{m}^*_{0,0}$ can be thought of as predicting the label based on *neither* feature (and is simply the average predicted label). The functions $\mathfrak{m}^*_{1,0}$ and $\mathfrak{m}^*_{0,1}$ (often called partial dependence functions) predict the label based on, respectively, only x or only y. In our case,

$$m_{0,0} = 0.46,$$

and the plots of $\mathfrak{m}^*_{1,0}$ and $\mathfrak{m}^*_{0,1}$ are given in Figure 12.3 (sometimes called in the literature partial dependence plot). Visually, for a given $x \in \mathbb{R}$, $\mathfrak{m}^*_{1,0}$ averages $\mathfrak{m}^*$ over a synthetic dataset of (x, y') where y' is given by the ground-truth dataset. As x moves to the right, the dataset in Figure 12.1 has more points with label 0, so the average value decreases. This occurs fairly gradually, so the plot of $\mathfrak{m}^*_{1,0}$ decreases fairly slowly. Conversely, for a given $y \in \mathbb{R}$, $\mathfrak{m}^*_{0,1}$ averages $\mathfrak{m}^*$ over a synthetic dataset of (x', y), where now x' is given by the ground-truth dataset. As y moves up, the dataset in Figure 12.1 has more points with label 1 so the average value increases. There is a fairly sharp transition for y near 0, so the plot of $\mathfrak{m}_{0,1}$ quickly increases near 0.

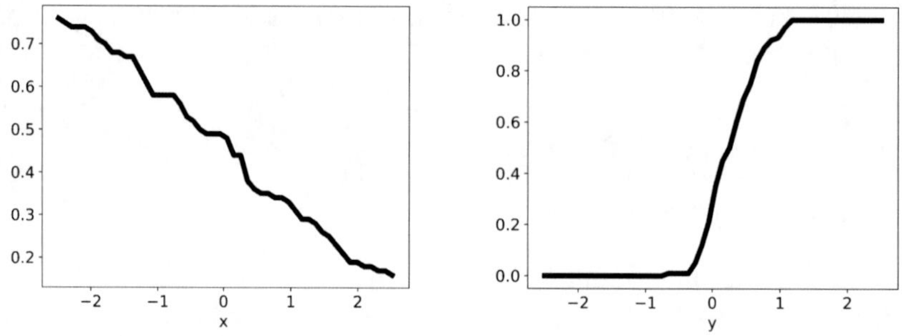

Figure 12.3. Partial dependence plots: plot of $\mathfrak{m}^*_{1,0}$ (left) and $\mathfrak{m}^*_{0,1}$ (right) data

We want to understand the accuracy of the models of (12.3) Borrowing from ideas of computing, let's now *overload* the $\mathfrak{m}^*_{i,j}$'s to again make them functions of both variables, so that we can apply $\mathcal{A}$ of (12.1). Let's define

$$\begin{aligned}
\mathfrak{m}^*_{1,0}(x,y) &\stackrel{\text{def}}{=} \mathfrak{m}^*_{1,0}(x),\\
\mathfrak{m}^*_{0,1}(x,y) &\stackrel{\text{def}}{=} \mathfrak{m}^*_{0,1}(y),\\
\mathfrak{m}^*_{0,0}(x,y) &\stackrel{\text{def}}{=} \mathfrak{m}^*_{0,0},
\end{aligned}$$

which means that $\mathfrak{m}^*_{i,j}$ is being evaluated as $\mathfrak{m}^*_{i,j}$ in the used features.

We can now define *accuracies* based on subsets of feature dimensions. Define

$$\begin{aligned}
\alpha(\{x,y\}) &\stackrel{\text{def}}{=} \mathcal{A}_{\mathcal{D}}(\mathfrak{m}^*_{1,1}),\\
\alpha(\{x\}) &\stackrel{\text{def}}{=} \mathcal{A}_{\mathcal{D}}(\mathfrak{m}^*_{1,0}),\\
\alpha(\{y\}) &\stackrel{\text{def}}{=} \mathcal{A}_{\mathcal{D}}(\mathfrak{m}^*_{0,1}),\\
\alpha(\emptyset) &\stackrel{\text{def}}{=} \mathcal{A}_{\mathcal{D}}(\mathfrak{m}^*_{0,0}).
\end{aligned}$$

We have

$$\begin{aligned}
\alpha(\{x,y\}) &= 0.92,\\
\alpha(\{x\}) &= 0.54,\\
\alpha(\{y\}) &= 0.83,\\
\alpha(\emptyset) &= 0.50.
\end{aligned}$$

We can now quantify how much each feature adds to various subsets of feature dimensions (not already containing that feature):

$$\begin{aligned}
\alpha(\{x\}) - \alpha(\emptyset) &= 0.54 - 0.50 = 0.04, \\
\alpha(\{x, y\}) - \alpha(\{y\}) &= 0.92 - 0.83 = 0.09, \\
\alpha(\{y\}) - \alpha(\emptyset) &= 0.83 - 0.50 = 0.32, \\
\alpha(\{x, y\}) - \alpha(\{x\}) &= 0.92 - 0.54 = 0.38.
\end{aligned}$$

In line with our intuition, y adds much more accuracy than x. The *Shapley value* in our case averages these over each feature:

$$\begin{aligned}
S_x &\stackrel{\text{def}}{=} \frac{1}{2}\left\{\{\alpha(\{x\}) - \alpha(\emptyset)\} + \{\alpha(\{x, y\}) - \alpha(\{y\})\}\right\}, \\
S_y &\stackrel{\text{def}}{=} \frac{1}{2}\left\{\{\alpha(\{y\}) - \alpha(\emptyset)\} + \{\alpha(\{x, y\}) - \alpha(\{x\})\}\right\}.
\end{aligned}$$

In our case,

$$\begin{aligned}
S_x &= \frac{1}{2}\{0.09 + 0.04\} = 0.07, \\
S_y &= \frac{1}{2}\{0.38 + 0.32\} = 0.35.
\end{aligned}$$

Namely, y adds about 35% predictive power, while x adds only about 7%,

$$S_y > S_x,$$

and we quantitatively have that y is a more important feature than x.

Let's start to convert our notation to problems involving more than $N = 2$ features. Let's write

$$S_x = \frac{1}{N}\left\{\frac{\alpha(\{x, y\}) - \alpha(\{y\})}{\binom{N-1}{|\{y\}|}} + \frac{\alpha(\{x\}) - \alpha(\emptyset)}{\binom{N-1}{|\emptyset|}}\right\}$$

with $N = 2$,

$$\binom{N-1}{|\emptyset|} = \binom{1}{0} = 1.$$

With $N = 3$ predictive variables x, y, and z, the Shapley value would be defined as

$$\begin{aligned} S_x &= \frac{1}{N}\left\{\frac{\alpha(\{x,y,z\}) - \alpha(\{y,z\})}{\binom{N-1}{|\{y,z\}|}} + \frac{\alpha(\{x,y\}) - \alpha(\{y\})}{\binom{N-1}{|\{y\}|}}\right. \\ &\qquad \left. + \frac{\alpha(\{x,z\}) - \alpha(\{z\})}{\binom{N-1}{|\{z\}|}} + \frac{\alpha(\{x\}) - \alpha(\emptyset)}{\binom{N-1}{|\emptyset|}}\right\} \\ &= \frac{1}{3}\left\{\frac{\alpha(\{x,y,z\}) - \alpha(\{y,z\})}{\binom{2}{2}} + \frac{\alpha(\{x,y\}) - \alpha(\{y\})}{\binom{2}{1}}\right. \\ &\qquad \left. + \frac{\alpha(\{x,z\}) - \alpha(\{z\})}{\binom{2}{1}} + \frac{\alpha(\{x\}) - \alpha(\emptyset)}{\binom{2}{0}}\right\} \\ &= \frac{1}{3}\left\{\frac{\alpha(\{x,y,z\}) - \alpha(\{y,z\})}{1} + \frac{\alpha(\{x,y\}) - \alpha(\{y\})}{2}\right. \\ &\qquad \left. + \frac{\alpha(\{x,z\}) - \alpha(\{z\})}{2} + \frac{\alpha(\{x\}) - \alpha(\emptyset)}{1}\right\}. \end{aligned}$$

Generally, if $\mathcal{F}$ is the collection of features, $|\mathcal{F}| = N$, and f a feature,

$$\begin{aligned} S_f &\stackrel{\text{def}}{=} \frac{1}{N} \sum_{F \subset \mathcal{F}\setminus\{f\}} \frac{\alpha(F \cup \{x\}) - \alpha(F)}{\binom{N-1}{|F|}} \\ &= \frac{1}{N} \sum_{n=0}^{N-1} \frac{1}{\binom{N-1}{n}} \sum_{\substack{F \subset \mathcal{F}\setminus\{f\} \\ |F| = n}} \{\alpha(F \cup \{x\}) - \alpha(F)\}, \end{aligned}$$

where the second sum has been organized by $|F|$; there are $\binom{N-1}{n}$ ways to choose a subset of $\mathcal{F} \setminus \{f\}$ of size n.

Let's finally note that the Shapley value is bounded by the maximal accuracy. If $\alpha \leq \bar{\alpha}$ ($\bar{\alpha} = 1$ in our calculation),

$$S_f \leq \frac{1}{N} \sum_{n=0}^{N-1} \frac{1}{\binom{N-1}{n}} \sum_{\substack{F \subset \mathcal{F}\setminus\{f\} \\ |F| = n}} \bar{\alpha} = \frac{1}{N} \sum_{n=0}^{N-1} \frac{1}{\binom{N-1}{n}} \binom{N-1}{n} \bar{\alpha} = \frac{1}{N} N\bar{\alpha} = \bar{\alpha}.$$

12.4. Feature Permutation versus Shapley Value

Let us now discuss how feature permutation, explored in Section 12.2, and Shapley value, explored in Section 12.3, compare with each other.

Both methods measure feature importance. Their main difference is that while feature permutation relies on the decrease in model performance, Shapley value is based on how much a feature adds to all possible predictions. Feature permutation does not include a direction, whereas Shapley values can be positive or negative depending on the influence on the predicted outcome.

Feature permutation is calculated on the entire dataset, whereas Shapley value shows how much a feature influences the prediction relative to the average outcome in the dataset.

As a more practical guide, one could use feature permutation to decide which features to keep and which ones are the most important to the accuracy of the model. Feature importance can also guide additional feature engineering. On the other hand, Shapley value can be used to understand which features influence predictions more, or how different values of a given feature affect predictions.

12.5. Brief Concluding Remarks

Up to now we have seen the main ingredients in the formulation of a deep learning algorithm. Namely, we have seen the basic formulation of a feed forward neural network in Chapter 5, backpropagation in Chapter 6, stochastic gradient descent algorithm in Chapter 7, and then the main principles in training, validation, and testing in Chapter 11, and feature selection in Chapter 12. Another important component that has contributed tremendously in the success of deep learning in practice is the realization that dependent and structured data require appropriate architectures. As such, in Chapter 13 we study recurrent neural networks and transformers that are widely used to model time series and sequential data. In Chapter 14 we study convolution neural networks that are widely used in image recognition problems.

12.6. Exercises

Exercise 12.1. Suppose we are building a binary classifier on points in $\mathbb{R}^2$. Consider a test set of three datapoints:

n	1	2	3
X_n	0.5	-2	1
Y_n	1	3	5
ℓ_n	1	0	0

This table means that the first datapoint in the set is $(0.5, 1)$ with label 1. Define the accuracy of the map $\mathfrak{m} : \mathbb{R}^2 \mapsto [0, 1]$ as

$$\mathcal{A}_{\mathcal{D}}(\mathfrak{m}) = 1 - \frac{1}{3}\sum_{n=1}^{3} |\mathfrak{m}(X_n, Y_n) - \ell_n|.$$

Suppose we have the classifier

$$\mathfrak{m}_0(x, y) = \begin{cases} 1, & \text{if } y > x^2 \\ 0, & \text{if } y \leq x^2. \end{cases}$$

In addition define

$$\hat{\mathfrak{m}}_{1,0}(x) = \frac{1}{3}\sum_{n=1}^{3} \mathfrak{m}_0(x, Y_n),$$

$$\hat{\mathfrak{m}}_{0,1}(y) = \frac{1}{3}\sum_{n=1}^{3} \mathfrak{m}_0(X_n, y),$$

$$\hat{\mathfrak{m}}_{0,0} = \frac{1}{3}\sum_{n=1}^{3} \mathfrak{m}_0(X_n, Y_n).$$

With these definitions at hand do the following:

(1) Compute $\hat{\mathfrak{m}}_{0,0}$, $\hat{\mathfrak{m}}_{1,0}(x)$, and $\hat{\mathfrak{m}}_{0,1}(y)$ for each x and y in the available dataset.

(2) Compute $\mathcal{A}_{\mathcal{D}}(\hat{\mathfrak{m}}_{0,0})$, $\mathcal{A}_{\mathcal{D}}(\hat{\mathfrak{m}}_{1,0})$, $\mathcal{A}_{\mathcal{D}}(\hat{\mathfrak{m}}_{0,1})$, and $\mathcal{A}_{\mathcal{D}}(\mathfrak{m}_0)$.

(3) Compute the Shapley value of x and the Shapley value of y.

Exercise 12.2. Suppose we are building a binary classifier on points in $\mathbb{R}^2$. Consider a test set of three datapoints

n	1	2	3
X_n	0.5	−2	1
Y_n	1	3	5
ℓ_n	1	0	1

This table means that the first datapoint in the set is $(0.5, 1)$ with label 1. Define the accuracy of the map $\mathfrak{m} : \mathbb{R}^2 \mapsto [0, 1]$ as

$$\mathcal{A}_{\mathcal{D}}(\mathfrak{m}) = 1 - \frac{1}{3}\sum_{n=1}^{3} |\mathfrak{m}(X_n, Y_n) - \ell_n|.$$

Suppose we have the classifier

$$\mathfrak{m}_0(x, y) = \begin{cases} 1, & \text{if } y > x^2 \\ 0, & \text{if } y \leq x^2. \end{cases}$$

In addition define

$$\hat{\mathfrak{m}}_{1,0}(x) = \frac{1}{3}\sum_{n=1}^{3} \mathfrak{m}_0(x, Y_n),$$

$$\hat{\mathfrak{m}}_{0,1}(y) = \frac{1}{3}\sum_{n=1}^{3} \mathfrak{m}_0(X_n, y),$$

$$\hat{\mathfrak{m}}_{0,0} = \frac{1}{3}\sum_{n=1}^{3} \mathfrak{m}_0(X_n, Y_n).$$

With these definitions in hand,

(1) Compute $\hat{\mathfrak{m}}_{0,0}$, $\hat{\mathfrak{m}}_{1,0}(x)$, and $\hat{\mathfrak{m}}_{0,1}(y)$ for each x and y in the available dataset.

(2) Compute $\mathcal{A}_{\mathcal{D}}(\hat{\mathfrak{m}}_{0,0})$, $\mathcal{A}_{\mathcal{D}}(\hat{\mathfrak{m}}_{1,0})$, $\mathcal{A}_{\mathcal{D}}(\hat{\mathfrak{m}}_{0,1})$, and $\mathcal{A}_{\mathcal{D}}(\mathfrak{m}_0)$.

(3) Compute the Shapley value of x and the Shapley value of y.

(4) Compare the results with those of Exercise 12.1.

Chapter 13

Recurrent Neural Networks for Sequential Data

13.1. Introduction

We next turn to applications of deep neural networks to *sequential data*. For example, sequential data may refer to data representing *time series*, text in *language*, etc. Let us for a moment focus on time series data. We want to predict a time series $\{Y_t\}_{t=1}^{\infty}$ of *outputs* (similar to labels) on the basis of a time series $\{X_t\}_{t=1}^{\infty}$ of *inputs* (similar to features). We want to do so with two thoughts in mind:

- We want to respect time (or generally speaking order in the sequence) as we cannot use future information in predicting present output.
- Real systems have some sort of *internal memory*.

Again, let's use an example dataset to motivate some thoughts. The first few lines of some sample data (see github) are in Table 13.1. See also Figure 13.1.

In Section 13.2 we go over the plant-observer paradigm, which leads to the basic recurrent neural networks: Jordan networks in Section 13.3 and Elman networks in Section 13.4. In Section 13.5.1 we discuss how backpropagation is applied to recurrent neural networks leading to the truncated backpropagation through time. Stability of recurrent neural networks is studied in Section 13.6 where we discuss conditions under which the basic recurrent neural network is stable in the sense that the output does not saturate. This motivates the

Table 13.1. Sample sequential data (an example of a time series)

t	X	Y
1	1.025	−0.041
2	0.893	−0.033
3	1.068	−0.029
4	1.035	−0.026
5	1.028	−0.024

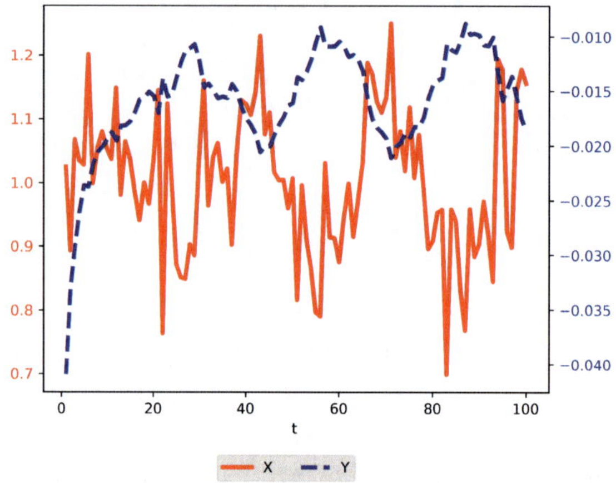

Figure 13.1. Sample sequential data (a graphical example of a time series)

construction of more advanced recurrent neural network architectures, such as gated recurrent networks (GRU), long-short term memory models (LSTM), and bidirectional recurrent neural networks, Section 13.7. Implementation details for *recurrent neural networks* (RNNs), such as dropout, batch normalization, and layer normalization, are discussed in Section 13.8. In Section 13.9 we change gears slightly and discuss the attention mechanism and the basic transformer architecture that has been very successful in sequential data related to large language models, and we discuss how it relates to the more advanced recurrent neural networks such as the LSTM.

13.2. The Plant-Observer Paradigm

We can organize our discussion by the *plant-observer* paradigm of systems theory. Suppose that the inputs drive a dynamical *plant*

$$Z_t = f(Z_{t-1}, X_t). \tag{13.1}$$

The observations are assumed to be some function of the state of the plant

$$Y_t = g(Z_t).$$

The plant Z may be very high-dimensional, reflecting the evolution of a large amount of *hidden* (or *latent*) information. If the plant dynamics (13.1) are linear, Z becomes an autoregressive process. The function g collapses all of this information to the observed values.

13.3. Jordan Networks

Let's assume that g is the identity map. Then $Y_t = Z_t$ and the plant-observer model reduces to

$$Y_t = f(Y_{t-1}, X_t).$$

Explicitly, we are representing the current output as a combination of the prior output and the current input:

$$(\text{feature}_t, \text{label}_t) = ((X_t, Y_{t-1}), Y_t).$$

This becomes a standard deep learning problem with *label* Y_t and *feature* (Y_{t-1}, X_t); see Table 13.2. We note in Table 13.2 that Y_{-1} does not exist, so we do not have a ground-truth pair $((X_0, Y_{-1}), Y_0)$ in our ground-truth dataset.

Table 13.2. Feature-label data for Jordan networks

		Feature	**Label**
	X	**lagged Y**	**Y**
t			
1	1.025	nan	−0.041
2	0.893	−0.041	−0.033
3	1.068	−0.033	−0.029
4	1.035	−0.029	−0.026
5	1.028	−0.026	−0.024

As a point of comparison for more complicated models, let's write out a one-layer model for a Jordan network. We might consider

$$\mathfrak{m}(X_t, Y_{t-1}; \theta) \stackrel{\text{def}}{=} \tanh(w_1 X_t + w_2 Y_{t-1} + b), \tag{13.2}$$

where $\theta \in \mathbb{R} \times \mathbb{R} \times \mathbb{R}$ is the parameter vector

$$\theta = (w_1, w_2, b).$$

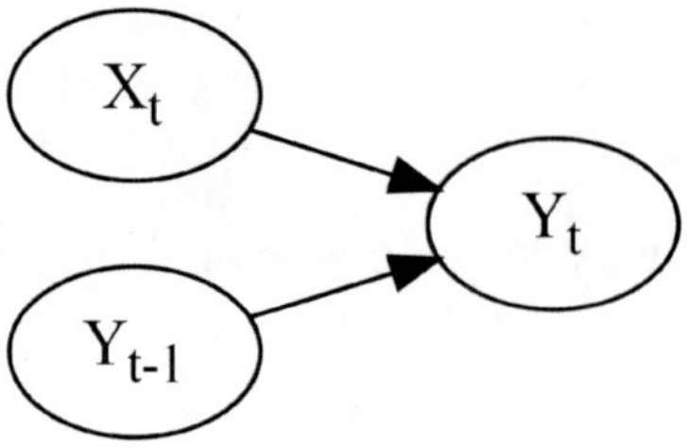

Figure 13.2. Jordan network

In Figure 13.2 we present schematically a Jordan network. Given (X_t, Y_{t-1}), $\mathfrak{m}(X_t, Y_{t-1})$ would be our prediction of Y_t. We can use the mean-square error

$$\Lambda(\theta) \stackrel{\text{def}}{=} \frac{1}{T}\sum_{t=1}^{T}\left(Y_t - \mathfrak{m}(X_t, Y_{t-1}; \theta)\right)^2, \qquad \theta = (w_1, w_2, b),$$

to find the best such model. More layers can easily be added.

13.4. Elman Networks

Elman networks implement the full plant-observer paradigm. Let's think through the analogue of (13.2). Let's consider models of the form

$$\begin{aligned} Z_t^{\mathfrak{m}}(\theta) &= \tanh(w_{zi}X_t + w_{zz}Z_{t-1}^{\mathfrak{m}}(\theta) + b_z) \\ Y_t^{\mathfrak{m}}(\theta) &= \tanh(w_o Z_t^{\mathfrak{m}}(\theta) + b_o) \end{aligned} \qquad t \in \{1, 2, \dots\}, \tag{13.3}$$

where

$$\theta = \begin{pmatrix} w_{zi} & w_{zz} & b_z & w_o & b_0 \end{pmatrix}$$

is in $\mathbb{R}^5$.

In Figure 13.3 we present schematically an Elman network.

Let's assume that

$$Z_0^{\mathfrak{m}}(\theta) = 0.1 \tag{13.4}$$

to start the plant process $Z^{\mathfrak{m}}$ at a specific nonzero value. Let's also take $T = 3$ as our time horizon; that will allow us to exactly write out several calculations.

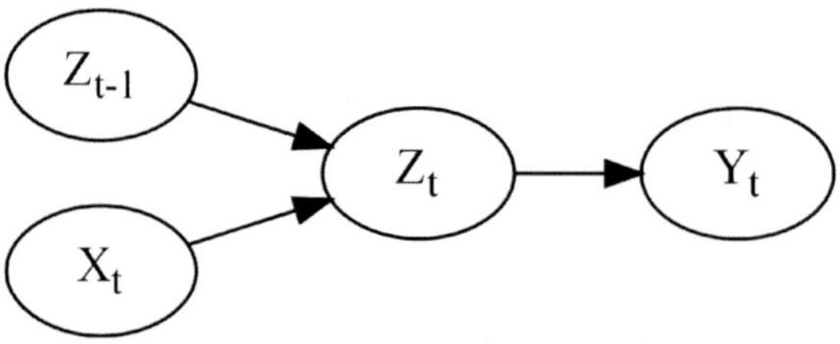

Figure 13.3. Elman network

Our loss function will be

$$\Lambda(\theta) \stackrel{\text{def}}{=} \frac{1}{T}\sum_{t=1}^{T}\left(Y_t^{\mathrm{m}}(\theta) - Y_t\right)^2. \tag{13.5}$$

As usual, we want to use *gradient descent* to minimize (13.5) over all θ's in $\mathbb{R}^5$. Let's think through how this would work with a concrete example. Consider

$$\theta = \begin{pmatrix} \overset{w_{zi}}{0.20} & \overset{w_{zz}}{-0.25} & \overset{b_z}{0.30} & \overset{w_o}{0.10} & \overset{b_o}{-0.15} \end{pmatrix}, \tag{13.6}$$

and let's compute $\nabla\Lambda(\theta)$. A concrete example may help us keep track of what is known and what needs to be calculated.

First, knowing the initial condition (13.4) and the parameters (13.6), we can reconstruct the plant Z using w_{zi}, w_{zz}, and b_z for $t \in \{1, 2, 3\}$:

$$\begin{aligned}
Z_t^{\mathrm{m}}(\theta) &= \tanh(w_{zi}X_t + w_{zz}Z_{t-1}^{\mathrm{m}}(\theta) + b_z) = \tanh(0.2X_t - 0.25Z_{t-1}^{\mathrm{m}}(\theta) + 0.3)\\
Z_0^{\mathrm{m}}(\theta) &= 0.1.
\end{aligned}$$

Explicitly, we may compute

$$\begin{aligned}
Z_0^{\mathrm{m}}(\theta) &= 0.1,\\
Z_1^{\mathrm{m}}(\theta) &\approx \tanh((-0.25)\times 0.1 + 0.2\times 1.025 + 0.3) \approx \tanh(0.48) \approx 0.446,\\
Y_1^{\mathrm{m}}(\theta) &\approx \tanh(0.1\times 0.446 - 0.15) \approx \tanh(-0.105) \approx -0.105,\\
Z_2^{\mathrm{m}}(\theta) &\approx \tanh((-0.25)\times 0.446 + 0.2\times 0.893 + 0.3) \approx \tanh(0.367) \approx 0.351,\\
Y_2^{\mathrm{m}}(\theta) &\approx \tanh(0.1\times 0.351 - 0.15) \approx \tanh(-0.115) \approx -0.114,\\
Z_3^{\mathrm{m}}(\theta) &\approx \tanh((-0.25)\times 0.351 + 0.2\times 1.068 + 0.3) \approx \tanh(0.426) \approx 0.402,\\
Y_3^{\mathrm{m}}(\theta) &\approx \tanh(0.1\times 0.402 - 0.15) \approx \tanh(-0.11) \approx -0.109.
\end{aligned}$$

See Table 13.3. As a point of reference, we then have, to three significant digits, that

$$\Lambda(\theta) \approx \frac{1}{3}\left\{(-0.041 + 0.105)^2 + (-0.033 + 0.114)^2 + (-0.029 + 0.109)^2\right\} \approx 0.006.$$

Table 13.3. Given and constructed values

	Ground Truth			Model
	X	**Y**	$\mathbf{Z}^{\mathrm{m}}(\theta)$	$\mathbf{Y}^{\mathrm{m}}(\theta)$
t				
0	nan	nan	0.100	nan
1	1.025	−0.041	0.446	−0.105
2	0.893	−0.033	0.351	−0.114
3	1.068	−0.029	0.402	−0.109

Differentiating with respect to w_o and b_o, we then have

$$\begin{aligned}
\frac{\partial \Lambda}{\partial w_o}(\theta) &= \frac{2}{T}\sum_{t=1}^{T}(Y_t^{\mathrm{m}} - Y_t)\frac{\partial Y_t^{\mathrm{m}}}{\partial w_o} \\
&= \frac{2}{T}\sum_{t=1}^{T}(Y_t^{\mathrm{m}} - Y_t)\tanh'(w_o Z_t^{\mathrm{m}} + b_o) Z_t^{\mathrm{m}}, \\
\frac{\partial \Lambda}{\partial b_o}(\theta) &= \frac{2}{T}\sum_{t=1}^{T}(Y_t^{\mathrm{m}} - Y_t)\frac{\partial Y_t^{\mathrm{m}}}{\partial b_o} \\
&= \frac{2}{T}\sum_{t=1}^{T}(Y_t^{\mathrm{m}} - Y_t)\tanh'(w_o Z_t^{\mathrm{m}} + b_o).
\end{aligned}$$

Explicitly,

$$\begin{aligned}
\frac{\partial \Lambda}{\partial w_o}(\theta) &\approx \frac{1}{3}\{(-0.041 + 0.105)\tanh'(0.1 \times 0.446 - 0.15) \times 0.446 \\
&\qquad + (-0.033 + 0.114)\tanh'(0.1 \times 0.351 - 0.15) \times 0.351 \\
&\qquad + (-0.029 + 0.109)\tanh'(0.1 \times 0.402 - 0.15) \times 0.402\} \\
&\approx \frac{1}{3}\{0.028 + 0.028 + 0.032\}, \\
&\approx 0.029, \\
\frac{\partial \Lambda}{\partial b_o}(\theta) &\approx \frac{1}{3}\{(-0.041 + 0.105)\tanh'(0.1 \times 0.446 - 0.15) \\
&\qquad + (-0.033 + 0.114)\tanh'(0.1 \times 0.351 - 0.15) \\
&\qquad + (-0.029 + 0.109)\tanh'(0.1 \times 0.402 - 0.15)\} \\
&\approx \frac{1}{3}\{0.063 + 0.08 + 0.079\} \\
&\approx 0.074.
\end{aligned}$$

How do we take derivatives of Λ with respect to w_{zi}, w_{zz}, and b_i? Again differentiating, we have

$$\begin{aligned}
\frac{\partial \Lambda}{\partial w_{zi}}(\theta) &= \frac{2}{3}\sum_{t=0}^{T}(Y_t^{\mathrm{m}} - Y_t)\frac{\partial Y_t^{\mathrm{m}}}{\partial w_{zi}} \\
&= \frac{2}{3}\sum_{t=0}^{T}(Y_t^{\mathrm{m}} - Y_t)\tanh'(w_o Z_t^{\mathrm{m}} + b_o)\, w_o \frac{\partial Z_t^{\mathrm{m}}}{\partial w_{zi}}, \\
\frac{\partial \Lambda}{\partial w_{zz}}(\theta) &= \frac{2}{3}\sum_{t=0}^{T}(Y_t^{\mathrm{m}} - Y_t)\frac{\partial Y_t^{\mathrm{m}}}{\partial w_{zz}} \\
&= \frac{2}{3}\sum_{t=0}^{T}(Y_t^{\mathrm{m}} - Y_t)\tanh'(w_o Z_t^{\mathrm{m}} + b_o)\, w_o \frac{\partial Z_t^{\mathrm{m}}}{\partial w_{zz}}, \\
\frac{\partial \Lambda}{\partial b_z}(\theta) &= \frac{2}{3}\sum_{t=0}^{T}(Y_t^{\mathrm{m}} - Y_t)\frac{\partial Y_t^{\mathrm{m}}}{\partial b_z} \\
&= \frac{2}{3}\sum_{t=0}^{T}(Y_t^{\mathrm{m}} - Y_t)\tanh'(w_o Z_t^{\mathrm{m}} + b_o)\, w_o \frac{\partial Z_t^{\mathrm{m}}}{\partial b_z}.
\end{aligned}$$

In turn,

$$\begin{aligned}
\frac{\partial Z_t^{\mathrm{m}}}{\partial w_{zi}} &= \tanh'(w_{zi}X_t + w_{zz}Z_{t-1}^{\mathrm{m}} + b_z)\left\{w_{zz}\frac{\partial Z_{t-1}^{\mathrm{m}}}{\partial w_{zi}} + X_t\right\}, \\
\frac{\partial Z_t^{\mathrm{m}}}{\partial w_{zz}} &= \tanh'(w_{zi}X_t + w_{zz}Z_{t-1}^{\mathrm{m}} + b_z)\left\{w_{zz}\frac{\partial Z_{t-1}^{\mathrm{m}}}{\partial w_{zz}} + Z_{t-1}^{\mathrm{m}}\right\}, \\
\frac{\partial Z_t^{\mathrm{m}}}{\partial b_z} &= \tanh'(w_{zi}X_t + w_{zz}Z_{t-1}^{\mathrm{m}} + b_z)\left\{w_{zz}\frac{\partial Z_{t-1}^{\mathrm{m}}}{\partial b_z} + 1\right\},
\end{aligned}$$

which can be organized as a matrix evolution

$$\begin{aligned}
\begin{pmatrix} \partial Z_t^{\mathrm{m}}/\partial w_{zi} \\ \partial Z_t^{\mathrm{m}}/\partial w_{zz} \\ \partial Z_t^{\mathrm{m}}/\partial b_z \end{pmatrix} &= \tanh'(w_{zi}X_t + w_{zz}Z_{t-1}^{\mathrm{m}} + b_z)\left\{w_{zz}\begin{pmatrix} \partial Z_{t-1}^{\mathrm{m}}/\partial w_{zi} \\ \partial Z_{t-1}^{\mathrm{m}}/\partial w_{zz} \\ \partial Z_{t-1}^{\mathrm{m}}/\partial b_z \end{pmatrix} + \begin{pmatrix} X_t \\ Z_{t-1}^{\mathrm{m}} \\ 1 \end{pmatrix}\right\}, \\
\begin{pmatrix} \partial Z_{-1}^{\mathrm{m}}/\partial w_{zi} \\ \partial Z_{-1}^{\mathrm{m}}/\partial w_{zz} \\ \partial Z_{-1}^{\mathrm{m}}/\partial b_z \end{pmatrix} &= \begin{pmatrix} 0 \\ 0 \\ 0 \end{pmatrix}.
\end{aligned}$$

Explicitly,

$$\begin{pmatrix} \partial Z_1^{\mathrm{m}}/\partial w_{zi} \\ \partial Z_1^{\mathrm{m}}/\partial w_{zz} \\ \partial Z_1^{\mathrm{m}}/\partial b_z \end{pmatrix} \approx \tanh(0.2 \times 1.025 + (-0.25) \times 0.1 + 0.3)$$

$$\times \left\{ \begin{pmatrix} 0.000 \\ 0.000 \\ 0.000 \end{pmatrix} \times (-0.25) + \begin{pmatrix} 1.025 \\ 0.100 \\ 1.000 \end{pmatrix} \right\}$$

$$\approx \tanh(0.48) \begin{pmatrix} 1.025 \\ 0.100 \\ 1.000 \end{pmatrix}$$

$$\approx \begin{pmatrix} 0.821 \\ 0.080 \\ 0.801 \end{pmatrix},$$

$$\begin{pmatrix} \partial Z_2^{\mathrm{m}}/\partial w_{zi} \\ \partial Z_2^{\mathrm{m}}/\partial w_{zz} \\ \partial Z_2^{\mathrm{m}}/\partial b_z \end{pmatrix} \approx \tanh(0.2 \times 0.893 + (-0.25) \times 0.446 + 0.3)$$

$$\times \left\{ \begin{pmatrix} 0.821 \\ 0.080 \\ 0.801 \end{pmatrix} \times (-0.25) + \begin{pmatrix} 0.893 \\ 0.446 \\ 1.000 \end{pmatrix} \right\}$$

$$\approx \tanh(0.367) \begin{pmatrix} 0.893 \\ 0.446 \\ 1.000 \end{pmatrix}$$

$$\approx \begin{pmatrix} 0.783 \\ 0.391 \\ 0.877 \end{pmatrix},$$

$$\begin{pmatrix} \partial Z_3^{\mathrm{m}}/\partial w_{zi} \\ \partial Z_3^{\mathrm{m}}/\partial w_{zz} \\ \partial Z_3^{\mathrm{m}}/\partial b_z \end{pmatrix} \approx \tanh(0.2 \times 1.068 + (-0.25) \times 0.351 + 0.3)$$

$$\times \left\{ \begin{pmatrix} 0.783 \\ 0.391 \\ 0.877 \end{pmatrix} \times (-0.25) + \begin{pmatrix} 1.068 \\ 0.351 \\ 1.000 \end{pmatrix} \right\}$$

$$\approx \tanh(0.426) \begin{pmatrix} 1.068 \\ 0.351 \\ 1.000 \end{pmatrix}$$

$$\approx \begin{pmatrix} 0.896 \\ 0.295 \\ 0.839 \end{pmatrix},$$

which then gives us

$$
\begin{aligned}
\frac{\partial \Lambda}{\partial w_{zz}}(\theta) &\approx \frac{1}{3}\big\{(-0.041+0.105)\tanh'(0.1\times 0.446-0.15)\times 0.1\times 0.08 \\
&\qquad +(-0.033+0.114)\tanh'(0.1\times 0.351-0.15)\times 0.1\times 0.391 \\
&\qquad +(-0.029+0.109)\tanh'(0.1\times 0.402-0.15)\times 0.1\times 0.295\big\} \\
&\approx \frac{1}{3}\{0.001+0.003+0.002\} \\
&\approx 0.002, \\
\frac{\partial \Lambda}{\partial w_{zi}}(\theta) &\approx \frac{1}{3}\big\{(-0.041+0.105)\tanh'(0.1\times 0.446-0.15)\times 0.1\times 0.821 \\
&\qquad +(-0.033+0.114)\tanh'(0.1\times 0.351-0.15)\times 0.1\times 0.783 \\
&\qquad +(-0.029+0.109)\tanh'(0.1\times 0.402-0.15)\times 0.1\times 0.896\big\} \\
&\approx \frac{1}{3}\{0.005+0.006+0.007\} \\
&\approx 0.006, \\
\frac{\partial \Lambda}{\partial b_{z}}(\theta) &\approx \frac{1}{3}\big\{(-0.041+0.105)\tanh'(0.1\times 0.446-0.15)\times 0.1\times 0.801 \\
&\qquad +(-0.033+0.114)\tanh'(0.1\times 0.351-0.15)\times 0.1\times 0.877 \\
&\qquad +(-0.029+0.109)\tanh'(0.1\times 0.402-0.15)\times 0.1\times 0.839\big\} \\
&\approx \frac{1}{3}\big\{0.064\tanh'(-0.105)\times 0.1\times 0.801 \\
&\qquad +0.081\tanh'(-0.115)\times 0.1\times 0.877 \\
&\qquad +0.08\tanh'(-0.11)\times 0.1\times 0.839\big\} \\
&\approx \frac{1}{3}\{0.005+0.007+0.007\} \\
&\approx 0.006.
\end{aligned}
$$

13.5. Training and Backpropagation for Recurrent Neural Networks

Let us recall the generic mean-square loss function introduced in (13.5). In Section 13.4 we investigated in a concrete example how the derivatives of the loss function Λ with respect to how the parameters in θ look for the standard Elman network. The goal of this section is to go over the idea behind the implementation of backpropagation for a generic recurrent neural network. Instead of considering the specific model (13.3), let us generalize this slightly and instead consider

$$
(13.7) \qquad \begin{aligned} Z_t^{\mathfrak{m}} &= \mathfrak{m}_Z(X_t, Z_{t-1}^{\mathfrak{m}}; \theta) \\ Y_t^{\mathfrak{m}} &= \mathfrak{m}_Y(X_t, Z_t^{\mathfrak{m}}; \theta) \end{aligned} \qquad t \in \{1, 2, \dots, T\}.
$$

We first note that the parameters that are included in the vector θ are common across all times $t \in \{1, 2, \dots, T\}$. As we shall now demonstrate, a recurrent neural network such as (13.7) can be realized as a very deep neural network with number of layers being T. Indeed, let us rewrite (13.7) in the form

$$
(13.8) \qquad \begin{aligned} Z_t^{\mathfrak{m}} &= \mathfrak{m}_Z(X_t, Z_{t-1}^{\mathfrak{m}}; \theta_t), \\ Y_t^{\mathfrak{m}} &= \mathfrak{m}_Y(X_t, Z_t^{\mathfrak{m}}; \theta_t), \\ \tilde{\Lambda}(\theta_1, \dots, \theta_T) &= \sum_{t=1}^{T} (Y_t^{\mathfrak{m}} - Y_t)^2, \\ \theta_t &= \theta, \quad t \in \{1, 2, \dots, T\}. \end{aligned}
$$

Note that (13.8) is a multi-layer neural network, where θ_t is the parameter vector for the tth layer. By the chain rule we have

$$
(13.9) \qquad \begin{aligned} \nabla_\theta \Lambda(\theta) &= \nabla_\theta \frac{1}{T} \sum_{t=1}^{T} (Y_t^{\mathfrak{m}} - Y_t)^2 \\ &= \frac{1}{T} \sum_{t=1}^{T} \nabla_{\theta_t} \tilde{\Lambda}(\theta_1, \dots, \theta_T) \\ &= \frac{1}{T} \sum_{t=1}^{T} \left[2(Y_t^{\mathfrak{m}} - Y_t) \frac{\partial \mathfrak{m}_Y(X_t, Z_t^{\mathfrak{m}}; \theta_t)}{\partial \theta} \right] + \frac{1}{T} \sum_{t=1}^{T-1} \left[\nabla_{Z_t^{\mathfrak{m}}} \tilde{\Lambda}(\theta_1, \dots, \theta_T) \frac{\partial Z_t^{\mathfrak{m}}}{\partial \theta_t} \right]. \end{aligned}
$$

Using the standard δ notation in backpropagation, if we now set

$$
\delta_t = \nabla_{Z_t^{\mathfrak{m}}} \tilde{\Lambda}(\theta_1, \dots, \theta_T),
$$

we shall have for $t = 1, \dots, T-1$, again by the chain rule, that

$$
\delta_t = 2(Y_{t+1}^{\mathfrak{m}} - Y_{t+1}) \frac{\partial \mathfrak{m}_Y(X_{t+1}, Z_{t+1}^{\mathfrak{m}}; \theta_{t+1})}{\partial Z^{\mathfrak{m}}} + \frac{\partial \mathfrak{m}_Z(X_{t+1}, Z_{t+1}^{\mathfrak{m}}; \theta_{t+1})}{\partial Z^{\mathfrak{m}}} \delta_{t+1},
$$

with $\delta_T = 0$. Note also that using the δ notation, (13.9) can be written as

$$
\begin{aligned}
\nabla_\theta \Lambda(\theta) &= \frac{1}{T}\sum_{t=1}^{T}\left[2\left(Y_t^{\mathrm{m}} - Y_t\right)\frac{\partial \mathfrak{m}_Y\left(X_t, Z_t^{\mathrm{m}};\theta_t\right)}{\partial\theta}\right] + \frac{1}{T}\sum_{t=1}^{T-1}\delta_t\frac{\partial Z_t^{\mathrm{m}}}{\partial\theta_t}\\
&= \frac{1}{T}\sum_{t=1}^{T}\left[2\left(Y_t^{\mathrm{m}} - Y_t\right)\frac{\partial \mathfrak{m}_Y\left(X_t, Z_t^{\mathrm{m}};\theta_t\right)}{\partial\theta}\right] + \frac{1}{T}\sum_{t=1}^{T-1}\delta_t\frac{\partial \mathfrak{m}_Z\left(X_t, Z_t^{\mathrm{m}};\theta_t\right)}{\partial\theta}\\
&= \frac{1}{T}\left[\left(2\left(Y_T^{\mathrm{m}} - Y_T\right)\frac{\partial \mathfrak{m}_Y\left(X_T, Z_T^{\mathrm{m}};\theta_T\right)}{\partial\theta}\right)\right.\\
&\qquad \left. + \sum_{t=1}^{T-1}\left(2\left(Y_t^{\mathrm{m}} - Y_t\right)\frac{\partial \mathfrak{m}_Y\left(X_t, Z_t^{\mathrm{m}};\theta_t\right)}{\partial\theta} + \delta_t\frac{\partial \mathfrak{m}_Z\left(X_t, Z_t^{\mathrm{m}};\theta_t\right)}{\partial\theta}\right)\right]. \qquad (13.10)
\end{aligned}
$$

Formula (13.10) gives the essence of the backpropagation algorithm in the case of a recurrent neural network; see also Exercise 13.3. One can see that the computational cost of this backpropagation algorithm, called backpropagation through time (BPTT), is of the order of T. This computational cost can be too costly to bear. For this reason, in typical applications the truncated backpropagation through time (tBPTT) algorithm is used instead of the BPTT algorithm, see Section 13.5.1.

13.5.1. Truncated backpropagation through time. As we just demonstrated, typical computational cost of the backpropagation algorithm based on (13.5) would be of order T, but if the sequence is long, then the computational cost would be too large to afford. In practice, the algorithm being used is the so-called *truncated backpropagation through time*, or tBPTT for short.

In particular, consider a truncation length $\tau \ll T$ (many times in practice $\tau = 1$) and set the objective function at the kth iteration to be

$$\Lambda_k(\theta_k) = \frac{1}{\tau}\sum_{t=\tau(k-1)+1}^{\tau k}\left(Y_t^{\mathrm{m}} - Y_t\right)^2,$$

where the normalization with $\frac{1}{\tau}$ is many times omitted in practice (as it is also the case for the $\frac{1}{T}$ normalization in (13.5)). The update equations take the form

$$
\begin{aligned}
Z_t^{\mathrm{m}} &= \mathfrak{m}_Z\left(X_t, Z_{t-1}^{\mathrm{m}};\theta_k\right)\\
Y_t^{\mathrm{m}} &= \mathfrak{m}_Y\left(X_t, Z_t^{\mathrm{m}};\theta_k\right)
\end{aligned}
\qquad t \in \{\tau(k-1)+1, \dots, \tau k\},
$$

and we remark that $Z^{\mathrm{m}}_{\tau(k-1)} = \theta_{k-2}$. Then, the SGD update for the model (13.3) becomes

$$\theta_{k+1} = \theta_k - \eta_k \nabla \Lambda_k(\theta_k).$$

Compared to traditional SGD, the computational cost is now of order τ. Even though tBPTT based SGD is biased, it has proven to work well in practice. Then, backpropagation is done similarly to (13.10) but bases the calculations on Λ_k rather than on Λ, which was the case for (13.10).

Next, we present the tBPTT algorithm, in pseudocode, in the case of truncation length $\tau = 1$, see Algorithm 1. Let us also consider the simplest case of an Elman network with a shallow neural network, a generic activation function σ for the memory and affine activation function for the output, i.e., going back to (13.3), we have the recurrent neural network model

$$\begin{aligned} Z_t^{\mathrm{m}} &= \sigma(w_{zi}X_t + w_{zz}Z_{t-1}^{\mathrm{m}} + b_z) \\ Y_t^{\mathrm{m}} &= w_o \cdot Z_t^{\mathrm{m}} + b_o \end{aligned} \qquad t \in \{1, 2, \dots\}.$$

Note that in general, $X \in \mathbb{R}^d$ is a vector, $w_{zi} \in \mathbb{R}^{N\times d}$, $w_{zz} \in \mathbb{R}^{N\times N}$ would be matrices, $Z_\cdot^{\mathrm{m}}, b_z, w_o \in \mathbb{R}^N$ would be vectors, and $Y_\cdot^{\mathrm{m}}, b_o$ would be one dimensional. Here, the vector of parameters is $\theta = (w_o, w_{zi}, w_{zz}, b_z, b_o)$. Let us assume that we initialize the parameters based on some distribution λ, i.e., at iteration $k = 0$ we have that $\theta_0 \sim \lambda$.

Algorithm 1 Online SGD with tBPTT for truncation length $\tau = 1$

procedure ▷ (Input parameters network size N, initial parameters distribution λ, running time T)
  Initialize: initial parameters $\theta_0 \sim \lambda$, initial memories $\forall i, Z_0^{\mathrm{m},i} = 0$, step $k = 0$
  while $k \le T$ **do**
    for all $i \in \{1, 2, \dots, N\}$ **do** ▷ Truncated forward propagation
      $Z_{k+1}^{\mathrm{m},i} \leftarrow \sigma\left(\sum_{j=1}^d w_{zi,k}^{i,j} X_k^j + \sum_{\ell=1}^N w_{zz,k}^{i,\ell} Z_k^{\mathrm{m},\ell} + b_{z,k}^i\right)$ ▷ Updating memory
    end for
    $Y_k^{\mathrm{m}} \leftarrow \sum_{i=1}^N w_{o,k}^i Z_{k+1}^{\mathrm{m},i} + b_{o,k}$ ▷ Updating output
    $\Lambda_k(\theta_k) = \frac{1}{2}(Y_k^{\mathrm{m}} - Y_k)^2$ ▷ Computing loss
    for all $i \in \{1, 2, \dots, N\}$ **do** ▷ Truncated backward propagation on $\Lambda_k(\theta_k)$
      $\Delta Z_{k+1}^{\mathrm{m},i} \leftarrow \sigma'\left(\sum_{j=1}^d w_{zi,k}^{i,j} X_k^j + \sum_{\ell=1}^N w_{zz,k}^{i,\ell} Z_k^{\mathrm{m},\ell} + b_{z,k}^i\right)$
      $b_{o,k+1} = b_{o,k} - \eta_k 2(Y_k^{\mathrm{m}} - Y_k)$
      $w_{o,k+1}^i = w_{o,k}^i - \eta_k 2(Y_k^{\mathrm{m}} - Y_k) Z_{k+1}^{\mathrm{m},i}$
      $w_{zi,k+1}^{i,j} = w_{zi,k}^{i,j} - \eta_k 2(Y_k^{\mathrm{m}} - Y_k) w_{o,k}^i \Delta Z_{k+1}^{\mathrm{m},i} X_k^j, \quad j = 1, \dots, d$
      $w_{zz,k+1}^{i,\ell} = w_{zz,k}^{i,\ell} - \eta_k 2(Y_k^{\mathrm{m}} - Y_k) w_{o,k}^i \Delta Z_{k+1}^{\mathrm{m},i} Z_k^{\mathrm{m},\ell}, \quad \ell = 1, \dots, N$
      $b_{z,k+1}^i = b_{z,k}^i - \eta_k 2(Y_k^{\mathrm{m}} - Y_k) w_{o,k}^i \Delta Z_{k+1}^{\mathrm{m},i}$
    end for
  end while
end procedure

Some bibliographical remarks are in order. In Algorithm 1 training is done *online*, i.e., we update the parameters every time we observe a new step of our sequences (X_k, Y_k). Convergence of algorithms like the online Algorithm 1, after appropriate scalings and as a number of hidden units and training steps $N, k \to \infty$, have been recently studied in [**CHLSS23**]. Scaling limits for recurrent neural networks when training is *offline* by continuous gradient descent after observing a fixed number of steps of the sequence $(X_k)_{k\geq 0}$ has been studied in [**ALM23**]. We will present detailed analysis of scaling limit results for feed forward neural networks in Chapters 19, 20, and, in the context of reinforcement learning, in Chapter 21.

13.6. Stability

Let's return to (13.3). Replacing tanh by the identity map, let's consider the *linear* evolution

$$Z_t^L = w_{zz}Z_{t-1}^L + w_{zi}X_t + b_o \qquad t \in \{1, 2 \dots\}. \tag{13.11}$$

This is a reasonable approximation of the dynamics of (13.3) if $Z^{\mathrm{m}} \approx 0$. Explicitly solving (13.11), we have that

$$Z_t^L = w_{zz}^t Z_0^L + \sum_{t'=1}^{t} w_{zz}^{t-t'} \{w_{zi}X_{t'} + b_o\}.$$

Thus

- If $|w_{zz}| < 1$, then $\lim_{n\nearrow\infty} |w_{zz}^n| = 0$, and the effects of past become negligible (and (13.11) is *stable*)
- If $|w_{zz}| > 1$, then $\lim_{n\nearrow\infty} |w_{zz}^n| = \infty$, and the effects of past become magnified (and (13.11) is *unstable*).

If $|w_{zz}| > 1$ and (13.11) is unstable, the tanh in the dynamics of Z^{m} of (13.3) is likely to *saturate*.

To find w_{zz}, we use gradient descent, which involves sensitivities. By comparison, differentiating (13.11) with respect to w_{zz}, we get

$$\frac{\partial Z_t^L}{\partial w_{zz}} = t w_{zz}^t Z_0^L + \sum_{t'=1}^{t} (t-t') w_{zz}^{t-t'-1} \{w_{zi}X_{t'} + b_z\}.$$

Since

$$\lim_{t\to\infty} |t w_{zz}^t| = \begin{cases} 0 & \text{if } |w_{zz}| < 1 \\ \infty & \text{if } |w_{zz}| > 1, \end{cases}$$

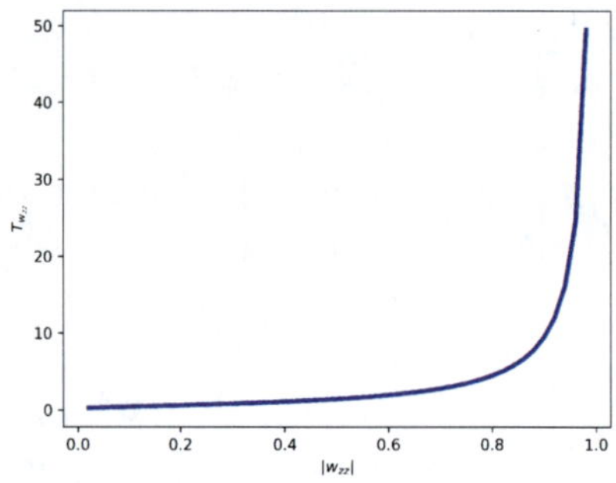

Figure 13.4. Time horizon of $|w_{zz}|^t$

the gradient $\partial Z_t^L/\partial w_{zz}$ is likely to

- *explode* as $t \nearrow \infty$ if $|w_{zz}| > 1$
- *vanish* as $t \nearrow \infty$ if $|w_{zz}| < 1$.

This is a particular challenge if one simply uses gradient descent to update w_{zz} to improve model fit. Even if $|w_{zz}| < 1$,

$$w_{zz} - \eta \frac{\partial \Lambda}{\partial w_{zz}}$$

may fail to be in $(-1, 1)$.

If Z^{m} is multidimensional, we can replace w_{zz} by a square matrix. In that case, we replace $|w_{zz}|$ with spectral radius of A.

If $|w_{zz}| < 1$, writing

$$\begin{aligned} |w_{zz}|^t &= (\exp[\ln |w_{zz}|])^t = \exp[t \ln |w_{zz}|] \\ &= \exp[-t \ln {}^1\!/\!_{|w_{zz}|}] = \exp\left[-\frac{t}{T_{w_{zz}}}\right], \end{aligned}$$

where

$$T_{w_{zz}} \stackrel{\text{def}}{=} \frac{1}{\ln {}^1\!/\!_{|w_{zz}|}},$$

we can think of $T_{w_{zz}}$ as a *time horizon* (see Figure 13.4); $|w_{zz}|^t \ll 1$ if and only if $t \gg T_{w_{zz}}$.

13.7. Advanced Architectures

Standard recurrent neural networks are difficult to train in order to get a stable time horizon of memory length; we want w_{zz} of (13.3) to be in $(-1, 1)$. One resolution for this is to replace w_{zz} with a logistic function S, which both take values in $(0, 1)$, and allows one to dynamically *gate* memory.

Let's replace (13.3) with

$$Z_t^{\mathrm{m}} = S(WX_t + B) \odot Z_{t-1}^{\mathrm{m}} + \tilde{f}(X_t)$$

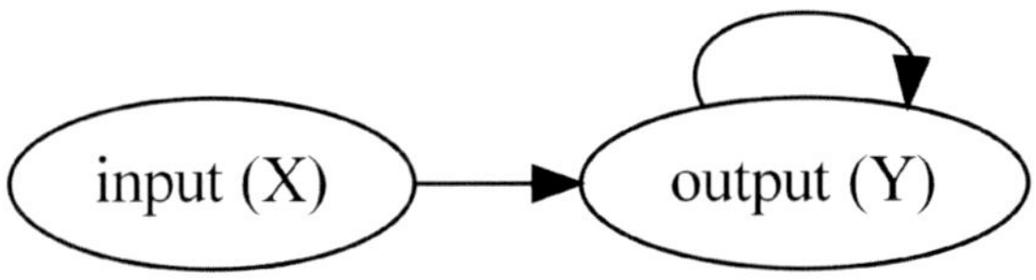

Figure 13.5. GRU architecture

for $t \in \{1, 2, \dots\}$. If $WX_t + B \gg 1$, then

$$Z_t^{\mathrm{m}} \approx Z_{t-1}^{\mathrm{m}} + \tilde{f}(X_t),$$

and we have strong memory. On the other hand, if $WX_t + B \ll -1$, then

$$Z_t^{\mathrm{m}} \approx \tilde{f}(X_t),$$

and the plant forgets the prior state. Since the logistic function takes values in $(0, 1)$, dependence on past should be close to stable.

$$Z_t^{\mathrm{m}} = \left\{\prod_{t'=1}^{t} S(WX_{t'} + B)\right\} Z_0^{\mathrm{m}} + \sum_{t'=1}^{t} \left\{\left\{\prod_{t''=t'}^{t} S(WX_{t''} + B)\right\}\right\} \tilde{f}(X_{t'}).$$

13.7.1. Gated Recurrent Units (GRU). In gated recurrent units (GRUs), we set the observer map to the identity ($Z_t^{\mathrm{m}} = Y_t$) and formally make Z_t^{m} a convex combination of Z_{t-1}^{m} and the effect of the input. In particular, we write

$$Y_t = \alpha(X_t) \odot Y_{t-1} + (1 - \alpha(X_t)) \odot \tilde{f}_{\mathrm{GRU}}(X_t, Y_{t-1}).$$

In Figure 13.5 we present schematically a GRU architecture.

A typical realization of the GRU architecture amounts to setting

$$\begin{aligned} \alpha(X_t) &= S\left(W^{(1)}X_t + B^{(1)}\right), \\ \tilde{f}_{\mathrm{GRU}}(X_t, Y_{t-1}) &= \tanh\left(W^{(2)}X_t + W^{(3)}S\left(W^{(4)}X_t + B^{(4)}\right)Y_{t-1} + B^{(2)}\right), \end{aligned}$$

resulting in the architecture

$$\begin{aligned} Y_t = {} & S\left(W^{(1)}X_t + B^{(1)}\right) \odot Y_{t-1} + \left(1 - S\left(W^{(1)}X_t + B^{(1)}\right)\right) \\ & \odot \tanh\left(W^{(2)}X_t + W^{(3)}S\left(W^{(4)}X_t + B^{(4)}\right)Y_{t-1} + B^{(2)}\right). \end{aligned}$$

Motivated by the analysis before, we see that if $S\left(W^{(1)}X_t + B^{(1)}\right) \approx 1$, then

$$Y_t \approx Y_{t-1},$$

and we have strong memory. On the other hand, if $S\left(W^{(1)}X_t + B^{(1)}\right) \approx 0$, then

$$Y_t \approx \tanh\left(W^{(2)}X_t + W^{(3)}S\left(W^{(4)}X_t + B^{(4)}\right)Y_{t-1} + B^{(2)}\right),$$

and the plant depends less on the prior state.

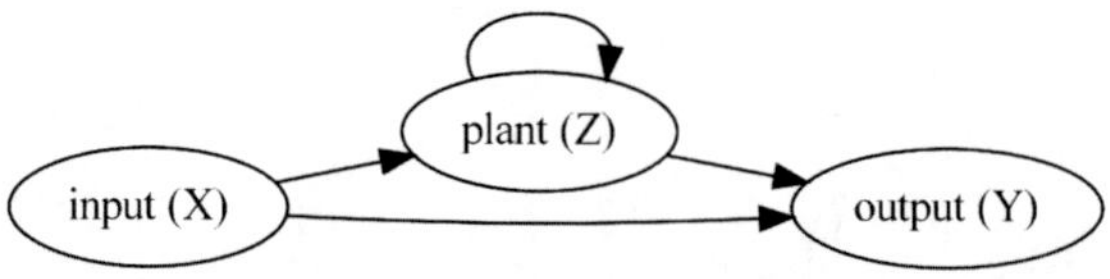

Figure 13.6. LSTM architecture

By setting $u_t = \alpha(X_t)$, one can write the GRU architecture in the following equivalent manner

$$\begin{aligned} u_t &= S\left(W^{(1)}X_t + B^{(1)}\right), \\ v_t &= \tanh\left(W^{(2)}X_t + W^{(3)}r_t \odot Y_{t-1} + B^{(2)}\right), \\ r_t &= S\left(W^{(4)}X_t + B^{(4)}\right), \\ Y_t &= u_t \odot Y_{t-1} + (1 - u_t) \odot v_t, \end{aligned}$$

where, r_t is called the reset gate, u_t is called the update gate, and Y_t is the memory. Of course, one can replace the logistic S and the tanh activation functions by any other activation functions.

13.7.2. Long-Short-Term Memory (LSTM). Long-short-term memory networks, similar to Elman networks, have a nonlinear observer function. In particular, we write

$$\begin{aligned} Z_t^{\mathrm{m}} &= \alpha_1(X_t) \odot Z_{t-1}^{\mathrm{m}} + \alpha_2(X_t) \odot \tilde{f}_{\mathrm{LSTM}}(X_t), \\ Y_t^{\mathrm{m}} &= \alpha_3(X_t) \tanh(Z_t^{\mathrm{m}}). \end{aligned}$$

In Figure 13.6 we present schematically the LSTM architecture.

A typical realization of the LSTM architecture amounts to setting

$$\begin{aligned} \alpha_1(X_t) &= S\left(W^{(1)}X_t + B^{(1)}\right), \\ \alpha_2(X_t) &= S\left(W^{(2)}X_t + B^{(2)}\right), \\ \tilde{f}_{\mathrm{LSTM}}(X_t) &= \tanh\left(W^{(3)}X_t + B^{(3)}\right), \\ \alpha_3(X_t) &= S\left(W^{(4)}X_t + B^{(4)}\right), \end{aligned}$$

resulting in the architecture

$$\begin{aligned} Z_t^{\mathrm{m}} &= S\left(W^{(1)}X_t + B^{(1)}\right) \odot Z_{t-1}^{\mathrm{m}} + S\left(W^{(2)}X_t + B^{(2)}\right) \tanh\left(W^{(3)}X_t + B^{(3)}\right), \\ Y_t^{\mathrm{m}} &= S\left(W^{(4)}X_t + B^{(4)}\right) \tanh(Z_t^{\mathrm{m}}). \end{aligned}$$

Note that the functions $\alpha_i(x) \in (0,1)$ for $i = 1,2,3$ are all logistic functions. As such, and similarly to the GRU case, if $\alpha_i(x) \approx 0$, the corresponding component will not be present in the model, or it will be fully present if $\alpha_i(x) \approx 1$.

The previous way of writing the LSTM architecture gives some intuition on the role of the different components. By setting $f_t = \alpha_1(X_t)$, $g_t = \alpha_2(X_t)$, and $q_t = \alpha_3(X_t)$, one can write the basic LSTM architecture in the following equivalent form

$$\begin{aligned}
f_t &= S\left(W^{(1)}X_t + B^{(1)}\right),\\
g_t &= S\left(W^{(2)}X_t + B^{(2)}\right),\\
Z_t^{\mathrm{m}} &= f_t \odot Z_{t-1}^{\mathrm{m}} + g_t \odot \tanh\left(W^{(3)}X_t + B^{(3)}\right),\\
q_t &= S\left(W^{(4)}X_t + B^{(4)}\right),\\
Y_t^{\mathrm{m}} &= q_t \odot \tanh(Z_t^{\mathrm{m}}),
\end{aligned}$$

where f_t is called the forget gate, g_t is called the input gate, Z_t^{m} is the memory, and Y_t^{m} is the hidden state (sometimes called the output). Of course, one can replace the logistic S and the tanh activation functions by other activation functions.

The LSTM in its most general form is written as

$$\begin{aligned}
f_t &= \sigma\left(W^{(1)}X_t + U^{(1)}Y_{t-1}^{\mathrm{m}} + B^{(1)}\right),\\
g_t &= \sigma\left(W^{(2)}X_t + U^{(2)}Y_{t-1}^{\mathrm{m}} + B^{(2)}\right),\\
r_t &= \tanh\left(W^{(3)}X_t + U^{(3)}Y_{t-1}^{\mathrm{m}} + B^{(3)}\right),\\
Z_t^{\mathrm{m}} &= f_t \odot Z_{t-1}^{\mathrm{m}} + g_t \odot r_t, \qquad (13.12)\\
q_t &= \sigma\left(W^{(4)}X_t + U^{(4)}Y_{t-1}^{\mathrm{m}} + B^{(4)}\right),\\
Y_t^{\mathrm{m}} &= q_t \odot \tanh(Z_t^{\mathrm{m}}),
\end{aligned}$$

where the matrices $U^{(1)}, U^{(2)}, U^{(3)}, U^{(4)}$ are also part of the parameter of the model θ that needs to be learned and σ can be a generic activation function. Also compared to the GRU architecture, the LSTM's forget gate and input gate are replaced by a single update gate in the GRU.

Remark 13.1. It is interesting to iterate the formula for Z_t^{m} in (13.12). We notice that we can write

$$\begin{aligned}
Z_t^{\mathrm{m}} &= f_t \odot Z_{t-1}^{\mathrm{m}} + g_t \odot r_t\\
&= \sum_{k=0}^{t}\left(g_k \odot \prod_{\ell=k+1}^{t} f_\ell\right) \odot r_k\\
&= \sum_{k=0}^{t} M_{k,t} \odot r_k, \qquad (13.13)
\end{aligned}$$

where we have set $M_{k,t} = g_k \odot \prod_{\ell=k+1}^{t} f_\ell$. Formula (13.13) shows that an LSTM memory state models long-distance context and it can be thought of as an (elementwise) weighted sum of a standard RNN state r_k. The weights are products of the input gate g_k and every future forget gate f_ℓ for $\ell \geq k+1$.

We will return to this interpretation of the LSTM in Section 13.9.3 where we compare the LSTM with the attention mechanism.

13.7.3. Bidirectional RNNs. In standard recurrent neural networks Z_t depends in a nonlinear manner on the data sequence $X_t, X_{t-1}, X_{t-2}, \ldots, X_1$, but not upon $X_{t+1}, X_{t+2}, \ldots, X_T$. This structure ignores potential important dependencies and data for a prediction at time t. In order to address this point, *bidirectional* RNNs were developed in [**SP97**]. In bidirectional RNNs, there are plants flowing in forward and backward time, both contributing to the observations:

$$\begin{aligned} \overrightarrow{Z}_t &= f_+(\overrightarrow{Z}_{t-1}, X_t), \quad t = 0, 1, \ldots, T, \\ \overleftarrow{Z}_t &= f_-(\overleftarrow{Z}_{t+1}, X_t), \quad t = T-1, T-2, \ldots, 0, \\ Y_t &= g(\overrightarrow{Z}_t, \overleftarrow{Z}_t, X_t), \quad t = 1, 2, \ldots, T. \end{aligned}$$

Here, $\overrightarrow{Z}_t$ is called the forward internal state and $\overleftarrow{Z}_t$ is called the backward internal state. Typically, $\overrightarrow{Z}_0$ and $\overleftarrow{Z}_T$ are initialized to be constants.

(Truncated) backpropagation through time is typically used to train bidirectional RNNs which at a high level read as follows:

- Calculate the forward in time direction $\overrightarrow{Z}_t$ for $t = 1, \ldots, T$.
- Calculate the backward in time direction $\overleftarrow{Z}_t$ for $t = T-1, T-2, \ldots, 0$.
- Calculate output layer Y_t for for $t = 0, 1, \ldots, T$.
- In the backpropagation step we calculate the gradients with respect to the model parameters.

Note that what makes the implementation feasible is that the forward internal states $\overrightarrow{Z}_t$ and backward internal states $\overleftarrow{Z}_t$ are independent. The same way one builds standard bidirectional RNNs one can also build a bidirectional GRU or a bidirectional LSTM.

13.8. Implementation Aspects for Recurrent Neural Networks

In this section we discuss several implementation aspects of recurrent neural networks, including the regularization method of dropout, Subsection 13.8.1, that we explored in Chapter 9; batch-normalization, Subsection 13.8.2, that we explored in Chapter 10; as well as layer-normalization, Subsection 13.8.3, which is similar to batch normalization but instead of normalizing per batch, we normalize per layer.

We mention here, without going into many details, that in recent years a number of techniques have been developed to address problems of possible saturation and vanishing grading problems associated to recurrent neural network architectures. Non-saturating recurrent neural networks proposed in [**CSV+19**] address the saturation problem by replacing saturating gates in LSTM or GRU models by using ReLU nonlinearities for activation functions (as opposed for example to the standard sigmoid and tanh activation functions used in LSTMs). Highway networks (see [**SGS15**] for the original paper and [**ZSKS17**] for its extension to recurrent neural network structures) combine identity functions with gates (similar to LSTMs and GRUs) and can help with addressing the vanishing gradient problem but also successfully build networks with many layers. Highway networks have also been successfully applied to numerically solving partial differential equations, see for example [**SS18**]. We leave further reading on non-saturating RNNs and highway networks to the interested reader.

13.8.1. Dropout in RNNs. As we discussed in Chapter 9 dropout is a regularization method that is widely used in deep learning. This naturally includes RNNs, see for instance [**SSB16**]. Let us recall that the memory state of a recurrent neural network reads as

$$Z_t^{\mathrm{m}} = \mathfrak{m}_Z\left(X_t, Z_{t-1}^{\mathrm{m}}; \theta\right), \qquad t \in \{1, \dots, T\}.$$

Recalling the dropout operator $\mathcal{D}$ from (9.1), the simplest, probably, way to use dropout is to instead consider

$$Z_t^{\mathrm{m}} = \mathfrak{m}_Z\left(X_t, \mathcal{D}(Z_{t-1}^{\mathrm{m}}); \theta\right), \qquad t \in \{1, \dots, T\},$$

where we recall that $\mathcal{D}(h) = h \odot \gamma$, where γ is a Bernoulli vector with success probability (componentwise) $p \in (0, 1)$.

As we explored in Section 9.5 we would use $\mathcal{D}(Z_{t-1}^{\mathrm{m}})$ during training and replace that by $Z_{t-1}^{\mathrm{m}} \odot \mathbb{E}(\gamma)$ during testing.

A further question we need to answer is whether the Bernoulli random vector γ would change with respect to t or not and whether it would be the same for all components of the memory state. This leads to the per-sequence and the per-step sampling.

To have a concrete architecture in mind consider the LSTM architecture (13.12). Per-sequence sampling is when sample dropout masks are applied once and then used in for every step in the entire sequence. Specifically, let $\gamma_1, \gamma_2, \gamma_3, \gamma_4$ be independent Bernoulli vectors and consider for $t = 1, \dots, T$ the architecture

$$\begin{aligned}
f_t &= \sigma\big(W^{(1)} X_t + \gamma_1 \odot U^{(1)} Y_{t-1}^{\mathrm{m}} + B^{(1)}\big), \\
g_t &= \sigma\big(W^{(2)} X_t + \gamma_2 \odot U^{(2)} Y_{t-1}^{\mathrm{m}} + B^{(2)}\big), \\
Z_t^{\mathrm{m}} &= f_t \odot Z_{t-1}^{\mathrm{m}} + g_t \odot \tanh\big(W^{(3)} X_t + \gamma_3 \odot U^{(3)} Y_{t-1}^{\mathrm{m}} + B^{(3)}\big), \\
q_t &= \sigma\big(W^{(4)} X_t + \gamma_4 \odot U^{(4)} Y_{t-1}^{\mathrm{m}} + B^{(4)}\big), \\
Y_t^{\mathrm{m}} &= q_t \odot \tanh(Z_t^{\mathrm{m}}).
\end{aligned}$$

Analogously, per-step sampling amounts to consider for $t = 1, \ldots, T$ independent Bernoulli vectors $\gamma_{1,t}, \gamma_{2,t}, \gamma_{3,t}, \gamma_{4,t}$ and then consider for $t = 1, \ldots, T$ the architecture

$$\begin{aligned}
f_t &= \sigma\big(W^{(1)}X_t + \gamma_{1,t} \odot U^{(1)}Y^{\mathrm{m}}_{t-1} + B^{(1)}\big)\\
g_t &= \sigma\big(W^{(2)}X_t + \gamma_{2,t} \odot U^{(2)}Y^{\mathrm{m}}_{t-1} + B^{(2)}\big)\\
Z^{\mathrm{m}}_t &= f_t \odot Z^{\mathrm{m}}_{t-1} + g_t \odot \tanh\big(W^{(3)}X_t + \gamma_{3,t} \odot U^{(3)}Y^{\mathrm{m}}_{t-1} + B^{(3)}\big)\\
q_t &= \sigma\big(W^{(4)}X_t + \gamma_{4,t} \odot U^{(4)}Y^{\mathrm{m}}_{t-1} + B^{(4)}\big)\\
Y^{\mathrm{m}}_t &= q_t \odot \tanh(Z^{\mathrm{m}}_t).
\end{aligned}$$

Another possibility is to apply dropout to the update gate. Namely replace in the classical LSTM model (13.12) the equation for the cell memory Z^{m}_t by

$$Z^{\mathrm{m}}_t = f_t \odot Z^{\mathrm{m}}_{t-1} + g_t \odot \gamma_t \odot \tanh\big(W^{(3)}X_t + U^{(3)}Y^{\mathrm{m}}_{t-1} + B^{(3)}\big),$$

where γ_t for $t = 1, \ldots, T$ are independent Bernoulli random vectors. This formulation has been shown to yield good results in practice, see [**SSB16**].

13.8.2. Batch-normalization in RNNs. As we have argued in Chapter 10, batch normalization is a useful technique that is typically applied to neural networks in order to accelerate training and improve accuracy. It is also applied to recurrent neural networks, see [**LPB$^+$16**].

Let us recall that for a minibatch $\mathcal{D}' \subset \mathcal{D}$, we define (we set $\varepsilon = 0$ for notational convenience)

$$\begin{aligned}
\mu(\mathcal{D}') &= \frac{1}{|\mathcal{D}'|}\sum_{x\in\mathcal{D}'} x,\\
\sigma^2(\mathcal{D}') &= \frac{1}{|\mathcal{D}'|}\sum_{x\in\mathcal{D}'} (x - \mu(\mathcal{D}'))^2,\\
\hat{x} &= \frac{x - \mu(\mathcal{D}')}{\sqrt{\sigma^2(\mathcal{D}')}}, \qquad \text{for } x \in \mathcal{D}'.
\end{aligned}$$

Then for parameters $\theta^{(2)} = (w, b)$ and for $p = (\mu(\mathcal{D}'), \sigma^2(\mathcal{D}'))$ let

$$T_p(x, \theta^{(2)}) = w\hat{x} + b.$$

In a recurrent neural network one can set

$$Z^{\mathrm{m}}_t = \sigma\big(T_p(WX_t + UZ^{\mathrm{m}}_{t-1} + B, \theta^{(2)}),\big),$$

or

$$Z^{\mathrm{m}}_t = \sigma\big(T_p(WX_t, \theta^{(2)}) + UZ^{\mathrm{m}}_{t-1} + B\big),$$

while there is some empirical evidence that the latter works better in practice, see [**LPB$^+$16**].

When applied to LSTM for example, the resulting architecture becomes

$$
\begin{aligned}
f_t &= \sigma\big(T_p(W^{(1)}X_t, \theta^{(2)}) + U^{(1)}Y^{\mathrm{m}}_{t-1} + B^{(1)}\big),\\
g_t &= \sigma\big(T_p(W^{(2)}X_t, \theta^{(2)}) + U^{(2)}Y^{\mathrm{m}}_{t-1} + B^{(2)}\big),\\
Z^{\mathrm{m}}_t &= f_t \odot Z^{\mathrm{m}}_{t-1} + g_t \odot \tanh\big(T_p(W^{(3)}X_t, \theta^{(2)}) + U^{(3)}Y^{\mathrm{m}}_{t-1} + B^{(3)}\big),\\
q_t &= \sigma\big(T_p(W^{(4)}X_t, \theta^{(2)}) + \gamma_{4,t} \odot U^{(4)}Y^{\mathrm{m}}_{t-1} + B^{(4)}\big),\\
Y^{\mathrm{m}}_t &= q_t \odot \tanh(Z^{\mathrm{m}}_t).
\end{aligned}
$$

Notice that in the architecture above the normalization parameters $\theta^{(2)}$ and the normalization statistics are shared across times t. If we wanted to have different normalization statistics for each time t, then it would be challenging to address datasets with variable length sequences. This issue is partially addressed by the method of layer-normalization discussed in Section 13.8.3.

13.8.3. Layer-normalization in RNNs. As hinted in the end of Section 13.8.2, one of the advantages of layer-normalization is that it is easy to apply in datasets with variable length sequences, see [**BKH16**].

In contrast to batch normalization, in layer normalization the mean and variance used for normalization is computed from all of the summed inputs to the neurons in a given layer during a single training case. In particular, let the hidden layer at time t be

$$\alpha_t = WX_t + UZ^{\mathrm{m}}_{t-1},$$

and let the ith-hidden unit be $\alpha_{i,t}$ with H the total number of hidden units. Then, to perform layer normalization in recurrent neural networks we replace Z^{m}_t by

$$
\begin{aligned}
Z^{\mathrm{m}}_t &= \sigma\left(\frac{W}{\sigma_t} \odot (\alpha_t - \mu_t) + B\right),\\
\mu_t &= \frac{1}{H}\sum_{i=1}^{H} \alpha_{i,t},\\
\sigma_t &= \sqrt{\frac{1}{H}\sum_{i=1}^{H} \left(\alpha_{i,t} - \mu_t\right)^2},
\end{aligned}
$$

where W is an additional parameter.

Apart from the easiness in implementation, layer normalization is invariant to rescalings of the input X_t and Z^{m}_{t-1}. This would typically result in more stable internal state dynamics and can be useful for out-of-sample data with different lengths.

Let us conclude this subsection by demonstrating the claimed invariance. Let ζ be a scalar and let us set $X^{\zeta}_t = \zeta X_t$ and $Z^{\zeta,\mathrm{m}}_{t-1} = \zeta Z^{\mathrm{m}}_{t-1}$. Then, define α^{ζ}_t,

μ_t^ζ, and σ_t^ζ analogously to α_t, μ_t, and σ_t with $(X_t^\zeta, Z_{t-1}^{\zeta,\mathrm{m}})$ replacing $(X_t, Z_{t-1}^{\mathrm{m}})$. A simple calculation shows that

$$\mu_t^\zeta = \zeta\mu_t, \text{ and } \sigma_t^\zeta = \zeta\sigma_t.$$

Hence, we directly get that

$$\begin{aligned} Z_t^{\zeta,\mathrm{m}} &= \sigma\left(\frac{W}{\sigma_t^\zeta} \odot (\alpha_t^\zeta - \mu_t^\zeta) + B\right) \\ &= \sigma\left(\frac{W}{\sigma_t} \odot (\alpha_t - \mu_t) + B\right) \\ &= Z_t^{\mathrm{m}}, \end{aligned}$$

proving the invariance claim.

13.9. Attention Mechanism and Transformers

So far in this chapter we have studied the basic recurrent neural network and its more advanced architectures, such as GRU and LSTM. Recurrent neural networks are designed to model sequential data, that may be time series data (e.g., energy prices), language translation, music composition, etc.

When it comes to certain applications though, such as language models, the basic RNN has certain shortcomings. The next hidden state Z_t is a function of the previous hidden state Z_{t-1} and the input for the current position X_t. In the context of language models, the current position amounts to the current word in the text. That way the network learns to use information from previous words in the sequence. However, due to the way that the basic RNN is built, the effect of previous words goes down as we move within the text in the forward direction. We saw a glimpse of this in the stability analysis of Section 13.6. However, language is more complicated. Specific words in the early part of the text can be very meaningful for understanding the context of later parts of the text.

This issue is alleviated to some extent with the use of more advanced RNN architectures such as GRU or LSTM. As we saw in Section 13.7, a GRU or LSTM neural network introduces additional memory cell states. The gates control the influence of the memory cells through the parameters that are being learned via some version of SGD.

However, as with the basic RNN, GRU, and LSTM, we face the issue of processing the data in a purely sequential manner. In order to apply the neural network on a new word vector X_t (typically given as a vector that appropriately maps the alphabet), we also need to know the effect of the network on the previous word vector X_{t-1}. That aspect hinders the ability to parallelize. This shortcoming becomes more evident in longer sequence lengths due to memory constraints across examples. We would like to not have to do things sequentially (to an extent mimicking how actual language works) and to be able to stack word vectors in a matrix and apply the appropriate neural network to the whole matrix at once.

Essentially, we would like our model to be able to pay attention to words that have appeared much earlier in the sequence (e.g., text). Some advances in that direction were made in the mid-2010s with the concept of the attention mechanism, but the real breakthrough came with the paper [**VSP+17**], where the authors recognized that the attention mechanism, if done correctly, does not need to be embedded in a recurrent neural network structure, and they proposed the transformer architecture which is a combination of attention mechanism and standard feed forward neural networks, Chapter 5, and layer normalization, Section 13.8.3.

The goal of this section to present the basic architecture and compare it with recurrent neural networks. In Section 13.9.1 we discuss the building block of the transformer architecture, which is the self-attention/attention mechanism. In Section 13.9.2 we present the basic transformer architecture. In Section 13.9.3 we discuss how the architectures of transformer and LSTM compare with each other.

13.9.1. Self-Attention/Attention Mechanism. A central component of the transformer architecture that we will present in Section 13.9.2 is the self-attention layer that we study in this section.

For the discussion that follows it will be useful to have language as an application domain of interest. The framework is more generically applicable, but having a concrete application domain will ease the presentation.

Consider a sequence of input word vectors $x_1, \dots, x_n \in \mathbb{R}^d$. Oftentimes in the literature, an element x_i of the sequence $\{x_i\}_{i\in\{1,\dots,n\}}$ is called a *token* and the whole sequence $\{x_i\}_{i\in\{1,\dots,n\}}$ is called a *prompt*. Tokenization refers to the process of obtaining a sequence of tokens, and embedding is the vector representation of a token.

Let X be the matrix with rows $x_1, \dots, x_n$. This means that $X \in \mathbb{R}^{n\times d}$.

Consider now three matrices to be learned through training, $W_{h,q} \in \mathbb{R}^{d\times d_k}$, $W_{h,k} \in \mathbb{R}^{d\times d_k}$, $W_{h,v} \in \mathbb{R}^{d\times d_v}$. Here $h = 1, \dots, H$ is the hth attention layer or *head*.

Let us then set

$$Q_h(X) = XW_{h,q}, \quad K_h(X) = XW_{h,k}, \quad V_h(x) = XW_{h,v}.$$

These linear operations correspond to queries, keys, and values, respectively. The idea for these names is that a word-vector x_i has a query that will be tested with the key word-vector x_j. If they are compatible, then their inner product would be large, in which case we look up the value of x_j. As such, the matrices $W_{h,q}, W_{h,k}, W_{h,v}$ are called *query*, *key*, and *value*, respectively.

Next, we define the matrix $A_h = [a_{h,ij}]_{i,j=1}^n$ where $a_{h,ij}$ are computed as a softmax function applied to the rows of a matrix-matrix product

$$A_h = S_{\text{softmax}}\left(\frac{Q_h(X)K_h^\top(X)}{\sqrt{d_k}}\right).$$

We recall that the output of a softmax function can be interpreted as probabilities. Let us now decipher what it means to apply the softmax function to a matrix. If, with some abuse of notation, we write

$$Q_h(x_i) = x_i W_{h,q}, \quad K_h(x_i) = x_i W_{h,k}, \quad V_h(x_i) = x_i W_{h,v}, \tag{13.14}$$

then $A_h = [a_{h,ij}]_{i,j=1}^n$ where $a_{h,ij}$

$$a_{h,ij} = S_{\text{softmax}}\left(\frac{\langle Q_h(x_i), K_h(x_j)\rangle}{\sqrt{d_k}}; i\right) = \frac{e^{\frac{\langle Q_h(x_i),K_h(x_j)\rangle}{\sqrt{d_k}}}}{\sum_{m=1}^n e^{\frac{\langle Q_h(x_i),K_h(x_m)\rangle}{\sqrt{d_k}}}}. \tag{13.15}$$

The scaling by $\frac{1}{\sqrt{d_k}}$ is to avoid numerical overflow. The elements of the matrix A, $a_{h,ij}$, are called self-attention weights and control how much x_i attends to x_j. Then, with $W_{h,o} \in \mathbb{R}^{d_k \times d}$, another matrix whose elements are to be learned through training, we define

$$z_i = \sum_{h=1}^{H} W_{h,o}^\top \sum_{j=1}^{n} a_{h,ij} V_h(x_j), \quad i = 1, \dots, n. \tag{13.16}$$

In the last display we sum up the value of each word-vector x_j, i.e., $V_h(x_j)$, in a way that is proportional to the compatibility of x_i and x_j via $a_{h,ij}$. Since we are interested in how much each element of the matrix X attends to each other, we call this the self-attention mechanism. We do this for all heads $h = 1, \dots, H$ to get the multihead self-attention z_i for $i = 1, \dots, n$. This process can be parallelized. Equations (13.14)–(13.16) refer to the multihead self-attention block. In Figures 13.7 and 13.8 we present schematic representations of the self-attention block (or mechanism) and of the multihead self-attention block, respectively.

One interesting property of the self-attention block is that it is permutation equivariant. Let us first define what this is.

Definition 13.2. Let $X \in \mathbb{R}^{n \times d}$ be a given matrix. Let π be the permutation operator of n objects and the permutation matrix associated with π, $L(\pi) = (e_{\pi_1}, \dots, e_{\pi_n}) \in \mathbb{R}^{n \times n}$, where e_{π_i} are one-hot vectors whose π_i element is 1 and

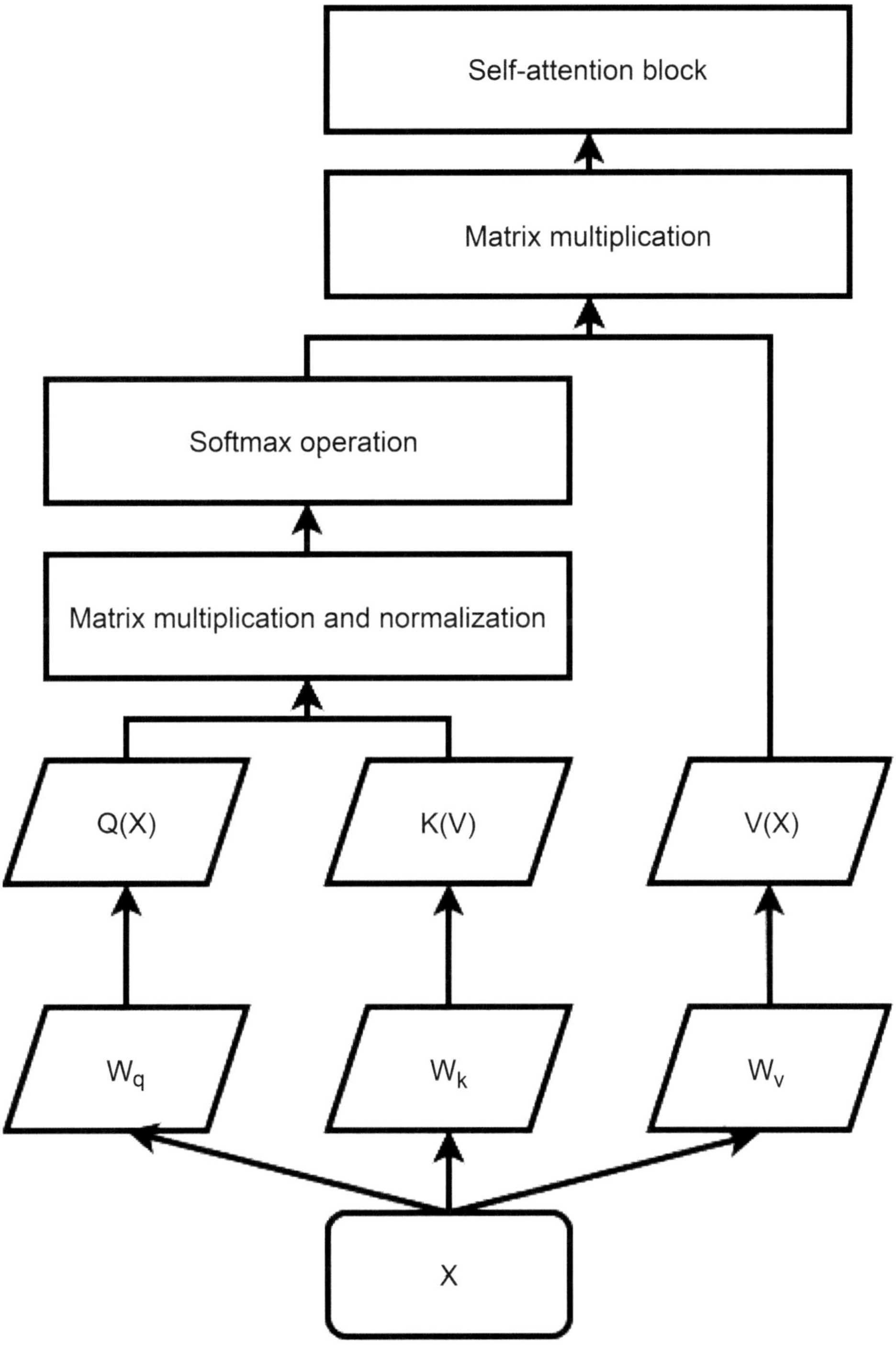

Figure 13.7. A schematic representation of a self-attention block

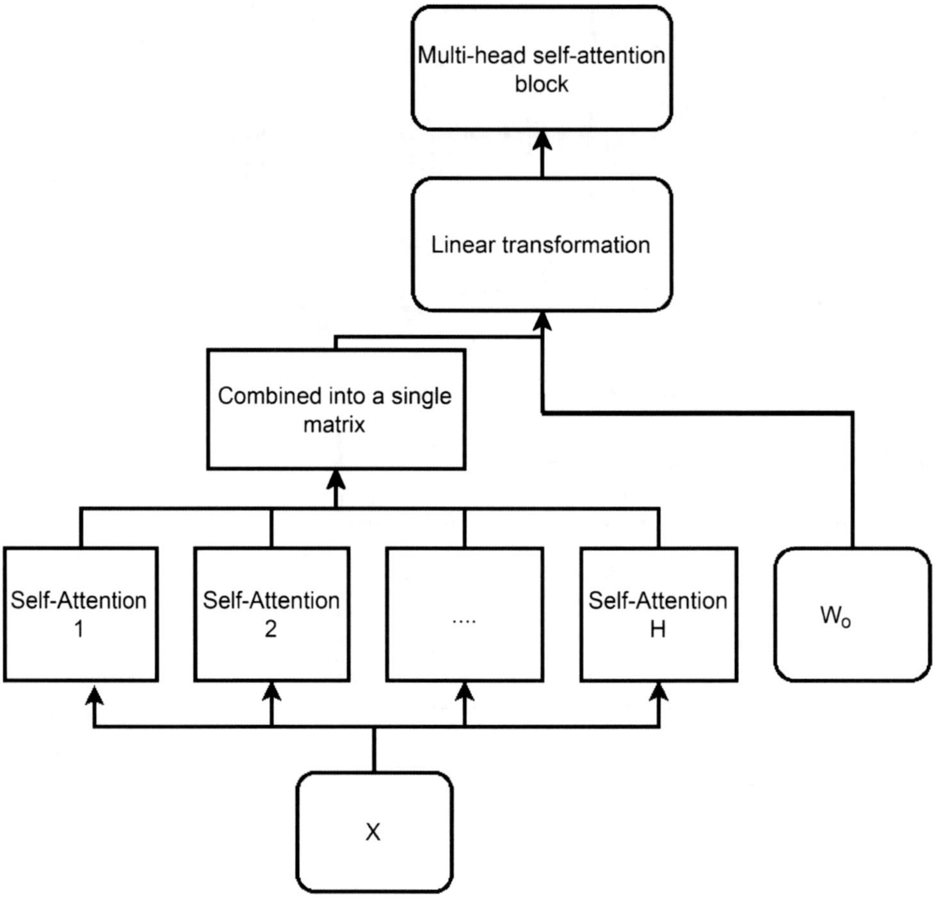

Figure 13.8. A schematic representation of a multihead self-attention block

the rest of the elements are 0. We will call the transformation $M(X;\pi) = L(\pi)X$ a permutation of X.

Definition 13.3. An operator $\mathcal{Z} : \mathbb{R}^{n\times d} \mapsto \mathbb{R}^{n\times d}$ is called permutation equivariant if for any $X \in \mathbb{R}^{n\times d}$ and for $M(X;\pi) = L(\pi)X$ a permutation operator, we have $\mathcal{Z}(M(X;\pi)) = \left[M\left(\mathcal{Z}^\top(X);\pi\right)\right]^\top$. The operator $\mathcal{Z}$ is called permutation invariant if for any $X \in \mathbb{R}^{n\times d}$, $\mathcal{Z}(M(X;\pi)) = \mathcal{Z}(X)$.

We will now show that the self-attention block operator is permutation equivariant. Consider for simplicity and without loss of generality the single-head attention, i.e., the case of $H = h = 1$, and set

$$\mathcal{Z}(X) = W_o^\top (W_v)^\top X^\top S_{\text{softmax}}\left(\frac{XW_qW_k^\top X^\top}{\sqrt{d_k}}\right).$$

Then, we have the following lemma.

Lemma 13.4. *We have that the self-attention block operator $\mathcal{Z}$ is permutation equivariant.*

Proof. We observe that

$$\begin{aligned}
&\mathcal{Z}(M(X;\pi)) \\
&= W_o^\top (W_v)^\top (M(X;\pi))^\top S_{\text{softmax}}\left(\frac{(M(X;\pi))W_q W_k^\top (M(X;\pi))^\top}{\sqrt{d_k}}\right) \\
&= W_o^\top (W_v)^\top X^\top L(\pi)^\top S_{\text{softmax}}\left(\frac{L(\pi)XW_q W_k^\top X^\top L(\pi)^\top}{\sqrt{d_k}}\right) \\
&= W_o^\top (W_v)^\top X^\top L(\pi)^\top L(\pi) S_{\text{softmax}}\left(\frac{XW_q W_k^\top X^\top}{\sqrt{d_k}}\right) L(\pi)^\top \\
&= W_o^\top (W_v)^\top X^\top S_{\text{softmax}}\left(\frac{XW_q W_k^\top X^\top}{\sqrt{d_k}}\right) L(\pi)^\top \\
&= \left[M\left(\mathcal{Z}^\top(X);\pi\right)\right]^\top .
\end{aligned}$$

To derive this, we used the fact that for the orthogonal matrix, $L(\pi)$ we have that $L(\pi)^\top L(\pi) = I$ and that for a given matrix A, we have

$$S_{\text{softmax}}(L(\pi)AL(\pi)^\top) = L(\pi)S_{\text{softmax}}(A)L(\pi)^\top. \qquad \square$$

Lemma 13.4 says something important. It suggests that the order in which we consider the input affects the output in the same way; for example the predictions change direction if we change the direction of the input. Said otherwise, the output is permuted the same way as the input is permuted.

In Exercise 13.8 we will see that in the attention-block case (not self-attention) the property of permutation invariance is true. The difference between attention and self-attention is that in the former we have the query $Q_h(X) = W_{h,q}$ instead of $Q_h(X) = XW_{h,q}$; i.e., the query does not depend on the input X, or said otherwise it is the same for all input instances.

13.9.2. Transformer architecture. The transformer architecture is composed by transformer blocks. A transformer block gets an input $X \in \mathbb{R}^{n\times d}$ and transforms it to another object, say $Y \in \mathbb{R}^{n\times d}$.

Before we give the definition of a transformer block, we recall the layer normalization (LN) operation from Section 13.8.3. For a vector $x \in \mathbb{R}^d$ and for vectors $w, b \in \mathbb{R}^d$, we define the layer normalization as in Section 13.8.3

$$LN(x;(w,b)) = \frac{w}{\sigma_x} \odot (x - \mu_x) + b,$$

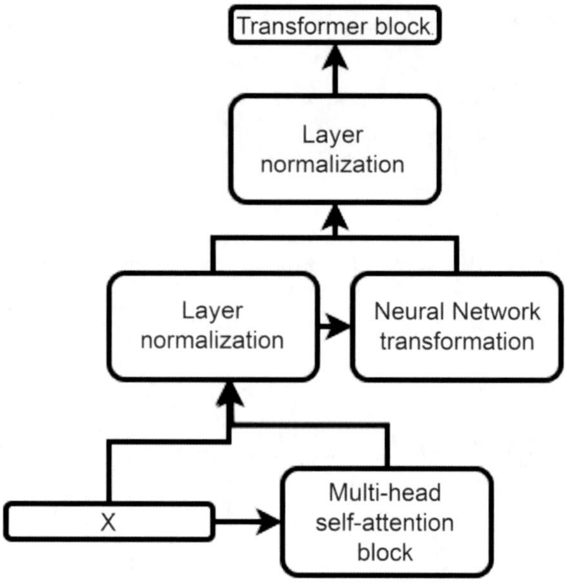

Figure 13.9. A schematic representation of a transformer block

where the vectors w, b are to be learned during the training process. For fixed w, b, one can think of the layer normalization here as a constraint on x being on an ellipsoid.

Now we are in position to give the definition of the basic transformer block. Let $\{x_i\}_{i\in\{1,...,n\}}$ be an input sequence and let z_i be the multihead self-attention block from (13.16). The basic transformer block is given by

$$\begin{aligned} \hat{z}_i &= LN(x_i + z_i; (w_1, b_1)), \\ \hat{y}_i &= W_3^{\mathsf{T}} \mathsf{ReLU}(W_2^T \hat{z}_i + B_2) \\ y_i &= LN(\hat{y}_i + \hat{z}_i; (w_4, b_4)), \quad i = 1, \dots, n, \end{aligned} \tag{13.17}$$

where $W_2 \in \mathbb{R}^{d \times d_f}$, $W_3 \in \mathbb{R}^{d_f \times d}$ are matrices and w_1, w_4, b_1, b_2, b_4 are vectors. In fact the parameter to be learned during the training phase is

$$\theta = (w_1, w_4, b_1, b_2, b_4, W_2, W_3, \{(W_{h,q}, W_{h,k}, W_{h,v}, W_{h,o}), h = 1, \dots, H\}).$$

So, in the end, we have obtained the model for the transformer block $Y = \mathfrak{m}(X; \theta)$. In Figure 13.9 we present a schematic representation of a transformer block.

A transformer is then a combination of transformer blocks. Different applications may use different variations of combinations of transformer blocks.

Remark 13.5. Note also that the model is flexible enough to accommodate different kind of neural network architectures. For example, referring back to

(13.17) one can generally set

$$\hat{y}_i = \tilde{m}(\hat{z}_i),$$

where $\tilde{m}$ is any kind of deep neural network architecture that takes $\hat{z}_i$ as its input, and not necessarily the shallow neural network with ReLU activation function that was considered in (13.17).

In [**VSP+17**] the problem of machine translation was studied and an encoder-decoder combination was used to define the transformer architecture. The definition of the transformer block for an encoder is (13.16) with the attention mechanism (13.15). For a decoder, the definition is slightly different and is based on what is called a *masked* multihead self-attention. In particular, in that case we still have (13.16), but with (13.15) replaced by

$$a_{h,ij} = \begin{cases} \frac{\exp\left(\frac{\langle Q_h(x_i), K_h(x_j)\rangle}{\sqrt{d_k}}\right)}{\sum_{m=1}^{i} \exp\left(\frac{\langle Q_h(x_i), K_h(x_m)\rangle}{\sqrt{d_k}}\right)}, & \text{for } j \leq i, \\ 0, & \text{otherwise.} \end{cases} \tag{13.18}$$

Note that in an encoder, the output z_i sees the whole sequence $x_1, \dots, x_n$. On the other hand, in a decoder, the output z_i sees only $x_1, \dots, x_i$ and it does not see $x_{1+1}, \dots, x_n$.

As a concrete example of a transformer let us describe the one presented in [**VSP+17**]. This architecture is composed by an encoder, a decoder and positional encoding. The *encoder* is a composition of L transformer blocks, each one with each own parameters, i.e., the output of the encoder is

$$\mathfrak{m}(\cdot; \theta_L) \circ \cdots \circ \mathfrak{m}(\cdot; \theta_1) \in \mathbb{R}^{n \times d}.$$

The *decoder* is again a composition of L transformer blocks, each one with its own parameters. However, one difference with the encoder is that we modify the transformer block as follows. After we have computed the $\hat{z}_i$ in (13.17), we then apply a masked multihead self-attention mechanism as in (13.18). The output of this operation then goes through a layer normalization and then we proceed as in (13.17), with a feed forward neural network followed by a subsequent layer normalization.

Lastly, we discuss what *position encoding* refers to. We note that up to now in our description, the relative order of the word-vectors x_i in a sentence did not play any role. It would be good for the model though to have some information on the relative positions of the words in the sentence. To achieve this, we add positional encoding to all inputs before feeding the encoder or the decoder with the information. Consider a sequence $x_i \in \mathbb{R}^d$, $i \in \{1, \dots, n\}$, of word embeddings and let $q_i \in \mathbb{R}^d$ be the positional encoding of the x_i word embedding for $i \in \{1, \dots, n\}$. While there are many different choices of position encoding in the literature (see for instance [**VSP$^+$17, GAG$^+$17**]), in [**VSP$^+$17**] the authors used the sinusoidal position encoding $q \in \mathbb{R}^{n \times d}$ where

$$q_i(2k) = \sin\left(\frac{i}{(10^4)^{2k/d}}\right) \text{ and } q_i(2k+1) = \cos\left(\frac{i}{(10^4)^{2k/d}}\right),$$

with $k \in \{1, \ldots, d/2 - 1\}$ and then transform the data to

$$\tilde{x} = \text{FeedForward}(x) + q.$$

Then $\tilde{x}$ is input to the initial stage of the encoder and decoder, i.e., at time zero of the algorithm. We lastly mention that in the original [**VSP$^+$17**] paper, $L = 6, H = 8, d_k = 64, d = 512$, and $d_f = 2048$ for the inner feed forward layer of (13.17). Another way to do positional encoding is to let q_i to be some random vector (random positional encoding) or a transformation that is learned through training.

13.9.3. Comparison of Transformer and LSTM. In this section we compare the attention mechanism (consequently the Transformer architecture) to the LSTM of Section 13.7. Let us recall the recursive formula for the memory state of LSTM (13.13).

Let us focus the discussion within the context of language models.

- The attention weights $a_{h,ij}$ are computed for all $i, j = 1, \ldots, n$. In the LSTM network the weights $M_{k,t}$ are computed only for $k \leq t$. This means that in the LSTM network an item of length t only attends to items of length $k \leq t$ and not to longer sequences.
- The recursive formula for the memory state of LSTM, (13.13), indeed shows that an LSTM tends to give more weight to recent words because the weights decay over time. Indeed, if the forget gate $f_\ell < 1$ for all $\ell \leq t$, then $M_{k_1,t} \leq M_{k_2,t}$ for $k_1 < k_2$. This is one difference with the attention mechanism of Section 13.9. The attention mechanism attends equally well to all items.
- Attention has a probabilistic interpretation through the softmax function, which gives probabilities of how much a given word attends to another word. In contrast, in LSTM the weights in the formula (13.13) may grow up to the length of the sequence.

The observations above suggest that the attention mechanism may have certain advantages in tasks such as text and language models over recurrent architectures like the LSTM. Indeed, in recent years variants of the transformer architecture have enjoyed many successes in large language models. For this purpose, bidirectional LSTMs have also been used in large language models, defined analogously to the bidirectional RNN structure of Section 13.7; two LSTMs are considered, one moving in the forward direction in text and the other one moving in the backward direction in text, effectively increasing the amount of information available to the network. We note that the attention mechanism can also be considered in conjunction with LSTMs where a given input datapoint is first processed by an LSTM before being fed into the attention mechanism.

Even though transformers are indeed very effective in handling long-range dependencies in data, they are also very costly in terms of memory usage and computational resources. Typically, transformers need more memory and computation power when compared to RNN based architectures, which stems from the fact that they extensively use the self-attention mechanism. Despite attention-based models and transformer architectures having advantages, RNNs and the related advanced architectures like GRU, LSTM, and bidirectional RNNs are a very powerful class of models for time-series data and generally sequential data analysis.

13.10. Brief Concluding Remarks

In this chapter we studied recurrent neural networks and transformers that are widely used to model time series and sequential data.

The attention mechanism was popularized with the paper [**VSP⁺17**], and since then, modifications of the basic Transformer architecture have found many applications in music generation [**HVU⁺19**], image generation [**PVU⁺18**], and more. Both LSTM based recurrent neural networks and various variants of the transformer architecture have found many applications in large language models, see for example [**DCLT19, RNSS18, RWC⁺19**],and [**BMR⁺20**]. LSTM such as recurrent neural network architectures have also found applications in solution of high-dimensional partial differential equations, see for example [**SS18**]. [**LLFZ18**] includes a discussion on the analogy between LSTM and the transformer's attention weights.

In Chapter 14, we study convolution neural networks that are widely used in image recognition problems.

13.11. Exercises

Exercise 13.1. Consider the one-dimensional dynamical system with system updates $Z_n = f(Z_{n-1}, \theta)$. For the following choices for f, find the hidden state of the system in the long run, $\lim_{n\to\infty} Z_n$.

(1) $f(z,\theta) = \tanh(\theta z)$ with $|\theta| < 1$.

(2) $f(z,\theta) = \sigma(\theta \cdot z)$, where σ is the logistic function.

(3) $f(z,\theta) = \sin(\theta \cdot z)$.

Exercise 13.2. In the context of Section 13.5 prove that tBPTT based SGD is a biased algorithm.

Exercise 13.3. Fill in the missing details in the chain rule derivations in Section 13.5 together with the details of the resulting backpropagation algorithm.

Exercise 13.4. Consider the recurrent neural network

$$H_t = w_h H_{t-1}$$
$$Z_t = w_z H_t,$$

with initial condition $H_{-1} = 1$. Provide formulas for

(1) Z_3,

(2) $\frac{\partial Z_3}{\partial w_z}$,

(3) $\frac{\partial Z_3}{\partial w_h}$,

in terms of w_h and w_z.

Exercise 13.5. Consider the single layer recurrent neural network

$$H_t = \tanh\left(w_{hh} H_{t-1} + w_{ih} X_t + b_h\right)$$
$$H_{-1} = h_0.$$

Compute the partial derivative $\frac{\partial H_t}{\partial h_0}$ for $t \in \{1, 2, \dots\}$.

Exercise 13.6. Consider a simple RNN with parameters $\theta = (a, b, c) \in \mathbb{R}^3$:

$$Z_{t+1} = a\sigma(bZ_t + cX_t),$$

where $\sigma(z) = \frac{e^z}{1+e^z}$. Suppose that initially $a = b = \frac{1}{2}$. What is $\frac{\partial Z_T}{\partial Z_t}$ where $T > t$? What happens when $T \to \infty$ and why is this an example of the vanishing gradient problem?

Exercise 13.7. Prove that for a permutation matrix $L(\pi)$ as in Definition 13.2 and $A \in R^{n \times n}$ a matrix, we have

$$S_{\text{softmax}}(L(\pi)AL(\pi)^\top) = L(\pi)S_{\text{softmax}}(A)L(\pi)^\top.$$

Exercise 13.8. Consider the attention block operator

$$\mathcal{Z}(X) = W_o^\top (W_v)^\top X^\top S_{\text{softmax}}\left(\frac{XW_q W_k^\top}{\sqrt{d_k}}\right).$$

Prove that it is permutation invariant. Why is the permutation invariance property useful for image classification problems?

Chapter 14

Convolution Neural Networks

14.1. Introduction

Is there an umbrella in Figure 14.1? You have probably seen enough umbrellas enough times to be able to recognize one! When you look at Figure 14.1, you scan over it and try to find a part of the image that matches the pattern you have in your mind for the umbrella.

Let's see if we can understand a deep learning approach to this. We start with a lot of *training* images, some of which contain (label 1) an umbrella, and others of which don't (label 0). We want to build a pattern representing the umbrella, and move this pattern around to see if we get a match.

Although two-dimensional images are the common application of *convolution neural networks* (CNN), we shall develop the ideas within the framework of one-dimensional signals. Almost all of the ideas can be developed, and the notation is simpler.

We begin in Section 14.2 with the simpler task of detecting a known reference signal. In Section 14.3 we elaborate what to do when we do not really know the reference signal. Auxiliary topics like stride and channels are presented in Section 14.4. Then, in Sections 14.5 and 14.6 we discuss details on the implementation of stochastic gradient descent for the case of single and multiple channels, respectively.

Figure 14.1. Sample image of an umbrella (photograph by the second author)

14.2. Detection of Known Signal

Let's start by trying to *detect* a known reference signal $\mathbf{R} = (R_n)_{n=0}^{N'-1}$ of Figure 14.2 (where $N' = 7$) in a larger observed signal of $\mathbf{X} = (X_n)_{n=0}^{N-1}$ (where $N = 25$). See Figure 14.3.

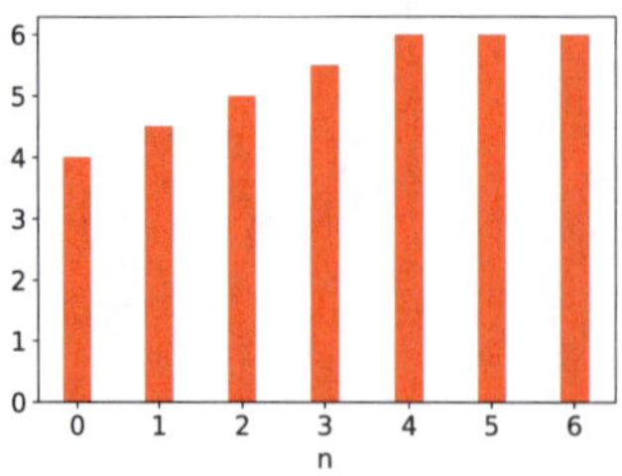

n	R
0	4.0
1	4.5
2	5.0
3	5.5
4	6.0
5	6.0
6	6.0

Figure 14.2. Reference signal R

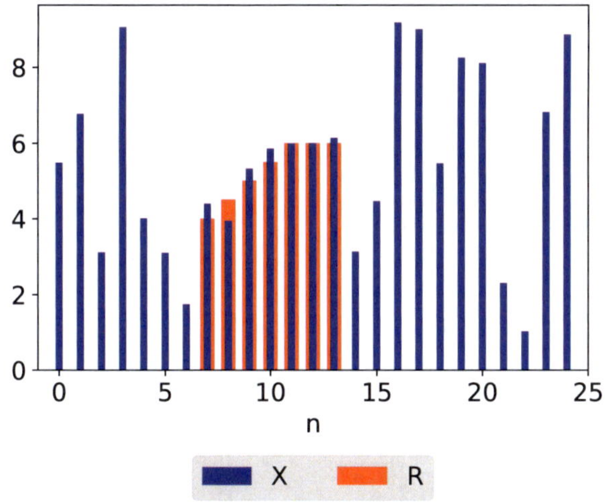

Figure 14.3. Typical observed signal **X**. The reference signal **R** starts at 7.

Let's think of this in feature-label space.

- *Feature* space will be $\mathbb{R}^N$ (i.e., $\mathbb{R}^{25}$); each element of feature space will be a signal of length 25.
- *Label* space will be $\{0, 1\}$
  - label 1 will correspond to **R** being present somewhere in the signal
  - label 0 will correspond to absence of **R** in the signal

Convolution allows us to efficiently search for the presence of the reference signal at all starting points. In our introductory two-dimensional example of Figure 14.1, convolution would have allowed us to search for all positions of the umbrella in two dimensions.

Let's formalize things a bit more by considering some ground-truth training data; more specifically, ten feature-label pairs as in Figure 14.4. The reference signal, **R**, when present, has been corrupted by noise. The reference signal can also start anywhere. For simplicity we will restrict the start n of the signal so that $n + (N' - 1) \leq (N - 1)$; i.e., $n \leq 24 - 6 = 18$, so that the entire signal is either present or not in the observation.

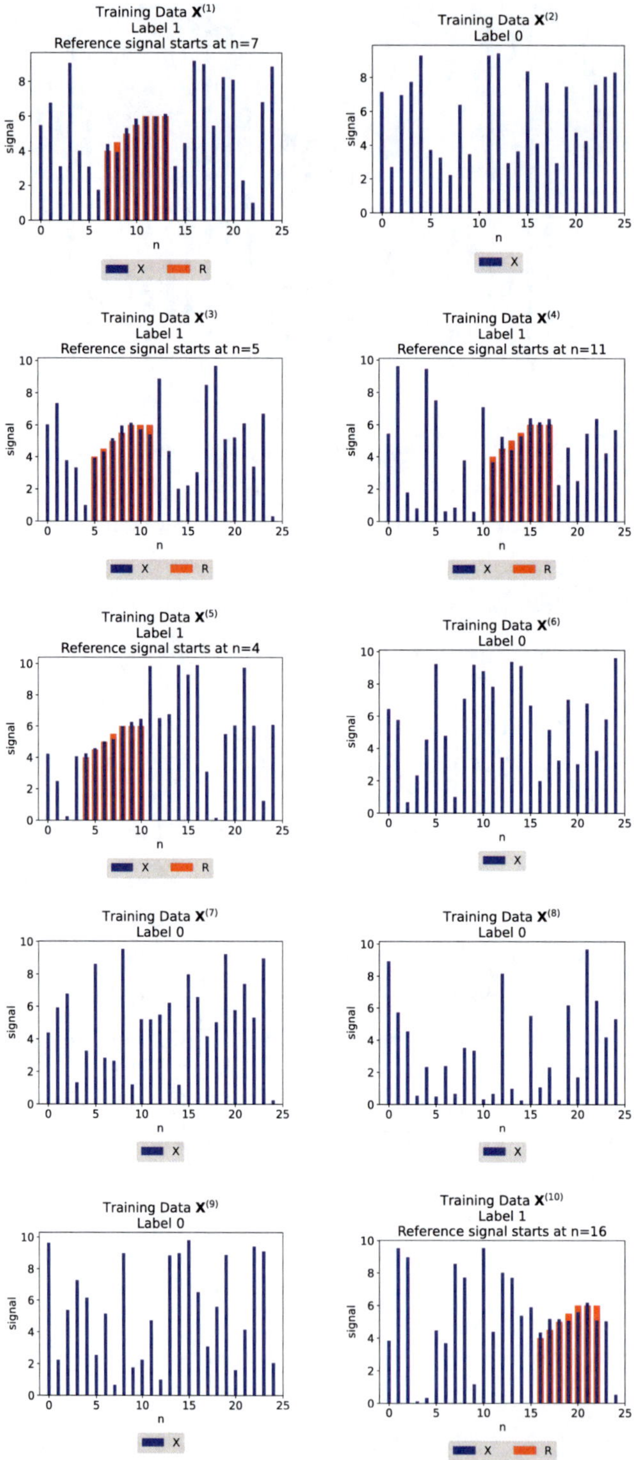

Figure 14.4. ground-truth training data on observed and reference signals, **X** and **R**, respectively

A somewhat natural way to try to find the reference signal is to compute correlations. For an observed signal $\mathbf{X} \in \mathbb{R}^{25}$, define the Pearson correlation (for the specific signal at hand)

$$\ell_n(\mathbf{X})$$

$$\stackrel{\text{def}}{=} \frac{\frac{1}{7}\sum_{n'=0}^{6}\left\{R_{n'} - \left(\frac{1}{7}\sum_{n'=0}^{6} R_{n'}\right)\right\}\left\{X_{n+n'} - \left(\frac{1}{7}\sum_{n'=0}^{6} X_{n+n'}\right)\right\}}{\sqrt{\frac{1}{7}\sum_{n'=0}^{6}\left\{R_{n'} - \left(\frac{1}{7}\sum_{n'=0}^{6} R_{n'}\right)\right\}^2}\sqrt{\frac{1}{7}\sum_{n'=0}^{6}\left\{X_{n+n'} - \left(\frac{1}{7}\sum_{n'=0}^{6} X_{n+n'}\right)\right\}^2}}$$

$$= \frac{\frac{1}{7}\sum_{n'=0}^{6} R_{n'}X_{n+n'} - \left(\frac{1}{7}\sum_{n'=0}^{6} R_{n'}\right)\left(\frac{1}{7}\sum_{n'=0}^{6} X_{n+n'}\right)}{\sqrt{\frac{1}{7}\sum_{n'=0}^{6} R_{n'}^2 - \left(\frac{1}{7}\sum_{n'=0}^{6} R_{n'}\right)^2}\sqrt{\frac{1}{7}\sum_{n'=0}^{6} X_{n+n'}^2 - \left(\frac{1}{7}\sum_{n'=0}^{6} X_{n+n'}\right)^2}}$$

between $(R_{n'})_{n'=0}^{6}$ and $(X_{n+n'})_{n'=0}^{6}$. If there were no noise, then $\ell_n(\mathbf{X}) = 1$ if $X_{n'+n} = R_{n'}$ for $n' \in \{0, 1, 2, \dots, 6\}$. We note, however, that the reverse is *not* true; if $\ell_n = 1$, we only know that there are $a > 0$ and $b \in \mathbb{R}$ such that

$$X_{n+n'} = aR_{n'} + b, \qquad n' \in \{0, 1, 2, \dots, 6\}$$

(i.e., $(R_{n'})_{n'=0}^{6}$ and $(X_{n+n'})_{n'=0}^{6}$ are colinear). See Figure 14.5. We do indeed see that

$$\overline{\ell}(\mathbf{X}) \stackrel{\text{def}}{=} \max\{\ell_n(\mathbf{X}) : n \in \{0, 1, \dots, 18\}$$

tends to be large when the signal is present and small when it isn't. We might try to detect $\mathbf{R}$ in a signal $\mathbf{X}$ by selecting a threshold ℓ and declaring

- $\mathbf{R}$ is present if $\overline{\ell}(\mathbf{X}) > \ell$
- $\mathbf{R}$ is absent if $\overline{\ell}(\mathbf{X}) < \ell$.

In other words, we can think of detecting $\mathbf{R}$ by applying logistic regression to $\overline{\ell}(\mathbf{X})$. This is an example of *feature engineering*.

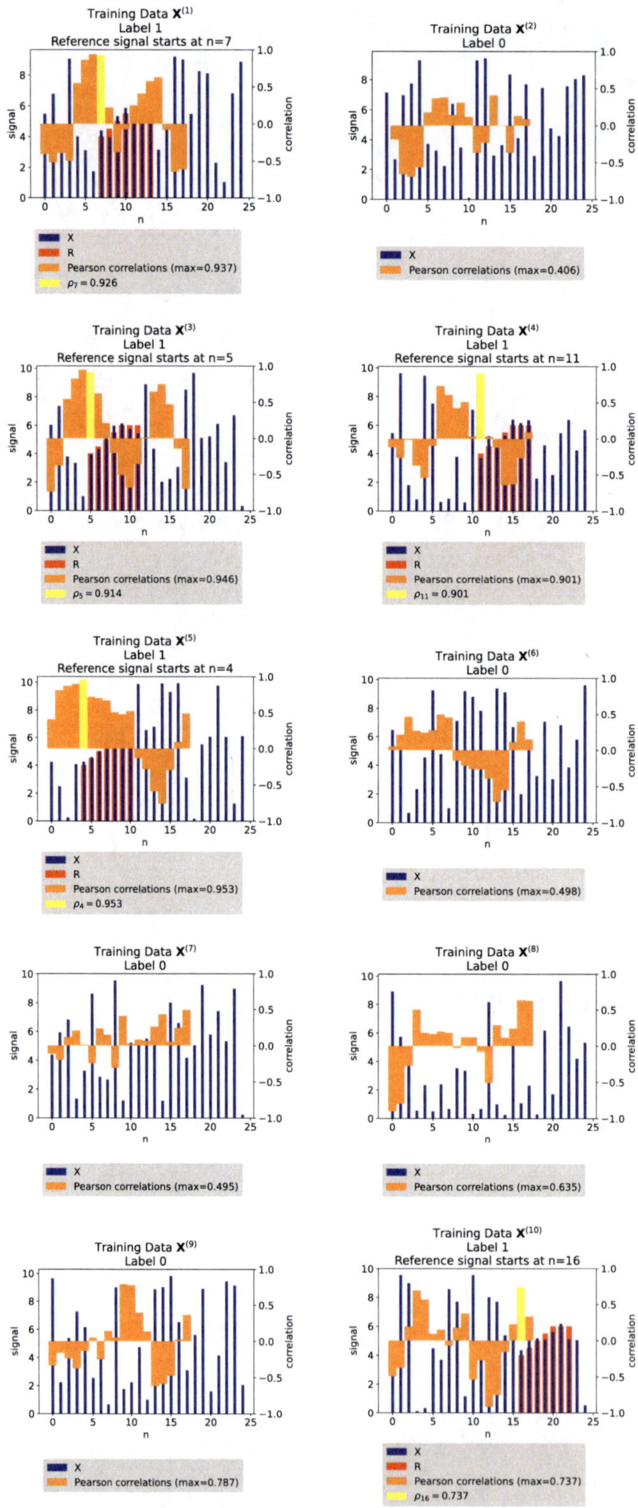

Figure 14.5. Observed **X** and reference **R** signals and the associated Pearson correlations

Let's rewrite the correlation coefficient;

$$\ell_n(\mathbf{X}) = w_n(\mathbf{X})(\mathbf{R} \star \mathbf{X})_n - b_n(\mathbf{X}), \tag{14.1}$$

where

$$w_n(\mathbf{X}) \stackrel{\text{def}}{=} \frac{\frac{1}{7}}{\sqrt{\frac{1}{7}\sum_{n'=0}^{6} R_{n'}^2 - \left(\frac{1}{7}\sum_{n'=0}^{6} R_{n'}\right)^2}\sqrt{\frac{1}{7}\sum_{n'=0}^{6} X_{n+n'}^2 - \left(\frac{1}{7}\sum_{n'=0}^{6} X_{n+n'}\right)^2}},$$

$$b_n(\mathbf{X}) \stackrel{\text{def}}{=} \frac{\left(\frac{1}{7}\sum_{n'=0}^{6} R_{n'}\right)\left(\frac{1}{7}\sum_{n'=0}^{6} X_{n+n'}\right)}{\sqrt{\frac{1}{7}\sum_{n'=0}^{6} R_{n'}^2 - \left(\frac{1}{7}\sum_{n'=0}^{6} R_{n'}\right)^2}\sqrt{\frac{1}{7}\sum_{n'=0}^{6} X_{n+n'}^2 - \left(\frac{1}{7}\sum_{n'=0}^{6} X_{n+n'}\right)^2}},$$

and

$$(\mathbf{R} \star \mathbf{X})_n \stackrel{\text{def}}{=} \sum_{n'=0}^{6} R_{n'} X_{n+n'}.$$

Instead of trying to detect $\mathbf{R}$ in $\mathbf{X}$ by using the engineered feature $\overline{\ell}(\mathbf{X})$, (14.1) suggests that we might instead try to detect $\mathbf{R}$ in $\mathbf{X}$ by applying logistic regression to the engineered feature (the *correlation*)

$$\max\{(\mathbf{R} \star \mathbf{X})_n \,:\, n \in \{0, 1, \ldots, 18\}\}.$$

Namely, let's consider training a model

$$\mathfrak{m}(\mathbf{X}, \theta) \stackrel{\text{def}}{=} S\left(w \max_n (\mathbf{R} \star \mathbf{X})_n + b\right)$$

(where, as usual, S is the logistic function) for the probability that the reference signal is present (and then voting to decide the label 1 or 0). As usual with logistic regression,

$$\theta = \begin{pmatrix} w \\ b \end{pmatrix}.$$

The training data is

$$\mathcal{D} \subset \mathbb{R}^{25} \times \{0, 1\}$$

and per-datapoint losses are

$$\lambda_{(\mathbf{X},y)}(\theta) = \ell_y\left(S\left(w \max_n (\mathbf{R} \star \mathbf{X})_n + b\right)\right), \qquad \theta = \begin{pmatrix} w \\ b \end{pmatrix} \in \mathbb{R}^2, \tag{14.2}$$

where, as in (3.7),

$$\ell_y(y') \stackrel{\text{def}}{=} y \ln \frac{y}{y'} + (1 - y) \ln \frac{1 - y}{1 - y'}.$$

As usual, the average loss is then

$$\Lambda(\theta) \stackrel{\text{def}}{=} \frac{1}{|\mathcal{D}|} \sum_{(\mathbf{X},y)\in\mathcal{D}} \lambda_{(\mathbf{X},y)}(\theta), \qquad \theta = \begin{pmatrix} w \\ b \end{pmatrix} \in \mathbb{R}^2. \tag{14.3}$$

Several side issues are worth some attention. Firstly, we have restricted the starting point of the signal $\mathbf{R}$, if present, to be such that the entirety of the signal is present (i.e., $n \le 18$). We can easily extend $(\mathbf{R} \star \mathbf{X})_n$ to other n by *zero-padding*; for any $n \in \mathbb{Z}$, we can define

$$(\mathbf{R} \star \mathbf{X})_n \stackrel{\text{def}}{=} \sum_{n'=0}^{6} R_{n'} X_{n+n'} \mathbf{1}_{\{0 \le n+n' \le 24\}}.$$

Secondly, the correlation $(\mathbf{R} \star \mathbf{X})$ can be rewritten as

$$\begin{aligned}(\mathbf{R} \star \mathbf{X})_n = \sum_{n'=0}^{N'-1} R_{n'} X_{n+n'} &= \sum_{m'=-N'+1}^{0} R_{-m'} X_{n-m'} = \sum_{m''=n}^{n+N'-1} R_{m''-n} X_{m''} \\ &= \sum_{m' \in \mathbb{Z}} K_{m'} X_{n-m'} = \sum_{m'' \in \mathbb{Z}} K_{n-m''} X_{m''}\end{aligned} \tag{14.4}$$

with

$$K_m = \begin{cases} R_{-m} & \text{if } 0 \le -m \le N'-1 \\ 0 & \text{else.} \end{cases}$$

The last expression in (14.4) is the *convolution* of $\mathbf{X}$ with $(K_m)_{m\in\mathbb{Z}}$.

Let's think through how we might minimize the average loss Λ of (14.3). Reusing a number of calculations from Chapter 3, we can compute the gradient of λ with respect to the vector θ;

$$\nabla \lambda_{(\mathbf{X},y)}(\theta) = (\ell_y \circ S)' \left(w \max_n (\mathbf{R} \star \mathbf{X})_n + b \right) \begin{pmatrix} \max_n (\mathbf{R} \star \mathbf{X})_n \\ 1 \end{pmatrix}. \tag{14.5}$$

14.3. Detection of Unknown Signal

Of course, in reality, we don't know the reference signal $\mathbf{R}$. Thinking back to Figure 14.1, we mentally have built a reference signal from experience. In other words, *we can think of* $\mathbf{R}$ *as part of the parameter vector.*

Let's work through this. Let's modify (14.2) to include $\mathbf{r} = (r_n)_{n=0}^{N'-1}$ as a parameter. Define

$$\lambda_{(\mathbf{X},y)}(\theta) = \ell_y \left(S \left(w \max_n (\mathbf{r} \star \mathbf{X})_n + b \right) \right), \qquad \theta = \begin{pmatrix} r_0 \\ r_1 \\ \vdots \\ r_{N'-1} \\ w \\ b \end{pmatrix} \in \mathbb{R}^{N'+2}.$$

The derivatives of $\lambda_{(\mathbf{X},y)}$ of (14.5) with respect to w and b are given in (14.5). To fully implement gradient descent, we need to work through the derivatives of $\max_n(\mathbf{r} \star \mathbf{X})_n$ with respect to the r_n's.

Let's first of all understand derivatives of the *maxpool* function

$$\mathsf{M}(x_1, x_2, x_3) \stackrel{\text{def}}{=} \max\{x_1, x_2, x_3\}, \qquad (x_1, x_2, x_3) \in \mathbb{R}^3.$$

For specificity, let's take the derivatives at $(x_1, x_2, x_3) = (-1, 5, 7)$. Explicitly,

$$\mathsf{M}(-1, 5, 7) = 7$$

and

$$\operatorname{argmax}(-1, 5, 7) = 3,$$

i.e., $x_3 > \max\{x_1, x_2\}$. For small ε,

$$\begin{aligned} \mathsf{M}(-1+\varepsilon, 5, 7) &= 7, \\ \mathsf{M}(-1, 5+\varepsilon, 7) &= 7, \\ \mathsf{M}(-1, 5, 7+\varepsilon) &= 7+\varepsilon, \end{aligned}$$

implying that

$$\begin{aligned} \frac{\partial \mathsf{M}}{\partial x_1}(-1, 5, 7) &= 0, \\ \frac{\partial \mathsf{M}}{\partial x_2}(-1, 5, 7) &= 0, \\ \frac{\partial \mathsf{M}}{\partial x_3}(-1, 5, 7) &= 1. \end{aligned}$$

More generally,

$$\frac{\partial \mathsf{M}}{\partial x_n}(x_1, x_2, x_3) = \begin{cases} 1 & \text{if } n = \operatorname{argmax}(x) \\ 0 & \text{if } n \neq \operatorname{argmax}(x) \end{cases} = \mathbf{1}_{\operatorname{argmax}(x)}(n).$$

The gradient of M is of course not defined at places where argmax is not unique:

$$M(-1, 7, 7+\varepsilon) = \begin{cases} 7+\varepsilon & \text{if } \varepsilon > 0 \\ 7 & \text{if } \varepsilon < 0 \end{cases}$$

so $\frac{\partial \mathsf{M}}{\partial x_3}(-1, 7, 7)$ doesn't exist. Generically, a floating point computation is unlikely to encounter such a case.

Secondly, let's compute derivatives of the correlation operator. For $\mathbf{r}$,

$$(\mathbf{r} \star \mathbf{X})_n \stackrel{\text{def}}{=} \sum_{n'=0}^{N'-1} r_{n'} X_{n+n'} = r_0 X_n + r_1 X_{n+1} \cdots r_{N'-1} X_{n+N'-1}.$$

Thus

$$\frac{\partial(\mathbf{r}\star\mathbf{X})_n}{\partial r_0} = X_n$$
$$\frac{\partial(\mathbf{r}\star\mathbf{X})_n}{\partial r_1} = X_{n+1}$$
$$\vdots$$
$$\frac{\partial(\mathbf{r}\star\mathbf{X})_n}{\partial r_{n'}} = X_{n+n'}.$$

Let's collect things together. We have

$$\nabla\lambda_{(\mathbf{X},y)}(\theta) = \begin{pmatrix} \frac{\partial\lambda_{(\mathbf{X},y)}}{\partial r_0}(\theta) \\ \frac{\partial\lambda_{(\mathbf{X},y)}}{\partial r_1}(\theta) \\ \vdots \\ \frac{\partial\lambda_{(\mathbf{X},y)}}{\partial r_{N'-1}}(\theta) \\ \frac{\partial\lambda_{(\mathbf{X},y)}}{\partial m}(\theta) \\ \frac{\partial\lambda_{(\mathbf{X},y)}}{\partial b}(\theta) \end{pmatrix} = (\ell_y\circ S)'\,(m\,\mathsf{M}(\mathbf{r}\star\mathbf{X})+b)\begin{pmatrix} m\frac{\partial\,\mathsf{M}(\mathbf{r}\star\mathbf{X})}{\partial r_0} \\ m\frac{\partial\,\mathsf{M}(\mathbf{r}\star\mathbf{X})}{\partial r_1} \\ \vdots \\ w\frac{\partial\,\mathsf{M}(\mathbf{r}\star\mathbf{X})}{\partial r_{N'-1}} \\ \mathsf{M}(\mathbf{r}\star\mathbf{X}) \\ 1 \end{pmatrix}$$

for

$$\theta = \begin{pmatrix} r_0 \\ r_1 \\ \vdots \\ r_{N'-1} \\ w \\ b \end{pmatrix} \in \mathbb{R}^{N'+2}.$$

Explicitly,

$$\begin{aligned}
\frac{\partial\,\mathsf{M}(\mathbf{r}\star\mathbf{X})}{\partial r_0} &= \sum_{n'}\frac{\partial\,\mathsf{M}}{\partial x_{n'}}(\mathbf{r}\star\mathbf{X})\frac{\partial(\mathbf{r}\star\mathbf{X})_n}{\partial r_0} \\
&= \frac{\partial(\mathbf{r}\star\mathbf{X})_{\operatorname{argmax}(\mathbf{r}\star\mathbf{X})}}{\partial r_0} = X_{\operatorname{argmax}(\mathbf{r}\star\mathbf{X})} \\
\frac{\partial\,\mathsf{M}(\mathbf{r}\star\mathbf{X})}{\partial r_1} &= \sum_{n'}\frac{\partial\,\mathsf{M}}{\partial x_{n'}}(\mathbf{r}\star\mathbf{X})\frac{\partial(\mathbf{r}\star\mathbf{X})_n}{\partial r_1} \\
&= \frac{\partial(\mathbf{r}\star\mathbf{X})_{\operatorname{argmax}(\mathbf{r}\star\mathbf{X})}}{\partial r_1} = X_{\operatorname{argmax}(\mathbf{r}\star\mathbf{X})+1} \\
&\vdots \\
\frac{\partial\,\mathsf{M}(\mathbf{r}\star\mathbf{X})}{\partial r_n} &= \sum_{n'}\frac{\partial\,\mathsf{M}}{\partial x_{n'}}(\mathbf{r}\star\mathbf{X})\frac{\partial(\mathbf{r}\star\mathbf{X})_n}{\partial r_n} \\
&= \frac{\partial(\mathbf{r}\star\mathbf{X})_{\operatorname{argmax}(\mathbf{r}\star\mathbf{X})}}{\partial r_n} = X_{\operatorname{argmax}(\mathbf{r}\star\mathbf{X})+n}.
\end{aligned}$$

Combining things together,

(14.6)

$$\nabla\lambda_{(\mathbf{X},y)}(\theta) = \begin{pmatrix} \frac{\partial\lambda_{(\mathbf{X},y)}}{\partial r_0}(\theta) \\ \frac{\partial\lambda_{(\mathbf{X},y)}}{\partial r_1}(\theta) \\ \vdots \\ \frac{\partial\lambda_{(\mathbf{X},y)}}{\partial r_{N'-1}}(\theta) \\ \frac{\partial\lambda_{(\mathbf{X},y)}}{\partial w}(\theta) \\ \frac{\partial\lambda_{(\mathbf{X},y)}}{\partial b}(\theta) \end{pmatrix} = (\ell_y \circ S)'\,(w\,\mathsf{M}(\mathbf{r}\star\mathbf{X}) + b) \begin{pmatrix} wX_{\operatorname{argmax}(\mathbf{r}\star\mathbf{X})} \\ wX_{\operatorname{argmax}(\mathbf{r}\star\mathbf{X})+1} \\ \vdots \\ wX_{\operatorname{argmax}(\mathbf{r}\star\mathbf{X})+N'-1} \\ \mathsf{M}(\mathbf{r}\star\mathbf{X}) \\ 1 \end{pmatrix}.$$

We can in fact easily interpret (14.6). Gradient descent for CNN tries to change reference patterns by the subpattern in the observations which is most likely the current estimate of the reference pattern. Of course, then this algorithm may *lock* onto the wrong pattern if the argmax is wrong.

CNNs are trying to do two things at once:

- Find the pattern of interest (add randomness to avoid locking into the wrong pattern).
- Find the map from observations to the label.

Note that it is reasonable to expect that longer patterns in shorter observations are typically better, because it is then easier to find the pattern of interest.

14.4. Auxiliary Thoughts

14.4.1. ReLU. Since in fact we are only interested in positive correlations, we might add a ReLU function (see Figure 14.6) after the convolution, but before the maxpool;

$$\mathsf{M}(\mathsf{ReLU}(\mathbf{r}\star\mathbf{X})).$$

The network will only act on parts of $\mathbf{X}$ which have *positive* correlation with $\mathbf{r}$ (but will get stuck if all correlations are negative; we want enough randomness in training to avoid that).

14.4.2. Stride. Returning to Section 14.2, if the reference signal $\mathbf{R}$ is (known to be) continuous; $\mathbf{R}\star\mathbf{X}$ should also be continuous. In the case of Figure 14.3, the explicit formula for $\mathbf{R}$ was

$$R_n = 6 + \frac{1}{2}\min\{n-4, 0\}, \qquad n \in \{0, 1, \dots, 6\},$$

implying that

$$|R_{n_1} - R_{n_2}| \le \frac{1}{2}|n_1 - n_2|, \qquad n_1, n_2 \in \{0, 1, \dots, 6\}.$$

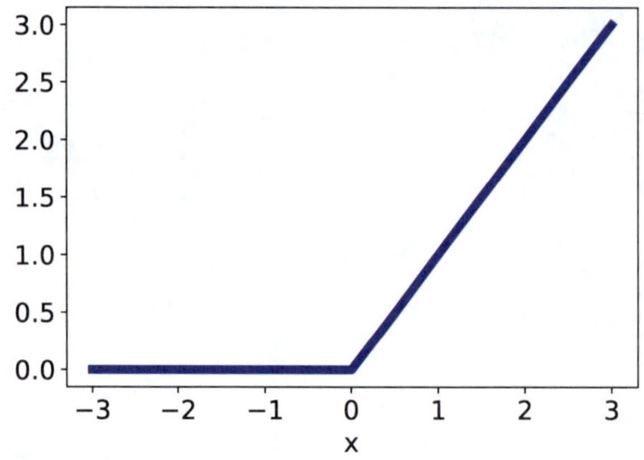

Figure 14.6. ReLU function

For any n_1 and n_2 in $\mathbb{Z}$, we then have that

$$\begin{aligned}(\mathbf{R}\star\mathbf{X})_{n_1}-(\mathbf{R}\star\mathbf{X})_{n_2} &= \sum_{n'\in\mathbb{Z}} R_{n'}X_{n_1+n'} - \sum_{n'} R_{n'}X_{n_2+n'}\\ &= \sum_{k\in\mathbb{Z}} R_{k-n_1}X_k - \sum_{k\in\mathbb{Z}} R_{k-n_2}X_k\\ &= \sum_{k\in\mathbb{Z}} \{R_{k-n_1}-R_{k-n_2}\}X_k\end{aligned}$$

and thus

$$|(\mathbf{R}\star\mathbf{X})_{n_1}-(\mathbf{R}\star\mathbf{X})_{n_2}| \le \frac{1}{2}|n_1-n_2|\left\{\sum_k |X_k|\right\}.$$

If $\mathbf{R}\star\mathbf{X}$ is sufficiently *continuous*, we might be able to approximate

$$\max_n(\mathbf{R}\star\mathbf{X})_n \qquad \text{and} \qquad \operatorname{argmax}_n(\mathbf{R}\star\mathbf{X})_n$$

with (stride $=2$ for example)

$$\max_{n\text{ even}}(\mathbf{R}\star\mathbf{X})_n \qquad \text{and} \qquad \operatorname{argmax}_{n\text{ even}}(\mathbf{R}\star\mathbf{X})_n.$$

Effectively, the stride parameter determines how many steps the calculation in the convolution operation shifts. For example, in the discussion above we took the stride to be equal to 2. A stride of 2 means that at each step there is a shift by two units (this could be a shift by two pixels if the object of study is an image for example) in the calculation of the convolution operation. To compare, a stride of 1 would mean that at each time step there is a shift by one unit.

We define multilayer convolution networks with arbitrary values for stride later on, in Section 14.7. The stride parameter is another hyperparameter that the user is choosing and can affect the performance of the algorithm. Generally

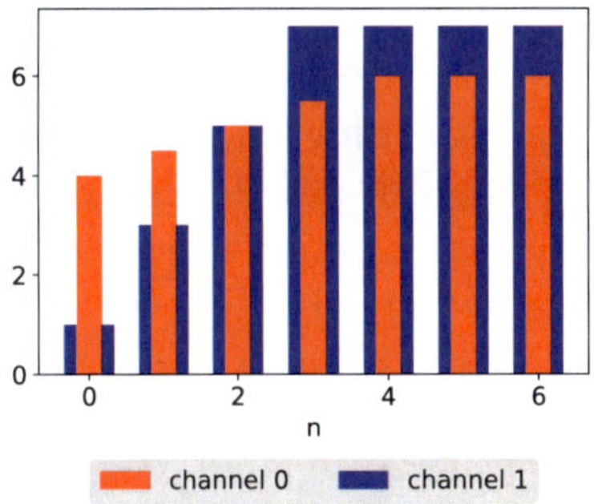

n	channel 0	channel 1
0	4.0	1
1	4.5	3
2	5.0	5
3	5.5	7
4	6.0	7
5	6.0	7
6	6.0	7

Figure 14.7. Reference signal with two channels

speaking, longer strides decrease computational time because they involve a smaller output dimension, but may also decrease accuracy.

Lastly, we mention that, in practice, we do not know **R**, and we need to adaptively find it, i.e., set **r** in place of **R** as was done in Section 14.3.

14.4.3. Channels. In practice, there may be several (i.e., K) *channels* (patterns) corresponding to the desired label;

- red, green, blue in images.
- voice ranges: bass, baritone, tenor, alto, mezzo-soprano, and soprano.

An example with $K = 2$ channels is in Figure 14.7.

Denoting the channels of pattern as $(\mathbf{r}^{(k)})_{k=1}^K$ and observation as $(\mathbf{X}^{(k)})_{k=1}^K$, we can construct the correlations

$$(\mathbf{r} \star \mathbf{X})_{k,n} = \sum_{n=0}^{N'-1} r_{k,n} X_n.$$

For each channel $n \in \mathbb{N}$ and $k \in \{1, 2, \dots, K\}$, where $\mathbf{r}^{(k)} = (r_{k,n})_{n=0}^{N'-1}$ and $\mathbf{X}^{(k)} = (X_{k,n})_{n \in \mathbb{N}}$. The argmax

$$\operatorname{argmax}_{n \in \mathbb{N}} \sum_{1 \le k \le K} |(\mathbf{r} \star \mathbf{X})_{k,n}|$$

will find the index $n \in \mathbb{N}$ at which the channel-averaged cross-correlation

$$\frac{1}{K} \sum_{1 \le k \le K} |(\mathbf{r} \star \mathbf{X})_{k,n}|$$

is maximum. The setting of multiple channels in convolution neural networks will be studied in more detail in Section 14.6.

14.5. SGD for Convolution Neural Networks with a Single Channel

We now generalize the motivational examples and setting studied in the previous sections. We investigate how stochastic gradient descent looks for a convolution neural network with a single hidden layer in a multidimensional setting. Let the input image be $\mathbf{X} \in \mathbb{R}^{d\times d}$ and an unknown signal (often also called filter) $\mathbf{r} \in \mathbb{R}^{k_y\times k_x}$. Convolutions will be taken with a stride of $s = 1$ and there will only be a single channel. Generalizations to the case of stride size $s > 1$, multiple channels, and multiple hidden layers will be considered later. We also do not include the bias term, to further simplify calculations.

We define a convolution of the matrix $\mathbf{X}$ with the filter $\mathbf{r}$ as the map $\mathbf{r} \star \mathbf{X}$: $\mathbb{R}^{k_y\times k_x} \times \mathbb{R}^{d\times d} \to \mathbb{R}^{(d-k_y+1)\times(d-k_x+1)}$ where

$$(\mathbf{r} \star \mathbf{X})_{i,q} = \sum_{m=0}^{k_y-1} \sum_{n=0}^{k_x-1} r_{m,n} X_{i+m,q+n}.$$

The hidden layer applies an elementwise nonlinearity $\sigma : \mathbb{R} \to \mathbb{R}$ to each element of the matrix $\mathbf{r} \star \mathbf{X}$. We define the variable $Z \in \mathbb{R}^{(d-k_y+1)\times(d-k_x+1)}$ and the hidden layer $H \in \mathbb{R}^{(d-k_y+1)\times(d-k_x+1)}$ where

$$\begin{aligned} H_{i,q} &= \sigma\big((Z)_{i,q}\big), \\ Z &= \mathbf{r} \star \mathbf{X}. \end{aligned}$$

Y is the label for the image $\mathbf{X}$ and takes values in the set $\mathcal{Y} = \{0, 1, \ldots, J-1\}$. The output of the network is simply the softmax function applied to a linear function of the hidden layer H:

$$\begin{aligned} \mathfrak{m}(x;\theta) &= S_{\text{softmax}}(U), \\ U_j &= W_{j,:,:} \cdot H + b_j, \end{aligned}$$

where $W \in \mathbb{R}^{J\times(d-k_y+1)\times(d-k_x+1)}$, $b \in \mathbb{R}^J$, $U \in \mathbb{R}^J$, and

$$W_{j,:,:} \cdot H = \sum_{i,q} W_{j,i,q} H_{i,q}.$$

Recall that the : notation in the place of an index signals summation with respect to that index.

The collection of parameters is $\theta = \{\mathbf{r}, W, b\}$. The cross-entropy error for a single data sample $(\mathbf{X}, Y)$ is

$$\ell := \ell_Y(\mathfrak{m}(\mathbf{X};\theta)) = -\log\Big(\mathfrak{m}_Y(\mathbf{X};\theta)\Big).$$

In order to implement the stochastic gradient descent algorithm, we must calculate $\nabla_\theta \ell$. We will next derive the backpropagation rule for single-layer convolution networks.

First, define

$$\delta_{i,q} := \frac{\partial \ell}{\partial H_{i,q}} = \sum_{j=0}^{J-1} \frac{\partial \ell}{\partial U_j} W_{j,i,q} = \frac{\partial \ell}{\partial U} \cdot W_{:,i,q}.$$

Recall that

$$\frac{\partial \ell}{\partial U} = -\big(e(Y) - \mathfrak{m}(\mathbf{X}; \theta)\big),$$
$$e(y) = \big(\mathbf{1}_{y=0}, \ldots, \mathbf{1}_{y=J-1}\big).$$

This of course immediately yields

$$\frac{\partial \ell}{\partial W_{j,:,:}} = \frac{\partial \ell}{\partial U_j} H,$$
$$\frac{\partial \ell}{\partial b} = \frac{\partial \ell}{\partial U}.$$

We must next derive the gradient with respect to the filter $\mathbf{r}$. By the chain rule,

$$\begin{aligned} \frac{\partial \ell}{\partial r_{i,q}} &= \sum_{m=0}^{d-k_y} \sum_{n=0}^{d-k_x} \delta_{m,n} \frac{\partial H_{m,n}}{\partial r_{i,q}} \\ &= \sum_{m=0}^{d-k_y} \sum_{n=0}^{d-k_x} \delta_{m,n} \sigma'(Z_{m,n}) X_{i+m,q+n} \\ &= X_{i:i+d-k_y, q:q+d-k_x} \cdot \big(\sigma'(Z) \odot \delta\big), \end{aligned}$$

where, with a slight abuse of notation, $\sigma'(Z)$ is the elementwise application of the nonlinearity $\sigma(\cdot)$, i.e.

$$\sigma'(Z) = \begin{bmatrix} \sigma'(Z_{0,0}) & \sigma'(Z_{0,1}) & \ldots & \sigma'(Z_{0,d-k_x}) \\ \sigma'(Z_{1,0}) & \sigma'(Z_{1,1}) & \ldots & \sigma'(Z_{1,d-k_x}) \\ \vdots & \vdots & \ddots & \vdots \\ \sigma'(Z_{d-k_y,0}) & \sigma'(Z_{d-k_y,1}) & \ldots & \sigma'(Z_{d-k_y,d-k_x}) \end{bmatrix}.$$

In particular, we have that

$$\Big(\mathbf{X} \star \big(\sigma'(Z) \odot \delta\big)\Big)_{i,q} = \sum_{m=0}^{d-k_y} \sum_{n=0}^{d-k_x} \big(\sigma'(Z) \odot \delta\big)_{m,n} X_{i+m,q+n} = \frac{\partial \ell}{\partial r_{i,q}}.$$

Therefore, from the definition of a convolution,

$$\frac{\partial \ell}{\partial \mathbf{r}} = \mathbf{X} \star \big(\sigma'(Z) \odot \delta\big).$$

This is a very nice result since the gradient with respect to the parameters also involves a convolution. That is, both the backward and forward steps in the backpropagation algorithm can be written in terms of a convolution.

Collecting our results, the stochastic gradient descent algorithm for updating θ is:

- Randomly select a new data sample $(\mathbf{X}, Y)$.
- Compute the forward step (Z, H, U, ℓ).
- Calculate the partial derivatives $(\frac{\partial \ell}{\partial U}, \delta, \frac{\partial \ell}{\partial \mathbf{r}})$.
- Update the parameters $\theta = \{\mathbf{r}, W, b\}$ with a stochastic gradient descent step:

$$b_{k+1} = b_k - \eta_k \frac{\partial \ell}{\partial U},$$
$$W_{j,\cdot,\cdot;k+1} = W_{j,\cdot,\cdot;k} - \eta_k \frac{\partial \ell}{\partial U_j} H, \quad j = 0, \dots, J,$$
$$\mathbf{r}_{k+1} = \mathbf{r}_k - \eta_k \Big(\mathbf{X} \star \big(\sigma'(Z) \odot \delta \big) \Big),$$

where η_k is the learning rate used at iteration k.

14.6. On Convolution Neural Networks with Multiple Channels

Now we study convolution neural networks with multiple channels, and for now we focus on the single layer case. The hidden layer now contains K *feature maps*. The number of feature maps K is often called the *number of channels*. By having multiple feature maps (instead of a single feature map), the network will be able to represent more complex relationships in the data.

The feature maps for the hidden layer are represented by a variable $H \in \mathbb{R}^{(d-k_y+1)\times(d-k_x+1)\times K}$. Each of the feature maps is produced by a convolution with a filter. The convolution layer has an array (or *stack*) of K filters where each filter is of size $k_y \times k_x$. The filters are given by the variable $\mathbf{r} \in \mathbb{R}^{d_y \times d_x \times K}$.

The hidden layer H is given by

$$H_{i,q,k} = \sigma\Big(\sum_{m=0}^{k_y-1} \sum_{n=0}^{k_x-1} r_{m,n,k} X_{i+m,q+n} \Big).$$

Therefore,

$$H_{:,:,k} = \sigma\Big(Z_{:,:,k} \Big),$$
$$Z_{:,:,k} = \mathbf{X}_{:,:} \star \mathbf{r}_{:,:,k}.$$

The output of the network is simply the softmax function applied to a linear function of the hidden layer H:

$$\mathfrak{m}(x;\theta) = S_{\text{softmax}}(U),$$
$$U_j = W_{j,:,:,:} \cdot H + b_j,$$

where $W \in \mathbb{R}^{J\times(d-k_y+1)\times(d-k_x+1)\times K}$, $b \in \mathbb{R}^J$, $U \in \mathbb{R}^J$, and $W_{j,:,:,:} \cdot H = \sum_{i,q,k} W_{j,i,q,k} H_{i,q,k}$ (the : notation in the place of an index signals summation with respect to that index). The collection of parameters is $\theta = \{\mathbf{r}, W, b\}$.

Define

$$\delta_{i,q,k} := \frac{\partial \ell}{\partial H_{i,q,k}} = \frac{\partial \ell}{\partial U} \cdot W_{:,i,q,k}.$$

The backpropagation algorithm is essentially the same as before, with

$$\nabla_{\mathbf{r}_{:,:,k}} \ell = X \star \left(\sigma'(Z_{:,:,k}) \odot \delta_{:,:,k}\right),$$

and

$$\frac{\partial \ell}{\partial b} = \frac{\partial \ell}{\partial U},$$
$$\frac{\partial \ell}{\partial W_{j,:,:,:}} = \frac{\partial \ell}{\partial U_j} H.$$

Let us now briefly discuss multi-layer convolution networks with multiple channels. The setting is that the input image $\mathbf{X} \in \mathbb{R}^{d\times d\times K^0}$ and that the ℓth convolution network contains K^ℓ channels (often called feature maps).

The first feature map is taken to be $H^0 = \mathbf{X}$. The ℓth hidden layer is $H^\ell \in \mathbb{R}^{d_y^\ell \times d_x^\ell \times K^\ell}$ and is given by

$$H^\ell_{i,q,k} = \sigma\left(\sum_{k'=0}^{K^{\ell-1}-1} \sum_{m=0}^{k_y^\ell - 1} \sum_{n=0}^{k_x^\ell - 1} r^\ell_{m,n,k,k'} H^{\ell-1}_{i+m,q+n,k'}\right).$$

The height d_y^ℓ and width d_x^ℓ of the feature maps in the ℓth layer depend on the height $d_y^{\ell-1}$ and width $d_x^{\ell-1}$ of the feature maps in the previous layer and on the filters $k_y^\ell \times k_x^\ell$. In particular, we have

$$d_y^\ell = d_y^{\ell-1} - k_y^\ell + 1,$$
$$d_x^\ell = d_x^{\ell-1} - k_x^\ell + 1.$$

In this case *zero-padding* means that we expand the matrices $H^{\ell-1}_{:,:,k}$ by adding P zeros on all sides to form a larger tensor

$$\hat{H}^{\ell-1} \in \mathbb{R}^{(d_y^{\ell-1}+2P)\times(d_x^{\ell-1}+2P)\times K^{\ell-1}}$$

and we have

$$H^\ell_{i,q,k} = \sigma\left(\sum_{k'=0}^{K^{\ell-1}-1} \sum_{m=0}^{k_y^\ell - 1} \sum_{n=0}^{k_x^\ell - 1} r^\ell_{m,n,k,k'} \hat{H}^{\ell-1}_{i+m,q+n,k'}\right).$$

Now H^ℓ has dimensions $(d_y^{\ell-1} - k_y^\ell + 2P + 1) \times (d_x^{\ell-1} - k_x^\ell + 2P + 1) \times K^\ell$.

Analogously, a convolution neural network with *stride s* would be

$$H^{\ell}_{i,q,k} = \sigma\left(\sum_{k'=0}^{K^{\ell-1}-1}\sum_{m=0}^{k^{\ell}_{y}-1}\sum_{n=0}^{k^{\ell}_{x}-1} r^{\ell}_{m,n,k,k'} H^{\ell-1}_{is+m,qs+n,k'}\right) \tag{14.7}$$

and H^{ℓ} has dimensions $\left(\lfloor\frac{d^{\ell-1}_{y}-k^{\ell}_{y}}{s}\rfloor+1\right)\times\left(\lfloor\frac{d^{\ell-1}_{x}-k^{\ell}_{x}}{s}\rfloor+1\right)\times K^{\ell}$.

Now that we have seen how convolution neural networks look in a general setting, let us conclude this section with a discussion on why convolution neural networks work well for image recognition problems. The first relevant observation is that a CNN has shared weights across layers and sparse interactions when compared to fully connected neural networks. Thus, it has much fewer parameters that must be learned when compared to a fully connected network; this naturally leads to potentially less overfitting. The second relevant observation is that a CNN learns all weights irrespective of where the specific object of interest is located in the image. The third relevant observation is that CNNs are invariant to translations. This translation invariance is a desired feature: think for example of a chair that has been photographed from different angles.

Let us now offer a short proof of why the latter statement of translation invariance is true. Consider an image $\mathbf{X} : \mathbb{Z}\times\mathbb{Z}\mapsto \mathbf{r}$ and set

$$Z_{i,q} = (\mathbf{r}\star\mathbf{X})_{i,q} = \sum_{m=0}^{k_y-1}\sum_{n=0}^{k_x-1} r_{m,n}X_{i+m,q+n}.$$

Consider the translation operator T defined by

$$\bar{X}_{i,q} = T(\mathbf{X})_{i,q} = X_{i-a,q-b}.$$

We notice that

$$\begin{aligned}(\mathbf{r}\star\bar{\mathbf{X}})_{i,q} &= \sum_{m=0}^{k_y-1}\sum_{n=0}^{k_x-1} r_{m,n}\bar{X}_{i+m,q+n}\\ &= \sum_{m=0}^{k_y-1}\sum_{n=0}^{k_x-1} r_{m,n}X_{i+m-a,q+n-b}\\ &= (\mathbf{r}\star\mathbf{X})_{i-a,q-b}\\ &= Z_{i-a,q-b}\\ &= T(Z)_{i,q}.\end{aligned}$$

Therefore, we have indeed established that

$$\mathbf{r}\star T(\mathbf{X}) = T(\mathbf{r}\star\mathbf{X}),$$

which demonstrates that shifting the data does not change the output of the convolution operator.

14.7. Brief Concluding Remarks

In this chapter we studied convolution neural networks. Convolutional neural networks have been very successful in image recognition problems. In the next chapter we discuss generative models, another very successful class of models that are widely used when one's goal is to generate data from a given distribution.

There is an increasingly wide range of applications of deep learning to image processing (typically building on convolutional neural networks); see [**GBC16**, **KWRL17**, **RHK23**] and the references therein.

14.8. Exercises

Exercise 14.1. For a reference signal R we have

n	0	1
R_n	1	2

We want to calibrate a convolution neural network of the form

$$\mathfrak{m}(x,\theta) = S\left(w \max_{n\in\{0,1,2\}} (\mathbf{R} \star x)(n) + b\right),$$

where $\mathbf{R}\star x$ is the correlation between $\mathbf{R}$ and x, $\theta = (w,b)^\top$ and S is the logistic function. Fix an observation X

n	0	1	2	3
X_n	1.2	2.3	0	−1

which contains the reference signal and some noise. Our goal is to calibrate the model $\mathfrak{m}$ to this datapoint.

(1) Compute $(\mathbf{R} \star X)(n)$ for $n \in \{0,1,2\}$.

(2) Compute $\mathfrak{m}(X,(0.5,0.1))$.

(3) If we let $\lambda(\theta) = H(1,\mathfrak{m}(X,\theta))$, where H is relative entropy, compute $\frac{\partial \lambda}{\partial w}(0.5,0.1)$ and $\frac{\partial \lambda}{\partial b}(0.5,0.1)$.

Exercise 14.2. Fix an observation X

n	0	1	2	3
X_n	1.2	2.3	0	−1

and consider the function

$$\mathfrak{m}(R_0,R_1) = \max_{n\in\{0,1,2\}} \{((R_0,R_1)\star X)(n)\}.$$

(1) Compute $\mathfrak{m}(1,2)$.

(2) Compute $\frac{\partial \mathfrak{m}}{\partial R_0}(1,2)$ and $\frac{\partial \mathfrak{m}}{\partial R_1}(1,2)$.

Exercise 14.3. What would be the formula for a $3-d$ convolution with stride, padding, and multiple output channels? The input is a $3-d$ image with K_{in} input channels, i.e., $X \in \mathbb{R}^{d_x \times d_y \times d_z \times K_{in}}$.

Exercise 14.4. Consider the convolution network with bias parameter

$$\begin{aligned}
Z &= \mathbf{R} \star \mathbf{X} + b^1, \\
H &= \sigma(Z), \\
U_j &= W_{j,:,:} \cdot H + b_j^2, \quad j = 0, 1, \ldots, J-1, \\
\mathfrak{m}(x;\theta) &= S_{\text{softmax}}(U).
\end{aligned}$$

For the cross-entropy error function, calculate $\frac{\partial \ell}{\partial b^1}$.

Exercise 14.5. Consider that the filter is given by

$$\mathbf{R} = \{\ldots, R_{-3}, R_{-2}, R_{-1}, R_0, R_1, R_2, R_3, \ldots\}.$$

Let the result of the convolution between the data $\mathbf{X}$ and the filter $\mathbf{R}$ be

$$Z_n = (\mathbf{R} \star \mathbf{X})_n = \sum_{n'=-\infty}^{\infty} R_{n'} X_{n+n'}.$$

What is Z_n in the case where the filter $\mathbf{R}$ has $R_j = 0$ for all $j \notin \{0,1\}$ and $R_0 = R_1 = \frac{1}{2}$? What about when the filter $\mathbf{R}$ has $R_j = 0$ for all $j \notin \{0,1,2,3\}$ and $R_0 = R_1 = R_2 = R_3 = \frac{1}{4}$? What do the signals resemble in both cases?

Chapter 15

Variational Inference and Generative Models

15.1. Introduction

In the problems of regression and classification that we have seen, the goal is to match the target data, which could be real-valued or categorical, respectively. In this chapter, we change gears and we consider generative models where the goal is to train models that create new data. Imagine for example that we have image data and the goal is to create a model that outputs similar images.

This problem is inherently probabilistic and what we are after is the probability distribution that generates the data that we are interested in, say images for instance. One of the main ingredients that we need in order to do so is a notion of distance between two probability distributions. In particular, consider two probability distributions μ and ν. There are many candidate options for *measuring* how close μ and ν are, for example total variation distance, Hellinger distance, integral probability metrics, Rényi divergence, Kullback-Leibler divergence, Wasserstein metric, and more, each one of them having their advantages and disadvantages. Two popular choices that we will use in this chapter are:

- Kullback-Leibler (KL) divergence defined by

$$\mathrm{KL}(\mu|\nu) = \int_{\mathcal{X}} \log \frac{\mu(dx)}{\nu(dx)} \mu(dx).$$

- Wasserstein r-metric defined by

$$W_r(\mu, \nu) = \inf_{\gamma \in \Pi(\mu,\nu)} (\mathbb{E}|X - Y|^r)^{1/r},$$

where $\gamma \in \Pi(\mu, \nu)$ is any coupling with marginals μ and ν, i.e. the $x-$marginal of γ is the measure μ and the y-marginal of γ is the measure ν, $(X, Y) \sim \gamma$ and $r \in [1, \infty]$.

Given a fixed distribution ν, a variational optimization problem with the loss function being the KL-divergence for instance would be

$$\mu^* = \text{argmin}_{\mu \in \mathcal{M}} \text{KL}(\mu|\nu),$$

and analogously for the Wasserstein metric. The choice for the set $\mathcal{M}$ is very important. If $\mathcal{M}$ is too big, then the problem becomes too hard. If it is too small, then it may not contain the distribution we would like to include.

With some abuse of notation, for probability distribution functions $\mu(dx)$ and $\nu(dx)$ that have densities $g(x)$ and $p(x)$, we shall write $\text{KL}(g|p)$ or $W_r(g, p)$ for KL-divergence and Wasserstein r-metric, respectively.

The rest of the chapter is organized as follows. In Section 15.2 we present one characteristic way of estimating probability density functions that is based on the Kullback-Leibler divergence and leads to the so-called Evidence Lower Bound and to the encoder-decoder paradigm. In Section 15.3 we introduce Generalized Adversarial Networks (GANs) that have been very influential generative models. In Section 15.4 we study optimization in GANs. In Section 15.5 we describe Wasserstein GANs, which is a class of GANs motivated by the Wasserstein metric.

15.2. Estimating Densities and the Evidence Lower Bound

In this section, we present one of the main approaches in learning densities of distributions generating the class of data we are interested in. A key component in such a consideration is the concept of a *latent variable*. Imagine, for example, that we want to draw a new car. Before drawing the new car, we need to decide its type (for example sedan/SUV/truck, etc.), its color, the background (is it moving on the road or is it parked), etc. These variables are called latent variables and will be denoted by z.

Then, after deciding on z, we will draw the car. However, two different people with the same choices for the latent variables z, will still draw two different paintings with cars. Namely, the process is inherently random. Therefore, what we are truly after is to obtain samples from the conditional distribution given z, say $p(x|z)$. Obtaining such samples is what is called the generative process. Also, since we do not know z, we assume a prior distribution on z, say $p(z)$.

We note that we are abusing terminology since both $p(x|z)$ and $p(z)$ are densities and not probability distribution functions. However, in this section we will do so without any further warning as working with densities makes the presentation simpler.

In practice, we have access to training data $\mathcal{D}_{\text{train}} = \{x_m\}_{m=1}^M$. We do not know the posterior probability density function $p(z|x)$ and we do not know the true latent variables z. So, it makes sense to look into the problem

$$\min_{g \in \mathcal{G}} \frac{1}{M} \sum_{m=1}^{M} \mathrm{KL}(g(z|x_m) | p(z|x_m)), \tag{15.1}$$

where the set $\mathcal{G}$ is the set that contains the possible probability densities of interest and hopefully contains p as well. Let us denote a minimizer by $g^*\left(z|\{x_m\}_{m=1}^M\right)$ emphasizing the conditioning on the training data $\mathcal{D}_{\text{train}} = \{x_m\}_{m=1}^M$.

We have assumed that the probability distributions we are working with have well-defined probability densities. Let us also assume momentarily for convenience that z and x_m are univariate random variables; see Remark 15.1 for the generalization to the multivariate case.

In this section we shall develop the basic principles of variational inference and its connection to neural networks using the encoder-decoder paradigm, building towards variational auto-encoders.

15.2.1. The Evidence Lower Bound. Before continuing to explore the properties of the optimization problem (15.1), let us rewrite the KL divergence in a more useful way. We notice that for any $m \in \{1, \dots, M\}$

$$\begin{aligned}
\mathrm{KL}(g(z|x_m)|p(z|x_m)) &= \mathbb{E}\left[\log \frac{g(Z|x_m)}{p(Z|x_m)}\right] \\
&= \mathbb{E}\left[\log g(Z|x_m)\right] - \mathbb{E}\left[\log p(Z|x_m)\right] \\
&= \mathbb{E}\left[\log g(Z|x_m)\right] - \mathbb{E}\left[\log p(Z, x_m)\right] + \log p(x_m),
\end{aligned}$$

where $Z \sim g(Z|x_m)$. However, by definition, we have that

$$\mathrm{KL}(g(z|x_m)|p(z|x_m)) \geq 0.$$

So, we obtain that

$$\log p(x_m) \geq \mathbb{E}\left[\log p(Z, x_m)\right] - \mathbb{E}\left[\log g(Z|x_m)\right]. \tag{15.2}$$

The left-hand side of the last display is the true log probability density function $\log p(x)$ that we are after when evaluated at the point $x = x_m$; oftentimes called the evidence. Thus, it makes sense to find a function g that maximizes the right-hand side of (15.2). The right-hand side of (15.2) is called the *evidence lower bound* (ELBO). In particular, we define

$$\begin{aligned} \mathrm{ELBO}(x_m|g) &= \mathbb{E}\left[\log \frac{p(Z,x_m)}{g(Z|x_m)}\right] \\ &= \mathbb{E}\left[\log p(Z,x_m)\right] - \mathbb{E}\left[\log g(Z|x_m)\right], \end{aligned} \tag{15.3}$$

where we recall that $Z \sim g(z|x_m)$. Hence, instead of working with (15.1), we work with the optimization problem

$$\max_{g\in\mathcal{G}} \frac{1}{M}\sum_{m=1}^{M} \mathrm{ELBO}(x_m|g),$$

and let a candidate maximizer be denoted by $g^*\left(z|\{x_m\}_{m=1}^M\right)$.

Next, we observe that we can write

$$\begin{aligned} \mathrm{ELBO}(x_m|g) &= \mathbb{E}\left[\log \frac{p(Z,x_m)}{g(Z|x_m)}\right] \\ &= \mathbb{E}\left[\log p(Z,x_m)\right] - \mathbb{E}\left[\log g(Z|x_m)\right] \\ &= \mathbb{E}\left[\log p(x_m|Z)\right] - \left(\mathbb{E}\left[\log g(Z|x_m)\right] - \mathbb{E}\left[\log p(Z)\right]\right) \\ &= \mathbb{E}\left[\log p(x_m|Z)\right] - \mathrm{KL}(g(Z|x_m)|p(Z)). \end{aligned}$$

It is interesting to note that in the last display, $\log p(x_m|z)$ is the log-likelihood of x_m given the variable z, whereas $\mathrm{KL}(g(z|x_m)|p(z))$ is the *error* being made by using $g(z|x_m)$ as a proxy for the prior distribution $p(z)$. We have arrived at the program

$$\begin{aligned} &\max_{g\in\mathcal{G}} \frac{1}{M}\sum_{m=1}^{M} \mathrm{ELBO}(x_m|g) \\ &= \max_{g\in\mathcal{G}} \frac{1}{M}\sum_{m=1}^{M} \left(\mathbb{E}\left[\log p(x_m|Z)\right] - \mathrm{KL}(g(Z|x_m)|p(Z))\right), \end{aligned} \tag{15.4}$$

where we recall that $Z \sim g(Z|x_m)$.

15.2.2. The encoder-decoder paradigm. Let us next discuss how to use neural networks to solve (15.4). How do we choose the set $\mathcal{G}$ over which the optimization is performed?

This brings us to the so-called encoder-decoder paradigm. We will parametrize the probability density distributions $g(x|z)$, $p(x|z)$, and $p(z)$, view them as neural networks, and optimize (15.4) over the parameters of these neural networks. To simplify the discussion below, let us focus on parametrizing $g(x|z)$, $p(x|z)$. Let θ_e and θ_d be parameters of neural networks and set $g(x|z) = g(z|x;\theta_e)$ and $p(x|z) = p(x|z;\theta_d)$. Then the goal is to do gradient descent or stochastic gradient descent on (15.4) in order to learn the unknown parameters θ_e and θ_d, namely

$$\max_{(\theta_e,\theta_d)\in\Theta} \frac{1}{M}\sum_{m=1}^{M} \left(\mathbb{E}\left[\log p(x_m|Z;\theta_d)\right] - \mathrm{KL}(g(Z|x_m;\theta_e)|p(Z))\right), \tag{15.5}$$

where we recall that $Z \sim g(Z|x_m;\theta_e)$. Denote by (θ_e^*, θ_d^*) a solution to this optimization problem.

However, we immediately have another problem to solve. In doing SGD on (15.4) or (15.5), we want to be able to compute quantities, such as $\nabla_{\theta_e}\mathbb{E}\left[f(Z)\right]$ where $Z \sim g(Z|x;\theta_e)$ for some function f. Note that the parameter with respect to which we want to differentiate also affects the distribution under which we sample Z, which is problematic.

To this end, notice first that, under smoothness assumptions on g, we can write (its derivation is left as Exercise 15.2)

$$\nabla_{\theta_e}\mathbb{E}\left[f(Z)\right] = \mathbb{E}\left[f(Z)\nabla_{\theta_e}\log g(Z|x;\theta_e)\right]. \tag{15.6}$$

In the latter expression $\nabla_{\theta_e}\log g(z|x;\theta_e)$ is typically called the score function of the probability distribution g. Note that this identity reduces the calculation of the gradient of the expectation of a test function f, to the calculation of expectation of f multiplied with the score function.

The score function formulation (15.6) allows us to compute what we want because given a sample $Z = z$, we can get $\nabla_{\theta_e}\log g(z|x;\theta_e)$ by standard back-propagation and then use Monte Carlo for the estimation of the expectation in (15.6). Namely, we can approximate

$$\nabla_{\theta_e}\mathbb{E}\left[f(Z)\right] \approx \frac{1}{K}\sum_{k=1}^{K} f(z_k)\nabla_{\theta_e}\log g(z_k|x;\theta_e), \tag{15.7}$$

where $z_k \sim g(z|x;\theta_e)$. However, it is known in the literature that the usual Monte Carlo gradient estimator of this quantity may have high variance, see **[PBJ12]**. This issue typically originates from the fact that $\nabla_{\theta_e}\log g(z|x;\theta_e) = \frac{\nabla_{\theta_e} g(z|x;\theta_e)}{g(z|x;\theta_e)}$ and the denominator can take very small values if the sample z is in the tail of the distribution (i.e., if it is rare).

This then brings us to an alternative way to solve this problem. In particular, let us directly assume that $g(z|x;\theta_e)$ is the density of some distribution. We then model its parameters (that will be functions of x) as neural networks with parameter θ_e. So, effectively, we model for example $Z \sim N(\mu(x), \sigma^2(x))$ and parametrize $\mu(x), \sigma^2(x)$ to be neural networks with parameter θ_e. In particular, for a given x and θ_e we set for example $Z = \mu(x;\theta_e) + \sigma(x;\theta_e)\epsilon$ where $\epsilon \sim N(0,1)$ which implies that the expectations are taken with respect to the standard normal distribution which does not involve the parameter θ_e of differentiation. This then allows us to move the differentiation inside the expectation. Namely, we now write

$$\nabla_{\theta_e}\mathbb{E}\left[f(Z)\right] = \mathbb{E}\left[\nabla_{\theta_e} f(\mu(x;\theta_e) + \sigma(x;\theta_e)\epsilon)\right],$$

where the expectation on the right is taken with respect to $\epsilon \sim N(0,1)$, and can be evaluated via Monte Carlo estimation. This is usually called in the literature the reparametrization trick ([**KW13**]), and is used routinely in generative modeling.

To be more precise, with θ_e being the parameter vector of the neural networks $\mu(x_m;\theta_e)$ and $\log\sigma^2(x_m;\theta_e)$, we set

$$g(z|x_m) = \text{ density of Normal}\left(\mu(x_m;\theta_e), \sigma^2(x_m;\theta_e)\right). \tag{15.8}$$

This is called the encoder and it models $g(z|x_m)$. How do we model $p(x_m|z)$? We have data $\mathcal{D}_{\text{train}} = \{x_m\}_{m=1}^M$ and $\{z_m\}_{m=1}^M$ provided by the encoder. The decoder models $p(x_m|z)$ as a neural network with parameters θ_d. For example, $p(x_m|z;\theta_d)$ here could be Bernoulli (in case of binary data), i.e.,

$$\log p(x_m|z;\theta_d) = x_m \log \mathfrak{m}(z;\theta_d) + (1-x_m)\log(1-\mathfrak{m}(z;\theta_d))$$

with $\mathfrak{m}(z;\theta_d)$ an appropriate neural network modeling probabilities. Alternatively, $p(x_m|z;\theta_d)$ could be Gaussian (in case of real-valued data),

$$p(x_m|z;\theta_d) = \text{ density of Normal}\left(\mu(z;\theta_d), \sigma^2(z;\theta_d)\right),$$

where $\mu(z;\theta_d)$ and $\log\sigma^2(z;\theta_d)$ are neural networks.

In applications, one typically models the prior probability density function of the latent variables $p(z)$ as the density of a Normal$(0,1)$ distribution (or a product of standard normal densities if there are more than one latent variable). Note that $p(z)$ could be also parametrized as a neural network. We will not do so here for simplicity but the framework allows us to do so.

So, effectively we have replaced (15.4) by

$$\max_{(\theta_e,\theta_d)\in\Theta} \frac{1}{M}\sum_{m=1}^{M} \left(\mathbb{E}\left[\log p(x_m|Z;\theta_d)\right] - \text{KL}(g(Z|x_m;\theta_e)|p(\epsilon))\right), \tag{15.9}$$

where $p(x_m|z;\theta_d)$ and $g(Z|x_m;\theta_e)$ are densities of random variables (from appropriate desired distributions as discussed above) parametrized as neural networks with parameters θ_d and θ_e, respectively. Z is generated based on (15.8) and $p(\epsilon)$ is the density of a Normal$(0,1)$ distribution. This is an example of a variational auto-encoder [**KW13**].

Note that (15.9) can be simplified further. For example, by direct computation, we have that (see Exercise 15.1)

$$\mathrm{KL}\big(N(\mu,\sigma^2)|N(0,1)\big) = \frac{\sigma^2+\mu^2-1-2\log\sigma^2}{2},$$

which means that in the case where we work with Gaussian distributions

$$\mathrm{KL}(g(Z|x_m;\theta_e)|p(\epsilon)) = \frac{\sigma^2(x_m;\theta_e)+\mu^2(x_m;\theta_e)-1-2\log\sigma^2(x_m;\theta_e)}{2}.$$

Then, in order to solve (15.9) one can perform standard stochastic gradient descent.

Remark 15.1. In this section, we assumed that z and x are univariate variables. If instead they are multivariate but independent, then the framework is similar due to the property of KL-divergence that

$$\mathrm{KL}\left(\prod_{k=1}^{K} g_k(Z_k)\Big|\prod_{k=1}^{K} p_k(Z_k)\right) = \sum_{k=1}^{K}\mathrm{KL}(g_k(Z_k)|p_k(Z_k)).$$

Exercise 15.3 establishes the validity of this identity.

Remark 15.2. Of course, instead of Gaussian models, one can use other distributions. The framework allows for that, but the calculations may be more involved.

Remark 15.3. The score function formulation and the reparametrization trick offer two different ways to compute the gradient of expectation of test functions. Both are popular formulations that have generated a lot of interest in recent years and both are used in deep learning applications (e.g., in generative modeling and in reinforcement learning). Both methods have advantages and disadvantages. As we already discussed, the estimation via the score function formulation is subject to high variance, whereas the estimation via the reparametrization trick will be less accurate when $g(x|z)$ has more than one mode because in that case a normal distribution will not be able to capture that.

15.3. Generative Adversarial Networks

In this section, we discuss *generative adversarial networks* (GANs), a clever reformulation of a generic generative model. GANs have found many applications in super-resolution of images, data simulation, semisupervised learning with unlabeled data, and much more.

GANs transform estimating a density (studied in Section 15.2) to a classification problem. In a sense GANs owe their empirical success to this property by leveraging the fact that deep neural networks work very well when it comes to classification problems.

15.3.1. Revisiting the Basic Classification Problem of Chapter 3. We introduce GANs by connecting them to the basic classification problem that we studied in Chapter 3 via logistic regression. In particular, consider the situation were the $\{0, 1\}$-valued label y corresponds to whether an image comes from the generator's distribution $\nu(dx)$, in which case $y = 0$, or from the nature's distribution $\mu(dx)$, in which case $y = 1$. Our data are images $\mathcal{D} = \{x_m\}_{m=1}^{2M}$ such that $\mathcal{D} = \mathcal{D}_0 \cup \mathcal{D}_1$, where

$$\begin{aligned} \mathcal{D}_0 &= \{x_m \sim \nu, \quad m = 1, \dots, M\}, \\ \mathcal{D}_1 &= \{x_m \sim \mu, \quad m = M+1, \dots, 2M\}. \end{aligned}$$

In this case we model the probability of *success* as a neural network

$$\mathbb{P}(y = 1|x) = \mathfrak{m}(x; \theta).$$

In its simplest form, and as used in Chapter 3, the model $\mathfrak{m}(x; \theta)$ could be the logistic function

$$\mathfrak{m}(x; \theta) = S(wx + b) = \frac{e^{wx+b}}{1 + e^{wx+b}} = \frac{1}{1 + e^{-(wx+b)}},$$

where $\theta = \binom{w}{b}$ are the parameters of the model taking values in the appropriate space Θ.

No matter what the actual choice of the model $\mathfrak{m}(x; \theta)$ is, the associated logistic loss function is

$$\Lambda(\theta) = -\frac{1}{M} \sum_{x \in \mathcal{D}_1} \log \mathfrak{m}(x; \theta) - \frac{1}{M} \sum_{x \in \mathcal{D}_0} \log(1 - \mathfrak{m}(x; \theta)),$$

and naturally the goal is to find $\theta^* = \operatorname{argmin}_\theta \Lambda(\theta)$.

Let us now abstract this formulation a little bit. The population loss function corresponding to $\Lambda(\theta)$ defined above is

$$\Lambda_{\text{pop}}(\mathfrak{m}) = -\mathbb{E}_\mu \log \mathfrak{m}(x) - \mathbb{E}_\nu \log(1 - \mathfrak{m}(x)), \tag{15.10}$$

where the subscript in the expectation operator denotes the probability distribution under which the expectation is being considered. Then, in theory we would have

$$\mathfrak{m}^* = \operatorname{argmin}_{\mathfrak{m}} \Lambda_{\text{pop}}(\mathfrak{m}).$$

With the goal of building some intuition, let us characterize the optimal point $\mathfrak{m}^*(x)$.

Lemma 15.4. *Assume that in the loss function defined in* (15.10)*, the measures* μ *and* ν *have continuous densities* p_μ *and* p_ν*, respectively, that are bounded away* *from zero. Then we have*

$$\mathfrak{m}^*(x) = \operatorname{argmin}_{\mathfrak{m}} \Lambda_{\text{pop}}(\mathfrak{m}) = \frac{p_\mu(x)}{p_\mu(x) + p_\nu(x)}.$$

In addition, the global minimum $\mathfrak{m}^*(x)$ is achieved if and only if $p_\mu(x) = p_\nu(x)$ in which case $\Lambda_{\text{pop}}(\mathfrak{m}^*) = \ln 4$.

Proof. A perturbation argument for the variational derivative $\frac{\delta \Lambda_{\text{pop}}(\mathfrak{m})}{\delta \mathfrak{m}}$ gives that

$$\frac{\delta \Lambda_{\text{pop}}(\mathfrak{m})}{\delta \mathfrak{m}} = -\frac{p_\mu(x)}{\mathfrak{m}(x)} + \frac{p_\nu(x)}{1 - \mathfrak{m}(x)}.$$

Setting this equal to zero and solving for $\mathfrak{m}(x)$ gives that indeed $\mathfrak{m}^*(x) = \frac{p_\mu(x)}{p_\mu(x)+p_\nu(x)}$. Note that

$$\frac{\delta^2 \Lambda_{\text{pop}}(\mathfrak{m})}{\delta \mathfrak{m}^2} = \frac{p_\mu(x)}{\mathfrak{m}^2(x)} + \frac{p_\nu(x)}{(1 - \mathfrak{m}(x))^2} \geq 0,$$

which implies that indeed $\mathfrak{m}^*(x)$ is a minimum.

To answer the second part of the lemma, we note that by symmetry the minimum of $\Lambda_{\text{pop}}(\mathfrak{m}^*)$ is achieved when the two terms balance, which happens when $\mathfrak{m}^*(x) = \frac{1}{2}$. This is true if and only if $p_\mu(x) = p_\nu(x)$. In that case, we immediately see that $\Lambda_{\text{pop}}\left(\frac{1}{2}\right) = \ln 4$. □

15.3.2. The Discriminator-Generator Framework. Lemma 15.4 shows that the ideal case would be when $\mathfrak{m}^*(x) = \frac{1}{2}$ for all points x, which happens when $p_\mu(x) = p_\nu(x)$. This means that in that case all samples are produced from the same true distribution. The latter starts being suggestive that we should be viewing the two terms on the right-hand side of $\Lambda_{\text{pop}}(\mathfrak{m})$ as *competing* with each other. The first term (corresponding to distribution μ) tries to find out whether a datapoint x comes from nature's (true) distribution, while the second term (corresponding to distribution ν) gives an output x which is supposed to be close to points in the true dataset.

This point of view brings us to the definition of GANs. Formally, a GAN consists of two neural networks:

- Discriminator (D): this works as a classifier. Given a datapoint x (e.g., an image), a number between $[0, 1]$ is produced which corresponds to the probability of the datapoint x being part of the dataset, i.e., coming from nature's distribution.
- Generator (G): this gives an output datapoint x, which is supposed to be close to images in the dataset.

To record the fact that we can really think of discriminator and generator as two different things, we reformulate $\Lambda_{\text{pop}}(\mathfrak{m})$ as $\Lambda_{\text{pop}}(\mathfrak{m}_D, \mathfrak{m}_G)$ where

$$\Lambda_{\text{pop}}(\mathfrak{m}_D, \mathfrak{m}_G) = -\mathbb{E}_\mu \log \mathfrak{m}_D(x) - \mathbb{E}_\nu \log(1 - \mathfrak{m}_G(x)),$$

where the subscripts D, G correspond to discriminator and generator, respectively.

Typically $\mathfrak{m}_D, \mathfrak{m}_G$ are neural networks with parameters θ_D and θ_G, respectively, and we choose

$$\begin{aligned}\mathfrak{m}_D &= \mathfrak{m}_D(x;\theta_D) = \mathfrak{m}(x;\theta_D),\\ \mathfrak{m}_G &= \mathfrak{m}_G(x;\theta_G) = \mathfrak{m}(g(z;\theta_G);\theta_D),\end{aligned}$$

where $g : \mathcal{Z} \mapsto \mathcal{X}$ maps latent space to data space (and it is also modeled as a neural network). Thus, we can actually write

$$\begin{aligned}&\Lambda_{\text{pop}}(\mathfrak{m}(\cdot;\theta_D), \mathfrak{m}(g(\cdot;\theta_D);\theta_G))\\ &\quad = -\mathbb{E}_\mu \log \mathfrak{m}_D(x) - \mathbb{E}_\nu \log(1 - \mathfrak{m}_G(x))\\ &\quad = -\mathbb{E}_\mu \log \mathfrak{m}(x;\theta_D) - \mathbb{E}_\nu \log(1 - \mathfrak{m}(g(z;\theta_G);\theta_D)).\end{aligned}$$

We may view the loss function as a function of θ_D, θ_G, and thus we write

$$\hat{\Lambda}_{\text{pop}}(\theta_D, \theta_G) = -\mathbb{E}_\mu \log \mathfrak{m}(x;\theta_D) - \mathbb{E}_\nu \log(1 - \mathfrak{m}(g(z;\theta_G);\theta_D)). \tag{15.11}$$

The objective of a GAN is twofold. The discriminator updates its weights θ_D to minimize the loss Λ_{pop}, whereas and generator updates its weights θ_G to maximize it. Then, we choose

$$(\theta_D^*, \theta_G^*) = \text{argmax}_{\theta_G} \, \text{argmin}_{\theta_D} \, \hat{\Lambda}_{\text{pop}}(\theta_D, \theta_G). \tag{15.12}$$

Remark 15.5. In this remark we directly connect the empirical loss function corresponding to (15.11) (this is (15.13) that we shall explore in Section 15.4) to the binary cross entropy (3.7) $\ell_y(y')$ that we defined in Chapter 3. Let us think of the true datapoints being in $\mathcal{D}_1$ (i.e., with label 1) and the fake datapoints being $\mathcal{D}_0 = \{g(z;\theta_G) : z \text{ generated points, say from } N(0,1)\}$ (i.e., with label 0). Create the labeled dataset

$$\mathcal{D}_{\text{labeled}} = \{\mathcal{D}_1 \times \{1\}\} \cup \{\mathcal{D}_0 \times \{0\}\}.$$

Assume that the classifier $\mathfrak{m}(x;\theta_D)$ is used by the discriminator. Let $y \in \{0,1\}$ denote the label of whether a point is fake or real, respectively. Then, we have

$$\begin{aligned}
\tilde{\Lambda}(\theta_D, \theta_G) &= \frac{1}{|\mathcal{D}_{\text{labeled}}|} \sum_{(x,y)\in\mathcal{D}_{\text{labeled}}} \ell_y\left(m(x;\theta_D)\right) \\
&= \frac{1}{|\mathcal{D}_{\text{labeled}}|} \sum_{x\in\mathcal{D}_1} \ell_1\left(m(x;\theta_D)\right) + \frac{1}{|\mathcal{D}_{\text{labeled}}|} \sum_{x\in\mathcal{D}_0} \ell_0\left(m(x;\theta_D)\right) \\
&= \frac{1}{|\mathcal{D}_{\text{labeled}}|} \Bigg[- \sum_{x\in\mathcal{D}_1} \log \mathfrak{m}(x;\theta_D) \\
&\qquad - \sum_{z\in\{\text{fake dataset}\}} \log(1 - \mathfrak{m}\left(\mathfrak{g}(Z;\theta_G);\theta_D)\right) \Bigg].
\end{aligned}$$

Namely the loss empirical function is directly connected to the average cross entropy associated with a classification problem.

An equivalent formulation is to define

$$\check{\Lambda}_{\text{pop}}(\theta_D, \theta_G) = \mathbb{E}_\mu \log \mathfrak{m}(x;\theta_D) + \mathbb{E}_\nu \log(1 - \mathfrak{m}(\mathfrak{g}(z;\theta_G);\theta_D)),$$

in which case $(\theta_D^*, \theta_G^*) = \text{argmin}_{\theta_G} \text{argmax}_{\theta_D} \check{\Lambda}_{\text{pop}}(\theta_D, \theta_G)$. Note that the two formulations are equivalent since $\hat{\Lambda}_{\text{pop}}(\theta_D, \theta_G) = -\check{\Lambda}_{\text{pop}}(\theta_D, \theta_G)$. However, we will work with the formulation based on $\hat{\Lambda}_{\text{pop}}(\theta_D, \theta_G)$ from (15.11), since, as we discussed in Remark 15.5, $\hat{\Lambda}_{\text{pop}}(\theta_D, \theta_G)$ is directly related to the basic cross entropy (3.7) $\ell_y(y')$ from logistic regression.

We shall discuss how to implement in practice the min-max problem (15.12).

15.4. Optimization in GANs

In practice, we have training data $\mathcal{D}_{\text{train}} = \{x_m\}_{m=1}^M$, where μ is the empirical distribution of the data $\mathcal{D}_{\text{train}}$ and the generator produces data from some model distribution ν, say Gaussian $N(0,1)$. So, typically, in practice we have the empirical loss function

(15.13)

$$\hat{\Lambda}(\theta_D, \theta_G) = -\frac{1}{|\mathcal{D}_{\text{train}}|} \sum_{x\in\mathcal{D}_{\text{train}}} \log \mathfrak{m}(x;\theta_D) - \mathbb{E}_{Z\sim\nu} \log(1 - \mathfrak{m}\left(\mathfrak{g}(Z;\theta_G);\theta_D)\right).$$

The goal is to obtain $\max_{\theta_G} \min_{\theta_D} \hat{\Lambda}(\theta_D, \theta_G)$. Standard SGD updates then read as

$$\begin{aligned}
\theta_{D,k+1} &= \theta_{D,k} - \eta \nabla_{\theta_{D,k}} \hat{\Lambda}(\theta_{D,k}, \theta_{G,k}), \\
\theta_{G,k+1} &= \theta_{G,k} + \eta \nabla_{\theta_{G,k}} \hat{\Lambda}(\theta_{D,k+1}, \theta_{G,k}),
\end{aligned}$$

where $\eta > 0$ is a learning rate. Notice the different sign (minus for the update of $\theta_{D,k+1}$ and plus for the update of $\theta_{G,k+1}$) in the gradient descent updates. The difference in the sign is not a typo! It is because we are solving a min-max problem.

Let us now study the derivatives $\nabla_{\theta_D}\hat{\Lambda}(\theta_D, \theta_G)$ and $\nabla_{\theta_G}\hat{\Lambda}(\theta_D, \theta_G)$. Direct calculation gives

$$\begin{aligned}\nabla_{\theta_D}\hat{\Lambda}(\theta_D, \theta_G) = &-\frac{1}{|\mathcal{D}_{\text{train}}|}\sum_{x\in\mathcal{D}_{\text{train}}}\left(\frac{1}{\mathfrak{m}(x;\theta_D)}\frac{\partial \mathfrak{m}(x;\theta_D)}{\partial\theta_D}\right)\\ &+\mathbb{E}_{Z\sim\nu}\left(\frac{1}{1-\mathfrak{m}\left(g(Z;\theta_G);\theta_D\right)}\frac{\partial \mathfrak{m}\left(g(Z;\theta_G);\theta_D\right)}{\partial\theta_D}\right)\end{aligned}$$

and

$$\begin{aligned}\nabla_{\theta_G}\hat{\Lambda}(\theta_D, \theta_G) &= \mathbb{E}_{Z\sim\nu}\left(\frac{1}{1-\mathfrak{m}\left(g(Z;\theta_G);\theta_D\right)}\frac{\partial \mathfrak{m}\left(g(Z;\theta_G);\theta_D\right)}{\partial\theta_G}\right)\\ &= \mathbb{E}_{Z\sim\nu}\left(\frac{1}{1-\mathfrak{m}\left(g(Z;\theta_G);\theta_D\right)}\frac{\partial \mathfrak{m}\left(g(Z);\theta_D\right)}{\partial g(Z)}\frac{\partial g(Z;\theta_G)}{\partial\theta_G}\right).\end{aligned}$$

In practice, we sample a minibatch of images $\{x_1, \ldots, x_q\}$ and we sample latent variables $\{z_1, \ldots, z_q\}$. Then

(1) We update the generator $g(\cdot;\theta_G)$ using the gradient updates and the minibatch being sampled.

(2) We update the discriminator $\mathfrak{m}(\cdot;\theta_D)$ using the corresponding gradient updates and the minibatch being sampled.

Remark 15.6. Training goes in cycles. In the early phases of the training process, the generator produces noisy data-images, but over time it becomes better at producing images that are closer to the real ones. The discriminator is trained initially on both fake and real data but as the generator gets better at its job, the discriminator has a harder time telling apart real from fake data. If training is successful, then by Lemma 15.4, the discriminator in the end (i.e., when $\max_{\theta_G}\min_{\theta_D}\hat{\Lambda}(\theta_D, \theta_G)$ has been achieved) produces probabilities of $1/2$. Namely, it cannot decide whether a sample is real or fake. When that happens, the discriminator is useless and can be discarded, leaving only the generator as being useful.

This is an inherently min-max problem, since the discriminator is minimizing the objective whereas the generator is maximizing the objective function. Training a GAN can become complicated with gradient methods especially because solving the min-max problem amounts to finding saddle points (so involving gradients makes it relatively possible to climb up the hill or fall down the hill). See Remark 15.7 in that direction.

Remark 15.7. Early in training the discriminator has an easy job to do whereas the generator has a hard job to do. In those early steps of training if the generator is not good enough, then we will have

$$\nabla_{\theta_G} \log(1 - \mathfrak{m}\,(g(Z;\theta_G);\theta_D)) = -\frac{\partial_{\theta_G} \mathfrak{m}\,(g(Z;\theta_G);\theta_D)}{1 - \mathfrak{m}\,(g(Z;\theta_G);\theta_D)} \approx 0.$$

Hence, in that case the job of the generator gets even harder! This often leads to the phenomenon called *mode collapse*, which means that the generator produces the same output for many different inputs; we will also discuss this in Section 15.5. This is why in practice oftentimes the loss function is modified to be

$$\hat{\Lambda}(\theta_D, \theta_G) = -\frac{1}{|\mathcal{D}_{\text{train}}|} \sum_{x \in \mathcal{D}_{\text{train}}} \log \mathfrak{m}(x;\theta_D) - \mathbb{E}_{Z\sim\nu} \log \mathfrak{m}\,(g(Z;\theta_G);\theta_D),$$

where again ν is some model distribution, say standard Gaussian for example, $N(0,1)$ (if one dimensional). The goal is the same, i.e., to reach

$$\max_{\theta_G} \min_{\theta_D} \hat{\Lambda}(\theta_D, \theta_G).$$

This choice of the loss function does not face the same vanishing gradient problem as the original formulation.

We conclude this section with an illustrative max-min problem.

Example 15.8. Let us define

$$f(x,y) = (x-2y)^2 - 7(y-1)^2,$$

and consider the problem $\max_{y\in\mathbb{R}} \min_{x\in\mathbb{R}} f(x,y)$. Think of x as the generator and y as the discriminator.

A straightforward computation shows that

$$f^*(y) = \min_{x\in\mathbb{R}} f(x,y) = -7(y-1)^2,$$

with the minimum achieved at $x = 2y$ and

$$\max_{y\in\mathbb{R}} f^*(y) = 0,$$

with the latter maximum achieved at $y = 1$. So, the saddle point is $(x,y) = (2,1)$.

If we were to solve this via SGD, we would have

$$\begin{aligned}
\frac{\partial f}{\partial x}(x,y) &= 2(x-2y),\\
\frac{\partial f}{\partial y}(x,y) &= 4(x-2y) - 14(y-1) = 4x - 22y + 14.
\end{aligned}$$

Consequently, with $\eta > 0$ a given learning rate, the SGD update equations would be

$$x_{k+1} = x_k - \eta(2x_k - 4y_k),$$
$$y_{k+1} = y_k + \eta(4x_{k+1} - 22y_k + 14).$$

In Exercise 15.3 we will see how the SGD algorithm actually converges to the target saddle point $(x, y) = (2, 1)$.

15.5. Wasserstein GANs

As we discussed in Remark 15.7 of Section 15.4, the gradient of the standard GAN with respect to the generator parameters can be near zero at least in the early phases of training, which may lead to the phenomenon of mode collapse. Mode collapse is when the generator produces the same output for main different inputs and it is a challenge that large-scale images will often phase.

A popular remedy to this problem is the *Wasserstein GAN* (WGAN) algorithm that was originally introduced in [**ACB17**]. Let us first recall the Wasserstein r-metric:

$$W_r(\mu, \nu) = \inf_{\gamma \in \Pi(\mu,\nu)} (\mathbb{E}|X - Y|^r)^{1/r},$$

where $\Pi(\mu, \nu)$ is any coupling with marginals μ and ν, $(X, Y) \sim \gamma$ and $r \in [1, \infty]$. Consider the case of $r = 1$.

By the Kantonovich-Rubinstein duality (see [**Vil09**]), we have

$$W_1(\mu, \nu) = \sup_{\|f\|_{\text{Lip}} \leq 1} \left\{ \mathbb{E}_{X\sim\mu}[f(X)] - \mathbb{E}_{Y\sim\nu}[f(Y)] \right\},$$

where $\{f : \|f\|_{\text{Lip}} \leq 1\}$ is the set of globally Lipschitz functions with Lipschitz constant bounded by one. Note that if $W_1(\mu, \nu) = 0$, then $\mu = \nu$, and if we have a sequence such that $\lim_{n\to\infty} W_1(\mu, \nu^n) = 0$, then ν^n will converge weakly to the measure μ. These properties make Wasserstein metric appealing for generative modeling.

In the world of GANs we have $Y = \mathrm{g}(Z; \theta_G)$ and μ would be the empirical distribution of the data $\mathcal{D}_{\text{train}}$. Then, if $X \sim \mu$ and $Y \sim \nu_{\theta_G}$, we get

$$W_1(\mu, \nu) = \sup_{\|f\|_{\text{Lip}} \leq 1} \left\{ \mathbb{E}_{X\sim\mu}[f(X)] - \mathbb{E}_{Z\sim\text{model}}[f(\mathrm{g}(Z; \theta_G))] \right\}.$$

Hence in the end, we are interested in the problem

$$\min_{\theta_G} W_1(\mu, \nu) = \min_{\theta_G} \sup_{\|f\|_{\text{Lip}} \leq 1} \left\{ \mathbb{E}_{X\sim\mu}[f(X)] - \mathbb{E}_{Z\sim\text{model}}[f(\mathrm{g}(Z; \theta_G))] \right\}.$$

However, optimizing over the whole space $\{f : \|f\|_{\text{Lip}} \leq 1\}$ is a difficult task. On the other hand we know by the uniform approximation theorems that neural networks $\mathfrak{m}(x; \theta)$ approximate continuous functions on compact sets, see Chapter 16. This approximation, leads us to look into the more practically relevant optmization problem

$$\min_{\theta_G} \sup_{\theta_D} \left\{ \mathbb{E}_{X\sim\mu}[\mathfrak{m}(X;\theta)] - \mathbb{E}_{Z\sim\text{model}}[\mathfrak{m}(g(Z;\theta_G);\theta_D)] \right\}, \tag{15.14}$$

which is analogous to (15.13).

However, $\mathfrak{m}(x;\theta)$ may not be of Lipschitz constant one. To this end, we observe that

$$\begin{aligned} J(\mu, \nu_{\theta_G}) &= \sup_{\|f\|_{\text{Lip}}\leq K} \left\{ \mathbb{E}_{X\sim\mu}[f(X)] - \mathbb{E}_{Z\sim\text{model}}[f(g(Z;\theta_G))] \right\} \\ &= K \sup_{\|f\|_{\text{Lip}}\leq K} \left\{ \mathbb{E}_{X\sim\mu}\left[\frac{1}{K} f(X)\right] - \mathbb{E}_{Z\sim\text{model}}\left[\frac{1}{K} f(g(Z;\theta_G))\right] \right\} \\ &= K \sup_{\|h\|_{\text{Lip}}\leq 1} \left\{ \mathbb{E}_{X\sim\mu}[h(X)] - \mathbb{E}_{Z\sim\text{model}}[h(g(Z;\theta_G))] \right\} \\ &= KW(\mu, \nu_{\theta_G}). \end{aligned}$$

Thus, we shall have

$$\nabla_{\theta_G} W(\mu, \nu_{\theta_G}) = \frac{1}{K} \nabla_{\theta_G} J(\mu, \nu_{\theta_G}),$$

which shows that the two problems are equivalent. Hence, (15.14) can indeed be used in place of (15.13), also validating Remark 15.7. In Algorithm 2 we present pseudocode for the WGAN algorithm where we also clip the θ_D parameter.

Algorithm 2 WGAN algorithm with SGD

procedure ▷ (Input parameters)
  Initialise: initial parameters $\theta_{G,0}, \theta_{D,0}$, clipping constant $c > 0$, learning rate η, step $k = 0$
  while Not yet converged **do**
    Sample X, Z
    $\tilde{\theta}_D \leftarrow \theta_D + \eta\left(\nabla_{\theta_D}\mathfrak{m}(X;\theta_D) - \nabla_{\theta_D}\mathfrak{m}(g(Z;\theta_G);\theta_D)\right)$
    $\theta_D \leftarrow \max\left(\min\left(\tilde{\theta}_D, c\right), -c\right)$
  end while
  Sample Z
  $\theta_G \leftarrow \theta_G - \eta\left(-\nabla_{\theta_G}\mathfrak{m}(g(Z;\theta_G);\theta_D)\right)$
end procedure

Generally speaking, clipping the estimated parameter amounts to fixing a threshold $c > 0$ and then replacing θ by $\max(\min(\theta, c), -c)$. It is easy to see that this operation constrains the resulting estimated parameter to be within the interval $[-c, c]$. It is a technique often used in practice to reduce the magnitude of the parameter by scaling it back to a given threshold if it becomes too large in norm. Oftentimes, the clipping idea is used directly on the gradient updates to scale them back to a given threshold if they become too large (effectively trying to mitigate the exploding gradient problem), see [**PMB13**].

At the same time, clipping introduces a discontinuity which can sometimes lead to undesired results and to difficulty with training. Adding gradient penalty terms (essentially enforcing a Lipschitz condition with a penalization) as an alternative to weight clipping in WGAN algorithm, was suggested in [**GAA$^+$17**] as a way to improve the convergence properties of the original WGAN algorithm.

15.6. Brief Concluding Remarks

Excellent sources for the auto-encoding variational Bayes procedure that we analyzed in Section 15.2 are [**KW13**] and [**DMBM17**]. [**Bis06**] also contains a nice related exposition to variational inference.

The generative adversarial networks (GANs) that we studied in Section 15.3 were originally introduced in [**GPAM$^+$14**] (see also the book [**GBC16**]) and since then have been tremendously influential. The Wasserstein GAN (WGAN) algorithm that we presented in Section 15.5 was originally introduced in [**ACB17**]. Some of the pitfalls (associated with critical weight clipping) of the original WGAN algorithm were empirically demonstrated in [**GAA$^+$17**], and adding a gradient penalty term was proposed to alleviate these pitfalls. Theoretically understanding this phenomenon is a subject of active research.

In this chapter we mainly discussed the Kullback-Leibler divergence and the Wasserstein metric as distance measures and used them in training generative adversarial networks. There are other distance measures, such as integral probability metrics and f-divergences, that sometimes are advantageous, especially when we aim to compare distributions which are not absolutely continuous with each other, see [**BDK$^+$22**] for details.

This chapter concludes Part 1. In Part 2, we go deeper into several topics, some that are more of a theoretical nature and some that are of a more computational nature.

15.7. Exercises

Exercise 15.1. Consider two independent random variables $X \sim N(\mu_x, \sigma_x^2)$ and $Y \sim N(\mu_y, \sigma_y^2)$. Find a formula for $\mathrm{KL}\big(N(\mu_x, \sigma_x^2)|N(\mu_y, \sigma_y^2)\big)$ in terms of the mean and variances of the random variables X and Y.

Exercise 15.2. Prove that under the proper assumptions relation

$$\nabla_{\theta_e}\mathbb{E}\left[f(Z)\right] = \mathbb{E}\left[f(Z)\nabla_{\theta_e}\log g(z|x;\theta_e)\right],$$

will hold.

Exercise 15.3. Prove that the statement of Remark 15.1 holds. Namely prove that if Z are multivariate but independent random variables, then

$$\mathrm{KL}\left(\prod_{k=1}^{K} g_k(Z_k) \Big| \prod_{k=1}^{K} p_k(Z_k)\right) = \sum_{k=1}^{K} \mathrm{KL}(g_k(Z_k)|p_k(Z_k)).$$

Derive how (15.9) will look in the multivariate case.

Exercise 15.4. Consider the setting of Example 15.8, namely let

$$f(x, y) = (x - 2y)^2 - 7(y - 1)^2$$

and consider the problem $\max_{y\in\mathbb{R}} \min_{x\in\mathbb{R}} f(x, y)$. Implement the stochastic gradient descent algorithm to show convergence to the saddle point. Assume starting point $(x, y) = (0, 0)$. Produce an $x - y$ plot showing the progress of the max-min problem towards the saddle point $(2, 1)$.

Exercise 15.5. Consider the loss function (15.13). Assume that $\mathfrak{m}(x; \theta_D) = \frac{e^{b(x;\theta_D)}}{1+e^{b(x;\theta_D)}}$.

(1) Write down $\nabla_{\theta_D} \hat{\Lambda}(\theta_D, \theta_G)$ in this case.
(2) Express this in terms of KL-divergence when the optimal $\mathfrak{m}^*(x)$ by Lemma 15.4 takes the form $\frac{e^{b(x)}}{1+e^{b(x)}}$.

Exercise 15.6. Consider the discriminator to be a sigmoid neural network of one hidden unit. Namely, let us set $\mathfrak{m}_D(x) = \mathfrak{m}(x; \theta_D) = \frac{e^{w\cdot x+b}}{1+e^{w\cdot x+b}}$ with $\theta_D = (w, b)$. Simplify the loss function (15.13) in that case.

Exercise 15.7. Consider a payoff function $\Lambda(x, y)$ which depends on the values x and y. Then, consider that we are interested in

$$(\hat{x}, \hat{y}) = \mathrm{argmax}_x \, \mathrm{argmin}_y \, \Lambda(x, y).$$

Obtain the point $(\hat{x}, \hat{y})$ as the long time behavior of a gradient descent method in continuous time of the joint variable $(x(t), y(t))$.

Exercise 15.8. Consider a loss function $\Lambda(x, y) = xy$ and consider the system

$$\begin{aligned} x_{k+1} &= x_k - \eta y_k, \\ y_{k+1} &= y_k + \eta x_k, \end{aligned}$$

where $\eta > 0$ is a learning rate. Find conditions under which (x_k, y_k) converges as $k \to \infty$.

Part 2

Advanced Topics and Convergence Results in Deep Learning

Transitioning from Part 1 to Part 2

As we mentioned in the introductory Chapter 1, this book is composed of two parts. In Part 1, we introduced the main tools for deep learning from a mathematical perspective based on a unified mathematical language. We often made connections with and drew motivation from classical statistics and machine learning topics. In Part 2 we dive into more advanced topics of deep learning. We build gradually.

We recall that the paradigm we have implemented in this book follows the diagram in the next figure.

Some of the components of the diagram are presented in Part 1 and some in Part 2, depending on the background that is needed. As we now transition from Part 1 to Part 2 of the book, let us elaborate in more detail in those components.

1. Motivating Learning: Part 1.

In Part 1, we first presented linear and logistic regression in Chapters 2 and 3, respectively, from the lens of optimization and in the language of deep learning. We then motivated how neural networks appear via kernels in Chapter 4. In Chapter 5 we visited the feed forward neural network architecture.

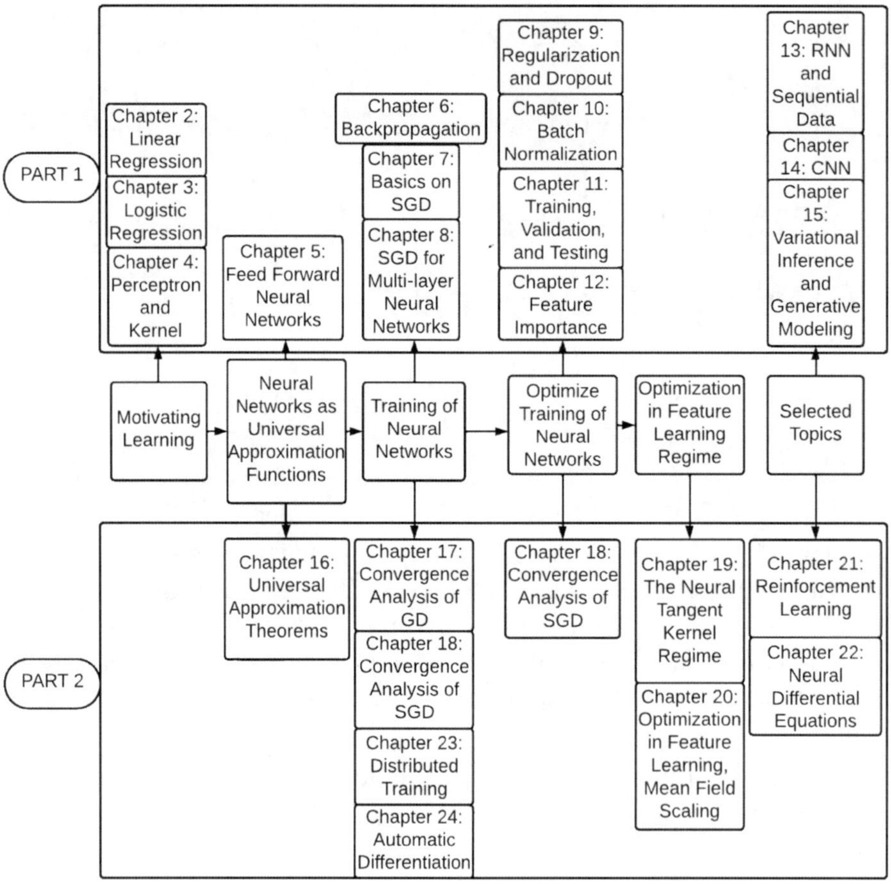

Graphical representation of how the book is organized

2. Neural Networks and Universal Approximation: Part 1 $\longrightarrow$ Part 2.

The development of Chapter 5 (Part 1) was tied to truth tables. This is motivated by the fact that neural networks are universal approximators. Namely, neural networks can approximate reasonable functions to good accuracy. The theory of uniform approximation of neural networks is well established, and we present it in Chapter 16 (Part 2). It is presented in Part 2 because its presentation requires more advanced mathematical tools.

3. Training of Neural Networks: Part 1 ⟶ Part 2.

Now that we know that neural networks are universal approximators and that they can be used to approximate a given function of interest,

How do we learn (estimate) the parameters in the neural network?

This is where the method of gradient descent (GD) comes in. Gradient descent involves differentiating the loss function of interest. Backpropagation presented in Chapter 6 (Part 1) enables us to do so in an efficient way. The backpropagagion algorithm can be implemented using automatic differentiation, which we discuss in more detail in Chapter 24 (Part 2). In practice, stochastic gradient descent (SGD) is being used, and we presented the principles of SGD in Chapters 7 and 8 (Part 1).

But an important question remains:

Does gradient descent or stochastic gradient descent converge?

The answer is yes, but it can be proven only under proper conditions. The mathematical theory for that is developed in Chapters 17 and 18 for GD and SGD, respectively (Part 2). This convergence theory is presented in Part 2 because it requires more advanced mathematical tools.

However, training of very large deep neural networks can become very expensive, making the need for computing power clear. Distributed learning is a way to help in that direction; this is described in Chapter 23 of Part 2 because it is conceptually and computationally more advanced.

4. Optimize Training of Neural Networks: Part 1 ⟶ Part 2.

Now that we have learned that neural networks are universal approximators and we know how to train them,

How do we optimize training?

There are several methods to improve training, for instance regularization methods (Chapter 9 in Part 1) and batch normalization (Chapter 10 in Part 1). Of course, we need to blend all of these with data and have a sense on how to tune things and what features are important; these topics were presented in Chapters 11 and 12 in Part 1.

Last, but not least, is there only the vanilla SGD algorithm, or are there more advanced algorithms? There are of course other optimization algorithms and we present those in Chapter 18 in Part 2 together with their properties. They are presented in Part 2 as their analysis requires sometimes more advanced mathematical tools.

5. Optimization in the Feature Learning Regime: Part 2.

The next natural question is:

Will the algorithms converge to the true answer, and will they perform well with unseen data? Under which conditions will that be the case?

Note that a priori it is not clear that this would have been the case (due to the training algorithm involved) even though neural networks are universal approximators. This brings us to the convergence results of Chapters 19 and 20 in Part 2. In Chapter 19 we study the neural tangent kernel (NTK) limit for neural networks which, since it is derived using a linearization, is referred to as the linear regime in this book. In Chapter 20 we discuss the nonlinear regime and the mean field scaling. The development of this theory requires more advanced exposure to stochastic processes convergence theory, so we present it in Part 2.

6. Selected Topics: Part 1 ⟶ Part 2.

The thematic units that we just described compose the main aspects of any deep learning algorithm. However, in deep learning there are a number of other topics of interest. We present a selection of these topics in either Part 1 or Part 2 depending on the level of mathematical and/or computational maturity required.

6A. Specialized Architectures (RNN, Transformer, CNN): Part 1. Dependent data, sequential data or image data may need more specialized architectures that take into account the nature of the data. This brings us to recurrent neural networks, transformers, and convolution neural networks which were presented in Chapters 13 and 14 in Part 1. Even though RNNs, transformers, and CNNs are more advanced architectures than feed forward neural networks, we present them towards the end of Part 1 because these more advanced architectures are still within reach conceptually.

6B. Variational Inference and Generative Modeling: Part 1. In Chapter 15 we present variational inference and generative adversarial networks which have been very successful frameworks to generate data from desired (often unknown) distributions; this is another very exciting area of research.

6C. Control Problems and Reinforcement Learning: Part 2. What if now we want to learn how to control a dynamical system to achieve a certain goal?

Can we use deep learning to control a dynamical system? Will the algorithm converge to the optimal control policy?

The answer is yes and this is the field of reinforcement learning that we describe in Chapter 21 in Part 2. In Chapter 21, we also present convergence properties of reinforcement learning algorithms with neural networks which is mainly why Chapter 21 is in Part 2 and not in Part 1. Reinforcement learning with neural networks is a very exciting area of research with a long history.

6D. Neural ODEs and SDEs: Part 2. In Chapter 22 we present the topic of neural differential equations, which essentially amounts to using neural networks to learn dynamical systems, another very exciting area of research. It is presented in Part 2 as its treatment oftentimes requires more advanced mathematical and conceptual tools.

Chapter 16

Universal Approximation Theorems

16.1. Introduction

In Chapter 5 we discussed feed forward neural networks. We demonstrated there, via explicit basic constructions, that indicators of rectangles can approximate generic functions.

In this chapter we demonstrate that this is part of a more general pattern for neural networks. One of the main mathematical questions is:

Given a target function $\bar{\mathfrak{m}}$, *is there a vector of parameter values* θ
and a neural network (shallow or deep) $\mathfrak{m}(x;\theta)$
that is close to $\bar{\mathfrak{m}}(x)$ *in an appropriate sense?*

One of the reasons neural networks work well in practice is because they are able to accurately approximate typical functions. In other words, neural networks are universal approximators. In this chapter we present some of the main results on the approximation properties of neural networks.

16.2. Basic Universal Approximation Theorems

In this section we focus on a single-layer neural network,

$$\mathfrak{m}(x;\theta) = \sum_{n=1}^{N} c^n \sigma(w^n \cdot x + b^n),$$

where $\theta = (c^n, w^n, b^n)_{n=1}^N \in \mathbb{R}^{(d+2)N}$ is the parameter to be estimated, N is the number of hidden units, σ is the activation function, and $x \in \mathbb{R}^d$ is the input data.

In this section, $\mathcal{M}(X)$ denotes the space of finite, signed measures on a space X. $C(X)$ is the space of continuous functions on X, and $L^2(X)$ is the space of square-integrable functions on X. Typically, we shall take $X = I_d = [0,1]^d$, the hypercube in d dimensions.

In Definition 16.1 we define what a discriminatory function is, which is a key concept in the theory of universal approximation of neural networks.

Definition 16.1. Consider a measure $\mu \in \mathcal{M}(I_d)$ and a function $\sigma : \mathbb{R} \mapsto \mathbb{R}$. Then σ is called *discriminatory* with respect to the measure μ if

$$\int_{I_d} \sigma(w \cdot x + b)\mu(dx) = 0$$

for every $(w, b) \in \mathbb{R}^{d+1}$ implies $\mu = 0$.

Definition 16.2. Consider a function $\sigma : \mathbb{R} \mapsto [0,1]$. Then, σ is said to be *sigmoidal* if

$$\lim_{x\to-\infty} \sigma(x) = 0, \qquad \lim_{x\to\infty} \sigma(x) = 1.$$

A question that arises is what kind of activation functions are indeed discriminatory? Proposition 16.3 shows that this is a quite common property, and in fact any continuous sigmoidal function is discriminatory in the sense of Definition 16.1.

Proposition 16.3. *Let σ be a given continuous sigmoidal function. Then σ is discriminatory with respect to all measures $\mu \in \mathcal{M}(I_d)$.*

Proof. Let us start by fixing a measure $\mu \in \mathcal{M}(I_d)$. Let σ be a continuous sigmoidal function such that

$$\int_{I_d} \sigma(w \cdot x + b)\mu(dx) = 0 \quad \text{for all } (w, b) \in \mathbb{R}^{d+1}.$$

By Definition 16.1, we would like to show that $\mu = 0$. For this purpose, we define the function $\sigma_\rho(x) = \sigma(\rho(w \cdot x + b) + q)$. Since σ is assumed to be sigmoidal, we will have

$$\lim_{\rho\to\infty} \sigma_\rho(x) = \begin{cases} 1, & \text{if } w \cdot x + b > 0 \\ \sigma(q), & \text{if } w \cdot x + b = 0 \\ 0, & \text{if } w \cdot x + b < 0. \end{cases}$$

Slightly abusing notation, let us now set

$$\sigma_\infty(x) = \begin{cases} 1, & \text{if } x \in B^+_{w,b} = \{x \in I_d \,:\, w \cdot x + b > 0\} \\ \sigma(q), & \text{if } x \in B^\circ_{w,b} = \{x \in I_d \,:\, w \cdot x + b = 0\} \\ 0, & \text{if } x \in B^-_{w,b} = \{x \in I_d \,:\, w \cdot x + b < 0\}, \end{cases}$$

and note that $\lim_{\rho\to\infty} \sigma_\rho(x) = \sigma_\infty(x)$. By a bounded convergence theorem we then have

$$\begin{aligned} 0 &= \lim_{\rho\to\infty} \int_{I_d} \sigma_\rho(x)\mu(dx) = \int_{I_d} \sigma_\infty(x)\mu(dx) \\ &= \mu(B^+_{w,b}) + \sigma(q)\mu(B^\circ_{w,b}). \end{aligned}$$

So, we have deduced that $\mu(B^+_{w,b}) + \sigma(q)\mu(B^\circ_{w,b}) = 0$. Taking now $q \to \infty$, we obtain $\mu(B^+_{w,b}) + \mu(B^\circ_{w,b}) = 0$. On the other hand, if we take $q \to -\infty$, we shall have that $\mu(B^+_{w,b}) = 0$. Thus, we get $\mu(B^\circ_{w,b}) = 0$. But $B^+_{w,b} = B^-_{-w,-b}$. Thus, μ vanishes on the half-planes of $\mathbb{R}^d$, which then implies that $\mu = 0$. Even though the latter conclusion is not so obvious here because the measure μ is a finite signed measure (not necessarily a positive measure), it effectively follows by the argument of Example 16.8.

All in all, we have shown that σ is discriminatory with respect to the arbitrarily chosen measure $\mu \in \mathcal{M}(I_d)$. □

When a function σ is discriminatory with respect to all measures $\mu \in \mathcal{M}(I_d)$, then we will say that σ is a discriminatory function. Proposition 16.3 shows that, at least, continuous sigmoidal functions are indeed discriminatory.

Proposition 16.4 shows that neural networks with continuous and discriminatory activation functions are dense in the space of continuous functions, i.e., they can approximate any given continuous function. In particular, Proposition 16.4 shows that for any given $g \in C(I_d)$, there exists a neural network such that for a given $\epsilon > 0$, we have

$$|g(x) - \mathfrak{m}(x;\theta)| < \epsilon \quad \text{for all } x \in I_d,$$

which is the definition of density of the space of shallow neural network functions in the space of continuous functions; see also Definition B.4.

Proposition 16.4. *Consider a continuous discriminatory function σ. Then functions of the form*

$$\mathfrak{m}(x;\theta) = \sum_{n=1}^{N} c^n \sigma(w^n \cdot x + b^n)$$

with $w^n \in \mathbb{R}^d$ and $c^n, b^n \in \mathbb{R}$ are dense in $C(I_d)$, where $I_d = [0,1]^d$.

Proof. We begin the proof by defining the space of all shallow neural networks

$$U = \left\{ \mathfrak{m} \,:\, \mathfrak{m}(x;\theta) = \sum_{n=1}^{N} c^n \sigma(w^n \cdot x + b^n), w^n \in \mathbb{R}^d, c^n, b^n \in \mathbb{R}, x \in I_d \right\}.$$

The assumption that σ is continuous, means that the space U is a linear subspace of $C(I_d)$. Let us now suppose that U is not dense in $C(I_d)$. Then, by Lemma B.6 there is a measure $\mu \in \mathcal{M}(I_d)$ such that

$$\sum_{n=1}^{N} \int_{I_d} c^n \sigma(w^n \cdot x + b^n) \mu(dx) = 0 \quad \text{for all } (c^n, w^n, b^n) \in \mathbb{R}^{d+2}.$$

By choosing the c^n's appropriately, we get that for all $(w^n, b^n) \in \mathbb{R}^{d+1}$, the relation holds

$$\int_{I_d} \sigma(w^n \cdot x + b^n) \mu(dx) = 0.$$

Since σ is discriminatory, we get that $\mu = 0$. This is immediately a contradiction, yielding the proposition. □

As we discussed, Proposition 16.4 shows that for any given $\mathrm{g} \in C(I_d)$, there exists $\mathfrak{m} \in U$ of the form of a neural network such that for all $\epsilon > 0$, we have that

$$\sup_{x \in I_d} |\mathrm{g}(x) - \mathfrak{m}(x;\theta)| < \epsilon.$$

The aforementioned results bring us to one of the first universal approximation theorems.

Theorem 16.5 ([**Cyb89**]). *Consider a continuous sigmoidal function σ. Then, the finite sums of the form $\mathfrak{m}(x;\theta) = \sum_{n=1}^{N} c^n \sigma(w^n \cdot x + b^n)$ with $w^n \in \mathbb{R}^d$ and $c^n, b^n \in \mathbb{R}$ are dense in $C(I_d)$.*

Proof. This is a direct consequence of Propositions 16.4 and 16.3. □

Another one of the early classical universal approximation theorems was derived in [**HSW89**].

Theorem 16.6 ([**HSW89**]). *Consider a continuous, nonconstant function $\sigma : \mathbb{R} \mapsto \mathbb{R}$. Then, the set*

$$U = \left\{ \mathfrak{m} \,:\, \mathfrak{m}(x;\theta) = \sum_{n=1}^{N} c^n \prod_{j=1}^{J_n} \sigma(w^{jn} \cdot x + b^{jn}), w^{jn} \in \mathbb{R}^d, c^n, b^{jn} \in \mathbb{R}, x \in I_d \right\}$$

is dense in $C(I_d)$.

Proof. Since σ is continuous, we can verify that the set U is an algebra on I_d, see Definition B.8.

In addition, U separates points in I_d. Since σ is nonconstant, there exist $u_1, u_2 \in \mathbb{R}$ such that $u_1 \neq u_2$ and $\sigma(u_1) \neq \sigma(u_2)$. Next, we pick points x, y in the hyperplanes $\{w \cdot x + b = u_1\}$ and $\{w \cdot y + b = u_2\}$, respectively. Then, we shall have that the function $\mathfrak{m}(z;\theta) = \sigma(w \cdot z + b)$ separates x and y. Indeed, we have that $\mathfrak{m}(x;\theta) = \sigma(w \cdot x + b) = \sigma(u_1)$ and $\mathfrak{m}(y;\theta) = \sigma(w \cdot y + b) = \sigma(u_2)$ with $\sigma(u_1) \neq \sigma(u_2)$.

Moreover, we have that U contains nonzero constants. Indeed, let b be such that $\sigma(b) \neq 0$ and choose $w = (0, 0, \dots, 0) \in \mathbb{R}^d$. Then $\mathfrak{m}(x;\theta) = \sigma(b) \neq 0$.

Hence, overall we have shown that U satisfies the assumptions of the Stone-Weierstrass theorem, Theorem B.9, implying that U is dense in $C(I_d)$. □

Theorems 16.5 and 16.6 address the uniform approximation properties of shallow neural networks in the supremum norm in $C(I_d)$. However, one may be interested in approximations of certain target functions in other norms. For example, one may be interested in approximating square integrable functions $g \in L^2(I_d)$, i.e., functions for which $\int_{I_d} |f(x)|^2 dx < \infty$. To illustrate this point, let us consider the following definition.

Definition 16.7. A function $0 \leq \sigma \leq 1$ is called *discriminatory* in L^2 if for $f \in L^2(I_d)$,

$$\int_{I_d} \sigma(w \cdot x + b) f(x) dx = 0$$

for every $(w, b) \in \mathbb{R}^{d+1}$ implies $f = 0$.

Before presenting the uniform approximation theorem in this case, let us see an example that not only is useful as a building block, but it also builds useful intuition.

Example 16.8. Let us prove that the indicator function $\sigma(x) = 1_{\{x \geq 0\}}$ is discriminatory in L^2. Indeed, let us consider $f \in L^2(I_d)$ and assume that for every $(w, b) \in \mathbb{R}^{d+1}$,

$$\begin{aligned} &\int_{I_d} \sigma(w \cdot x + b) f(x) dx = 0 \\ \Longrightarrow &\int_{\{x : w \cdot x + b \geq 0\} \cap I_d} f(x) dx = 0 \\ \Longrightarrow &\int_{\{x : w \cdot x \geq -b\} \cap I_d} f(x) dx = 0. \end{aligned}$$

Since, the last relation is true for every $(w, b) \in \mathbb{R}^{d+1}$, it motivates the idea that the integral of $f(x)$ is zero over all intervals for $w \cdot x$, and by linearity of the integral, it will also be zero over unions of disjoint intervals. Let us make this idea precise.

We view $\int_{I_d} \sigma(w \cdot x + b) f(x)dx$ as the functional,

$$\mathfrak{F}(g) = \int_{I_d} g(x)f(x)dx = \int_{I_d} 1_{\{w\cdot x \geq -b\}} f(x)dx = \int_{\{x:w\cdot x\geq -b\}\cap I_d} f(x)dx$$

for $g(x) = 1_{\{w\cdot x\geq -b\}}$. Then, since this is true for all $b \in \mathbb{R}$ and since the integral operator has the additive property, if U_i are disjoint intervals in $\mathbb{R}^d$, then $\mathfrak{F}(H) = 0$ for $H(x) = \sum_{i=1}^{K} u_i 1_{U_i}(x)$ for $u_i \in \mathbb{R}$. H is a simple function, and we know that simple functions are dense in the set of bounded functions. Therefore, by density, if g is a bounded function, we will also have that $\mathfrak{F}(g) = 0$.

Let us now calculate the Fourier transform of f. We have by definition

$$\begin{aligned}
\hat{f}(u) &= \int_{I_d} e^{i(u\cdot x)} f(x)dx \\
&= \int_{I_d} (\cos(u \cdot x) + i \sin(u \cdot x))\, f(x)dx \\
&= \mathfrak{F}(\cos) + i\mathfrak{F}(\sin) \\
&= 0,
\end{aligned}$$

because $\cos(x), \sin(x)$ are both bounded functions. Here, $i^2 = -1$ is the standard imaginary number. Since the inverse Fourier transform of f is zero for all $u \in \mathbb{R}^d$, then $f = 0$ almost everywhere. This completes the derivation.

Then, we have the following theorem.

Theorem 16.9. *Let σ be discriminatory in L^2 according to Definition* 16.7. *Then, the set*

$$U = \left\{ \mathfrak{m} : \mathfrak{m}(x; \theta) = \sum_{n=1}^{N} c^n \sigma(w^n \cdot x + b^n), c^n, b^n \mathbb{R}, w^n \in \mathbb{R}^d, x \in I_d \right\}$$

is dense in $L^2(I_d)$. Namely, if $f \in L^2(I_d)$, then for every $\epsilon > 0$, there is $\mathfrak{m} \in U$ such that

$$\int_{I_d} |f(x) - \mathfrak{m}(x; \theta)|^2 dx < \epsilon.$$

Proof. The proof of this result is an application of the celebrated Riesz representation theorem, Theorem B.1. Indeed, let us assume by contradiction that the set U is not dense in $L^2(I_d)$. By a reformulation of Lemma B.6, we get that there will be a linear bounded functional H such that $H \neq 0$ on $L^2(I_d)$, but with

$H = 0$ on U. Hence, by the Riesz representation theorem, Theorem B.1 there is some $f \in L^2(I_d)$ such that

$$H(h) = \int_{I_d} h(x)f(x)dx,$$

and in addition $\|H\| = \|f\|_2$.

In addition, for any $\mathfrak{m} \in U$ we shall have that $H(\mathfrak{m}) = 0$. This means that

$$\int_{I_d} \sigma(w \cdot x + b)f(x)dx = 0.$$

Since σ is discriminatory according to Definition 16.7, we get that $f = 0$. This, however, would mean that $\|H\| = 0$ which would be a contradiction since we have assumed that $H \neq 0$ on $L^2(I_d)$. □

Since the fundamental works of [**Cyb89, HSW89**] there have been many universal approximation theorems, and we will not cover all of them. Nevertheless, we have gotten a good taste of the basic universal approximations theorems here. In addition, we visit some more recent results in Section 16.3 using ReLU activation functions.

We conclude this section with a short detour on error bounds. In particular, we have seen that shallow neural networks are universal approximator functions in $C(I_d)$ and in $L^2(I_d)$.

But how good really is such an approximation? To get a taste of such a result, we will visit one of the classical results in this direction by [**Bar94**]. In order to state this result, we first need to discuss some properties of target functions having a Fourier representation. In particular, assume that the target function $\bar{\mathfrak{m}}$ has the inverse Fourier representation (with $\hat{\bar{\mathfrak{m}}}$ the Fourier transform),

$$\bar{\mathfrak{m}}(x) = \int_{\mathbb{R}^d} e^{iu \cdot x}\hat{\bar{\mathfrak{m}}}(u)du \quad \text{for } x \in \mathbb{R}^d.$$

Let us now assume that $u\hat{\bar{\mathfrak{m}}}(u)$ is integrable and let us set

$$D(\bar{\mathfrak{m}}) = \int_{\mathbb{R}^d} \|u\|_1 |\hat{\bar{\mathfrak{m}}}(u)|du,$$

where $\|u\|_1 = \sum_i |u_i|$. Then, we have the following result, which we present without proof referring the interested reader to [**Bar94**] for its proof.

Theorem 16.10 ([**Bar94**]). *Let σ be a sigmoidal activation function, i.e.,*

$$\lim_{x \to -\infty} \sigma(x) = 0 \quad \textit{and} \quad \lim_{x \to +\infty} \sigma(x) = 1.$$

Let $\bar{\mathfrak{m}}$ be a target function with $D(\bar{\mathfrak{m}}) < \infty$ and consider $\mu \in \mathcal{M}(I_d)$. Then, for any $N \in \mathbb{N}$ there exists a function $\mathfrak{m}^N(x;\theta) = \sum_{n=1}^N c^n\sigma(w^n \cdot x + b^n)$ such that

$$\left(\int_{I_d} |\bar{\mathfrak{m}}(x) - \mathfrak{m}^N(x;\theta)|^2\mu(dx)\right)^{1/2} \leq \frac{D(\bar{\mathfrak{m}})}{\sqrt{N}}.$$

Theorem 16.10 says that the approximation error in the L^2 norm is of the order of $1/\sqrt{N}$.

16.3. Universal Approximation Results Using ReLU Activation Functions

In this section, we visit some more recent universal approximation theorems using deep neural networks based on ReLU activation functions.

Let us start by defining what a ReLU *deep neural network* (DNN) is. First, we recall that a one-dimensional ReLU function simply is $\mathsf{ReLU}(x) = \max(x, 0)$ for $x \in \mathbb{R}$. Similarly if $x = (x_1, \dots, x_d) \in \mathbb{R}^d$, then we define

$$\mathsf{ReLU}(x) = (\mathsf{ReLU}(x_1), \dots, \mathsf{ReLU}(x_d)).$$

Let us also establish some notation for linear transformations. With $x \in R^d$, let us set for $k = 1, \dots, n$, $T_k(x) = w^k \cdot x + b^k = \sum_{i=1}^d w^{ik}x_i + b^k$ and define the mapping $T : \mathbb{R}^d \mapsto \mathbb{R}^n$ to be

$$T(x) = (T_1(x), \dots, T_n(x)).$$

Now, we can define what a ReLU DNN is.

Definition 16.11. A ReLU DNN is a function $\mathfrak{m} : \mathbb{R}^d \mapsto \mathbb{R}^n$ of the form

$$\mathfrak{m}(x;\theta) = (T_{L+1} \circ \mathsf{ReLU} \circ T_L \circ \cdots \circ T_2 \circ \mathsf{ReLU} \circ T_1)(x;\theta), \tag{16.1}$$

where $x \in \mathbb{R}^d$, $T_1 : \mathbb{R}^d \mapsto \mathbb{R}^{\ell_1}$, $T_k : \mathbb{R}^{\ell_{k-1}} \mapsto \mathbb{R}^{\ell_k}$ for $k = 2, \dots, L$, $T_{L+1} : \mathbb{R}^{\ell_L} \mapsto \mathbb{R}^n$. Here $\ell_i \in \mathbb{N}$ for $i = 1, \dots, L$ represents the widths of the hidden layers, $L+1$ is the depth of the network, and θ is the vector with all parameters $(w_1, \dots, w_{L+1}, b_1, \dots, b_{L+1})$ in the affine transformations $T_k(x)$.

The depth of such a ReLU DNN is $L + 1$, the width is $\max\{\ell_1, \dots, \ell_L\}$ and the size is $\sum_{i=1}^L \ell_i$. L is the number of hidden layers.

A basic property of ReLU DNNs is that compositions and additions of ReLU DNNs yield a ReLU DNN. Indeed, we have the following lemma.

Lemma 16.12 ([**ABMM18**]). *The set of* ReLU *DNNs defined by Definition* 16.11 *is closed under the operations of addition and composition. Indeed,*

(1) *Let $\mathfrak{m}_1, \mathfrak{m}_2 : \mathbb{R}^d \mapsto \mathbb{R}^n$ be two* ReLU *DNNs with depth $L + 1$ and sizes s_1 and s_2, respectively. Then $\mathfrak{m}_1 + \mathfrak{m}_2$ is a* ReLU *DNN with depth $L + 1$ and size $s_1 + s_2$.*

(2) *Let* $\mathfrak{m}_1 : \mathbb{R}^d \mapsto \mathbb{R}^n$ *and* $\mathfrak{m}_2 : \mathbb{R}^m \mapsto \mathbb{R}^d$ *be two* ReLU *DNNs with depths* $L_1 + 1$ *and* $L_2 + 1$, *respectively, and sizes* s_1 *and* s_2, *respectively. Then, the composition function* $\mathfrak{m}_1 \circ \mathfrak{m}_2$ *is a* ReLU *DNN with depth* $L_1 + L_2 + 1$ *and size* $s_1 + s_2$.

Proof. Both parts follow directly by the structure of a ReLU DNN via Definition 16.11. The first part follows by combining the coordinates of the outputs while the second part follows by noting that composition of affine transformations is an affine transformation. □

Lemma 16.13 ([**ABMM18**]). *Let* $\mathfrak{m}_1, \dots, \mathfrak{m}_k : \mathbb{R}^d \mapsto \mathbb{R}^1$ *be* k ReLU *DNNs with depths* $L_i + 1$ *and sizes* s_i *for* $i = 1, \dots, k$, *respectively. Then* $\mathfrak{m} : \mathbb{R}^d \mapsto \mathbb{R}_1$ *defined as* $\mathfrak{m}(x) = \max\{\mathfrak{m}_1(x), \dots, \mathfrak{m}_k(x)\}$ *can be written as a* ReLU *DNN with depth at most* $\max\{L_1, \dots, L_k\} + \lceil \log_2(k) \rceil + 1$ *and size at most* $\sum_{i=1}^k s_i + 4(2k - 1)$.

Proof. The proof proceeds by induction. The case $k = 1$ is trivial (and note that a special case when $k = 2$ is shown in Exercise 16.4). For $k \geq 2$ define the functions $f_1 = \max\{\mathfrak{m}_1, \dots, \mathfrak{m}_{\lfloor \frac{k}{2} \rfloor}\}$ and $f_2 = \max\{\mathfrak{m}_{\lfloor \frac{k}{2} \rfloor + 1}, \dots, \mathfrak{m}_k\}$.

Let us first show the big-picture idea of the proof and then go into the more detailed computations. Define the vector-valued function $F : \mathbb{R}^d \mapsto \mathbb{R}^2$ by $F(x) = (f_1(x), f_2(x))$ and the function $T : \mathbb{R}^2 \mapsto \mathbb{R}$ by $T(x_1, x_2) = \max\{x_1, x_2\}$. By the second part of Lemma 16.12 we have that $\mathfrak{m} = T \circ F$ is indeed a ReLU DNN.

It remains to clarify the appropriate depth and size formulas. Let us go back to f_1, f_2. The induction hypothesis is that the claimed formulas for depth and size hold for all $m < k$, and we want to prove the claim for the given k. Note that both f_1 and f_2 represent a maximum of not more than $k_1 = \lfloor \frac{k}{2} \rfloor$ and $k_2 = \lceil \frac{k}{2} \rceil$ terms, respectively (both $k_1, k_2 < k$). So, due to the induction hypothesis,

- f_1 is a ReLU DNN with depth at most $\max\{L_1, \dots, L_{k_1}\} + \lceil \log_2(k_1) \rceil + 1$ and size at most $\sum_{i=1}^{k_1} s_i + 4(2k_1 - 1)$.
- f_2 is a ReLU DNN with depth at most $\max\{L_{k_1+1}, \dots, L_k\} + \lceil \log_2(k_2) \rceil + 1$ and size at most $\sum_{i=k_1+1}^{k} s_i + 4(2k_2 - 1)$.

This means that $F(x) = (f_1(x), f_2(x))$ can be written as a ReLU DNN with depth at most $\max\{L_1, \dots, L_k\} + \lceil \log_2(k_2) \rceil + 1$ and size at most $\sum_{i=1}^k s_i + 4(2k - 2)$. Next note that T can be written as a ReLU DNN with two layers and size 4. The proof now is concluded by using the second part of Lemma 16.12 for the formulas for depth and size for the composition $T \circ F$. □

Then, we have the following theorem.

Theorem 16.14 ([**ABMM18**]). *Any continuous piecewise linear function* $\mathbb{R}^d \mapsto \mathbb{R}$ *can be represented by a* ReLU *DNN of at most* $\lceil \log_2(d+1) \rceil + 1$ *depth, and any* ReLU *DNN* $\mathbb{R}^d \mapsto \mathbb{R}$ *represents a continuous piecewise linear function.*

Proof. The fact that any ReLU DNN $\mathbb{R}^d \mapsto \mathbb{R}$ represents a continuous piecewise linear function follows immediately by Definition 16.11. Indeed, the composition of continuous piecewise linear functions is a continuous piecewise linear function.

The converse now is trickier and it is based on Theorem 1 of [**WS05**], saying that every continuous piecewise linear function $f : \mathbb{R}^d \mapsto \mathbb{R}$ can be written in the form

$$f(x) = \sum_{j=1}^{p} s_j \max_{i \in S_j} g_i(x),$$

where $s_j \in \{-1, 1\}$, $g_1, \ldots, g_k$ are affine functions and subsets $S_j \subset \{1, \ldots, k\}$ for $k \in \mathbb{N}$ and $j = 1, \ldots, p$ (not necessarily disjoint) where each S_j is of maximum cardinality $d + 1$.

Then, we notice that the functions $\max_{i \in S_j} g_i$ are piecewise linear convex functions. In addition, each one of the functions $\max_{i \in S_j} g_i$ has at most $d + 1$ affine pieces. Hence, the formula for $f : \mathbb{R}^d \mapsto \mathbb{R}$ above says that any continuous piecewise linear function can be thought of as a linear combination of piecewise linear convex functions with at most $d + 1$ pieces.

Recall that $\{g_i\}$ are affine functions. By Lemma 16.13 each of the maximums $\max_{i \in S_j} g_i$ (each one with at most $d + 1$ terms) can be represented by a ReLU DNN with at most $\lceil \log_2(d+1) \rceil + 1$ depth. By Lemma 16.12 we have that additions of ReLU DNNs each one of depth $\lceil \log_2(d+1) \rceil + 1$ at most, can be represented by a ReLU DNN with at most depth $\lceil \log_2(d+1) \rceil + 1$, completing the proof of the theorem. □

Theorem 16.15 ([**ABMM18**]). *Let* $1 \leq q < \infty$. *Consider a function* $f \in L^q(\mathbb{R}^d)$, *i.e.,* f *is such that* $\|f\|_q = \left(\int_{\mathbb{R}^d} |f(x)|^q dx\right)^{1/q} < \infty$. *Then, there is a* ReLU *DNN with at most* $\lceil \log_2(d+1) \rceil + 1$ *hidden layers that approximates* f *in* L^q *to arbitrary accuracy.*

Proof. By classical density results (see for example [**RF10**]) the space of continuous piecewise linear functions is dense in $L^q(\mathbb{R}^d)$ for any $1 \leq q < \infty$. This means that given $f \in L^q(\mathbb{R}^d)$ and some given $\epsilon > 0$ there is a continuous piecewise linear function, say $h : \mathbb{R}^d \mapsto \mathbb{R}$, such that

$$\|f(x) - h(x)\|_q \leq \epsilon.$$

But, by Theorem 16.14 we have that any continuous piecewise linear function $\mathbb{R}^d \mapsto \mathbb{R}$ can be represented by a ReLU DNN of at most $\lceil \log_2(d+1) \rceil + 1$ depth. This concludes the proof. □

We conclude this section with a result by [**Han19**] that showcases aspects of the tradeoffs between deep-and-narrow ReLU DNN and shallow-and-wide ReLU neural networks.

Theorem 16.16 ([**Han19**]). *Consider a* ReLU *neural network* $\mathfrak{m} : [0,1]^d \mapsto \mathbb{R}$ *with input dimension d, output dimension* 1*, and with a single layer of width n. Then, there exists another* ReLU *DNN defined as in Definition* 16.11 *which computes the same function and has input dimension d, output dimension* 1*,* $n+2$ *hidden layers, each one with width* $d+2$.

This result is interesting especially in the case of $n > d$. It basically says that the width goes from n to $d + 2$ but at the expense of an increase in the number of hidden layers from 1 to $n + 2$.

Proof. Let T_k for $k = 1, \dots, n$ be the affine functions computed by the hidden neurons in the single layer of f. Namely, we set $T_k(x) = \sum_{i=1}^d w^{ik} x_i + b^k$. The neural network $\mathfrak{m}(x)$ (we ignore for notational convenience to explicitly denote the dependence on the parameter θ) can be represented as

$$\mathfrak{m}(x) = \text{ReLU}\left(b + \sum_{k=1}^{n} c^k \text{ReLU}(T_k(x))\right).$$

By continuity and since $[0,1]^d$ is a compact set, there is $\Gamma > 0$ large enough so that for all $k = 1, \dots, n$ and $x \in [0,1]^d$,

$$\Gamma + \sum_{i=1}^{k} c^i \,\text{ReLU}(T_i(x)) > 0.$$

Consider now the affine transformations

$$\begin{aligned}
\tilde{T}_1(x) &= (x, T_1(x), \Gamma),\\
\tilde{T}_{n+2}(x,y,z) &= z - \Gamma + b,\\
\tilde{T}_j(x,y,z) &= (x, T_j(x), z + c^{j-1} y), \quad j = 2, \dots, n+1.
\end{aligned}$$

These are the affine transformations that will define the new ReLU DNN per Definition 16.11. Let us confirm this. For the kth layer ($k \le n+1$), the activation is

$$\begin{aligned}
\tilde{\mathfrak{m}}_{(k)}(x) &= (\text{ReLU} \circ \tilde{T}_k \circ \cdots \circ \tilde{T}_2 \circ \text{ReLU} \circ \tilde{T}_1)(x)\\
&= \left(x, \text{ReLU}(T_k(x)), \Gamma + \sum_{i=1}^{k-1} c^i \text{ReLU}(T_i(x))\right).
\end{aligned}$$

So, for the last layer we shall have

$$\begin{aligned}\tilde{\mathfrak{m}}_{(n+2)}(x) &= \mathsf{ReLU} \circ \tilde{T}_{n+2} \circ (\tilde{\mathfrak{m}}_{(n+1)})(x) \\ &= \mathsf{ReLU} \circ \tilde{T}_{n+2}\left(x, \mathsf{ReLU}(T_{n+1}(x)), \Gamma + \sum_{i=1}^{n} c^i \mathsf{ReLU}(T_i(x))\right) \\ &= \mathsf{ReLU}(\Gamma + \sum_{i=1}^{n} c^i \mathsf{ReLU}(T_i(x)) - \Gamma + b) \\ &= \mathfrak{m}(x),\end{aligned}$$

which concludes the proof. □

Smooth functions can approximate to any given accuracy positive continuous functions and therefore the same is true for ReLU DNNs with a single hidden layer, see for example [**MP16**] for a related discussion. With that in mind Theorem 16.16 directly gives rise to the following result.

Theorem 16.17. *Consider $f : [0,1]^d \mapsto \mathbb{R}_+$ to be a positive and bounded continuous function. Then, there exists a* ReLU *DNN with input dimension d, output dimension* 1*, and width of hidden layer being $d + 2$ that approximates f to arbitrary accuracy.*

We conclude this section with a result showing that for positive, continuous, piecewise linear, convex functions f, the width upper bound is $d + 1$.

Theorem 16.18 ([**Han19**]). *Let $f : [0,1]^d \mapsto \mathbb{R}_+$ be the function computed by a* ReLU *neural network with arbitrarily given width. Assume in addition that f is convex. Then, there are positive affine functions $g_i : [0,1]^d \mapsto \mathbb{R}$ such that we can write*

$$f(x) = g(x) = \max_{1 \le i \le N} g_i(x),$$

where g is a positive convex function. In addition, there exists a feed forward ReLU *DNN $\mathfrak{m} : [0,1]^d \mapsto \mathbb{R}_+$ with hidden layers width $d + 1$ and depth N that computes f exactly.*

Proof. The representation of f as the maximum of positive affine functions follows by Theorem 16.14.

We want to show that f can be computed by a ReLU DNN that has hidden-layer width $d + 1$ and depth N. For $x = (x_1, \dots, x_d) \in \mathbb{R}^d$, $x_{d+1} \in \mathbb{R}^1$, and $i = 1, \dots, N$, let us define

$$T_i : \mathbb{R}^{d+1} \mapsto \mathbb{R}^{d+1}, \text{ with } T_i(x, x_{d+1}) = (x, g_i(x) + x_{d+1})$$

and

$$\tilde{T}_i : \mathbb{R}^{d+1} \mapsto \mathbb{R}^{d+1}, \text{ with } \tilde{T}_i(x, x_{d+1}) = (x, -g_i(x) + x_{d+1}).$$

Next we compute that for $x \in \mathbb{R}_+^d$ and for any function $\kappa : \mathbb{R}_+^d \mapsto \mathbb{R}$,

$$\begin{aligned} T_i \circ \mathsf{ReLU} \circ \tilde{T}_i(x, \kappa(x)) &= T_i(x, \max\{\kappa(x) - g_i(x), 0\}) \\ &= (x, g_i(x) + \max\{\kappa(x) - g_i(x), 0\}) \\ &= (x, \max\{\kappa(x), g_i(x)\}). \end{aligned}$$

Defining now $\mathcal{G}_i = T_i \circ \mathsf{ReLU} \circ \tilde{T}_i$, the latter relation implies that under the mapping $\mathcal{G}_i$, the graph of $x \mapsto \kappa(x)$ is the graph of $x \mapsto \max\{\kappa(x), g_i(x)\}$ when viewed as function on $\mathbb{R}_+^d$.

Now that we defined $\mathcal{G}_1, \ldots, \mathcal{G}_N$, let us also set

$$\begin{aligned} \mathcal{G}_0(x) &= \mathsf{ReLU}(x, 0), \\ \mathcal{G}_{N+1}(x, x_{d+1}) &= \mathsf{ReLU}(x_{d+1}). \end{aligned}$$

After these definitions, we are ready to construct the ReLU DNN representation of $f(x) = g(x) = \max_{1 \le i \le N} g_i(x)$. In particular, we see that we can write

$$\max_{1 \le i \le N} g_i(x) = (\mathcal{G}_{N+1} \circ \mathcal{G}_N \circ \cdots \circ \mathcal{G}_1 \circ \mathcal{G}_0)(x),$$

which has input dimension d, hidden layer width $d + 1$, and depth N. □

16.4. Brief Concluding Remarks

In this chapter we presented the main uniform approximation theorems. The main references here are [**Cyb89, HSW89, Hor91, KH91**] for classical results and [**GWFM$^+$13, ABMM18, Han19, Yar17, SH17**] for some more recent developments using rectified liner units (ReLU) as activation functions. The ability of ReLU DNNs to represent continuous piecewise linear functions and related uniform approximation results was observed in [**GWFM$^+$13**]. Later on, [**ABMM18**] improved upon those results with an upper bound on the depth of such ReLU DNNs. [**Han19**] found width and depth upper bounds for ReLU DNN representations of positive continuous piecewise linear functions. In [**Yar17**] it is shown that deep ReLU networks can have advantages when it comes to approximation of smooth functions compared to shallow neural networks. In [**SH17**] the author shows that the depth (number of layers) of the neural network architecture is important when it comes to ReLU activation functions. It is shown in [**SH17**] that, for any network architecture satisfying a certain condition, one can obtain good approximation rates. See also [**Cal20**] for a more extensive exposition to universal approximation theorems. The proofs in Section 16.2 are based on [**Cyb89**] and [**HSW89**]. The proofs in Section 16.3 are based on [**ABMM18**] and [**Han19**].

16.5. Exercises

Exercise 16.1. Prove that the logistic function $\sigma(x) = \frac{e^x}{1+e^x}$ is discriminatory in the L^2-sense.

Exercise 16.2. Consider the Heaviside function,

$$H(x) = \begin{cases} 1, & x \geq 0 \\ 0, & x < 0. \end{cases}$$

(1) Show that the function $H(x)$ is discriminatory in the L^2-sense.

(2) Show that any function $f \in L^2(I_d)$ can be approximated by a one-layer perceptron model of the form $\mathfrak{m}(x;\theta) = \sum_{n=1}^{N} c^n H(w^n \cdot x + b^n)$ with $\theta = \{(c^n, w^n, b^n)_{n=1}^N\}$ for N large enough.

Exercise 16.3. Let f be a continuous function $f : [a,b] \mapsto \mathbb{R}$. Prove that for every $\epsilon > 0$ there is an equidistant partition $a = x_0 < x_1 < \cdots < x_N = b$ such that a piecewise linear function f^* that passes through the points $\{x_n, f(x_n)\}_{n=1}^N$ satisfies

$$\sup_{x\in[a,b]} |f(x) - f^*(x)| < \epsilon.$$

Exercise 16.4. Consider the function $f(x_1,x_2) = \max\{x_1,x_2\}$. By using the representation $\max\{x_1,x_2\} = \frac{x_1+x_2}{2} + \frac{|x_1-x_2|}{2}$, show in a constructive manner that there are $p \in \mathbb{N}$, $c^i, w^{1,i}, w^{2,i} \in \mathbb{R}$ so that

$$f(x_1,x_2) = \sum_{i=1}^{p} c^i \,\mathsf{ReLU}(w^{1,i}x_1 + w^{2,i}x_2).$$

Exercise 16.5. Prove that the function $f(x_1,x_2) = \max\{x_1,x_2\}$, with $x_1 \in \mathbb{R}$ and $x_2 \geq 0$, is implementable by a ReLU DNN that has hidden layer width 2, depth 2 and output dimension 1. Namely, identify the linear transformations T_1, T_2, T_3 so that f can be written in the form (16.1) and establish that identity.

Exercise 16.6. Consider a one-layer neural network with ReLU activation function,

$$\mathfrak{m}(x;\theta) = \sum_{n=1}^{N} c^n \mathsf{ReLU}(x + \theta^n) + b.$$

Let $\bar{\mathfrak{m}} \in C([a,b];\mathbb{R})$ be a target function. Show that for every $\epsilon > 0$, there exist c^n, θ^n and $N \geq 1$ such that

$$\sup_{x\in[a,b]} |\mathfrak{m}(x;\theta) - \bar{\mathfrak{m}}(x)| < \epsilon.$$

Exercise 16.7. Let $y = \bar{\mathfrak{m}}(x)$. The universal approximation theorem states that for every $\epsilon > 0$ there exist a neural network $\mathfrak{m}(x;\theta)$ and a parameter choice θ^* such that $\mathbb{E}_{(X,Y)}[\|Y - \mathfrak{m}(X;\theta^*)\|] < \epsilon$. If we estimate θ using stochastic gradient descent, will it converge to θ^*, and why?

Chapter 17

Convergence Analysis of Gradient Descent

17.1. Introduction

As we have seen in the preceding chapters, we must search over classes of deep neural networks to find parameters which best fit our ground-truth data. A *loss* function helps us compare choices of parameters and (informally) leads to the best parameters. *Gradient descent algorithms*, which (informally, again) iteratively improve upon parameter choices, provide the mathematical underpinning of numerical methods to minimize the loss function.

In this chapter we present the classical theory of gradient descent-type algorithms. In particular, we will quantify the convergence and performance of gradient descent optimization. This will allow us to understand tradeoffs between various types of gradient descent algorithms.

Our goal in this chapter is to present the main results with an eye towards the developments in Chapter 18 where we study convergence properties of stochastic gradient descent, which is what is typically being used in deep learning algorithms.

The results of this chapter are *informative*. In reality, loss functions are oftentimes unknown and very high-dimensional and may have degeneracies. In Section 17.2 we discuss the gradient flow and convergence properties under convexity assumptions. Convergence results in the nonconvex case are discussed in Section 17.3. Accelerated gradient descent methods such as Polyak's and Nesterov's methods are discussed in Section 17.4.

17.2. Convergence Properties under Convexity Assumptions

Consider a loss function $\Lambda \in \mathcal{C}^1$ and the ordinary differential equation

$$\dot{\theta} = -\nabla\Lambda(\theta), \quad \theta(0) = \theta_0. \tag{17.1}$$

An easy calculation shows that

$$\frac{d}{dt}\Lambda(\theta(t)) = \dot{\theta}(t)\nabla\Lambda(\theta(t)) = -\left\|\nabla\Lambda(\theta(t))\right\|_2^2 \leq 0.$$

Thus, we have that $\frac{d}{dt}\Lambda(\theta(t)) \leq 0$ which means that as t increases, $\Lambda(\theta(t))$ decreases. This is a good thing! The relation (17.1) is gradient flow and Λ acts as a Lyapunov function for the dynamical system (17.1). However, the fact that $\frac{d}{dt}\Lambda(\theta(t)) \leq 0$ does not guarantee convergence. It only says that it is non-increasing.

Now let us discretize (17.1) to get

$$\theta_{k+1} = \theta_k - \eta\nabla\Lambda(\theta_k), \tag{17.2}$$

where η is the discretization step of the gradient flow and it is also called the *learning rate*. It can also be viewed as the timestep size in a discretization of the gradient flow. In fact, relation (17.2) is called gradient descent (GD). If we know the structure of Λ, generally speaking, a good choice for the learning rate η is one that does not overshoot the minimum θ^*, but we will come back to that point shortly.

17.2.1. Convexity and Convergence Criteria.

Consider a loss function $\Lambda : \mathbb{R}^d \mapsto \mathbb{R}$, $\alpha \in [0,1]$ and $\theta, \theta' \in \mathbb{R}^d$. Let's start with some definitions.

Definition 17.1 (Convexity). We say that the function Λ is convex if

$$\Lambda(\alpha\theta + (1-\alpha)\theta') \leq \alpha\Lambda(\theta) + (1-\alpha)\Lambda(\theta'),$$

with $\alpha \in (0,1)$, or equivalently if $\Lambda \in \mathcal{C}^1$

$$\Lambda(\theta') \geq \Lambda(\theta) + \langle\nabla\Lambda(\theta), \theta' - \theta\rangle.$$

Note that the second inequality above can be derived from the first inequality by rearranging the terms and letting $\alpha \to 0$.

Definition 17.2 (Strict convexity). We say that the function Λ is strictly convex if

$$\Lambda(\alpha\theta + (1-\alpha)\theta') < \alpha\Lambda(\theta) + (1-\alpha)\Lambda(\theta'),$$

with $\alpha \in (0,1)$, or equivalently if $\Lambda \in \mathcal{C}^1$

$$\Lambda(\theta') > \Lambda(\theta) + \langle\nabla\Lambda(\theta), \theta' - \theta\rangle.$$

Definition 17.3 (Strong convexity). We say that the function Λ is strongly convex if there exists $0 < \gamma < \infty$ such that $\Lambda(\theta) - \frac{\gamma}{2}\|\theta\|_2^2$ is convex. Oftentimes, we shall call refer to this as γ-strong convexity.

Let us now discuss some facts and conditions related to optimality.

- We say that θ is a local minimum if for every θ' in the neighborhood of θ, $\Lambda(\theta) \leq \Lambda(\theta')$.
- Local minima are global minima for convex functions.
- Global minimum is unique for strictly convex functions. Indeed, let θ, θ' be two candidate global minima, and apply the definition of strict convexity to $\theta'' = \frac{\theta+\theta'}{2}$ to get $\Lambda(\theta'') < \frac{1}{2}(\Lambda(\theta) + \Lambda(\theta')) = \Lambda(\theta)$. This immediately gives us a contradiction.
- If θ is a local minimum and $\theta \mapsto \Lambda(\theta)$ is once differentiable, then $\nabla\Lambda(\theta) = 0$. The latter condition is sufficient for a global minimum if Λ is convex.

Remark 17.4. Note that if Λ is convex the following hold for all θ, θ':

$$\begin{aligned}\Lambda(\theta') &\geq \Lambda(\theta) + \langle \nabla\Lambda(\theta), \theta' - \theta\rangle\,,\\ \Lambda(\theta) &\geq \Lambda(\theta') + \langle \nabla\Lambda(\theta'), \theta - \theta'\rangle\,.\end{aligned}$$

By adding these two expressions we get that

$$\langle \nabla\Lambda(\theta) - \nabla\Lambda(\theta'), \theta - \theta'\rangle \geq 0.$$

This is called monotonicity of the gradients and it says that the gradient of Λ and θ change in the same direction.

17.2.2. Newton's Method. Let $\Lambda \in \mathcal{C}^2$ and let θ_0 be an initial guess of a minimizer. Assume that Λ is convex around θ_0. Using a Taylor series expansion, we have

$$\Lambda(\theta) \approx \Lambda(\theta_0) + (\theta - \theta_0)\nabla\Lambda(\theta_0) + \frac{1}{2}(\theta - \theta_0)^2\nabla^2\Lambda(\theta_0)(\theta - \theta_0),$$

where the third-order term has been ignored.

At the same time, we also have by a Taylor series expansion again (ignoring the error term),

$$\nabla\Lambda(\theta) \approx \nabla\Lambda(\theta_0) + \nabla^2\Lambda(\theta_0)(\theta - \theta_0). \tag{17.3}$$

Recall that at a minimizer θ^* of the objective function the gradient $\nabla\Lambda(\theta^*) = 0$. Then, substituting θ^* for θ in the equation above yields

$$\nabla^2\Lambda(\theta_0)\theta^* \approx -\nabla\Lambda(\theta_0) + \nabla^2\Lambda(\theta_0)\theta_0.$$

If in addition, we have that the matrix $\nabla^2\Lambda(\theta_0)$ is invertible (i.e., it does not have zero eigenvalues), then

$$\theta^* = \theta_0 - \left(\nabla^2\Lambda(\theta_0)\right)^{-1} \nabla\Lambda(\theta_0)$$

gives the minimum of the quadratic approximation (17.3) to $\Lambda(\theta)$. This observation motivates the update,

$$\theta_{k+1} = \theta_k - \left(\nabla^2\Lambda(\theta_k)\right)^{-1} \nabla\Lambda(\theta_k),$$

which is called Newton's method.

Newton's method converges to the minimum of Λ faster than gradient descent does, but it requires the computation of the Hessian at each iteration, which can be a very costly step.

17.2.3. Convergence Rate Results for Gradient Descent. Let us now try to answer some basic questions:

- How quickly does gradient descent converge?
- Can we pick learning rates without overshooting?
- How many iterations do we need in order to be within some given distance of the minimum?

We visited some of these questions to a certain extent in the case of logistic regression in Chapter 3 for original versus normalized data. In this section we will consider these questions in a more general context. To answer these questions, we first impose certain assumptions to at least be able to discuss those questions in stylized settings. We assume that the parameter space is a Euclidean space, $\theta \in \Theta$ with $\Theta = \mathbb{R}^d$.

Assumption 17.5. We assume that the loss function Λ is $L_\circ$-Lipschitz in the sense that for all $\theta, \theta' \in \Theta$,

$$|\Lambda(\theta) - \Lambda(\theta')| \leq L_\circ \|\theta - \theta'\|_2.$$

Assumption 17.6. We assume that the gradient of the loss function $\nabla\Lambda$ is L-Lipschitz in the sense that for all $\theta, \theta' \in \Theta$,

$$\|\nabla\Lambda(\theta) - \nabla\Lambda(\theta')\|_2 \leq L\|\theta - \theta'\|_2.$$

Notice that Assumption 17.6 implies that

$$\langle \nabla\Lambda(\theta) - \nabla\Lambda(\theta'), \theta - \theta' \rangle \leq L\|\theta - \theta'\|_2^2.$$

Convex optimization and convergence of gradient descent for convex problems is a classical topic in the literature, see for example [**Ber03, Nes04, Nes07**]. Below we present some of the main results of the literature, building towards the convergence results for stochastic gradient descent, which are discussed in Chapter 18. In this section we shall see two things:

- In Lemma 17.11 we show that for gradient descent under Assumption 17.6 with learning rate sufficiently small compared to the Lipshcitz constant ($\eta L \leq 1$), we have that
$$\Lambda(\theta_k) - \Lambda(\theta^*) \leq \frac{1}{2\eta k}\|\theta_0 - \theta^*\|_2^2,$$
where θ^* is a global minimum of the loss function $\Lambda(\cdot)$. This result means that gradient descent will result in the loss function converging to its minimum value as the number of iterations $k \to \infty$.
- If in addition, we assume that the loss function Λ is strongly convex (Definition 17.3), then we get a rate of convergence. In particular, as we shall see in Lemma 17.14, if the learning rate is even smaller ($\eta < \frac{2}{L+\gamma}$), then for $\lambda = 1 - \eta\frac{2\gamma L}{L+\gamma} < 1$, we have that
$$\|\theta_k - \theta^*\|_2^2 \leq \lambda^k \|\theta_0 - \theta^*\|_2^2.$$

We first present a few preliminary results that will naturally lead to these conclusions.

Lemma 17.7. *Under Assumption* 17.6 *we have for all* $\theta, \theta' \in \Theta$,
$$|\Lambda(\theta') - \Lambda(\theta) - \langle \nabla\Lambda(\theta), \theta' - \theta\rangle| \leq \frac{L}{2}\|\theta' - \theta\|_2^2.$$

In particular, we have that $\Lambda(\theta') - \Lambda(\theta) \leq \langle \nabla\Lambda(\theta), \theta' - \theta\rangle + \frac{L}{2}\|\theta' - \theta\|_2^2$.

Proof. Using Taylor's theorem, we can calculate that
$$\Lambda(\theta') - \Lambda(\theta) = \int_0^1 \langle \nabla\Lambda(\theta + \rho(\theta' - \theta)), \theta' - \theta\rangle \, d\rho.$$

Subtracting $\langle \nabla\Lambda(\theta), \theta' - \theta\rangle$ from both terms yields
$$\Lambda(\theta') - \Lambda(\theta) - \langle \nabla\Lambda(\theta), \theta' - \theta\rangle = \int_0^1 \langle \nabla\Lambda(\theta + \rho(\theta' - \theta)) - \nabla\Lambda(\theta), \theta' - \theta\rangle \, d\rho.$$

Therefore, we obtain
$$\begin{aligned}
&|\Lambda(\theta') - \Lambda(\theta) - \langle \nabla\Lambda(\theta), \theta' - \theta\rangle| \\
&\leq \int_0^1 \|\nabla\Lambda(\theta + \rho(\theta' - \theta)) - \nabla\Lambda(\theta)\|_2 \, \|\theta' - \theta\|_2 \, d\rho \\
&\leq L\left(\int_0^1 \rho d\rho\right)\|\theta' - \theta\|_2^2 \\
&= \frac{L}{2}\|\theta' - \theta\|_2^2,
\end{aligned}$$
completing the proof. □

Lemma 17.8. *Let Assumption* 17.6 *hold. Consider the setting of gradient descent and choose a learning rate* η *such that* $\eta L \leq 1$. *Then, we have*

$$\Lambda(\theta_{k+1}) - \Lambda(\theta_k) \leq -\frac{\eta}{2}\|\nabla\Lambda(\theta_k)\|_2^2.$$

Proof. By Lemma 17.7 we have that

$$\Lambda(\theta') - \Lambda(\theta) \leq \langle \nabla\Lambda(\theta), \theta' - \theta\rangle + \frac{L}{2}\|\theta' - \theta\|_2^2.$$

Hence, using the update equation $\theta_{k+1} = \theta_k - \eta\nabla\Lambda(\theta_k)$ from the GD algorithm, we get

$$\begin{aligned}
\Lambda(\theta_{k+1}) - \Lambda(\theta_k) &\leq \langle \nabla\Lambda(\theta_k), \theta_{k+1} - \theta_k\rangle + \frac{L}{2}\|\theta_{k+1} - \theta_k\|_2^2 \\
&= -\eta\|\nabla\Lambda(\theta_k)\|_2^2 + \frac{L\eta^2}{2}\|\nabla\Lambda(\theta_k)\|_2^2 \\
&\leq -\eta\|\nabla\Lambda(\theta_k)\|_2^2 + \frac{\eta}{2}\|\nabla\Lambda(\theta_k)\|_2^2 \\
&= -\frac{\eta}{2}\|\nabla\Lambda(\theta_k)\|_2^2,
\end{aligned}$$

where we used the assumption $\eta L \leq 1$ to derive the third line. □

Lemma 17.9. *Let Assumption* 17.6 *hold and let* θ^* *be the global minimum of the convex loss function* $\Lambda(\cdot)$. *Consider the setting of gradient descent and choose learning rate* η *such that* $\eta L \leq 1$. *Then, we have*

$$\Lambda(\theta_{k+1}) - \Lambda(\theta^*) \leq \frac{1}{2\eta}\left(\|\theta_k - \theta^*\|_2^2 - \|\theta_{k+1} - \theta^*\|_2^2\right).$$

Proof. We begin by expanding the square

$$\|\theta_k - \theta^* - \eta\nabla\Lambda(\theta_k)\|_2^2 = \|\theta_k - \theta^*\|_2^2 + \|\eta\nabla\Lambda(\theta_k)\|_2^2 - 2\langle \eta\nabla\Lambda(\theta_k), \theta_k - \theta^*\rangle,$$

which then leads to the identity

$$\langle \nabla\Lambda(\theta_k), \theta_k - \theta^*\rangle = -\frac{1}{2\eta}\|\theta_k - \theta^* - \eta\nabla\Lambda(\theta_k)\|_2^2 + \frac{1}{2\eta}\|\theta_k - \theta^*\|^2 + \frac{\eta}{2}\|\nabla\Lambda(\theta_k)\|_2^2.$$

Then, Lemma 17.8 (which uses the assumption $\eta L \le 1$) and the aforementioned formula for $\langle \nabla\Lambda(\theta_k), \theta_k - \theta^* \rangle$ yield

$$\begin{aligned}
\Lambda(\theta_{k+1}) - \Lambda(\theta^*) &\le \Lambda(\theta_k) - \Lambda(\theta^*) - \frac{\eta}{2}\|\nabla\Lambda(\theta_k)\|_2^2 \\
&\le \langle \nabla\Lambda(\theta_k), \theta_k - \theta^* \rangle - \frac{\eta}{2}\|\nabla\Lambda(\theta_k)\|_2^2 \\
&= -\frac{1}{2\eta}\|\theta_k - \theta^* - \eta\nabla\Lambda(\theta_k)\|_2^2 + \frac{1}{2\eta}\|\theta_k - \theta^*\|_2^2 \\
&\quad + \frac{\eta}{2}\|\nabla\Lambda(\theta_k)\|_2^2 - \frac{\eta}{2}\|\nabla\Lambda(\theta_k)\|_2^2 \\
&= -\frac{1}{2\eta}\|\theta_{k+1} - \theta^*\|_2^2 + \frac{1}{2\eta}\|\theta_k - \theta^*\|_2^2,
\end{aligned}$$

completing the proof of the lemma. Note that the second line above uses the definition of convexity and the last line uses the gradient descent equation $\theta_{k+1} = \theta_k - \eta\nabla\Lambda(\theta_k)$. □

This leads to Lemmas 17.10 and 17.11 which demonstrate that gradient descent makes progress towards the minimum.

Lemma 17.10. *Let Assumption* 17.6 *hold and let θ^* be global minimum of the convex loss function $\Lambda(\cdot)$. Consider the setting of gradient descent and pick a learning rate η such that $\eta L \le 1$. Then, we have that*

$$\|\theta_{k+1} - \theta^*\|_2^2 \le \|\theta_k - \theta^*\|_2^2.$$

Proof. The proof follows directly by Lemma 17.9 because, since θ^* is the global minimizer, $\Lambda(\theta_{k+1}) - \Lambda(\theta^*) \ge 0$. □

Lemma 17.11. *Let Assumption* 17.6 *hold. Consider the gradient descent update equation $\theta_{k+1} = \theta_k - \eta\nabla\Lambda(\theta_k)$ with learning rate η such that $\eta L \le 1$. Then, we have that*

$$\Lambda(\theta_k) - \Lambda(\theta^*) \le \frac{1}{2\eta k}\|\theta_0 - \theta^*\|_2^2.$$

Proof. By Lemma 17.9 we have that

$$\Lambda(\theta_i) - \Lambda(\theta^*) \le \frac{1}{2\eta}\left(\|\theta_{i-1} - \theta^*\|_2^2 - \|\theta_i - \theta^*\|_2^2\right).$$

Averaging over $i \in \{1, \ldots, k\}$, we then obtain

$$\begin{aligned}
\frac{1}{k}\sum_{i=1}^{k}\Lambda(\theta_i) - \Lambda(\theta^*) &\le \frac{1}{2\eta k}\sum_{i=1}^{k}\left(\|\theta_{i-1} - \theta^*\|_2^2 - \|\theta_i - \theta^*\|_2^2\right) \\
&\le \frac{1}{2\eta k}\|\theta_0 - \theta^*\|_2^2,
\end{aligned}$$

where we have used a telescoping series and the fact that the final term $-\|\theta_k - \theta^*\|_2^2 \leq 0$.

Since $\Lambda(\theta_k) \leq \Lambda(\theta_i)$ for all $i \in \{1, \ldots, k\}$ (see Lemma 17.8) we have

$$\Lambda(\theta_k) \leq \frac{1}{k}\sum_{i=1}^{k} \Lambda(\theta_i).$$

Therefore, we obtain

$$\Lambda(\theta_k) - \Lambda(\theta^*) \leq \frac{1}{2\eta k}\|\theta_0 - \theta^*\|_2^2,$$

completing the proof of the lemma. □

Remark 17.12. The conclusion from Lemma 17.11 is that to find θ^* so that $\Lambda(\theta_k) - \Lambda(\theta^*) \leq \epsilon$ for a convex loss function Λ, then we need $\mathcal{O}(1/\epsilon)$ steps.

Lemma 17.13. *Let Assumption* 17.6 *hold. Assume that Λ is strongly convex, i.e., there exists $\gamma > 0$ so that $\Lambda(\theta) - \frac{\gamma}{2}\|\theta\|_2^2$ is convex. Let the learning rate η be chosen such that $\eta < \frac{2}{L+\gamma}$. Then, we have that*

$$\|\theta_{k+1} - \theta^*\|_2^2 \leq \left(1 - \eta\frac{2\gamma L}{L+\gamma}\right)\|\theta_k - \theta^*\|_2^2.$$

Proof. Using the algorithm $\theta_{k+1} = \theta_k - \eta\nabla\Lambda(\theta_k)$ and expanding the square, we get

$$\begin{aligned}\|\theta_{k+1} - \theta^*\|_2^2 &= \|\theta_k - \eta\nabla\Lambda(\theta_k) - \theta^*\|_2^2 \\ &= \|\theta_k - \theta^*\|_2^2 + \eta^2\|\nabla\Lambda(\theta_k)\|_2^2 - 2\eta\left\langle\nabla\Lambda(\theta_k), \theta_k - \theta^*\right\rangle.\end{aligned}$$

For the last term, we use the definition of strong convexity of Λ to get that

$$\langle\nabla\Lambda(\theta_k), \theta_k - \theta^*\rangle \geq \frac{1}{L+\gamma}\|\nabla\Lambda(\theta_k)\|_2^2 + \gamma\frac{L}{L+\gamma}\|\theta_k - \theta^*\|_2^2.$$

Hence, we can continue the expression for $\|\theta_{k+1} - \theta^*\|_2^2$ to get

$$\begin{aligned}\|\theta_{k+1} - \theta^*\|_2^2 &\leq \left(1 - 2\eta\gamma\frac{L}{L+\gamma}\right)\|\theta_k - \theta^*\|_2^2 + \eta\left(\eta - \frac{2}{L+\gamma}\right)\|\nabla\Lambda(\theta_k)\|_2^2 \\ &\leq \left(1 - \eta\frac{2\gamma L}{L+\gamma}\right)\|\theta_k - \theta^*\|_2^2,\end{aligned}$$

where to get the last inequality we used the assumption $\eta - \frac{2}{L+\gamma} < 0$. This concludes the proof of the lemma. □

Lemma 17.14. *Let Assumption* 17.6 *hold and assume that Λ is strongly convex.*
Let $\eta < \frac{2}{L+\gamma}$ and define $\lambda = 1 - \eta\frac{2\gamma L}{L+\gamma} < 1$. Then, we have that

$$\|\theta_k - \theta^*\|_2^2 \leq \lambda^k\|\theta_0 - \theta^*\|_2^2.$$

Proof. It follows by iteratively applying Lemma 17.13. □

Remark 17.15. Note that strong convexity yields faster convergence than plain convexity. Let us set $\eta = \frac{2}{L+\gamma}$. Using the inequality

$$\lambda^k = \left(1 - \eta\frac{2\gamma L}{L+\gamma}\right)^k = \left(1 - 4\frac{\gamma L}{(L+\gamma)^2}\right)^k \leq e^{-4k\frac{\gamma L}{(L+\gamma)^2}},$$

Lemma 17.14 yields the bound

$$\|\theta_k - \theta^*\|_2^2 \leq e^{-4k\frac{\gamma L}{(L+\gamma)^2}} \|\theta_0 - \theta^*\|_2^2.$$

Suppose we want error ϵ. Then, with convexity we will need $\mathcal{O}(1/\epsilon)$ steps. On the other hand with strong convexity we will need $\mathcal{O}(\log(1/\epsilon))$ steps.

Also, if we again consider $\eta = \frac{2}{L+\gamma}$, then going back to Lemma 17.14, we have that

$$\lambda = 1 - \eta\frac{2\gamma L}{L+\gamma} = \left(\frac{\frac{L}{\gamma} - 1}{\frac{L}{\gamma} + 1}\right)^2.$$

Note that λ is a decreasing function of $\frac{L}{\gamma}$. Hence the convergence is faster when $\frac{L}{\gamma}$ is small. Large $\frac{L}{\gamma}$ means that some directions of the loss function Λ are highly curved whereas others are flat. So picking a small scalar η would not work equally well for all regions.

If step size η is too large, then the algorithm will overshoot in highly curved regions. If, on the other hand, η is too small, the progress of gradient descent will be slow.

All in all, the number $\frac{L}{\gamma}$ is important! We may also recognize the parameter $\frac{L}{\gamma} = \frac{\text{largest eigenvalue}}{\text{smallest eigenvalue}} > 1$ as the condition number for the Hessian of the loss function Λ.

17.3. Convergence in the Absence of Convexity Assumptions

Let us now return to the gradient descent algorithm (17.2)

$$\theta_{k+1} = \theta_k - \eta\nabla\Lambda(\theta_k)$$

and study convergence properties of gradient descent when $\Lambda(\theta)$ is nonconvex. As we shall see, we will need to choose the learning rate η to decrease in a specific way. In particular, the learning rate must satisfy the conditions $\sum_{k=1}^{\infty} \eta_k = \infty$ and $\sum_{k=1}^{\infty} \eta_k^2 < \infty$ (see [**BT00**]). This turns out to be a good learning schedule, and we will return to this in Section 18.3.3 where we will study stochastic gradient descent. We note that an example of a learning rate satisfying these conditions is $\eta_k = \frac{C_0}{C_1+C_2 k}$ for finite constants $0 < C_0, C_1, C_2 < \infty$.

We shall prove that with those choices for learning rates and, if we further assume that Assumption 17.6 holds and that there is some $L_1 < \infty$ such that $\|\nabla\Lambda(\theta)\|_2 \leq L_1$, then

$$\lim_{k\to\infty} \Lambda(\theta_k) \text{ exists, and that}$$
$$\lim_{k\to\infty} \|\nabla\Lambda(\theta_k)\|_2 = 0.$$

Note that we are not claiming that $\lim_{k\to\infty}\theta_k$ exists. Building towards the aforementioned result, we first write

$$\begin{aligned}\Lambda(\theta_{k+1}) &= \Lambda(\theta_k) - \eta_k \int_0^1 \langle \nabla\Lambda(\theta_k - s\eta_k \nabla\Lambda(\theta_k)), \nabla\Lambda(\theta_k)\rangle\, ds\\ &= \Lambda(\theta_k) - \eta_k \|\nabla\Lambda(\theta_k)\|_2^2\\ &\quad - \eta_k \int_0^1 \langle \nabla\Lambda(\theta_k - s\eta_k \nabla\Lambda(\theta_k)) - \Lambda(\theta_k), \nabla\Lambda(\theta_k)\rangle\, ds.\end{aligned}$$

Then, if we set $\xi_{k+1} = -\eta_k \int_0^1 \langle \nabla\Lambda(\theta_k - s\eta_k\nabla\Lambda(\theta_k)) - \Lambda(\theta_k), \nabla\Lambda(\theta_k)\rangle\, ds$, we can obtain the relation

$$\Lambda(\theta_{k+1}) - \Lambda(\theta_k) + \eta_k\|\nabla\Lambda(\theta_k)\|_2^2 = \xi_{k+1}.$$

Using the definition of ξ_{k+1} and Cauchy-Schwarz inequality (see Appendix B), we obtain

$$\begin{aligned}|\xi_{k+1}| &\leq \eta_k \int_0^1 \|\nabla\Lambda(\theta_k - s\eta_k\nabla\Lambda(\theta_k)) - \Lambda(\theta_k)\|_2 \|\nabla\Lambda(\theta_k)\|_2 ds\\ &\leq \eta_k^2 L \|\nabla\Lambda(\theta_k)\|_2^2 \int_0^1 s ds\\ &\leq \eta_k^2 L L_1^2,\end{aligned}$$

where in the last inequality we used the assumed Lipschitz property of $\nabla\Lambda$ from Assumption 17.6 as well as the global boundedness assumption $\|\nabla\Lambda(\theta)\|_2 \leq L_1$.

Since we have further assumed that $\sum_{k=1}^\infty \eta_k^2 < \infty$, we immediately obtain that $\sum_{k=1}^\infty |\xi_{k+1}| < \infty$, which is due to the fact that $\mathbb{R}$ is a complete metric space which gives that $\lim_{k\to\infty}\sum_{k'=1}^k \xi_{k'}$ exists. Let us set $\Xi_k = \sum_{k'=1}^k \xi_{k'}$. Then, we rewrite

$$\Lambda(\theta_{k+1}) - \Lambda(\theta_k) + \eta_k\|\nabla\Lambda(\theta_k)\|_2^2 = \Xi_{k+1} - \Xi_k.$$

We know that $\Xi_\infty = \lim_{k\to\infty} \Xi_k$ exists. By definition, this means that that for a given $\epsilon > 0$, there exists an $K(\epsilon)$ such that for $k \geq K(\epsilon)$,

$$|\Xi_k - \Xi_\infty| \leq \epsilon/2.$$

Therefore, for $K(\epsilon) \leq k_1 \leq k_2$, a telescoping series argument gives

$$\begin{aligned}\Lambda(\theta_{k_2}) - \Lambda(\theta_{k_1}) &\leq |(\Xi_{k_2} - \Xi_\infty) - (\Xi_{k_1} - \Xi_\infty)| \\ &\leq \epsilon.\end{aligned}$$

Taking first $k_2 \to \infty$ and then $k_1 \to \infty$ yields that

$$\limsup_{k_2\to\infty} \Lambda(\theta_{k_2}) \leq \liminf_{k_1\to\infty} \Lambda(\theta_{k_1}) + \epsilon.$$

Taking now $\epsilon \to 0$ we obtain

$$\limsup_{k\to\infty} \Lambda(\theta_k) \leq \liminf_{k\to\infty} \Lambda(\theta_k),$$

and since, trivially the reverse direction $\liminf_{k\to\infty} \Lambda(\theta_k) \leq \limsup_{k\to\infty} \Lambda(\theta_k)$, automatically holds, we obtain that indeed $\lim_{k\to\infty} \Lambda(\theta_k)$ exists.

Let us next prove that $\lim_{k\to\infty} \|\nabla\Lambda(\theta_k)\|_2 = 0$. We recall the relation

$$\Lambda(\theta_{k+1}) - \Lambda(\theta_k) + \eta_k \|\nabla\Lambda(\theta_k)\|_2^2 = \Xi_{k+1} - \Xi_k.$$

Let us suppose that $\zeta_- = \liminf_{k\to\infty} \|\nabla\Lambda(\theta_k)\|_2 > 0$. Then, there is some $K(\zeta_-) > 0$ such that for $k \geq K(\zeta_-)$ we shall have

$$\|\nabla\Lambda(\theta_k)\|_2 > \frac{1}{2}\zeta_-.$$

Using a telescoping series summation, we obtain

$$\Lambda(\theta_k) - \Lambda(\theta_{K(\zeta_-)}) + \sum_{k'=K(\zeta_-)}^{k-1} \eta_{k'} \|\nabla\Lambda(\theta_{k'})\|_2^2 = \Xi_k - \Xi_{K(\zeta_-)}.$$

Rearranging the latter relation gives

$$\begin{aligned}\sum_{k'=K(\zeta_-)}^{k-1} \eta_{k'} \left(\frac{1}{2}\zeta_-\right)^2 &\leq \sum_{k'=K(\zeta_-)}^{k-1} \eta_{k'} \|\nabla\Lambda(\theta_{k'})\|_2^2 \\ &\leq \left(\Lambda(\theta_k) - \Lambda(\theta_{K(\zeta_-)})\right) + \left(\Xi_k - \Xi_{K(\zeta_-)}\right) \\ &\leq \left|\Lambda(\theta_k) - \Lambda(\theta_{K(\zeta_-)})\right| + \left|\Xi_k - \Xi_{K(\zeta_-)}\right|.\end{aligned}$$

Therefore we have that

$$\sum_{k'=K(\zeta_-)}^{k-1} \eta_{k'} \leq \frac{1}{\left(\frac{1}{2}\zeta_-\right)^2} \left|\Lambda(\theta_k) - \Lambda(\theta_{K(\zeta_-)})\right| + \left|\Xi_k - \Xi_{K(\zeta_-)}\right|.$$

Since now we have already established that $\lim_{k\to\infty}\Lambda(\theta_k)$ and $\lim_{k\to\infty}\Xi_k$ exist and since by assumption $\sum_{k=1}^{\infty}\eta_k=\infty$, we obtain a contradiction under the assumption $\zeta_->0$. Hence, we have obtained that

$$\liminf_{k\to\infty}\|\nabla\Lambda(\theta_k)\|_2=0.$$

To conclude the proof, it remains to show that $\limsup_{k\to\infty}\|\nabla\Lambda(\theta_k)\|_2=0$. Let us on the contrary assume that

$$\zeta_+=\limsup_{k\to\infty}\|\nabla\Lambda(\theta_k)\|>0.$$

Since $\liminf_{k\to\infty}\|\nabla\Lambda(\theta_k)\|_2=0$, we have that $\|\nabla\Lambda(\theta_k)\|_2$ has an infinite number of oscillations above $\frac{2}{3}\zeta_+$ and below $\frac{1}{3}\zeta_+$. This means that there are $0\le k_1^-<k_1^+<k_2^-<k_2^+<\cdots$ such that

- $\lim_{\ell\to\infty}k_\ell^-=\lim_{\ell\to\infty}k_\ell^+=\infty$.
- $\|\nabla\Lambda(\theta_{k_\ell^-})\|_2<\frac{1}{3}\zeta_+$.
- $\frac{1}{3}\zeta_+\le\|\nabla\Lambda(\theta_k)\|\le\frac{2}{3}\zeta_+$ for $k_\ell^-<k<k_\ell^+$.
- $\|\nabla\Lambda(\theta_{k_\ell^+})\|_2>\frac{2}{3}\zeta_+$.

We then write

$$\begin{aligned}\sum_{k'=k_\ell^-}^{k_\ell^+-1}\left(\|\nabla\Lambda(\theta_{k'+1})\|_2-\|\nabla\Lambda(\theta_{k'})\|_2\right)&=\|\nabla\Lambda(\theta_{k_\ell^+})\|_2-\|\nabla\Lambda(\theta_{k_\ell^-})\|_2\\&\ge\frac{2}{3}\zeta_+-\frac{1}{3}\zeta_+\\&=\frac{1}{3}\zeta_+.\end{aligned}$$

On the other hand

$$\begin{aligned}\|\nabla\Lambda(\theta_{k+1})\|_2-\|\nabla\Lambda(\theta_k)\|_2&=\|\nabla\Lambda(\theta_{k+1})-\nabla\Lambda(\theta_k)+\nabla\Lambda(\theta_k)\|_2-\|\nabla\Lambda(\theta_k)\|_2\\&\le\|\nabla\Lambda(\theta_{k+1})-\nabla\Lambda(\theta_k)\|_2\\&\le L\|\theta_{k+1}-\theta_k\|_2\\&\le L\eta_k\|\nabla\Lambda(\theta_k)\|_2\\&\le LL_1\eta_k.\end{aligned}$$

Therefore, we have obtained that

$$\frac{1}{3}\zeta_+\le LL_1\sum_{k'=k_\ell^-}^{k_\ell^+-1}\eta_{k'}.$$

Note that

$$\sum_{k'=k_\ell^-}^{k_\ell^+-1} \eta_{k'} \|\nabla\Lambda(\theta_{k'})\|_2^2 \leq -\left(\Lambda(\theta_{k_\ell^+}) - \Lambda(\theta_{k_\ell^-})\right) + \left(\Xi_{k_\ell^+} - \Xi_{k_\ell^-}\right)$$

$$\leq \left|\Lambda(\theta_{k_\ell^+}) - \Lambda(\theta_{k_\ell^-})\right| + \left|\Xi_{k_\ell^+} - \Xi_{k_\ell^-}\right|$$

and that

$$\sum_{k'=k_\ell^-+1}^{k_\ell^+-1} \eta_{k'} \|\nabla\Lambda(\theta_{k'})\|_2^2 \geq \sum_{k'=k_\ell^-+1}^{k_\ell^+-1} \eta_{k'} \left(\frac{1}{3}\zeta_+\right)^2.$$

So, we have

$$\sum_{k'=k_\ell^-}^{k_\ell^+-1} \eta_{k'} \left(\frac{1}{3}\zeta_+\right)^2 \leq \left|\Lambda(\theta_{k_\ell^+}) - \Lambda(\theta_{k_\ell^-})\right| + \left|\Xi_{k_\ell^+} - \Xi_{k_\ell^-}\right| + \eta_{k_\ell^-}\left(\frac{1}{3}\zeta_+\right)^2,$$

which leads to

$$\sum_{k'=k_\ell^-}^{k_\ell^+-1} \eta_{k'} \leq \frac{1}{\left(\frac{1}{3}\zeta_+\right)^2}\left(\left|\Lambda(\theta_{k_\ell^+}) - \Lambda(\theta_{k_\ell^-})\right| + \left|\Xi_{k_\ell^+} - \Xi_{k_\ell^-}\right|\right) + \eta_{k_\ell^-}.$$

So we have

$$\frac{1}{3}\zeta_+ \leq LL_1 \sum_{k'=k_\ell^-}^{k_\ell^+-1} \eta_{k'}$$

$$\leq \frac{LL_1}{\left(\frac{1}{3}\zeta_+\right)^2}\left(\left|\Lambda(\theta_{k_\ell^+}) - \Lambda(\theta_{k_\ell^-})\right| + \left|\Xi_{k_\ell^+} - \Xi_{k_\ell^-}\right| + \eta_{k_\ell^-}\left(\frac{1}{3}\zeta_+\right)^2\right).$$

Since $\lim_{k\to\infty} \Lambda(\theta_k)$ and $\lim_{k\to\infty} \Xi_k$ exist and $\lim_{k\to\infty} \eta_k = 0$, we have

$$\lim_{k\to\infty} \left(\left|\Lambda(\theta_{k_\ell^+}) - \Lambda(\theta_{k_\ell^-})\right| + \left|\Xi_{k_\ell^+} - \Xi_{k_\ell^-}\right|\right) = 0.$$

However, this is a contradiction to the assumption that there are an infinite number of oscillations above $\frac{2}{3}\zeta_-$ and below $\frac{1}{3}\zeta_-$. The latter then leads to a contradiction to $\limsup_{k\to\infty} \|\nabla\Lambda(\theta_k)\|_2 > 0$. Therefore, we have indeed obtained that

$$\limsup_{k\to\infty} \|\nabla\Lambda(\theta_k)\|_2 = 0.$$

So, all in all, we have obtained the following result.

Theorem 17.16. *Assume that Assumption* 17.6 *holds and that there is some* $L_1 < \infty$ *such that* $\|\nabla\Lambda(\theta)\|_2 \leq L_1$. *Assume that the learning rate* η_k *is chosen so that* $\sum_{k=1}^{\infty}\eta_k = \infty$ *and* $\sum_{k=1}^{\infty}\eta_k^2 < \infty$. *Then, we have*

$$\lim_{k\to\infty} \Lambda(\theta_k) \text{ exists, and that}$$
$$\lim_{k\to\infty} \|\nabla\Lambda(\theta_k)\|_2 = 0.$$

We note that Theorem 17.16 proves convergence to critical points of the loss function in the case of the standard gradient descent algorithm. In Chapter 18, we will revisit this issue in the case of stochastic gradient descent, see Theorem 18.10 therein.

17.4. Accelerated Gradient Descent Methods

17.4.1. Polyak's method. Polyak's idea constitutes giving particles velocity; see [**Pol67**]. Let $v_k = \theta_{k+1} - \theta_k$ be the velocity (change) in θ. Then, we can write $v_k = \theta_{k+1} - \theta_k = -\eta\nabla\Lambda(\theta_k)$. What about now employing Newton's second law of motion and giving particles some velocity?

We will see two derivations of Polyak's method. Both derivations attempt to account for velocity (change) in θ albeit they have a different starting point.

In the first derivation, we can write as an approximation to the derivative for η small

$$-\nabla\Lambda(\theta_k) \approx \frac{\theta_{k+1} - 2\theta_k + \theta_{k-1}}{\eta},$$

which, by rearranging, leads to (making the $\approx$ sign to be an $=$ sign)

$$\theta_{k+1} = \theta_k - \eta\nabla\Lambda(\theta_k) + (\theta_k - \theta_{k-1}).$$

Now augment the latter by an additional hyperparameter ρ as

$$\theta_{k+1} = \theta_k - \eta\nabla\Lambda(\theta_k) + \rho(\theta_k - \theta_{k-1}).$$

The latter can be written equivalently as

$$\begin{aligned} u_{k+1} &= (1+\rho)\theta_k - \rho\theta_{k-1}, \\ \theta_{k+1} &= u_k - \eta\nabla\Lambda(\theta_k) \end{aligned}$$

or, equivalently,

$$\begin{aligned} u_{k+1} &= \rho u_k - \nabla\Lambda(\theta_k), \\ \theta_{k+1} &= \theta_k + \eta u_{k+1}. \end{aligned} \tag{17.4}$$

The second derivation (that is perhaps mathematically better motivated) of Polyak's momentum method is as follows. Let us start with gradient descent

Figure 17.1. An irregular road that naturally has many local minimums and local maximums and many possible different directions of motion. This motivates the idea of considering particles with velocity to explore loss functions with such characteristics. (Photograph by the second author.)

for the loss function $\Lambda(\theta)$

$$\theta_{k+1} = \theta_k - \eta \nabla \Lambda(\theta_k).$$

To avoid small fluctuations in the landscape of Λ (think for example of Figure 17.1), we replace $\nabla \Lambda(\theta)$ by the exponential moving average

$$\theta_{k+1} = \theta_k - \eta \sum_{k'=0}^{k} \rho^{k-k'}(1-\rho)\nabla \Lambda(\theta_{k'}). \tag{17.5}$$

At first sight this formula seems strange, but it relies on a rigorous mathematical result, oftentimes referred to as the Tauberian theorem. The Tauberian theorem connects exponential averaging and regular averaging. Note that $\Lambda(\theta)$ is a regular average. This is Lemma 18.17 which is presented and proven in Chapter 18. Lemma 18.17 says that if $\{\xi_k\}_{k\in\mathbb{N}}$ is a bounded sequence such that

$$\bar{\xi} \stackrel{\text{def}}{=} \lim_{k \nearrow \infty} \frac{1}{k} \sum_{k'=1}^{k} \xi_{k'}$$

is well defined, then we have that

$$\lim_{\rho \nearrow 1} (1-\rho) \sum_{k'=0}^{\infty} \rho^{k'} \xi_{k'} = \bar{\xi}.$$

Hence, one can indeed motivate (17.5) as another approximation to a regular average. The next step is to add an auxiliary equation for the evolution of the velocity of θ. For this purpose, we define

$$u_{k+1} = (1-\rho) \sum_{k'=0}^{k} \rho^{k-k'} \nabla\Lambda(\theta_{k'}),$$

with $u_0 = 0$. Then we have that

$$\begin{aligned} u_{k+1} &= (1-\rho) \sum_{k'=0}^{k-1} \rho^{k-k'} \nabla\Lambda(\theta_{k'}) + (1-\rho)\nabla\Lambda(\theta_k) \\ &= \rho(1-\rho) \sum_{k'=0}^{k-1} \rho^{k-1-k'} \nabla\Lambda(\theta_{k'}) + (1-\rho)\nabla\Lambda(\theta_k) \\ &= \rho u_k + (1-\rho)\nabla\Lambda(\theta_k). \end{aligned}$$

We have that if $\rho \approx 1$, then $u_{k+1} \approx u_k$ (representing the memory), whereas if $\rho \approx 0$, then $u_{k+1} \approx \nabla\Lambda(\theta_k)$ (update). Hence, we have arrived at the equations

$$\begin{aligned} u_{k+1} &= \rho u_k + (1-\rho)\nabla\Lambda(\theta_k), \\ \theta_{k+1} &= \theta_k - \eta u_{k+1} \end{aligned}$$

with $u_0 = 0$. We can of course rescale and define $u'_k = -u_k \frac{1}{(1-\rho)}$, in which case we get (for $\rho \neq 1$)

$$\begin{aligned} u'_{k+1} &= \rho u'_k - \nabla\Lambda(\theta_k), \\ \theta_{k+1} &= \theta_k + \eta' u'_{k+1}, \end{aligned}$$

where we define the new learning rate $\eta' = \eta(1-\rho)$. Essentially the latter is the same as (17.4).

Note that by defining $\hat{u}_k = \eta' u'_k$ the latter can also be written equivalently as

$$\begin{aligned} \hat{u}_{k+1} &= \rho \hat{u}_k - \eta' \nabla\Lambda(\theta_k), \\ \theta_{k+1} &= \theta_k + \hat{u}_{k+1}. \end{aligned} \tag{17.6}$$

Remark 17.17. An analysis similar to the plain GD method shows that if we want an error of order ϵ, then Polyak's method requires $\mathcal{O}\left(1/\sqrt{\epsilon}\right)$ steps for convex loss functions Λ and $\mathcal{O}\left(\sqrt{\frac{L}{\gamma}}\log(1/\epsilon)\right)$ steps for strongly convex loss functions Λ. However, Polyak's method may be unstable and not converge because when the iterates overshoot the global minimum, the inertia is different than the gradient.

17.4.2. Nesterov's method. This is an improvement of the classical momentum's Polyak method which tries to remove the oscillations when we are close to the global minimum, see [**Nes83, Nes04**]. The idea is to include damping (i.e., friction) in the motion. In particular, the effective force reduces the velocity without slowing down the weights much. To be exact, let us recall Polyak's method in the form (17.6) (ignore now the hat and prime notations):

$$\begin{aligned} u_{k+1} &= \rho u_k - \eta\nabla\Lambda(\theta_k), \\ \theta_{k+1} &= \theta_k + u_{k+1}. \end{aligned}$$

The idea is to approximate the next step in the gradient. Therefore, we replace $\nabla\Lambda(\theta_k)$ by an approximation of the next (probably better) point

$$\begin{aligned} \nabla\Lambda(\theta_{k+1}) = \nabla\Lambda(\theta_k + u_{k+1}) &= \nabla\Lambda\left(\theta_k + \rho u_k - \eta\nabla\Lambda(\theta_k)\right) \\ &\approx \nabla\Lambda(\theta_k + \rho u_k). \end{aligned}$$

Nesterov's idea in the latter calculation was that $\nabla\Lambda(\theta_k + \rho u_k)$ gives better performance near the minimum where $\nabla\Lambda = 0$. Note that the term $\nabla\Lambda(\theta_k + \rho u_k)$ computes the gradient at the current velocity which prevents overshooting in the neighborhood of the global minimum. So, we have the algorithm

$$\begin{aligned} \theta_{k+1} &= \theta_k + u_{k+1}, \\ u_{k+1} &= \rho u_k - \eta\nabla\Lambda(\theta_k + \rho u_k). \end{aligned}$$

If we now define $\hat{\theta}_k = \theta_k + \rho u_k$, then we get the alternative representation of Nesterov's algorithm

$$\begin{aligned} \hat{\theta}_{k+1} &= \hat{\theta}_k + \rho(u_{k+1} - u_k) + u_{k+1}, \\ u_{k+1} &= \rho u_k - \eta\nabla\Lambda(\hat{\theta}_k). \end{aligned}$$

Some related remarks are in order.

Remark 17.18. A typical choice is to set $\rho = \frac{\sqrt{m}-1}{\sqrt{m}+1}$ where $m = \frac{L}{\gamma}$ is the condition number. In practice we do not know the condition's number m. A typical value is $\rho = 0.9$ in most deep learning libraries. This choice is motivated as follows. If $m = 1$, then $\rho = 0$ and no inertia is needed. If $m \gg 1$, the Hessian of the loss function is badly conditioned and a value of $\rho \sim 1$ would be needed. So, choosing a value for ρ close to 1 is well motivated.

17.5. Brief Concluding Remarks

There are many excellent sources that cover in depth the classical topic of convergence theory for the gradient descent algorithm and related optimization results; see for example [**Ber03, Nes04, Nes07**] and the lecture notes [**Cha22**], which also partially motivated aspects of the presentation and proofs of lemmas of Section 17.2.3 on convergence rates for gradient descent. In this chapter, our goal was to present the main results paving the path towards Chapter 18 where we study convergence properties of stochastic gradient descent (SGD), which is the foundation of many deep learning algorithms.

17.6. Exercises

Exercise 17.1. Consider the algorithm $x_{k+1} = x_k - \eta\nabla\Lambda(x_k)$ with $\eta > 0$ and Λ a continuously differentiable function. Show that

(1) If the sequence $\{x_k\}$ converges, then $\nabla\Lambda(x_k) \to 0$.

(2) If the series $\sum_{k=0}^{\infty} \nabla\Lambda(x_k)$ converges, then the sequence $\{x_k\}$ converges.

Exercise 17.2. Consider the classical gradient descent problem with learning rate $\eta > 0$,

$$x_{k+1} = x_k - \eta\nabla\Lambda(x_k),$$

where Λ a continuously differentiable function.

(1) Let $k = n\Delta_t$ and $\eta = \lambda\Delta_t$. Show that as $\Delta_t \to 0$, the continuous time formulation of the gradient descent algorithm is the ordinary differential equation $\dot{x}_t = -\lambda\nabla\Lambda(x_t)$.

(2) Let $\Lambda(x) = \frac{1}{2}\|Ax - b\|_2^2$, where $A \in R^{m_1 \times m_2}$ and $b \in \mathbb{R}^{m_1}$. Show that the corresponding continuous time problem converges to the least squares solution $(ATA)^{-1}A^{\top}b$ as $t \to \infty$.

Exercise 17.3. Let $\Lambda : \mathbb{R}^m \mapsto \mathbb{R}$ be a convex function with minimum at $\theta^* \in R^m$. Assume that $\theta(t)$ solves the gradient flow ODE,

$$\dot{\theta} = -\nabla\Lambda(\theta),$$
$$\theta(0) = \theta_0.$$

Show that

$$\Lambda(\theta(t)) - \Lambda(\theta^*) \leq \frac{\|\theta_0 - \theta^*\|_2^2}{t}.$$

Exercise 17.4. Assume that $\Lambda(\theta)$ is strongly convex. Prove that the gradient descent update always decreases the objective function.

Exercise 17.5. Assume that the gradient of the loss function $\nabla\Lambda$ is L-Lipschitz, and let θ^* be the global minimum of the loss function $\Lambda(\cdot)$. Prove that for any $\theta \in \Theta$,

$$\Lambda(\theta^*) - \Lambda(\theta) \le -\frac{1}{2L}\|\nabla\Lambda(\theta)\|_2^2.$$

Chapter 18

Convergence Analysis of Stochastic Gradient Descent

18.1. Introduction

Gradient descent, whose convergence we analyzed in Chapter 17, calculates the gradient on all available data samples at each optimization iteration and thus becomes computationally expensive when the number of data samples is large. For large datasets, gradient descent becomes computationally infeasible since a single optimization iteration may require calculating the loss gradient for millions or even billions of data samples. This motivates the method of *stochastic gradient descent*, which at each optimization iteration only calculates the gradient of the loss on a randomly selected subset of the dataset. For large datasets, stochastic gradient descent has a substantially lower computational cost than gradient descent. Stochastic gradient descent can be viewed as using a noisy, stochastic estimate of the direction of steepest descent for the true objective function (which is evaluated on the entire dataset).

The key idea is at each optimization iteration (i.e., for each parameter update) to use a different *randomly* selected subset of the dataset. Thus, after the stochastic gradient descent algorithm has run for long enough, most likely all data samples in the overall dataset would have been used multiple times, and thus on average the effect should be the same as that of using the full dataset in gradient descent. Due to the fact that we randomly select a subset of the full dataset for every parameter update, the algorithm now is called stochastic gradient descent (SGD) instead of gradient descent (GD).

We discussed SGD in Chapters 7 and 8 without studying its theoretical aspects. In this chapter we elaborate on the convergence properties of SGD, we examine how SGD compares with classical GD, and we also present and discuss more sophisticated SGD methods such as SGD with momentum, AdaGrad, RMSProp, ADAM, and AdaMax.

18.2. Preliminary calculations

Assume we have data (X, Y) taking values in $\mathbb{R}^d \times \mathbb{R}$ and a model $\mathfrak{m}(x;\theta)$ where θ is the parameter of the model we want to estimate. We use the notation $\lambda_{(x,y)}(\theta)$ to measure how close the model's prediction $\mathfrak{m}(x;\theta)$ is to the actual observation y. We assume that we can sample data $\mathcal{D} = \{(x_m, y_m)_{m=1}^M\}$ from the distribution $\mathbb{P}$ of (X, Y). The loss function is

$$\Lambda(\theta) = \frac{1}{|\mathcal{D}|} \sum_{(x,y)\in\mathcal{D}} \lambda_{(x,y)}(\theta).$$

Oftentimes we may write $\Lambda_{\mathcal{D}}(\theta)$ to emphasize the dataset $\mathcal{D}$. Then, we will naturally choose $\theta^* = \operatorname{argmin}_{\theta\in\Theta} \Lambda(\theta)$. We recall that $\lambda_{(x,y)}(\theta)$ can be thought of as the per-data-sample loss and Λ as the average loss.

As we discussed in Section 17.2, gradient descent takes steps in the direction of steepest descent, recall (17.2)

$$\theta_{k+1} = \theta_k - \eta\nabla\Lambda(\theta_k).$$

Notice now that a Taylor series expansion yields (ignoring the higher-order error term)

$$\begin{aligned}\Lambda(\theta_{k+1})-\Lambda(\theta_k) &\approx \langle\nabla\Lambda(\theta_k),(\theta_{k+1}-\theta_k)\rangle+\frac{1}{2}(\theta_{k+1}-\theta_k)^\top\nabla^2\Lambda(\theta_k)(\theta_{k+1}-\theta_k)\\ &= -\eta\|\nabla\Lambda(\theta_k)\|^2 + \frac{1}{2}\eta^2\left(\nabla\Lambda(\theta_k)\right)^\top\nabla^2\Lambda(\theta_k)\nabla\Lambda(\theta_k).\end{aligned}$$

This shows that if η is sufficiently small, then $\Lambda(\theta_{k+1}) - \Lambda(\theta_k) \leq 0$. This indicates that—if η is selected to be sufficiently small—then the objective function loss for the gradient descent training algorithm is monotone decreasing. That is, the training algorithm is making progress towards a minimizer.

Next we notice that for a given finite dataset $\mathcal{D}$ we shall have that

$$\nabla_\theta\Lambda_{\mathcal{D}}(\theta) = \nabla_\theta\frac{1}{|\mathcal{D}|}\sum_{(x,y)\in\mathcal{D}}\lambda_{(x,y)}(\theta) = \frac{1}{|\mathcal{D}|}\sum_{(x,y)\in\mathcal{D}}\left[\nabla_\theta\lambda_{(x,y)}(\theta)\right].$$

If now $\mathcal{D}'$ is a randomly selected subset of $\mathcal{D}$, then we write

$$\Lambda_{\mathcal{D}'}(\theta) \stackrel{\text{def}}{=} \frac{1}{|\mathcal{D}'|}\sum_{(x,y)\in\mathcal{D}'}\lambda_{(x,y)}(\theta)$$

to emphasize that the computed loss function is on the randomly selected subset of $\mathcal{D}$, i.e., on $\mathcal{D}'_k \subset \mathcal{D}$. SGD ignores the expectation and follows a noisy (still unbiased) descent direction

$$\theta_{k+1} = \theta_k - \eta \nabla_\theta \Lambda_{\mathcal{D}'_k}(\theta_k),$$

where a new random data subset $\mathcal{D}'_k$ is selected uniformly-at-random at each parameter update iteration k.

The term $-\eta \nabla_\theta \Lambda_{\mathcal{D}'_k}(\theta_k)$ is an unbiased estimate of the direction of steepest descent for the objective function $\Lambda(\theta_k)$. The fact that it is unbiased follows from the assumption that $(x_m, y_m) \in \mathcal{D}'_k$ are i.i.d. samples from $\mathbb{P}$. Indeed, notice that under the i.i.d. assumption and given that at each given iteration we first choose the dataset $\mathcal{D}'_k$ and then we apply the gradient operator, we shall have for a given dataset $\mathcal{D}'_k \subset \mathcal{D}$ that

$$\begin{aligned}
\mathbb{E}\left[\nabla_\theta \Lambda_{\mathcal{D}'_k}(\theta)\right] &= \mathbb{E}\left[\frac{1}{|\mathcal{D}'_k|} \sum_{(x,y)\in\mathcal{D}'_k} \left[\nabla_\theta \lambda_{(x,y)}(\theta)\right]\right] \\
&= \mathbb{E}\left[\nabla_\theta \lambda_{(X,Y)}(\theta)\right] = \mathbb{E}\left[\nabla_\theta \Lambda_{\mathcal{D}}(\theta)\right].
\end{aligned}$$

Online learning typically corresponds to choosing $\mathcal{D}'$ with cardinality $|\mathcal{D}'| = 1$. Before we proceed with the analysis of SGD let us summarize some practical messages which we will develop from the analysis that will follow.

Remark 18.1.

(1) GD needs to compute gradients for each data sample in the dataset at every iteration, which is very computationally expensive for large datasets. SGD is computationally cheaper and is typically advantageous when the size of the dataset M is large. When M is large, we would sample a much smaller subset $|\mathcal{D}'| \ll M$ at each iteration.

(2) The learning rate $\eta = \eta_k$ determines the size of the parameter update step. It needs to decay appropriately in order to average out the noise in the SGD step.

(3) Typical requirements are $\sum_{k=1}^{\infty} \eta_k = \infty$ and $\sum_{k=1}^{\infty} \eta_k^2 < \infty$; see [**RM51**, **BT00**]. As an example consider the choice $\eta_k = \frac{C_0}{C_1 + C_2 k}$ where $0 < C_0, C_1, C_2 < \infty$ are constants.

(4) A choice that is typically used in practice is to define a priori constants $0 < C_0, C_1, C_2, C_3, C_4 < \infty$ and then consider the step function

$$\eta_k = \begin{cases} C_0, & k \leq C_1 \\ 0.1C_0, & C_1 < k \leq C_2 \\ 0.01C_0, & C_2 < k \leq C_3 \\ 0.001C_0, & C_2 < k \leq C_4. \end{cases}$$

(5) Note that too small of a learning rate leads to slow convergence, whereas too large of a learning rate means that the algorithm oscillates and overshoots.

Note that GD can use larger learning rates than SGD. In regards to SGD, this can be partially addressed by using minibatches, which use batches of random samples to reduce the noise. In particular, a minibatch is created by letting some $M_\circ \ll M$, a randomly selected subset $\mathcal{D}' \subset \mathcal{D}$ of cardinality $|\mathcal{D}'| = M_\circ$ and setting

$$\nabla \Lambda_{\mathcal{D}'}(\theta) = \frac{1}{|\mathcal{D}'|} \sum_{(x,y)\in\mathcal{D}'} \nabla_\theta \lambda_{(x,y)}(\theta). \tag{18.1}$$

To simplify notation and emphasize the cardinality of the randomly selected subset $\mathcal{D}' \subset \mathcal{D}$, we shall write $G^{(M_\circ)}(\theta_k) = \nabla \Lambda_{\mathcal{D}'_k}(\theta)$ in the SGD update equation for θ

$$\theta_{k+1} = \theta_k - \eta G^{(M_\circ)}(\theta_k). \tag{18.2}$$

We emphasize that at each iteration k of the algorithm in (18.2) a new random subset $\mathcal{D}'_k \subset \mathcal{D}$ of cardinality $M_\circ$ is selected.

Note that $G^{(M_\circ)}(\theta)$ is less noisy than $G^{(1)}(\theta)$ for $M_\circ > 1$. Indeed, we can calculate

$$\begin{aligned}
\operatorname{Var}\big(G^{(M_\circ)}(\theta_k)|\theta_k\big) &= \operatorname{Var}\left(\frac{1}{|\mathcal{D}'_k|} \sum_{(x,y)\in\mathcal{D}'} \nabla_\theta \lambda_{(x,y)}(\theta_k)\Big|\theta_k\right) \\
&= \frac{1}{|\mathcal{D}'|} \operatorname{Var}\Big(\nabla_\theta \lambda_{(x_j,y_j)}(\theta_k)\Big|\theta_k\Big) \\
&< \operatorname{Var}\Big(\nabla_\theta \lambda_{(x_j,y_j)}(\theta_k)\Big|\theta_k\Big) \\
&= \operatorname{Var}\big(G^{(1)}(\theta_k)|\theta_k\big),
\end{aligned}$$

where the data sample (x_j, y_j) is selected uniformly at random from the dataset $\mathcal{D}$. The above inequality shows that the minibatch SGD gradient estimate is less noisy (lower variance) than the classical SGD gradient estimate ($M_\circ = 1$).

Remark 18.2 (Training epochs). Related to the latter statement for the variance is also the notion of an epoch. An epoch, similar to the minibatch size, is another hyperparameter to be chosen. The number of epochs refers to the number of times the learning algorithm goes through the entire training dataset. Therefore, having completed one epoch means that each sample in the training dataset has been used in the algorithm. Naturally, one epoch can be composed by one or more minibatches and the number of epochs completed during training is typically large. Rigorous minibatch learning may take a long time to see all datapoints. On the other hand, unbiased estimation of the gradient of the loss function sort of depends on rigorous sampling of the data. Standard SGD with a large number of epochs tries to find a reasonable compromise between these two situations.

Essentially, we randomly partition $\mathcal{D} = \bigcup_{k=1}^{K} \mathcal{D}'_k$ with $|\mathcal{D}'_k| \approx |\mathcal{D}|/K$. Then, for each $n \in \mathbb{N}$ and $k \in \{1, 2, \dots, K\}$, we apply SGD

$$\theta_{nK+k} = \theta_{nK+k-1} - \eta \nabla \Lambda_{\mathcal{D}'_k}(\theta_{nK+k-1}).$$

By breaking gradient descent into epochs, after completing one epoch, each sample in the training dataset has had the opportunity to be used in the algorithm. However, we do note a bias issue. Even though, $\nabla \Lambda_{\mathcal{D}'_1}(\theta)$ is an unbiased estimate of $\nabla \Lambda_{\mathcal{D}(\theta)}$, the remaining $\nabla \Lambda_{\mathcal{D}'_k}(\theta)$ will be biased because $\mathcal{D}'_1$ has already been chosen.

This issue of bias can be addressed by simply running SGD using uniform-at-random sampling (with replacement) from the dataset $\mathcal{D}$ and considering an epoch to be the number of minibatch SGD updates such that the total number of data samples used equals the size of the overall datasets $|\mathcal{D}|$. Note that it is not guaranteed that in a single epoch every data sample in $\mathcal{D}$ will have been trained on though.

18.3. Convergence Results for SGD

18.3.1. Convergence of SGD for Convex Loss Functions. Let us consider a convex loss function $\Lambda(\theta)$ per Definition 17.1. Recall that $\lambda_{(x_m, y_m)}(\theta)$ denotes the per-data-sample loss.

Before proceeding with the convergence results for SGD for strongly convex loss functions, let us perform some initial calculations motivated by our analysis of standard GD. We recall that the standard update in GD is given by (17.2):

$$\theta_{k+1} = \theta_k - \eta \nabla \Lambda(\theta_k).$$

Substituting the above equation into Lemma 17.7 yields

$$\begin{aligned} \Lambda(\theta_{k+1}) - \Lambda(\theta_k) &\leq -\eta \|\nabla \Lambda(\theta_k)\|_2^2 + \frac{L\eta^2}{2} \|\nabla \Lambda(\theta_k)\|_2^2 \\ &= -\eta \left(1 - \frac{L\eta}{2}\right) \|\nabla \Lambda(\theta_k)\|_2^2. \end{aligned} \tag{18.3}$$

This relation means that if η is sufficiently small and, specifically, if $\eta < 2/L$, then GD will typically be making progress towards the global minimum. However, this is not necessarily guaranteed for SGD. In the case of plain SGD, the update is given by (18.2) with $|\mathcal{D}'| = 1$, i.e.,

$$\begin{aligned}\theta_{k+1} &= \theta_k - \eta G^{(1)}(\theta_k) \\ &= \theta_k - \eta \nabla \lambda_{(x_{i_k}, y_{i_k})}(\theta_k),\end{aligned}$$

where the index i_k is randomly sampled at the kth iteration. Substituting $\theta' = \theta_{k+1}$, $\theta = \theta_k$, and the aforementioned SGD update into Lemma 17.7 and then taking a conditional expectation produces the following lemma.

Lemma 18.3. *Under Assumption* 17.6 *we have that*

$$\begin{aligned}&\mathbb{E}\left[\Lambda(\theta_{k+1}) - \Lambda(\theta_k)|\theta_k\right] \\ &\quad\le \eta\left(\frac{L\eta}{2}\mathbb{E}\left[\|\nabla\lambda_{(x_{i_k},y_{i_k})}(\theta_k)\|_2^2|\theta_k\right] - \left\langle \nabla\Lambda(\theta_k), \mathbb{E}\left[\nabla\lambda_{(x_{i_k},y_{i_k})}(\theta_k)|\theta_k\right]\right\rangle\right) \\ &\quad= \eta\left(\frac{L\eta}{2}\mathbb{E}\left[\|\nabla\lambda_{(x_{i_k},y_{i_k})}(\theta_k)\|_2^2|\theta_k\right] - \|\nabla\Lambda(\theta_k)\|_2^2\right),\end{aligned}$$

where the expectation is taken under the index i_k sampled by the SGD algorithm at the kth iteration and we have used the fact that $\nabla\Lambda(\theta) = \mathbb{E}\left[\nabla\lambda_{(x_i,y_i)}(\theta)\right]$ (i.e., the stochastic gradient estimates are unbiased).

Lemma 18.3 shows that without extra assumptions, SGD updates may not monotonically decrease the value of the average loss function. In the last line, the first term (which is positive) may potentially be larger than the second term for a *fixed learning rate* η. However, it should be highlighted that there always does exist an $\eta > 0$ (which will depend upon θ_k) such that the last line is negative when $\|\nabla\Lambda(\theta_k)\|_2^2 > 0$. Furthermore, we observe that η must be smaller when $\|\nabla\Lambda(\theta_k)\|_2^2$ is smaller (which typically means θ_k is closer to a local minimizer) to guarantee that the last line is negative. This suggests that we should reduce the learning rate η during training. Specifically, the learning rate should be η_k, a function of the number of parameter update iterations, where $\eta_k \to 0$ as $k \to \infty$. Consequently, as θ_k approaches a local minimizer (and $\|\nabla\Lambda(\theta_k)\|_2^2$ becomes smaller), the learning rate will also become smaller.

Let us now try to improve upon Lemma 18.3. We will use the following assumptions.

Assumption 18.4. Let us assume the following.

- $\nabla\Lambda(\theta) = \mathbb{E}\left[\nabla\lambda_{(x_i,y_i)}(\theta)\right]$, i.e., that stochastic gradients are unbiased.
- There are constants $\kappa_1, \kappa_2 < \infty$ such that

  $$\mathbb{E}\left\|\nabla\lambda_{(x_i,y_i)}(\theta)\right\|_2^2 \le \kappa_1 + \kappa_2\|\nabla\Lambda(\theta)\|_2^2,$$

  i.e., that the second moments of SGD and GD are comparable in magnitude.

Some comments are in order. The first part of Assumption 18.4 will hold if the sampling distribution of the index is uniform, which is the standard sampling method for SGD and which we have previously used in our analysis, including in Lemma 18.3. The second part of Assumption 18.4 states that the second moments of the stochastic gradient are of a magnitude similar to the square of the full gradient. Combining the second assumption now with Lemma 18.3, we immediately have the following result.

Lemma 18.5. *Under Assumptions* 17.6 *and* 18.4 *we have that*

$$\mathbb{E}\left[\Lambda(\theta_{k+1}) - \Lambda(\theta_k)|\theta_k\right] \leq \eta\left(\eta\frac{L\kappa_1}{2} - \left(1 - \eta\frac{L\kappa_2}{2}\right)\|\nabla\Lambda(\theta_k)\|_2^2\right),$$

where the expectation is taken under the index i_k sampled by the SGD at the kth iteration.

Lemma 18.5 shows that in order for the right-hand side to be negative (i.e., in order for SGD to be making progress towards the global minimum), we need to select the step size η in such a way that the combined term $\eta\frac{L\kappa_1}{2} - \left(1 - \eta\frac{L\kappa_2}{2}\right)\|\nabla\Lambda(\theta_k)\|_2^2$ is negative.

In addition, note that Lemma 18.5 shows that if $\kappa_1 = 0$ and $\kappa_2 = 1$, i.e., in the absence of stochasticity, we recover the result of standard GD, i.e., the decay (18.3). If the stochastic gradient is noisy, i.e., if $\kappa_1 \neq 0$, then there is not necessarily a monotonic decay of the difference $\mathbb{E}\left[\Lambda(\theta_{k+1}) - \Lambda(\theta_k)|\theta_k\right]$ because for every iteration $k \in \mathbb{N}$, there is the positive term $\eta^2\frac{K\kappa_1}{2}$ at the right-hand side bound.

As it is then shown in [**BCN18**], under Assumptions 17.6 and 18.4 stronger results can be obtained. Note that even though we assume in the next lemmas that $\kappa_2 \geq 1$ from Assumption 18.4, this is done with loss of generality, as one can always use a larger upper bound if necessary. We next present the related theory.

Lemma 18.6. *Let Assumptions* 17.6 *and* 18.4 *with $\kappa_2 \geq 1$ hold, and in addition assume that Λ is γ-strongly convex (Definition* 17.3*). Assume that the learning rate is chosen according to $\eta \leq \frac{1}{L\kappa_2}$, where L is the Lipschitz constant from Assumption* 17.6. *Then we have the bound*

$$\mathbb{E}\left[\Lambda(\theta_{k+1})\right] - \Lambda(\theta^*) \leq \frac{\eta L\kappa_1}{2\gamma} + (1 - \eta\gamma)^k\left(\Lambda(\theta_0) - \Lambda(\theta^*) - \frac{\eta L\kappa_1}{2\gamma}\right).$$

The aforementioned upper bound converges to $\frac{\eta L\kappa_1}{2\gamma}$ as $k \to \infty$.

Proof of Lemma 18.6. First, we notice that by γ-strong convexity, we have for all $\theta, \theta' \in \Theta$

$$\Lambda(\theta') \geq \Lambda(\theta) + \langle\nabla\Lambda(\theta), \theta' - \theta\rangle + \frac{\gamma}{2}\|\theta' - \theta\|^2.$$

As a function of θ', the right-hand side (a quadratic function) has the unique minimizer

$$\hat{\theta} = \theta - \frac{1}{\gamma}\nabla\Lambda(\theta),$$

further yielding

$$\Lambda(\theta') \geq \Lambda(\theta) - \frac{1}{2\gamma}\|\nabla\Lambda(\theta)\|_2^2. \tag{18.4}$$

Set $\Lambda^* = \Lambda(\theta^*)$. By Lemma 18.5 and γ-strong convexity (in particular, applying (18.4) with $\theta' = \theta^*$ and $\theta = \theta_k$), we have for all $k \in \mathbb{N}$

$$\begin{aligned}\mathbb{E}\left[\Lambda(\theta_{k+1}) - \Lambda(\theta_k)|\theta_k\right] &\leq -\eta\left(1 - \eta\frac{L\kappa_2}{2}\right)\|\nabla\Lambda(\theta_k)\|_2^2 + \eta^2\frac{L\kappa_1}{2} \\ &\leq -\frac{\eta}{2}\|\nabla\Lambda(\theta_k)\|_2^2 + \eta^2\frac{L\kappa_1}{2} \\ &\leq -\eta\gamma(\Lambda(\theta_k) - \Lambda^*) + \eta^2\frac{L\kappa_1}{2}.\end{aligned}$$

The next step is to subtract Λ^* from both sides and then to take an expectation. Doing so, we obtain

$$\mathbb{E}\left[\Lambda(\theta_{k+1}) - \Lambda^*\right] \leq (1 - \eta\gamma)(\mathbb{E}\left[\Lambda(\theta_k) - \Lambda^*\right]) + \eta^2\frac{L\kappa_1}{2}.$$

This yields

$$\begin{aligned}\mathbb{E}\left[\Lambda(\theta_{k+1}) - \Lambda^*\right] - \frac{\eta L\kappa_1}{2\gamma} &\leq (1 - \eta\gamma)(\mathbb{E}\left[\Lambda(\theta_k) - \Lambda^*\right]) + \eta^2\frac{L\kappa_1}{2} - \frac{\eta L\kappa_1}{2\gamma} \\ &= (1 - \eta\gamma)\left(\mathbb{E}\left[\Lambda(\theta_k) - \Lambda^*\right] - \frac{\eta L\kappa_1}{2\gamma}\right).\end{aligned} \tag{18.5}$$

Recall now that we have chosen $\eta \leq \frac{1}{L\kappa_2}$. This means that

$$0 < \eta\gamma \leq \frac{\gamma}{L\kappa_2} \leq \frac{\gamma}{L} \leq 1,$$

the latter being true because we assumed $\kappa_2 \geq 1$ and because it must hold that $\gamma \leq L$ (recall that γ is the strong convexity constant of Λ whereas L is the global Lipschitz constant for the gradient of Λ). Therefore, (18.5) is a contraction and the result follows by applying (18.5) iteratively in $k \in \mathbb{N}$. This concludes the proof of the lemma. □

Lemma 18.6 shows that we must select η to be small in order to hope to eventually reach the global minimum. However, if $\eta > 0$ is chosen to be very small, then the algorithm will take a long time to converge, and we will still be $\mathcal{O}(\eta)$ away from the global minimum.

So, the question is: how can we guarantee that the SGD algorithm converges? The answer is to let the learning rate η gradually decrease over time, see Remark 18.1. We will specifically discuss the necessity for decay in the learning rate in the next Section 18.3.3. We will prove in Lemma 18.7 that SGD converges to the global minimizer for strongly convex functions if the learning rate appropriately decays as the number of update steps $\to \infty$.

Lemma 18.7. *Let Assumptions* 17.6 *and* 18.4 *with* $\kappa_2 \geq 1$ *hold and in addition assume that* Λ *is* γ*-strongly convex (Definition* 17.3*). Assume that the learning rate is chosen according to* $\eta = \eta_k = \frac{C_0}{C_1+k}$ *with* $\gamma C_0 > 1$, $C_1 > 0$, *and* $\eta_1 \leq \frac{1}{L\kappa_2}$, *where L is the Lipschitz constant from Assumption* 17.6*. Then, the following bound holds:*

$$\mathbb{E}\left[\Lambda(\theta_k)\right] - \Lambda(\theta^*) \leq \frac{\tau}{C_1 + k},$$

where $\tau = \max\left\{\frac{C_0^2 L\kappa_1}{2(\gamma C_0 - 1)}, (C_1 + 1)(\mathbb{E}[\Lambda(\theta_1)] - \Lambda(\theta^*))\right\}$.

Proof of Lemma 18.7. The proof of this lemma is similar to the proof of Lemma 18.6. Set $\Lambda^* = \Lambda(\theta^*)$. By Lemma 18.5 and γ-strong convexity, we have for all $k \in \mathbb{N}$

$$\begin{aligned}
\mathbb{E}\left[\Lambda(\theta_{k+1}) - \Lambda(\theta_k)|\theta_k\right] &\leq -\eta_k\left(1 - \eta_k\frac{L\kappa_2}{2}\right)\|\nabla\Lambda(\theta_k)\|_2^2 + \eta_k^2\frac{L\kappa_1}{2} \\
&\leq -\frac{\eta_k}{2}\|\nabla\Lambda(\theta_k)\|_2^2 + \eta_k^2\frac{L\kappa_1}{2} \\
&\leq -\eta_k\gamma(\Lambda(\theta_k) - \Lambda^*) + \eta_k^2\frac{L\kappa_1}{2},
\end{aligned}$$

where in the last display we used (18.4) with $\theta' = \theta^*$ and $\theta = \theta_k$.

The next step is to subtract Λ^* from both sides of the latter expression followed by taking expectation. We then obtain

$$\mathbb{E}\left[\Lambda(\theta_{k+1}) - \Lambda^*\right] \leq (1 - \eta_k\gamma)(\mathbb{E}\left[\Lambda(\theta_k) - \Lambda^*\right]) + \eta_k^2\frac{L\kappa_1}{2}.$$

Next, we proceed with an induction argument. The statement holds for $k = 1$ directly by the definition of τ. Then, let us assume that it holds for some integer k greater than 1 and prove it for $k + 1$. We have

$$\begin{aligned}
\mathbb{E}\left[\Lambda(\theta_{k+1}) - \Lambda^*\right] &\leq \left(1 - \frac{C_0\gamma}{C_1 + k}\right)\frac{\tau}{C_1 + k} + \frac{C_0^2 L\kappa_1}{2(C_1 + k)^2} \\
&= \frac{C_1 + k - C_0\gamma}{(C_1 + k)^2}\tau + \frac{C_0^2 L\kappa_1}{2(C_1 + k)^2} \\
&= \frac{C_1 + k - 1}{(C_1 + k)^2}\tau + \left[-\frac{C_0\gamma - 1}{(C_1 + k)^2}\tau + \frac{C_0^2 L\kappa_1}{2(C_1 + k)^2}\right] \\
&\leq \frac{\tau}{C_1 + k + 1},
\end{aligned}$$

where in the latter calculation we used that the term in the bracket

$$\left[-\frac{C_0\gamma - 1}{(C_1+k)^2}\tau + \frac{C_0^2 L\kappa_1}{2(C_1+k)^2}\right] \leq 0$$

due to the definition of τ and the fact that

$$(C_1+k)^2 \geq (C_1+k+1)(C_1+k-1).$$

This concludes the proof of the lemma. □

Some remarks on Lemmas 18.6 and 18.7 are in order. First, it is interesting to note that the choice of the learning rate $\eta = \eta_k = \frac{C_0}{C_1+k}$ satisfies the conditions $\sum_{k=1}^{\infty} \eta_k = \infty$ and $\sum_{k=1}^{\infty} \eta_k^2 < \infty$. (In fact, a slightly more general learning rate of $\eta_k = \frac{C_0}{C_1+C_2 k}$ could be used.) This turns out to be a good learning rate schedule. We already demonstrated this in the case of gradient descent in Theorem 17.16 and we will return to this in Section 18.3.3 for the case of stochastic gradient descent.

Second, the strong convexity constant $\gamma > 0$ is seen to play an important role for both statements in Lemmas 18.6 and 18.7. However, it affects the step size in different ways in the two lemmas. For the case of constant stepsize of Lemma 18.6 the stepsize is not affected by γ, even though the optimality gap is. On the other hand, in the vanishing stepsize case of Lemma 18.7 the choice of the learning rate is affected by γ. Indeed, we have chosen $\eta_k = \frac{C_0}{C_1+k}$ with $C_0 > 1/\gamma$.

Third, in the case of vanishing learning rate of Lemma 18.7, the choice of the initial point affects the optimality gap through the parameter τ via the term $\Lambda(\theta_1) - \Lambda(\theta^*)$. However, with an appropriate choice of the learning schedule, the effect of this term can be diminished. We refer the reader to [**BCN18**] for more details on this issue.

Fourth, let us comment on the effect of choosing a minibatch of size $M_\circ > 1$ as seen in (18.1)–(18.2). Note that in that case, Assumption 18.4 changes to $\mathbb{E}\,\|\nabla\rho_i(\theta)\|_2^2 \leq \frac{\kappa_1}{K} + \frac{\kappa_2}{M_\circ}\|\nabla\Lambda(\theta)\|_2^2$. This then leads to the statement of Lemma 18.6 becoming

$$\mathbb{E}\,[\Lambda(\theta_{k+1})] - \Lambda(\theta^*) \leq \frac{\eta L\kappa_1}{2\gamma M_\circ} + (1-\eta\gamma)^k\left(\Lambda(\theta_0) - \Lambda(\theta^*) - \frac{\eta L\kappa_1}{2\gamma M_\circ}\right)$$

and the requirement for the learning rate to change to $\eta \leq \frac{M_\circ}{L\kappa_s}$. This means that a larger choice of the learning rate is allowed. If we do choose the largest allowed constant rate, the term $\frac{\eta L\kappa_1}{2\gamma M_\circ}$ will not be affected by $M_\circ$. However, the term $(1-\eta\gamma)^k$ will be affected by it and in particular increasing the learning rate to $\frac{M_\circ}{L\kappa_s}$ will lead to faster decay of that term by a factor of $M_\circ$. However, we need to keep in mind that the choice of the minibatch size also requires $M_\circ$ calculations of the gradient.

18.3.2. Convergence of SGD for Nonconvex Loss Functions. As is typically the case, many important machine learning models lead to nonconvex optimization problems. In particular, neural networks are nonconvex. Even though the analysis of nonconvex functions trained with SGD is more complicated than the analysis in the convex case, one can still obtain meaningful results. We will follow the presentation of Section 18.3.1 by first presenting bounds for a constant learning rate and then for a decreasing learning rate.

Lemma 18.8. *Let Assumptions* 17.6 *and* 18.4 *hold. Assume that the learning rate is chosen according to* $\eta \leq \frac{1}{L\kappa_2}$*, where* L *is the Lipschitz constant from Assumption* 17.6. *In addition, assume that the sequence of iterates* θ_k *is contained in an open set over which* Λ *is bounded below by* Λ^*. *Then we have the bound*

$$\mathbb{E}\left[\frac{1}{M_\circ}\sum_{m=1}^{M_\circ}\|\nabla\Lambda(\theta_m)\|_2^2\right] \leq \eta L\kappa_1 + \frac{2(\Lambda(\theta_1)-\Lambda^*)}{M_\circ\eta}.$$

The aforementioned upper bound converges to $\eta L\kappa_1$ *as* $M_\circ \to \infty$.

Proof of Lemma 18.8. As in Lemma 18.6, we have for all $m \in \mathbb{N}$,

$$\begin{aligned}\mathbb{E}\left[\Lambda(\theta_{m+1})\right] - \mathbb{E}\left[\Lambda(\theta_m)\right] &\leq -\eta\left(1-\eta\frac{L\kappa_2}{2}\right)\mathbb{E}\|\nabla\Lambda(\theta_m)\|_2^2 + \eta^2\frac{L\kappa_1}{2}\\ &\leq -\frac{\eta}{2}\mathbb{E}\|\nabla\Lambda(\theta_m)\|_2^2 + \eta^2\frac{L\kappa_1}{2}.\end{aligned}$$

The result now follows by summing over all $m \in \{1,\dots,M_\circ\}$ and using the assumption that the sequence of iterates θ_m is contained in an open set over which Λ is bounded below by Λ^*. This completes the proof of the lemma. □

Lemma 18.9. *Let Assumptions* 17.6 *and* 18.4 *hold. In addition, assume that the sequence of iterates* θ_k *is contained in an open set over which* Λ *is bounded below by* Λ^*. *Assume that the sequence of learning rates satisfies* $\sum_{m=1}^{\infty}\eta_m = \infty$ *and* $\sum_{m=1}^{\infty}\eta_m^2 < \infty$. *Then, we have*

$$\lim_{M_\circ\to\infty}\mathbb{E}\left[\sum_{m=1}^{M_\circ}\eta_m\|\nabla\Lambda(\theta_m)\|_2^2\right] < \infty,$$

and consequently

$$\lim_{M_\circ\to\infty}\mathbb{E}\left[\frac{1}{\sum_{m=1}^{M_\circ}\eta_m}\sum_{m=1}^{M_\circ}\eta_m\|\nabla\Lambda(\theta_m)\|_2^2\right] = 0.$$

Proof of Lemma 18.9. The condition $\sum_{m=1}^{\infty} \eta_m^2 < \infty$ implies that $\eta_m \to 0$, so we may choose $\eta_m \leq \frac{1}{L\kappa_2}$. Then, proceeding as in the proof of Lemma 18.8,

$$\begin{aligned}\mathbb{E}\left[\Lambda(\theta_{m+1})\right] - \mathbb{E}\left[\Lambda(\theta_m)\right] &\leq -\eta_m\left(1 - \eta_m \frac{L\kappa_2}{2}\right)\mathbb{E}\|\nabla\Lambda(\theta_m)\|_2^2 + \eta_m^2 \frac{L\kappa_1}{2} \\ &\leq -\frac{\eta_m}{2}\mathbb{E}\|\nabla\Lambda(\theta_m)\|_2^2 + \eta_m^2 \frac{L\kappa_1}{2}.\end{aligned}$$

Next, we sum over $m \in \{1, \ldots, M_\circ\}$ to obtain

$$\Lambda^* - \mathbb{E}\left[\Lambda(\theta_1)\right] \leq -\frac{1}{2}\sum_{m=1}^{K} \eta_m \mathbb{E}\|\nabla\Lambda(\theta_m)\|_2^2 + \frac{L\kappa_1}{2}\sum_{m=1}^{M_\circ} \eta_m^2.$$

Rearranging this statement leads to the first claim of the lemma. The second claim follows by the first statement and the fact that $\sum_{m=1}^{\infty} \eta_m = \infty$. This concludes the proof of the lemma. □

Lemma 18.9 leads to the following important theorem.

Theorem 18.10 ([**BT00, BCN18**]). *Let Assumptions* 17.6 *and* 18.4 *hold. In addition, assume that the sequence of iterates* θ_k *is contained in an open set over which* Λ *is bounded below by* Λ^*. *Assume that the learning rate is chosen according to* η_k *satisfying* $\sum_{k=1}^{\infty} \eta_k = \infty$ *and* $\sum_{k=1}^{\infty} \eta_k^2 < \infty$. *Then, we have*

$$\liminf_{k\to\infty} \mathbb{E}\left[\|\nabla\Lambda(\theta_k)\|_2^2\right] = 0.$$

Note that this result includes a limit infimum result. Under stronger conditions this can be that reduced to $\lim_{k\to\infty} \mathbb{E}\left[\|\nabla\Lambda(\theta_k)\|_2^2\right] = 0$. This is Theorem 18.11 and was proven in [**BT00**] and under a somewhat different set of conditions in [**BCN18**].

18.3.3. Why Should the Learning Rate Decrease? Let us now demonstrate with a formal argument as to why the learning rate needs to decrease. We recall that we have already seen instances of this phenomenon in the cases of linear and logistic regression, Chapters 2 and 3, where we show that overshooting occurs if the learning rate is too large. The fully rigorous treatment in the general case of this result can be found in [**BT00**]. For simplicity we will focus on the case of plain SGD where $|\mathcal{D}'| = 1$. We have

$$\begin{aligned}\theta_{k+1} &= \theta_k - \eta_k \nabla_\theta \lambda_{(x_k,y_k)}(\theta_k) \\ &= \theta_k - \eta_k \mathbb{E}\left[\nabla_\theta \lambda_{(X,Y)}(\theta_k)\right] + \eta_k\left(\mathbb{E}\left[\nabla_\theta \lambda_{(X,Y)}(\theta_k)\right] - \nabla_\theta \lambda_{(x_k,y_k)}(\theta_k)\right) \\ &= \theta_k - \eta_k \nabla_\theta \Lambda(\theta_k) + \eta_k\left(\mathbb{E}\left[\nabla_\theta \lambda_{(X,Y)}(\theta_k)\right] - \nabla_\theta \lambda_{(x_k,y_k)}(\theta_k)\right).\end{aligned}$$

Let us set $R_k = \left(\mathbb{E}\left[\nabla_\theta \lambda_{(X,Y)}(\theta_k)\right] - \nabla_\theta \lambda_{(x_k,y_k)}(\theta_k)\right)$ which represents the randomness in SGD. We would like to prove that in the appropriate sense

$\lim_{k\to\infty} R_k = 0$. Now, let us assume that $\lambda \in \mathcal{C}_b^{2,2}$. Using Taylor expansion we obtain for some $k_1 > k$,

$$\Lambda(\theta_{k_1}) - \Lambda(\theta_k) = -\sum_{i=k}^{k_1} \eta_i \|\nabla_\theta \Lambda(\theta_i)\|_2^2 + \sum_{i=k}^{k_1} \eta_i \left(\nabla_\theta \Lambda(\theta_i)\right)^\top R_i$$
$$+ \sum_{i=k}^{k_1} \eta_i^2 \times (\text{higher order terms}).$$

If $\sum_{i=0}^{\infty} \eta_i^2 < \infty$, then we have that for every $\epsilon > 0$ there is a $k < \infty$ such that $\sum_{i=k}^{\infty} \eta_i^2 < \epsilon$. Assuming that the higher order terms are bounded, we then obtain that for this chosen k

$$\sum_{i=k}^{k_1} \eta_i^2(\text{higher order terms}) \leq C \sum_{i=k}^{k_1} \eta_i^2 \leq C\epsilon.$$

In addition, we shall have that

$$\begin{aligned}
&\mathbb{E}\left(\left(\sum_{i=k}^{k_1} \eta_i \left(\nabla_\theta \Lambda(\theta_i)\right)^\top R_i\right)^2\right) \\
&= \sum_{i=k}^{k_1} \eta_i^2 \mathbb{E}\left(\left(\nabla_\theta^\top \Lambda(\theta_i)\left(\mathbb{E}_{X,Y}\left(\nabla_\theta \lambda_{(X,Y)}(\theta_k) - \nabla_\theta \lambda_{(x_i,y_i)}(\theta_i)\right)\right)\right)^2\right) \\
&\quad + \sum_{i=k}^{k_1} \sum_{j=k, j\neq i}^{k_1} \eta_i \eta_j \mathbb{E}\left(\nabla_\theta \Lambda(\theta_i)\left(\mathbb{E}_{X,Y}\left(\nabla_\theta \lambda_{(X,Y)}(\theta_i) - \nabla_\theta \lambda_{(x_i,y_i)}(\theta_i)\right)\right.\right. \\
&\quad \left.\left.\times \nabla_\theta \Lambda(\theta_j)\left(\mathbb{E}_{X,Y}\left(\nabla_\theta \lambda_{(X,Y)}(\theta_j) - \nabla_\theta \lambda_{(x_j,y_j)}(\theta_j)\right)\right)\right)\right) \\
&\leq C \sum_{i=k}^{k_1} \eta_i^2 \\
&\leq C\epsilon.
\end{aligned}$$

In the calculation above we used the tower property of expectation to realize that

$$\mathbb{E}\left(\mathbb{E}_{X,Y}\left(\nabla_\theta \lambda_{(X,Y)}(\theta_j) - \nabla_\theta \lambda_{(x_j,y_j)}(\theta_j)\right) | \mathcal{F}_j\right) = 0,$$

where $\mathcal{F}_j$ is the filtration at time j, i.e., all of the information on the random variables at steps $0, 1, \ldots, j$.

Thus, by Chebyshev's inequality we have that for every $\epsilon, \delta > 0$ there is $k_1 > k$ so that

$$\mathbb{P}\left[\left(\sum_{i=k}^{k_1} \eta_i \left(\nabla_\theta \Lambda(\theta_i)\right)^\top R_i\right)^2 \geq \epsilon\right] < \delta.$$

In fact, by the Borel-Cantelli lemma, one actually has

$$\sum_{i=k}^{\infty} \eta_i \left(\nabla_\theta \Lambda(\theta_i)\right)^\top R_i \to 0 \quad \text{almost surely.}$$

Thus, if k is large enough, we have

$$\Lambda(\theta_{k_1}) - \Lambda(\theta_k) \approx -\sum_{i=k}^{k_1} \eta_i \|\nabla_\theta \Lambda(\theta_i)\|_2^2 < 0.$$

In fact, if $\|\nabla_\theta \Lambda(\theta_i)\|_2^2 \geq \lambda > 0$, then we have that

$$\Lambda(\theta_{k_1}) - \Lambda(\theta_k) \leq -\lambda \sum_{i=k}^{k_1} \eta_i \to -\infty.$$

But, by assumption $\theta \mapsto \Lambda(\theta)$ is bounded function, so the last display cannot happen. Thus, there is $i > k$ so that $\|\nabla_\theta \Lambda(\theta_i)\|_2^2 < \lambda$. What we just derived is a heuristic derivation of the following classical result.

Theorem 18.11 ([**BT00**]). *Assume $\nabla_\theta \Lambda(\theta)$ is globally Lipschitz and bounded. Assume $\Lambda(\theta)$ is bounded and that the learning rate is such that $\sum_{k=1}^{\infty} \eta_k = \infty$ and $\sum_{k=1}^{\infty} \eta_k^2 < \infty$. Then we have that*

$$\lim_{k \to \infty} \|\nabla_\theta \Lambda(\theta_k)\|_2 = 0 \quad \text{almost surely.}$$

Remark 18.12. Neural networks do not typically satisfy the assumptions of Theorem 18.11 because the gradient of the loss function will typically be neither globally Lipschitz nor globally bounded. Regardless, SGD on neural networks has been proven to be very effective in practice and the message of the theorem has empirically been shown to be valid, i.e., the learning rates should progressively decrease for convergence to occur but not too fast.

18.4. Comparing SGD with GD

18.4.1. SGD Is Like GD with Noise. Let us now investigate how SGD and GD are related to each other. Comparing the analysis in Sections 17.2.3 and 18.3 legitimately raises the question as to why we should use stochastic gradient descent versus gradient descent. After all, if one has a strongly convex function to minimize, GD is much faster to minimize it than SGD. Typically, in order to reach an ϵ neighborhood of the global minimum, GD requires $\mathcal{O}(1/\epsilon)$ steps whereas SGD typically requires $\mathcal{O}(1/\epsilon^2)$ steps.

When making these comparisons however we should keep in mind that the empirical loss we are minimizing consists of M terms. So, taking M into account, GD actually requires $\mathcal{O}(M/\epsilon)$ steps whereas plain SGD still requires $\mathcal{O}(1/\epsilon^2)$ steps.

Thus, if $\epsilon \ll 1/M$, then GD is superior to SGD. However, if $\epsilon \gg 1/M$, then SGD wins! Thus, if the sample size is large and we cannot afford to have a tiny ϵ accuracy, then SGD will typically be less costly and preferable to GD. This is one of the main reasons why SGD is well suited for machine learning.

Let us next investigate a more direct relation between *GD* and *SGD*. We will see that SGD is basically GD with noise. Let us consider for now a fixed learning rate η. Let us recall that

$$\theta_{k+1} = \theta_k - \eta \nabla_\theta \frac{1}{M} \sum_{m=1}^{M} \lambda_{(x_m, y_m)}(\theta_k).$$

Now notice that

$$\begin{aligned}\mathbb{E}\left[\theta_{k+1} - \theta_k | \theta_k\right] &= -\eta \mathbb{E}\left[\nabla_\theta \frac{1}{M} \sum_{m=1}^{M} \lambda_{(x_m, y_m)}(\theta_k) \Big| \theta_k\right] \\ &= -\eta \nabla_\theta \Lambda(\theta_k).\end{aligned}$$

The last display shows that the average change in weights of θ_k is proportional to the full gradient $\nabla_\theta \Lambda(\theta_k)$, which is exactly what GD gives.

What about the change in the variance? We have

$$\begin{aligned}\mathrm{Var}[\theta_{k+1} - \theta_k | \theta_k] &= \frac{\eta^2}{M^2} \sum_{m=1}^{M} \mathrm{Var}\left[\nabla_\theta \ell_{y_m}(\mathfrak{m}(x_m; \theta_k)) | \theta_k\right] \\ &= \frac{\eta^2}{M} \mathrm{Var}[\nabla_\theta \Lambda(\theta_k) | \theta_k],\end{aligned}$$

which follows because we sample uniformly which in turn means that the random variables $\ell_{y_m}(\mathfrak{m}(x_m; \theta_k))$ are i.i.d.

The previous discussion motivates us to model the transition probabilities $\mathbb{P}(\theta_{k+1}|\theta_k)$ as a Gaussian random variable. In particular,

$$\theta_{k+1} = \theta_k + \xi_k, \text{ where } \xi_k \sim N\left(-\eta \nabla_\theta \Lambda(\theta_k), \frac{\eta^2}{M} \mathrm{Var}[\nabla_\theta \Lambda(\theta_k) | \theta_k]\right).$$

So, on average SGD is like GD plus noise that decreases as M increases! Now, this allows us to approximate

$$\theta_{k+1} = \theta_k - \eta \nabla_\theta \Lambda(\theta_k) + \frac{\eta}{\sqrt{M}} \sqrt{\mathrm{Var}[\nabla_\theta \Lambda(\theta_k)]} Z_k, \text{ where } Z_k \sim N(0,1) \text{ i.i.d.}$$

This observation is the basis for continuous time approximation of SGD by stochastic differential equations (see Appendix A for a quick discussion on stochastic differential equations and further bibliographical remarks). In particular, we have the following result.

Theorem 18.13 (**[HLLL19]**). *Let $T < \infty$ be fixed and let us assume that Λ is such that $\sum_{|a|\leq 7} |D^a\Lambda|_\infty < C < \infty$ for some finite constant $C < \infty$ ($D^a\Lambda$ is the ath-order derivatives of the Λ function). Let (x_k, y_k) be a sequence of i.i.d. random variables sampled from some distribution $(X, Y) \sim \mathbb{P}$. Let $\tilde{\Theta}$ be the solution to the stochastic differential equation*

$$\dot{\tilde{\Theta}}(t) = \left(-\nabla\Lambda(\tilde{\Theta}(t)) - \frac{\eta}{4}\nabla|\nabla\Lambda(\tilde{\Theta}(t))|^2\right) + \sqrt{\eta}\sqrt{\mathrm{Var}_{(X,Y)}\left[\nabla\lambda_{(X,Y)}(\tilde{\Theta}(t))\right]}\dot{W}(t),$$

where $W(t)$ is a standard multidimensional Brownian motion. Then with $p = 2$ for any $\phi \in \mathcal{C}_b^{2(p+1)}$, there is $C < \infty$ and $\eta_0 > 0$ such that

$$|\mathbb{E}\phi(\theta_k) - \mathbb{E}\phi(\tilde{\Theta}(k\eta))| \leq C\eta^p \textit{ for all } k \leq T/\eta \textit{ and } \eta \in (0, \eta_0),$$

where θ_k satisfies $\theta_k = \theta_{k-1} - \eta\nabla\lambda_{(x_k, y_k)}(\theta_{k-1})$.

This theorem says that $\tilde{\Theta}$ approximates the sequence $\{\theta_k\}$ with weak order 2. As a matter of fact if we ignore the term $-\frac{\eta}{4}\nabla|\nabla\Lambda(\tilde{\Theta}(t))|^2$, the approximation above yields the same result but with $p = 1$, i.e., a weak error of order 1 approximation instead of 2. With or without the correction term, we can view $\tilde{\Theta}$ as the SGD diffusion approximation to the discrete algorithm for θ_k.

18.4.2. Momentum methods and SGD. In Section 17.4 we explored momentum kinds of methods for gradient descent. As was discussed there, momentum methods can be expected to lead to acceleration of gradient descent methods due to the use of inertia of particles. The natural question is whether momentum methods would be expected to accelerate convergence of SGD as they do with GD. It turns out that the answer to this question is not necessarily yes.

To this end we recall that SGD is a noisy approximation of the full gradient of the dataset. This means that the gradient will always be incorrect in SGD and as such one does not expect that the velocity in the next iteration will be accurate. In fact, it is shown in [**LB20, KNJK18**] that a standard application of either Polyak's or Nesterov's momentum methods does not always lead to acceleration. For instance, as demonstrated in [**LB20**] in the case of the mean-square error loss problem (a strongly convex problem), Nesterov's momentum method when applied to SGD leads to the estimate

$$|\mathbb{E}\left[\Lambda(\theta_k)\right] - \Lambda(\theta^*)| \leq e^{-k\frac{C}{m}}|\Lambda(\theta_0) - \Lambda(\theta^*)|$$

for k large enough, where $m = \frac{L}{\gamma}$ is the condition number and $C > 0$ is some constant. Comparing this to what one gets in the standard gradient descent problem for the same problem

$$|\mathbb{E}[\Lambda(\theta_k)] - \Lambda(\theta^*)| \le e^{-k\frac{1}{m}} |\Lambda(\theta_0) - \Lambda(\theta^*)|,$$

we conclude that the only effect of momentum method to SGD is the multiplicative factor for C. It is worthwhile to mention here that standard GD without momentum would lead to the bound

$$|\mathbb{E}[\Lambda(\theta_k)] - \Lambda(\theta^*)| \le e^{-k\frac{1}{\sqrt{m}}} |\Lambda(\theta_0) - \Lambda(\theta^*)|,$$

which means that momentum does accelerate GD, since when $m > 1$, $\frac{1}{\sqrt{m}} > \frac{1}{m}$. Even though, we will not investigate this in further depth in this book, we mention the following for completeness.

- Modifications to the momentum method to provably accelerate SGD have been recently proposed, see for example [**LB20**].
- A standard way to accelerate stochastic optimization methods is through the use of control variates, see [**RSB12**].

Momentum methods, such as Nesterov's method, are used in practice in conjunction with SGD to train deep neural networks with great success. So, the natural question is why do they work so well when the theory suggests that one should not necessarily see acceleration? One answer to this question is that while training deep neural networks, the SGD gradients are very close to the GD gradients (see [**CS17**]) for which we know that Nesterov's method leads to acceleration. In particular, as discussed in [**CS17**], many deep learning applications to datasets lead to weights of the neural network that have similar gradients with each other. So, even though stochastic gradients are computed on different batches of gradients, they are very similar to each other and thus to the full gradient. We refer the interested reader to [**CS17**] for a more detailed discussion on this.

18.4.3. GD and SGD for Linear Regression. In Chapter 2 we visited linear regression. We return now to this topic making connections with gradient descent and stochastic gradient descent. Consider the simple setting of the one-dimensional least square regression problem.

$$\Lambda(\theta) = \frac{1}{M}\sum_{m=1}^{M}(wx_m - y_m)^2, \quad \theta = w \in \mathbb{R}.$$

It is relatively easy to see that each term of this sum is minimized at $w_m^* = \frac{y_m}{x_m}$. Let us now set $w_{\min}^* = \min\{w_m^*, m = 1, \dots, M\}$ and $w_{\max}^* = \max\{w_m^*, m = 1, \dots, M\}$. If the initialization w_0 is such that

$$w_0 > w_{\max}^* \quad \text{or} \quad w_0 < w_{\min}^*,$$

then after a few number of steps of SGD, say k number of steps, we will eventually have that $w_k \in (w_{\min}^*, w_{\max}^*)$. The region $(w_{\min}^*, w_{\max}^*)$ is sometimes referred to in the literature as the confusion zone.

As soon as $w_k \in (w_{\min}^*, w_{\max}^*)$ then there is no convergence for SGD. For a fixed learning rate and because the weights are sampled uniformly, the weights will move either to the left or to the right of the confusion zone depending on which value was used to compute the gradient. This is because SGD samples a different point at each iteration. Consequently, SGD will oscillate in the confusion zone.

However, the objective function used in linear regression is convex, as the sum of convex functions, and there is a unique global minimum. The global minimum is

$$w^* = \frac{\sum_{m=1}^M x_m y_m}{\sum_{m=1}^M x_m^2},$$

which we recognize as the least squares estimator. In contrast to SGD, GD (which uses the gradient evaluated on the entire dataset) will converge to the global minimum!

18.5. Variants of Stochastic Gradient Descent

In this section we discuss some popular variants of stochastic gradient descent.

18.5.1. AdaGrad. AdaGrad (adaptive gradient) was introduced by [**DHS11**]. We define

$$G_k = \sum_{i=1}^{k} \nabla\Lambda(\theta_i)\,(\nabla\Lambda(\theta_i))^\top,$$

and then we consider the update

$$\theta_{k+1} = \theta_k - \eta G_k^{-1/2} \nabla\Lambda(\theta_k).$$

In high dimensions $G_k^{-1/2}$ is hard to compute. Therefore typically the update is done using only the diagonal elements of the matrix. In particular, the diagonal elements of G_k are $(G_k)_{j,j} = \sum_{i=1}^k ((\nabla\Lambda(\theta_i))_j)^2$ and we have

$$\theta_{k+1} = \theta_k - \eta \cdot \operatorname{diag}\left(G_k^{-1/2}\right) \nabla\Lambda(\theta_k).$$

Note that essentially AdaGrad is an adaptive learning algorithm where the adaptive learning rate is $\eta \cdot \operatorname{diag}\left(G_k^{-1/2}\right)$.

AdaGrad was conceived in order to deal with sparse and unbalanced data. So, it makes sense to use different learning rates for different coordinates. Also note that AdaGrad has a vanishing effective learning rate $\eta \cdot \operatorname{diag}\left(G_k^{-1/2}\right)$, so it will typically converge.

Remark 18.14. When applied in practice, AdaGrad usually takes the following equivalent form

$$\begin{aligned} u_{k+1} &= u_k + \nabla\Lambda(\theta_k) \odot \nabla\Lambda(\theta_k), \\ \theta_{k+1} &= \theta_k - \eta \frac{1}{\epsilon + \sqrt{u_{k+1}}} \odot \nabla\Lambda(\theta_k), \end{aligned}$$

where division and square root are applied elementwise and $\epsilon > 0$ is there to avoid instabilities in the denominator when a region of small gradients is reached by the algorithm.

18.5.2. RMSProp. RMSProp (root-mean square propagation) was developed by [**TH12**]. A drawback of AdaGrad, presented in Section 18.5.1, is that it treats all past gradients equally. It would make sense to use decaying weights for past gradients. One relatively simple way to do so is as follows.

Let $\delta \in (0, 1)$ be the factor controlling the exponential forgetting rate. Then, with $\alpha \odot \beta = (\alpha_1\beta_1, \dots, \alpha_d\beta_d)$ as the elementwise multiplication of two d-dimensional vectors α, β, we set

$$\begin{aligned} u_{k+1} &= \delta u_k + (1-\delta)\nabla\Lambda(\theta_k) \circ \nabla\Lambda(\theta_k), \\ \theta_{k+1} &= \theta_k - \eta \cdot u_{k+1}^{-1/2} \odot \nabla\Lambda(\theta_k). \end{aligned}$$

Note that by iterating the update rule for u_{k+1}, we have

$$u_{k+1} = \delta^{k+1} u_0 + (1-\delta) \sum_{j=1}^{k+1} \delta^{k+1-j} \nabla\Lambda(\theta_j) \odot \nabla\Lambda(\theta_j).$$

Since we have

$$\delta^{k+1} + (1-\delta) \sum_{j=1}^{k+1} \delta^{k+1-j} = 1,$$

we get that u_{k+1} is the weighted average of u_0 and all the elements of $\nabla\Lambda(\theta_j) \circ \nabla\Lambda(\theta_j)$ until step $k+1$. Notice that the most recent updates are more heavily weighted than the earlier updates.

Remark 18.15. When applied in practice, RMSProp usually takes the following equivalent form

$$u_{k+1} = \delta u_k + (1-\delta)\nabla\Lambda(\theta_k) \circ \nabla\Lambda(\theta_k),$$
$$\theta_{k+1} = \theta_k - \eta\frac{1}{\sqrt{\epsilon + u_{k+1}}} \odot \nabla\Lambda(\theta_k),$$

where division and square root are applied elementwise and $\epsilon > 0$ is there to avoid instabilities in the denominator when a region of small gradients is reached by the algorithm.

As an example, let us consider the convergence properties of RMSProp in the case of $\Lambda(\theta) = \frac{1}{2}\|\theta\|_2^2$. For simplicity, we shall focus on the one-dimensional case and we will consider $\theta_0 > 0$. In this case, the update rule for u_k becomes

$$u_{k+1} = \delta^{k+1}u_0 + (1-\delta)\sum_{j=1}^{k+1} \delta^{k+1-j}|\theta_j|^2.$$

The update rule for θ_{k+1} becomes

$$\theta_{k+1} = \theta_k\left(1 - \eta u_{k+1}^{-1/2}\right).$$

The latter relation immediately implies that if

$$0 < \min_{k\in\mathbb{N}}\left(1-\eta u_{k+1}^{-1/2}\right) \leq \max_{k\in\mathbb{N}}\left(1-\eta u_{k+1}^{-1/2}\right) < \zeta < 1,$$

then we will have that

$$\theta_k \in (0, \theta_0\zeta^k),$$

which subsequently implies (recall that we have claimed that $\zeta < 1$) that

$$\lim_{k\to\infty} \theta_k = 0,$$

i.e., we have established convergence to the global minimum of the loss function $\Lambda(\theta) = \frac{1}{2}\|\theta\|^2$.

It remains to discuss when the claim

$$0 < \min_{k\in\mathbb{N}}\left(1-\eta u_{k+1}^{-1/2}\right) \leq \max_{k\in\mathbb{N}}\left(1-\eta u_{k+1}^{-1/2}\right) < \zeta < 1$$

holds. Note that this requirement is equivalent to requiring that

$$\eta^2 < \min_{k\in\mathbb{N}} |u_k| \leq \max_{k\in\mathbb{N}} |u_k| < \left(\frac{\eta}{1-\zeta}\right)^2$$

holds.

First, let us address why $\eta^2 < \min_{k\in\mathbb{N}} |u_k|$ can be assumed to hold. If $\theta_k \to 0$, then there is nothing to prove. So, let us assume that θ_k does not converge to 0. This means that there are $\tau > 0$ and k_0 such that for all $k > k_0$, we have

$|\theta_k| \geq \tau > 0$. Without loss of generality, we may in fact assume that for all $k > 0$, we have $|\theta_k| \geq \tau > 0$ (by potentially adjusting τ). Then, we obtain

$$\begin{aligned} u_{k+1} &= \delta^{k+1} u_0 + (1-\delta) \sum_{j=1}^{k+1} \delta^{k+1-j} |\theta_j|^2 \\ &\geq \delta^{k+1} u_0 + \tau^2 (1-\delta) \sum_{j=1}^{k+1} \delta^{k+1-j} \\ &= \delta^{k+1} u_0 + \tau^2 (1 - \delta^{k+1}) \\ &= \delta^{k+1}(u_0 - \tau^2) + \tau^2 \\ &\geq \tau^2, \end{aligned}$$

if $(u_0 - \tau^2) > 0$. So, we will then obtain that if $\eta^2 < \tau^2$, then we indeed have that for all $k \in \mathbb{N}$, $\eta^2 < \tau^2 < u_{k+1}$. This calculation shows that even in the case where it is assumed that θ_k does not converge to 0, then if u_0 is sufficiently large and η is sufficiently small, we will indeed have that $\eta^2 < u_k$.

The inequality $\max_{k\in\mathbb{N}} |u_k| < \left(\frac{\eta}{1-\zeta}\right)^2$ is equivalent to requiring that the sequence u_k is uniformly bounded in $k \in \mathbb{N}$. This is effectively a consequence of the first inequality. Indeed, since we can indeed choose things so that $0 < \eta^2 < u_k$, assuming that $\theta_0 > 0$, we shall have that

$$\theta_{k+1} \leq \theta_k \leq \theta_{k-1} \leq \cdots \leq \theta_0.$$

Thus, if $\theta_0 < K$, we shall have that

$$\begin{aligned} u_{k+1} &= \delta^{k+1} u_0 + (1-\delta) \sum_{j=1}^{k+1} \delta^{k+1-j} |\theta_j|^2 \\ &\leq \delta^{k+1} u_0 + K^2 (1-\delta) \sum_{j=1}^{k+1} \delta^{k+1-j} \\ &= \delta^{k+1} u_0 + K^2 (1 - \delta^{k+1}) \\ &\leq u_0 + 2K^2, \end{aligned}$$

providing a uniform upper bound for u_{k+1} and thus proving the claim.

Let us conclude this subsection with a useful, general purpose result that can be of use beyond RMSProp. As a matter of fact, a number of algorithms, such as RMSProp, depend on *exponential moving averages*. Suppose that $\{\xi_n\}_{n\in\mathbb{N}}$ is a bounded sequence of real numbers. Fix $\delta \in (0,1)$ and define

$$\begin{aligned} \Xi_0 &= 0, \\ \Xi_{n+1} &= \delta \Xi_n + (1-\delta)\xi_{n+1}, \qquad n \in \{0, 1, \dots\}, \end{aligned}$$

where Ξ_{n+1} is a convex combination of its prior value (memory) and the new data ξ. As $\delta \searrow 0$, Ξ forgets is prior value (no memory), while if $\delta \nearrow 1$, Ξ has full memory, and the new data is discarded. It is easy to check that we explicitly have

$$\Xi_n = (1-\delta) \sum_{n'=1}^{n} \delta^{n-n'} \xi_{n'}$$

for all $n \in \mathbb{N}$. The sequence Ξ is often referred to as the exponential moving average of the ξ_n's.

Let's investigate this. A continuous-time *Tauberian* theorem suggests the connection between exponential moving averages and regular averages.

Lemma 18.16. *Assume that $f \in B(\mathbb{R}_+)$ (i.e., f is a bounded function on $\mathbb{R}_+$) is such that*

$$\bar{f} \stackrel{def}{=} \lim_{T \nearrow \infty} \frac{1}{T} \int_{t=0}^{T} f(t)dt$$

is well defined. Then

$$\lim_{\lambda \searrow 0} \lambda \int_{t=0}^{\infty} e^{-\lambda t} f(t)dt = \bar{f}.$$

Proof. Define

$$F(T) \stackrel{\text{def}}{=} \int_{t=0}^{T} f(t)dt$$

for $T > 0$. Integrating by parts,

$$F(T)e^{-\lambda T} = -\lambda \int_{t=0}^{T} e^{-\lambda t} F(t)dt + \int_{t=0}^{T} e^{-\lambda t} f(t)dt.$$

By assumption, $\|f\| \stackrel{\text{def}}{=} \sup_{t \in (0,\infty)} |f(t)|$ is finite; thus

$$\frac{|F(T)|}{T} \leq \|f\| \tag{18.6}$$

for all $T > 0$. Consequently,

$$\lim_{T \nearrow \infty} F(T)e^{-\lambda T} = 0.$$

Rearranging and multiplying by λ and then changing the variable of integration, we have that

$$\lambda \int_{t=0}^{\infty} e^{-\lambda t} f(t)dt = \lambda^2 \int_{t=0}^{\infty} e^{-\lambda t} F(t)dt = \int_{t=0}^{\infty} \frac{F(t/\lambda)}{t/\lambda} t e^{-t} dt.$$

In light of (18.6), dominated convergence implies the result. □

This implies a corresponding result for exponential moving averages.

Lemma 18.17. *Assume that* $\{\xi_n\}_{n\in\mathbb{N}}$ *is a bounded sequence such that*

$$\bar{\xi} \stackrel{def}{=} \lim_{N\nearrow\infty} \frac{1}{N} \sum_{n=1} \xi_n$$

is well defined. Then

$$\lim_{\delta\nearrow 1}(1-\delta) \sum_{n=0}^{\infty} \delta^n \xi_n = \bar{\xi}.$$

Proof. Let's rewrite the sum as an integral and make the change of variables $\lambda = -\ln\delta$ (so that $\delta = e^{-\lambda}$). Then

$$(1-\delta)\sum_{n=0}^{\infty}\delta^n\xi_n = (1-\delta)\int_{t=0}^{\infty}\delta^{\lfloor t\rfloor}\xi_{\lfloor t\rfloor}dt = (1-e^{-\lambda})\int_{t=0}^{\infty}e^{-\lambda\lfloor t\rfloor}\xi_{\lfloor t\rfloor}dt.$$

The asymptotic $\delta \nearrow 1$ is equivalent to $\lambda \searrow 0$. Rearranging so that we can use the previous result, we write

$$\begin{aligned}(1-\delta)\sum_{n=0}^{\infty}\delta^n\xi_n &= (1-e^{-\lambda})\int_{t=0}^{\infty}e^{-\lambda t}\exp[-\lambda(\lfloor t\rfloor - t)]\,\xi_{\lfloor t\rfloor}dt \\ &= \frac{1-e^{-\lambda}}{\lambda}\left\{\lambda\int_{t=0}^{\infty}e^{-\lambda t}\xi_{\lfloor t\rfloor}dt + \lambda\int_{t=0}^{\infty}e^{-\lambda t}\{\exp[\lambda(t-\lfloor t\rfloor)]-1\}\,\xi_{\lfloor t\rfloor}dt\right\}.\end{aligned}$$

By the above standard Tauberian theorem,

$$\lim_{\lambda\searrow 0}\lambda\int_{t=0}^{\infty}e^{-\lambda t}\xi_{\lfloor t\rfloor}dt = \lim_{T\nearrow\infty}\frac{1}{T}\int_{t=0}^{T}\xi_{\lfloor t\rfloor}dt = \bar{\xi}.$$

We of course also have that

$$\lim_{\lambda\searrow 0}\frac{1-e^{-\lambda}}{\lambda} = 1.$$

By assumption, $\|\xi\| \stackrel{\text{def}}{=} \sup_{n\in\mathbb{N}}|\xi_n|$ is finite. Using this and rescaling,

$$\left|\lambda\int_{t=0}^{\infty}e^{-\lambda t}\{\exp[\lambda(t-\lfloor t\rfloor)]-1\}\,\xi_{\lfloor t\rfloor}dt\right| \le \|\xi\|\int_{s=0}^{\infty}e^{-s}\left|\exp[s-\lambda\lfloor s/\lambda\rfloor]-1\right|ds.$$

Since $\lfloor x\rfloor \le x \le \lfloor x\rfloor + 1$ for all $x \ge 0$,

$$\lfloor s/\lambda\rfloor \le s/\lambda < \lfloor s/\lambda\rfloor + 1;$$

multiplying by λ and subtracting,

$$0 \le s - \lambda\lfloor s/\lambda\rfloor \le \lambda.$$

Hence, we obtain

$$\left|\exp[s-\lambda\lfloor s/\lambda\rfloor]-1\right| = \exp[s-\lambda\lfloor s/\lambda\rfloor]-1 \le e^{\lambda}-1,$$

which then leads to

$$\begin{aligned}\overline{\lim_{\lambda\searrow 0}}\left|\lambda\int_{t=0}^{\infty}e^{-\lambda t}\left\{\exp[\lambda(t-\lfloor t\rfloor)]-1\right\}\xi_{\lfloor t\rfloor}dt\right| &\le \overline{\lim_{\lambda\searrow 0}}\|\xi\|\int_{s=0}^{\infty}e^{-s}\left(e^{\lambda}-1\right)ds\\ &= \overline{\lim_{\lambda\searrow 0}}\|\xi\|\left(e^{\lambda}-1\right)\\ &= 0.\end{aligned}$$

Combining things together, we get the claim. □

18.5.3. ADAM. ADAM (adaptive learning method) was introduced in [**KB15**]. ADAM combines RMSProp (Section 18.5.2) and the momentum method. ADAM uses exponential moving averages to estimate first and second moments of the gradient and then applies bias corrections. We initialize $m_0 = u_0 = 0$, consider the exponential decay rates $\delta_1, \delta_2 \in (0,1)$, and define the updates

$$\begin{aligned}m_{k+1} &= \delta_1 m_k + (1-\delta_1)\nabla\Lambda(\theta_k),\\ u_{k+1} &= \delta_2 u_k + (1-\delta_2)\nabla\Lambda(\theta_k)\odot\nabla\Lambda(\theta_k).\end{aligned}$$

The moments m_{k+1} and u_{k+1} can be thought of as biased estimates of the first and second moments of $\nabla\Lambda(\theta)$ dictated by exponential moving averaging. In particular, by iterating the updates above we get

$$\begin{aligned}m_{k+1} &= (1-\delta_1)\sum_{i=1}^{k+1}\delta_1^{k+1-i}\nabla\Lambda(\theta_i),\\ u_{k+1} &= (1-\delta_2)\sum_{i=1}^{k+1}\delta_2^{k+1-i}\nabla\Lambda(\theta_i)\odot\nabla\Lambda(\theta_i).\end{aligned}$$

Next, we take expectations on the left and right-hand side of the formulas above. As done in [**KB15**], assuming that the first and second moments of $\nabla\Lambda(\theta_i)$ are stationary, we would get that

$$\begin{aligned}\mathbb{E}m_{k+1} &= \mathbb{E}\left[(1-\delta_1)\sum_{i=1}^{k+1}\delta_1^{k+1-i}\nabla\Lambda(\theta_i)\right] = (1-\delta_1^{k+1})\mathbb{E}\nabla\Lambda(\theta_{k+1}),\\ \mathbb{E}u_{k+1} &= \mathbb{E}\left[(1-\delta_2)\sum_{i=1}^{k+1}\delta_2^{k+1-i}\nabla\Lambda(\theta_i)\odot\nabla\Lambda(\theta_i)\right]\\ &= (1-\delta_2^{k+1})\mathbb{E}\left[\nabla\Lambda(\theta_{k+1})\odot\nabla\Lambda(\theta_{k+1})\right].\end{aligned}$$

This calculation motivates the introduction of the bias-corrected first and second raw moment estimates,

$$\hat{m}_{k+1} = \frac{1}{1-\delta_1^{k+1}} m_{k+1},$$
$$\hat{u}_{k+1} = \frac{1}{1-\delta_2^{k+1}} u_{k+1}.$$

With that in mind the final recursive formula for θ_{k+1} is defined to be

$$\theta_{k+1} = \theta_k - \eta \frac{\hat{m}_{k+1}}{\sqrt{\hat{u}_{k+1}} + \epsilon},$$

where $\epsilon > 0$ is there to prevent dividing by zero. Typical values suggested by the authors of [**KB15**] for the parameters are $\delta_1 = 0.9$, $\delta_2 = 0.999$, $\epsilon = 10^{-8}$, and $\eta = 10^{-3}$. ADAM is very popular for neural networks.

18.5.4. AdaMax. AdaMax (adaptive maximum method) was introduced in [**KB15**]. AdaMax is a variant of ADAM 18.5.3 using the infinity norm. Before presenting the method, let us first comment on the connection with ADAM. In ADAM, the update of u_{k+1} is

$$u_{k+1} = (1-\delta_2) \sum_{i=1}^{k+1} \delta_2^{k+1-i} \nabla\Lambda(\theta_i) \odot \nabla\Lambda(\theta_i).$$

This can be thought of as the update rule for individual weights scaling their gradients inversely proportional to a scaled L^2 norm of the individual current and past gradients. One can envision generalizing the L^2 norm based update rule to an L^p norm based update rule. As we will see next if we take $p \to \infty$ one obtains a simple and stable algorithm, [**KB15**].

In particular, let us consider

$$u_{k+1}^p = (1-\delta_2^p) \sum_{i=1}^{k+1} \delta_2^{p(k+1-i)} \|\nabla\Lambda(\theta_i)\|_p^p.$$

Next, we see how this update behaves as $p \to \infty$. We get

$$\begin{aligned}
u_{k+1} &= \lim_{p\to\infty} \left(u_{k+1}^p\right)^{1/p} = \lim_{p\to\infty} \left((1-\delta_2^p) \sum_{i=1}^{k+1} \delta_2^{p(k+1-i)} \|\nabla\Lambda(\theta_i)\|_p^p \right)^{1/p} \\
&= \lim_{p\to\infty} \left(\sum_{i=1}^{k+1} \left(\delta_2^{(k+1-i)} \|\nabla\Lambda(\theta_i)\|_p\right)^p \right)^{1/p} \\
&= \max\{\delta_2^k \|\nabla\Lambda(\theta_1)\|_\infty, \delta_2^{k-1} \|\nabla\Lambda(\theta_2)\|_\infty, \dots, \delta_2 \|\nabla\Lambda(\theta_k)\|_\infty, \|\nabla\Lambda(\theta_{k+1})\|_\infty\}.
\end{aligned}$$

The latter now can be written as

$$u_{k+1} = \max\{\delta_2 u_k, \|\nabla\Lambda(\theta_{k+1})\|_\infty\}$$

with $u_0 = 0$ and there is no need to correct for initialization bias in this case.

Thus summarizing, we initialize $m_0 = u_0 = 0$, consider the exponential decay rates $\delta_1, \delta_2 \in (0,1)$ as in Section 18.5.3, and define the updates

$$m_{k+1} = \delta_1 m_k + (1-\delta_1)\nabla\Lambda(\theta_k),$$
$$u_k = \max\{\delta_2 u_{k-1}, \|\nabla\Lambda(\theta_k)\|_\infty\}.$$

Then the parameters are updated as

$$\theta_{k+1} = \theta_k - \eta\frac{1}{1-\delta_1^{k+1}}\frac{m_{k+1}}{u_k}.$$

18.6. Brief Concluding Remarks

The proofs of Theorems 18.6, 18.7, 18.8, and 18.9 are based on [**BCN18**]. Some of the first main convergence results appeared in [**BT00**]. The book chapter [**Bot12**] contains much practical advice on how to implement and make use of SGD in practical applications. Momentum methods with SGD are discussed in [**LB20, KNJK18**]. The variants of SGD, such as AdaGrad, RMSProp, ADAM, and AdaMax, all of which are routinely used in practice, were introduced in [**DHS11**], [**TH12**], [**KB15**], and [**KB15**], respectively. The books [**BPM90, KY03**] cover a range of topics in stochastic approximation and adaptive algorithms in general.

18.7. Exercises

Exercise 18.1. Consider the stochastic gradient descent update

$$x_{t+1} = x_t - \eta(\nabla\Lambda(x_t) + \sigma\xi_t),$$

where $\eta < 1$, $\xi_t \sim N(0,1)$ i.i.d., and $\Lambda(x) = \frac{1}{2}\|x\|_2^2$. Show that as $t \to \infty$, x_t converges in distribution to a $N\left(0, \frac{\eta^2\sigma^2}{2\eta-\eta^2}\right)$. Can you interpret the effect of the injected noise $\sigma\xi_t$?

Exercise 18.2. What are the effects of the different hyperparameters in the RMSProp algorithm? What is the advtantage of the RMSProp algorithm over the AdaGrad algorithm?

Exercise 18.3. Suppose we randomly initialize a neural network $\mathfrak{m}(x;\theta)$ and train it for a long time with the final parameter estimate being θ^1. Suppose, we again randomly initialize the same neural network model $\mathfrak{m}(x;\theta)$ and train it again with the final parameter estimate being θ^2. Will θ^1 and θ^2 be similar and why?

Exercise 18.4. Suppose we randomly train K neural network models

$$\mathfrak{m}(x;\theta^1),\ldots,\mathfrak{m}(x;\theta^K)$$

using stochastic gradient descent. Furthermore, each time we train a new model the initial parameter is randomly initialized. Therefore, the trained $\theta^1,\ldots,\theta^K$ are i.i.d.

Is it correct to use the ensemble model $\mathfrak{m}\left(x;\frac{1}{K}\sum_{k=1}^K \theta^k\right)$, and why?

Exercise 18.5. Consider an example where the neural network will not train (i.e., the gradient with respect to at least one of the parameters will always be zero). It is sufficient to consider mean-square error, a single data sample (x,y), and a single ReLU hidden unit.

Exercise 18.6. Provide the details of the proof of Lemma 18.3.

Chapter 19

The Neural Tangent Kernel Regime

19.1. Introduction

The uniform approximation theory for neural networks that we explored in Chapter 16 shows that artificial neural networks with sufficiently many hidden units can approximate any reasonable function. With that in mind, we can ask the following natural questions:

- To where does the optimization of artificial neural networks converge?
- How does the loss surface of artificial neural networks look?
- Why do neural networks typically generalize well? Namely, why do neural networks tend to yield good results on unseen data?

A number of results in the literature (see [**CHM$^+$15, PDGB14, PB17**]) suggest that

- Finding *bad minima*, i.e., local minima with a much higher cost than global minima, is a low probability event as the number of hidden units increase.
- For large size networks, good local minima are equivalent, yield similar performance, and are easier to find than bad local minima.

In this chapter and in Chapter 20, we will visit the questions posed above through the lens of scaling limits for neural networks. As we shall see, proper scaling of the neural network can lead to a well-defined limit as the number of hidden units grows to infinity. In turn, this limit can be analyzed and offer valuable information on how both the limiting loss function and the limit of the scaled neural network behave.

Before we dive into the different scaling limits, we motivate why appropriate scalings make sense through an investigation of the weight initialization scheme suggested by [**GB10**] in Section 19.2. The neural tangent kernel (also called the linear) regime describes the limit of the trained neural network that we get when the parameters are initialized motivated by a square-root weight initialization. The neural tangent kernel (NTK) is presented in Section 19.3. In practice during the training phase, time is discrete and SGD is used. The exact mathematical analysis, convergence to, and convergence properties of the NTK limit (the linear regime) are then presented in Sections 19.4–19.7. In Chapter 20 we will study a different scaling regime, the mean field scaling regime (also called the nonlinear regime).

19.2. Weight Initialization

In this section, we discuss weight initialization. Weight initialization plays an important role in avoiding (to a certain degree) the vanishing and exploding gradient problems. The discussion that follows is largely heuristic, but it is indicative of how one can think about this issue and also represents what is often being done in practice.

In particular, we will go over the mathematics behind the so-called Xavier initialization [**GB10**], which has been very influential for weight initialization during training of neural networks; see also Remark 19.1.

Let $N_{\ell-1}$ be the number of hidden units in the $(\ell-1)$-th layer. Let us start by considering the standard feed forward neural network

$$H_j^{(\ell)} = \sigma\left(\sum_{i=1}^{N_{\ell-1}} w_{i,j}^{(\ell)} H_i^{(\ell-1)} + b_j^{(\ell)}\right).$$

Let us set $Z_j^{(\ell-1)} = \sum_{i=1}^{N_{\ell-1}} w_{i,j}^{(\ell)} H_i^{(\ell-1)} + b_j^{(\ell)}$.

Let us see why randomness at every layer in the initialization is needed and why proper scaling makes sense. We will make use of the following approximation. Consider a smooth function g and a square integrable random variable X. A linear approximation of X about $\mathbb{E}X$, using a first order Taylor expansion, suggests

$$\operatorname{Var}(g(X)) \approx (g'(\mathbb{E}X))^2 \operatorname{Var} X. \tag{19.1}$$

For the sake of this heuristic discussion, let us assume momentarily that $w_{ij}^{(\ell)}, b_j^{(\ell)}$ are deterministic. We would then obtain

$$\text{Var}\left(H_j^{(\ell)}\right) \approx \left(\sigma'\left(\sum_{i=1}^{N_{\ell-1}} w_{i,j}^{(\ell)} \mathbb{E} H_i^{(\ell-1)} + b_j^{(\ell)}\right)\right)^2 \left(\sum_{i=1}^{N_{\ell-1}} \left(w_{ij}^{(\ell)}\right)^2 \text{Var}(H_j^{(\ell-1)})\right).$$

Applying now the Cauchy-Schwarz inequality and assuming that $\|\sigma'\| \leq C < \infty$, we get

$$\sum_{j}^{N_{\ell-1}} \left(\text{Var}\left(H_j^{(\ell)}\right)\right)^2 \leq C^4 \sum_{i,j}^{N_{\ell-1}} \left(w_{ij}^{(\ell)}\right)^4 \sum_{i=1}^{N_{\ell-1}} \left(\text{Var}(H_j^{(\ell-1)})\right)^2.$$

Thus, if $\sum_j^{N_{\ell-1}} \left(w_{ij}^{(\ell)}\right)^4$ is small, then $\sum_j^{N_{\ell-1}} \left(\text{Var}\left(H_j^{(\ell)}\right)\right)^2$ may decrease when ℓ increases. So, in that case after passing through a few layers the signal becomes insignificant.

If, on the other hand, $|w_{ij}^{(\ell)}|$ are large in magnitude, then with $\sigma(x) = x$ we get that (assume momentarily that $w_{ij}^{(\ell)}, b_j^{(\ell)}$ are deterministic)

$$\text{Var}\left(H_j^{(\ell)}\right) = \sum_{i=1}^{N_{\ell-1}} \left(w_{ij}^{(\ell)}\right)^2 \text{Var}\left(H_i^{(\ell-1)}\right),$$

which suggests that in that case $\text{Var}\left(H_j^{(\ell)}\right)$ can increase with ℓ.

If $\sigma(x) = \frac{e^x}{e^x+1}$ is the sigmoid function, then large $|w_{ij}^{(\ell)}|$ means large $\sum_{i=1}^{N_{\ell-1}} w_{ij}^{(\ell)} H_i^{(\ell-1)}$, in which case σ becomes saturated leading to the vanishing gradient problem.

This then brings up the question of how do we initialize the weights in a way that would avoid the issues just described. The idea is to find weight values for which the variance remains fairly unchanged as the signal passes through each layer.

The analysis will be done in two steps, the forward pass and the backward pass. The general assumptions we will make here are the following.

- All inputs, all layers, and all weights at initialization are independent and identically distributed.
- The inputs are normalized with zero means, i.e., $\mathbb{E} H_i^{(0)} = 0$. All weights have mean zero at initialization, i.e., $\mathbb{E} w_{i,j}^{(\ell)} = 0$.
- The activation function σ is an odd function ($\sigma(-x) = -\sigma(x)$) such that $\sigma'(0) = 1$. For example $\sigma(x) = \tanh(x)$ satisfies these constraints.
- To simplify the algebra, let us also assume that the biases are zero, i.e., $b_j^{\ell} = 0$ for all j, ℓ.

19.2.1. Forward Pass. We start by applying formula (19.1) to the activation function σ. Since σ is assumed to be an odd function with $\sigma'(0) = 1$, we may assume that $\sigma(x) \approx x$ close to zero. Recall that $w_{i,j}^{(\ell)} H_i^{(\ell-1)}$ will be by assumption zero at initialization.

We have the following calculations

$$
\begin{aligned}
\operatorname{Var}\left(H_j^{(\ell)}\right) &\approx \left(\sigma'\left(\sum_{i=1}^{N_{\ell-1}} \mathbb{E}\left(w_{i,j}^{(\ell)} H_i^{(\ell-1)}\right)\right)\right)^2 \operatorname{Var}\left(\sum_{i=1}^{N_{\ell-1}} w_{ij}^{(\ell)} H_j^{(\ell-1)}\right) \\
&\approx \operatorname{Var}\left(\sum_{i=1}^{N_{\ell-1}} w_{ij}^{(\ell)} H_j^{(\ell-1)}\right) \qquad (\sigma(x) \approx x \text{ around } x = 0) \\
&= \sum_{i=1}^{N_{\ell-1}} \operatorname{Var}\left(w_{ij}^{(\ell)} H_j^{(\ell-1)}\right) \qquad (w_{ij}^{(\ell)}, H_j^{(\ell-1)} \text{ are independent}) \\
&= \sum_{i=1}^{N_{\ell-1}} \left[\left(\mathbb{E}\left(w_{ij}^{(\ell)}\right)\right)^2 \operatorname{Var}\left(H_j^{(\ell-1)}\right) + \operatorname{Var}\left(w_{ij}^{(\ell)}\right)\left(\mathbb{E}\left(H_j^{(\ell-1)}\right)\right)^2 \right. \\
&\qquad \left. + \operatorname{Var}\left(w_{ij}^{(\ell)}\right) \operatorname{Var}\left(H_j^{(\ell-1)}\right)\right] \\
&\qquad \text{(variance of product formula for two random variables)} \\
&= \sum_{i=1}^{N_{\ell-1}} \left[\operatorname{Var}\left(w_{ij}^{(\ell)}\right) \operatorname{Var}\left(H_j^{(\ell-1)}\right)\right] \\
&\qquad \left(\text{since by assumption } \mathbb{E}\left(w_{ij}^{(\ell)}\right) = \mathbb{E}\left(H_j^{(\ell-1)}\right) = 0\right) \\
&= N_{\ell-1} \left[\operatorname{Var}\left(w_{ij}^{(\ell)}\right) \operatorname{Var}\left(H_j^{(\ell-1)}\right)\right] \qquad \text{(i.i.d. assumption).}
\end{aligned}
$$

Requiring now that the variance of the different hidden layers is the same, we immediately obtain that we should choose

$$
\operatorname{Var}\left(w_{ij}^{(\ell)}\right) = \frac{1}{N_{\ell-1}}.
$$

19.2.2. Backward Pass. The starting point is the desire to maintain the variance of the gradient of the cost as it propagates through layers, in particular, we would like to have

$$
\operatorname{Var}\left(\frac{\partial \Lambda}{\partial Z^{(\ell-1)}}\right) = \operatorname{Var}\left(\frac{\partial \Lambda}{\partial Z^{(\ell)}}\right).
$$

Starting with the chain rule, we have the following computations:

$$
\begin{aligned}
\operatorname{Var}\left(\frac{\partial \Lambda}{\partial Z_j^{(\ell-1)}}\right) &= \operatorname{Var}\left(\sum_{i=1}^{N_\ell} w_{ij}^{(\ell)} \frac{\partial \Lambda}{\partial Z_i^{(\ell)}} \sigma'(Z_i^{(\ell-1)})\right) \quad \text{(chain rule)} \\
&\approx \sum_{i=1}^{N_\ell} \operatorname{Var}\left(w_{ij}^{(\ell)} \frac{\partial \Lambda}{\partial Z_i^{(\ell)}}\right) \\
&\quad \text{(by independence, } \sigma(x) \approx x \text{ around } x = 0 \text{ and } \mathbb{E}(Z_i^{(\ell-1)}) = 0) \\
&= \sum_{i=1}^{N_\ell} \left[\left(\mathbb{E}\left(w_{ij}^{(\ell)}\right)\right)^2 \operatorname{Var}\left(\frac{\partial \Lambda}{\partial Z_i^{(\ell)}}\right) + \operatorname{Var}\left(w_{ij}^{(\ell)}\right)\left(\mathbb{E}\left(\frac{\partial \Lambda}{\partial Z_i^{(\ell)}}\right)\right)^2 \right. \\
&\quad \left. + \operatorname{Var}\left(w_{ij}^{(\ell)}\right) \operatorname{Var}\left(\frac{\partial \Lambda}{\partial Z_i^{(\ell)}}\right)\right] \\
&\quad \text{(variance of product formula for two random variables)} \\
&= \sum_{i=1}^{N_\ell} \left[\operatorname{Var}\left(w_{ij}^{(\ell)}\right) \operatorname{Var}\left(\frac{\partial \Lambda}{\partial Z_i^{(\ell)}}\right)\right] \\
&\quad \text{(by assuming } \mathbb{E}\left(w_{ij}^{(\ell)}\right) = \mathbb{E}\left(\frac{\partial \Lambda}{\partial Z_i^{(\ell)}}\right) = 0) \\
&= N_\ell \left[\operatorname{Var}\left(w_{ij}^{(\ell)}\right) \operatorname{Var}\left(\frac{\partial \Lambda}{\partial Z_i^{(\ell)}}\right)\right] \quad \text{(i.i.d. assumption).}
\end{aligned}
$$

So eventually requiring that $\operatorname{Var}\left(\frac{\partial \Lambda}{\partial Z^{(\ell-1)}}\right) = \operatorname{Var}\left(\frac{\partial \Lambda}{\partial Z^{(\ell)}}\right)$ leads to the choice

$$\operatorname{Var}\left(w_{ij}^{(\ell)}\right) = \frac{1}{N_\ell}.$$

19.2.3. Conclusions and Motivation for Scaling Limits. The forward and backward calculations suggest that the variance of weights should be chosen to be inversely proportional to the number of hidden units in the given layer and in the previous layer too.

Even though these two expressions do not agree in the case $N_\ell \neq N_{\ell-1}$, one can replace them by the requirement that the variance of the weights is proportional to the average of the two, i.e.,

$$\operatorname{Var}\left(w_{ij}^{(\ell)}\right) = \frac{2}{N_\ell + N_{\ell-1}},$$

in which case the essence of the message coming from these calculations is still maintained. Namely, a principled choice for the weights is to initialize them so that their variance is inversely proportional to the number of layers of the current layer and/or of the previous layer. Even though these heuristics may not be solving the vanishing or exploding gradient problems, they are at least improving upon this issue.

For example, the following are then viable choices based on this framework.

- If w_{ij} are Gaussian, then we are led to choose
$$w_{ij} \sim N\left(0, \frac{2}{N_\ell + N_{\ell-1}}\right).$$
- If w_{ij} are uniform, then we are led to choose
$$w_{ij} \sim \text{Uniform}\left[-\sqrt{\frac{6}{N_{\ell-1} + N_{\ell-1}}}, \sqrt{\frac{6}{N_{\ell-1} + N_{\ell-1}}}\right].$$
The latter calculation stems from the fact that if $W \sim \text{Uniform}(-a, a)$, then $\text{Var}(W) = \frac{1}{3}a^2$ and then we set $\frac{1}{3}a^2 = \frac{2}{N_\ell + N_{\ell-1}}$ and solve for a.

Remark 19.1. In addition to the Xavier initialization that we presented here, another popular initialization scheme is the so-called He initialization [**HZRS15**] (sometimes also referred to Kaiming initialization). The He initialization suggests the choice of $\frac{1}{2}N_\ell \,\text{Var}\left(w_{ij}^{(\ell)}\right) = 1$, which leads to a zero-mean Gaussian initialization with standard deviation $\sqrt{2/N_\ell}$. The biases $b_j^{(\ell)}$ are initialized at zero. The difference between He initialization and Xavier initialization is that while the derivation for the Xavier initialization is based on the linear activation function, the derivation of He initialization takes into account ReLU activation functions.

The next natural question to pose is how a neural network that incorporates such initialization behaves in the limit as $N_{\ell-1}, N_\ell \to \infty$. We visit this question in the next section and the analysis gives rise to what is referred to in the literature as the neural tangent kernel, [**JGH18**]. As we shall see in the next section, weight initialization can be thought of being equivalent to an appropriate scaling of the neural network. This way of thinking gives rise to other kinds of scaling, such as the mean field scaling that we will explore in Chapter 20.

19.3. The Linear Asymptotic Regime: Neural Tangent Kernel

The *neural tangent kernel* (NTK) was initially introduced in [**JGH18**] and has been quite influential on how one can think about limiting theory and related analysis for training neural networks. The motivation for the formulation that follows comes from the weight initialization scalings that we saw in Section 19.2. In order to simplify the presentation and introduce the ideas in a more pedagogical setting, we restrict our attention to the shallow neural network case. An induction argument then directly gives the limit in the case of deep neural networks.

In order to introduce the ideas, consider the simplest possible setting of a shallow neural network where we set

$$\mathfrak{m}^N(x;\theta) = \frac{1}{\sqrt{N}} \sum_{n=1}^{N} C^n \sigma(W^n \cdot x),$$

with $C^n \in \mathbb{R}$, $W^n \in \mathbb{R}^d$, $x \in \mathbb{R}^d$, and $\sigma(\cdot) : \mathbb{R} \to \mathbb{R}$. The number of hidden units is N. We note that even though we do not include the bias term, i.e., we consider the model $\mathfrak{m}^N(x;\theta) = \frac{1}{\sqrt{N}} \sum_{n=1}^{N} C^n \sigma(W^n \cdot x)$ instead of $\mathfrak{m}^N(x;\theta) = \frac{1}{\sqrt{N}} \sum_{n=1}^{N} C^n \sigma(W^n \cdot x + b^n)$, we do so without loss of generality. Indeed, one can always consider the vector x to have the first element being equal to 1 which then immediately incorporates the bias term.

This formulation is slightly different from the discussion in Section 19.2, where one would typically have no scaling of $\frac{1}{\sqrt{N}}$ and instead sample C, W from mean zero normal distribution with variance of order $\frac{1}{N}$. We do note here that scaling the variance of C by $1/N$ without any scaling in the neural network is the same as scaling the neural network by $1/\sqrt{N}$ and sampling C from mean zero normal distribution with unit variance. The current formulation though is key in order to obtain a consistent asymptotic behavior as $N \to \infty$.

This section demonstrates how the neural network behaves as $N \to \infty$ and shows how the NTK arises. In order to motivate things, we shall treat time in this section as being continuous, even though in reality time evolves discretely. In this section we also use gradient descent. In Section 19.4 we switch gears and we will rigorously derive the evolution in discrete time and using stochastic gradient descent instead of gradient descent.

Let us assume that we have M datapoints and that the loss function is

$$\Lambda^N(\theta) = \frac{1}{2}\frac{1}{M} \sum_{m=1}^{M} \left(y_m - \mathfrak{m}^N(x_m;\theta)\right)^2.$$

Assuming that parameters $\theta_t = (C_t^1, \ldots, C_t^N, W_t^1, \ldots, W_t^N) \in \mathbb{R}^{N\times(1+d)}$ evolve in continuous time based on gradient descent, we have the following update equations:

$$
\begin{aligned}
\dot{C}_t^i &= \frac{\eta}{\sqrt{N}} \frac{1}{M} \sum_{m=1}^{M} \left(y_m - \mathfrak{m}^N(x_m;\theta_t)\right) \sigma(W_t^n \cdot x_m), \\
\dot{W}_t^i &= \frac{\eta}{\sqrt{N}} \frac{1}{M} \sum_{m=1}^{M} \left(y_m - \mathfrak{m}^N(x_m;\theta_t)\right) C_t^n \sigma'(W_t^n \cdot x_m) x_m, \\
\mathfrak{m}^N(x;\theta_t) &= \frac{1}{\sqrt{N}} \sum_{n=1}^{N} C_t^n \sigma(W_t^n \cdot x),
\end{aligned}
\tag{19.2}
$$

with $\eta > 0$ a given learning rate. At time $t = 0$ we initialize the parameters θ_0 in an i.i.d. fashion, from some distribution μ_0 with at least two finite moments.

Let us next define the empirical measure sitting on the learned parameters θ_t,

$$
\mu_t^N = \frac{1}{N} \sum_{n=1}^{N} \delta_{\theta_t^n}.
$$

Differentiating now in time the model $\mathfrak{m}^N(x;\theta_t)$ and using the update equations from (19.2), we obtain

$$
\begin{aligned}
\frac{d}{dt}\mathfrak{m}^N(x;\theta_t) &= \frac{1}{\sqrt{N}} \sum_{n=1}^{N} \left[\dot{C}_t^n \sigma(W_t^n \cdot x) + C_t^n \sigma'(W_t^n \cdot x) \dot{W}_t^n \cdot x\right] \\
&= \frac{\eta}{N} \sum_{n=1}^{N} \frac{1}{M} \sum_{m=1}^{M} \left(y_m - \mathfrak{m}^N(x_m;\theta_t)\right) \\
&\quad \times \left[\sigma(W_t^n \cdot x_m)\sigma(W_t^n \cdot x) + |C_t^n|^2 \sigma'(W_t^n \cdot x_m)\sigma'(W_t^n \cdot x) x_m \cdot x\right] \\
&= \frac{\eta}{M} \sum_{m=1}^{M} \left(y_m - \mathfrak{m}^N(x_m;\theta_t)\right) \\
&\quad \times \left\langle \sigma(w \cdot x_m)\sigma(w \cdot x) + c^2 \sigma'(w \cdot x_m)\sigma'(w \cdot x) x_m \cdot x, \mu_t^N \right\rangle.
\end{aligned}
$$

If we now set

$$
A(x, x'; \mu) = \left\langle \sigma(w \cdot x')\sigma(w \cdot x) + c^2 \sigma'(w \cdot x')\sigma'(w \cdot x) x' \cdot x, \mu \right\rangle,
$$

we see that for a given measure μ, the matrix A with elements $A(x_i, x_j; \mu)$ for $i, j = 1, \ldots, M$ is a symmetric and positive semidefinite matrix, see Corollary 19.6. Hence it defines a kernel and is the basis for the NTK. In particular, we can write

$$
\frac{d}{dt}\mathfrak{m}^N(x;\theta_t) = \frac{\eta}{M} \sum_{m=1}^{M} \left(y_m - \mathfrak{m}^N(x_m;\theta_t)\right) A(x, x_m; \mu_t^N).
$$

At time $t = 0$, i.e., at initialization, the i.i.d. assumption on the random variables θ_0 shows that for a given x (this is the standard central limit theorem for i.i.d. random variables, see Appendix A),

$$\mathfrak{m}^N(x;\theta_0) \stackrel{d}{\to} N(0, \langle |c\sigma(w\cdot x)|^2, \mu_0\rangle).$$

Let us denote the limit Gaussian distribution $N(0, \langle |c\sigma(w\cdot x)|^2, \mu_0\rangle)$ by $\mathcal{G}(x)$. For a finite dataset $\mathcal{D}$, we shall write $\mathcal{G}$ for the vector with elements $\mathcal{G}(x)$.

The next step is to investigate the behavior of the empirical measure μ^N as $N \to \infty$. We claim that μ_t^N actually converges to the distribution at initialization, i.e., $\mu_t^N \to \mu_0$ as $N \to \infty$. Namely, we claim that as N gets large, the distribution of the parameters remains very close to their distribution at initialization. To see this, let us fix a smooth and bounded function $g \in C_b^1(\mathbb{R}^{d+1})$ and study the evolution in time of the pairing $\langle g, \mu_t^N\rangle$. Using (19.2), we have

$$\begin{aligned}
\frac{d}{dt}\langle g, \mu_t^N\rangle &= \frac{d}{dt}\frac{1}{N}\sum_{n=1}^N g(C_t^n, W_t^n) \\
&= \frac{1}{N}\sum_{n=1}^N \left[\partial_c g(C_t^n, W_t^n)\dot{C}_t^n + \nabla_w g(C_t^n, W_t^n)\cdot \dot{W}_t^n\right] \\
&= \frac{\eta}{N^{3/2}}\sum_{n=1}^N \frac{1}{M}\sum_{m=1}^M (y_m - \mathfrak{m}^N(x_m;\theta_t)) \\
&\quad \times [\partial_c g(C_t^n, W_t^n)\sigma(W_t^n\cdot x_m) + \nabla_w g(C_t^n, W_t^n)\cdot C_t^n\sigma'(W_t^n\cdot x_m)x_m] \\
&= \frac{\eta}{\sqrt{N}}\frac{1}{M}\sum_{m=1}^M (y_m - \mathfrak{m}^N(x_m;\theta_t)) \\
&\quad \times \langle \partial_c g(c,w)\sigma(w\cdot x_m) + \nabla_w g(c,w)\cdot c\sigma'(x\cdot x_m)x_m, \mu_t^N\rangle.
\end{aligned}$$

Assume now that the activation function is $\sigma \in C_b^1$. The latter expression and boundedness of the components on its right-hand side (which can be derived similarly to the corresponding statement in discrete time, see Section 19.5.1) immediately show that as $N \to \infty$

$$\langle g, \mu_t^N\rangle \to \langle g, \mu_0\rangle.$$

Hence, we then get as $N \to \infty$ that for fixed x, x'

$$A(x, x'; \mu_t^N) \to A(x, x'; \mu_0).$$

This last statement says that for large N, the scaling of the neural network by $1/\sqrt{N}$ leads to the kernel A being constant over time and to the distribution of the trained parameters being close to their distribution at initialization.

Let us now set $\mathfrak{m}^N = \left(\mathfrak{m}^N(x_1), \dots, \mathfrak{m}^N(x_M)\right)$. The preceding calculations show that $\mathfrak{m}^N \to \mathfrak{m}$ as $N \to \infty$ where $\mathfrak{m}$ satisfies

$$d\mathfrak{m}_t = \frac{\eta}{M} A\left(\hat{Y} - \mathfrak{m}_t\right) dt,$$
$$\mathfrak{m}_0 = \mathcal{G},$$

with $\hat{Y} = (y_1, \dots, y_M)$.

Therefore, $\mathfrak{m}_t$ is the solution to a continuous-time gradient descent algorithm which minimizes the quadratic objective function,

$$J(\hat{Y}, \mathfrak{m}_t) = \frac{1}{2}(\hat{Y} - \mathfrak{m}_t)^\top A(\hat{Y} - \mathfrak{m}_t).$$

Therefore, even though the prelimit optimization problem is nonconvex, the neural network's limit will minimize a quadratic objective function.

Then as we shall see in Theorem 19.4, given that under the proper assumptions A is positive definite by Corollary 19.6, we have that

$$\mathfrak{m}_t = \hat{Y} + \left(\mathcal{G} - \hat{Y}\right) e^{-At},$$

showing that $\mathfrak{m}_t \to \hat{Y}$ as $t \to \infty$ exponentially fast. That is, in the limit of large numbers of hidden units and many training steps, the neural network model converges to a global minimum with zero training error. Namely, in the limit as $N \to \infty$ and $t \to \infty$, the algorithm recovers the true (at least in-sample) data.

19.4. The Linear Asymptotic Regime in the Discrete Time Case

In this section we rigorously derive the NTK analyzing the algorithm in discrete time which is also what is actually implemented in practice. We also show this result for the stochastic gradient descent algorithm.

To simplify the discussion we shall consider the simplest possible setting where we set

$$\mathfrak{m}^N(x; \theta) = \frac{1}{\sqrt{N}} \sum_{n=1}^{N} C^n \sigma(W^n \cdot x),$$

where $C^n \in \mathbb{R}$, $W^n \in \mathbb{R}^d$, $x \in \mathbb{R}^d$, and $\sigma(\cdot) : \mathbb{R} \to \mathbb{R}$. The number of hidden units is N and the output is scaled by a factor $\frac{1}{\sqrt{N}}$ (the widely used Xavier initialization [**GB10**]). We note that the analysis that follows can be generalized to feed forward neural networks of arbitrary depth with a little bit more additional work.

The objective function is

$$\Lambda^N(\theta) = \frac{1}{2}\frac{1}{M}\sum_{m=1}^{M}\left(y_m - \mathfrak{m}^N(x_m;\theta)\right)^2,$$

where $y_m \in \mathbb{R}$, $x_m \in \mathbb{R}^d$, and the parameters $\theta = (C^1, \dots, C^N, W^1, \dots, W^N) \in \mathbb{R}^{N\times(1+d)}$. For notational convenience, we may refer to $\mathfrak{m}^N(x;\theta)$ as $\mathfrak{m}^N(x)$ in our analysis below.

The model parameters θ are trained using stochastic gradient descent. The parameter updates are given by

$$C^n_{k+1} = C^n_k + \frac{\eta^N_k}{\sqrt{N}}(y_k - \mathfrak{m}^N_k(x_k))\sigma(W^n_k \cdot x_k),$$

$$W^n_{k+1} = W^n_k + \frac{\eta^N_k}{\sqrt{N}}(y_k - \mathfrak{m}^N_k(x_k))C^n_k\sigma'(W^n_k \cdot x_k)x_k,$$

$$\mathfrak{m}^N_k(x) = \frac{1}{\sqrt{N}}\sum_{n=1}^{N} C^n_k\sigma(W^n_k \cdot x), \tag{19.3}$$

for $k = 0, 1, \dots, \lfloor TN \rfloor$ where $T > 0$. We use η^N_k as the learning rate (which may depend upon N, k or both). The data samples are (x_k, y_k) and are assumed to be i.i.d. samples from a distribution $\pi(dx, dy)$.

We impose the following assumption.

Assumption 19.2. We have the following:

- The activation function $\sigma \in C^2_b(\mathbb{R})$, i.e., σ, is twice continuously differentiable and bounded with bounded derivatives.
- The randomly initialized parameters (C^n_0, W^n_0) are i.i.d., mean-zero random variables from a distribution $\mu_0(dc, dw)$.
- The random variable C^n_0 has compact support and $\langle \|w\|, \mu_0\rangle < \infty$.
- The sequence of data samples (x_k, y_k) is i.i.d. from the probability distribution $\pi(dx, dy)$.
- There is a fixed dataset of M data samples $(x_m, y_m)_{m=1}^M$ and therefore $\pi(dx, dy) = \frac{1}{M}\sum_{m=1}^M \delta_{(x_m, y_m)}(dx, dy)$.

Note that the last assumption also implies that $\pi(dx, dy)$ has compact support.

We will study the limiting behavior of the network output $\mathfrak{m}^N_k(x)$ for $x \in \mathcal{D} = \{x_1, \dots, x_M\}$ as the number of hidden units N and stochastic gradient descent steps $k = \lfloor Nt \rfloor$ simultaneously become large.

In particular, we view the network output $\mathfrak{m}^N_{\lfloor Nt \rfloor}(x)$ as a stochastic process. We shall prove that it converges in distribution to the solution of a random ODE as $N \to \infty$.

For this purpose, we define the empirical measure

$$\nu^N_k = \frac{1}{N} \sum_{n=1}^{N} \delta_{C^n_k, W^n_k},$$

and we observe that $\mathfrak{m}^N_k$ can be written as the inner-product

$$\mathfrak{m}^N_k(x) = \left\langle c\sigma(w \cdot x), \sqrt{N}\nu^N_k \right\rangle.$$

For each fixed for $x \in \mathcal{D}$, Assumption 19.2 allows us to use the standard central limit theorem (see Appendix A) to obtain that as $N \to \infty$,

$$\mathfrak{m}^N_0(x) \xrightarrow{d} \mathcal{G}(x),$$

where $\mathcal{G} \in \mathbb{R}^M$ is the Gaussian random variable $N(0, \langle |c\sigma(w \cdot x)|^2, \mu_0 \rangle)$. For the same reasons, the standard law of large numbers also gives

$$\nu^N_0 \xrightarrow{p} \nu_0 \equiv \mu_0.$$

Next, we define the scaled processes

$$\begin{aligned} h^N_t &= \mathfrak{m}^N_{\lfloor Nt \rfloor}, \\ \mu^N_t &= \nu^N_{\lfloor Nt \rfloor}, \end{aligned}$$

where $\mathfrak{m}^N_k = (\mathfrak{m}^N_k(x_1), \ldots, \mathfrak{m}^N_k(x_M))$, $h^N_t = (h^N_t(x_1), \ldots, h^N_t(x_M))$, where we set $h^N_t(x) = \mathfrak{m}^N_{\lfloor Nt \rfloor}(x)$.

We observe that (μ^N_t, h^N_t) is a pair of stochastic processes. We are interested in its behavior as $N \to \infty$. For this purpose, we will study its convergence in distribution as $N \to \infty$ in the space $D_E([0, T])$ where $E = \mathcal{M}(\mathbb{R}^{1+d}) \times \mathbb{R}^M$. $D_E([0, T])$ is the Skorokhod space (see Section A.4 for definitions) and $\mathcal{M}(S)$ is the space of probability measures on S.

We prove in this chapter that a neural network with Xavier initialization and trained with stochastic gradient descent converges in distribution to a random ODE as the number of units and training steps become large. In addition, the convergence analysis will also address several interesting questions:

- The results provide a rigorous convergence guarantee for Xavier initialization (i.e., the $\frac{1}{\sqrt{N}}$ normalization factor), which is almost universally used in deep learning models. A priori it is unclear if the neural network $\mathfrak{m}^N_k(x)$ will converge as $N \to \infty$ since, for $k > 0$, the $C^n \sigma(W^n \cdot x)$ is correlated with $C^j \sigma(W^j \cdot x)$ and therefore a limit may not exist. If a limit did not exist, this would imply that the neural network model could have poor numerical behavior for large N. We prove that a limit does exist.

- Although the prelimit problem of optimizing a neural network with respect to its parameters is nonconvex (and therefore the neural network may converge to a local minimum), the limit equation minimizes a quadratic objective function when viewed as a function of the limit empirical measure of parameters.
- We show that the matrix in the limiting quadratic objective function is positive definite, and therefore the neural network (in the limit) will converge to a global minimum with zero loss on the training set.

The main convergence results are presented below and their proofs follow in Section 19.7.

Theorem 19.3. *Assume that Assumption* 19.2 *holds and choose the learning rate to be* $\eta_k^N = \frac{\eta}{N}$ *for* $0 < \eta < \infty$, *a fixed constant. Then, the process* (μ_t^N, h_t^N) *converges in distribution in the space* $D_E([0,T])$[1] *as* $N \to \infty$ *to* (μ_t, h_t) *which satisfies, for every* $f \in C_2^b(\mathbb{R}^{1+d})$, *the random ODE*

$$
\begin{aligned}
h_t(x) &= h_0(x) + \eta \int_{\mathcal{X}\times\mathcal{Y}} (y - h_t(x'))\langle \sigma(w\cdot x)\sigma(w\cdot x'), \mu_t\rangle \pi(dx', dy)dt \\
&\quad + \eta \int_{\mathcal{X}\times\mathcal{Y}} (y - h_t(x'))\left\langle c^2\sigma'(w\cdot x')\sigma'(w\cdot x)x\cdot x', \mu_t\right\rangle \pi(dx', dy)dt, \\
h_0(x) &= \mathcal{G}(x), \\
\langle f, \mu_t\rangle &= \langle f, \mu_0\rangle .
\end{aligned} \tag{19.4}
$$

Recall that $\mathcal{G} \in \mathbb{R}^M$ is a Gaussian random variable with elements

$$
\mathcal{G}(x) \sim N(0, \left\langle |c\sigma(w\cdot x)|^2, \mu_0\right\rangle).
$$

In addition, note that $\bar{\mu}_t$ in the limit equation (19.4) is a constant, i.e., $\mu_t = \mu_0$ for $t \in [0,T]$. Therefore, (19.4) reduces to

$$
\begin{aligned}
h_t(x) &= h_0(x) + \eta \int_0^t \int_{\mathcal{X}\times\mathcal{Y}} (y - h_s(x'))\langle \sigma(w\cdot x)\sigma(w\cdot x'), \mu_0\rangle \pi(dx', dy)ds \\
&\quad + \eta \int_0^t \int_{\mathcal{X}\times\mathcal{Y}} (y - h_s(x'))\left\langle c^2\sigma'(w\cdot x')\sigma'(w\cdot x)x\cdot x', \mu_0\right\rangle \pi(dx', dy)ds, \\
h_0(x) &= \mathcal{G}(x).
\end{aligned} \tag{19.5}
$$

[1] $D_E([0,T])$ is the set of maps from $[0,T]$ into E which are right-continuous and which have left-hand limits. Here, we have $E = \mathcal{M}(\mathbb{R}^{1+d}) \times \mathbb{R}^M$ and $\mathcal{M}(\mathbb{R}^{1+d})$ is the space of probability measures in $\mathbb{R}^{1+d}$ (see also Section A.4).

Since (19.5) is a linear equation in $C_{\mathbb{R}^M}([0,T])$, the solution h_t is unique.

Equation (19.5) can be written more compactly in terms of the matrix $A \in \mathbb{R}^{M\times M}$ where

$$A(x,x') = \frac{\eta}{M}\langle \sigma(w\cdot x)\sigma(w\cdot x'), \mu_0\rangle + \frac{\eta}{M}\left\langle c^2\sigma'(w\cdot x')\sigma'(w\cdot x)x\cdot x', \mu_0\right\rangle,$$

where $x, x' \in \mathcal{D}$. A is finite-dimensional since we fixed a training set of size M in the beginning. A is called in the literature the NTK and notice that it is fixed in time, i.e., it does not change dynamically. Note that we can write

$$h_t(x) = \mathcal{G}(x) + \int_0^t A(x,x')(y - h_t(x'))\pi(dx', dy)ds.$$

Then, (19.5) becomes

$$dh_t = A\left(\hat{Y} - h_t\right)dt,$$
$$h_0 = \mathcal{G},$$

where $\hat{Y} = (y_1, \ldots, y_M)$.

Therefore, h_t is the solution to a continuous-time gradient descent algorithm which minimizes a quadratic objective function.

$$\frac{dh_t}{dt} = -\frac{1}{2}\nabla_h J(\hat{Y}, h_t),$$
$$J(y,h) = (y-h)^\top A(y-h),$$
$$h_0 = \mathcal{G}.$$

Therefore, even though the prelimit optimization problem is nonconvex, the neural network's limit will minimize a quadratic objective function.

The question of global convergence then translates to whether $h_t \to \hat{Y}$ as $t \to \infty$ or not. That is, in the limit of large numbers of hidden units and many training steps, does the neural network model converge to a global minimum with zero training error? The answer to this question is yes!

Theorem 19.4 shows that indeed we have that $h_t \to \hat{Y}$ as $t \to \infty$ if A is positive definite. Then Corollary 19.6 demonstrates that, under reasonable hyperparameter choices and if the data samples are in distinct directions (see [**Ito96**]), A will be positive definite.

Theorem 19.4. *If A is positive definite, then*

$$h_t \to \hat{Y} \quad \text{as} \quad t \to \infty.$$

Proof. Consider the transformation $\tilde{h}_t = h_t - \hat{Y}$. Then,

$$d\tilde{h}_t = -A\tilde{h}_t dt,$$
$$\tilde{h}_0 = \mathcal{G} - \hat{Y}.$$

Then, $\tilde{h}_t \to 0$ (and consequently $h_t \to \hat{Y}$) as $t \to \infty$ if A is positive definite. □

Under the proper assumptions the matrix A is positive definite. This is the content of Corollary 19.6. Before stating that result, we need to introduce the notion of data samples being in distinct directions following [**Ito96**].

Definition 19.5 (Distinct directions). For $x \in \mathbb{R}^d$ nonzero, define the line $L_x = \{y \in \mathbb{R}^d \,:\, y = tx, t \in \mathbb{R}\}$. The vectors $x^{(i)}$ are said to be in distinct directions if they are not zero and if the lines $L_{x^{(i)}}$ meet at the origin only.

Corollary 19.6. *Assume Assumption* 19.2. *A sufficient condition for A to be positive definite is* $\sigma(\cdot)$ *is non-polynomial and slowly increasing (i.e.,* $\lim_{x\to\infty} \frac{\sigma(x)}{x^a} = 0$ *for every* $a > 0$*),* μ_0 *is positive when evaluated on sets of positive Lebesgue measure, and the data samples* $x^{(i)}$ *are in distinct directions, per Definition* 19.5.

Examples of activation units $\sigma(\cdot)$ satisfying the conditions in Corollary 19.6 include sigmoid functions and hyperbolic tangent functions. Using a normal distribution for the initialization of the parameters in the neural network is a common choice in practice (covered by the requirements of Corollary 19.6).

Remark 19.7. For presentation purposes we have not explicitly denoted the bias term in the model. However, it is clear that this can be handled by requiring the first component of the vector x to be equal to one for example. This would result in the neural network taking the form

$$\mathfrak{m}^N(x;\theta) = \frac{1}{\sqrt{N}} \sum_{n=1}^{N} C^n \sigma(W^n \cdot x + b^n).$$

19.5. Preliminary Bounds and Existence of a Limit

19.5.1. Preliminary Bounds. Before we begin with the preliminary bounds, let us first define a notation that will be frequently used in this and in the subsequent chapters.

Definition 19.8. For a sequence of random variables $\{A_N\}_{N\in\mathbb{N}}$ and a sequence of real numbers $\{\beta_N\}_{N\in\mathbb{N}}$ we will write:

- $A_N = \mathcal{O}_p(\beta_N)$ if A_N/β_N is stochastically bounded. This means that for arbitrary $\epsilon > 0$, there is $M < \infty$ and some $N_o < \infty$ large enough so that
$$\mathbb{P}\left(\left|\frac{A_N}{\beta_N}\right| > M\right) < \epsilon \quad \text{for all } N > N_o.$$
- $A_N = \mathcal{O}(\beta_N)$ if A_N/β_N is bounded. This means that there exists a finite constant $C_o < \infty$, which is independent of N so that
$$|A_N| \leq C_o\beta_N \quad \text{for all } N \in \mathbb{N}.$$

The first a priori bounds we establish involve the parameters (C^n_k, W^i_k).

Lemma 19.9. *Let $T < \infty$ be given. There is a universal constant $C_o < \infty$, such that for all $n \in \mathbb{N}$ and all k with $k/N \leq T$,*

$$|C^n_k| < C_o < \infty,$$
$$\mathbb{E}\,\|W^n_k\| < C_o < \infty.$$

Proof. The unimportant finite constant $C_o < \infty$ may change from line to line. Recall the choice $\eta^N_k = \frac{\eta}{N}$ for $0 < \eta < \infty$ a constant. We first observe that

$$\begin{aligned}|C^n_{k+1}| &\leq |C^n_k| + \eta N^{-3/2}\,|y_k - \mathfrak{m}^N_k(x_k)|\,|\sigma(W^n_k \cdot x_k)|\\ &\leq |C^n_k| + \frac{C_o|y_k|}{N^{3/2}} + \frac{C_o}{N^2}\sum_{n=1}^{N}|C^n_k|,\end{aligned}$$

where the last inequality follows from the definition of $\mathfrak{m}^N_k(x)$ and the uniform boundedness assumption on $\sigma(\cdot)$.

Then, we subsequently obtain that

$$\begin{aligned}|C^n_k| &= |C^n_0| + \sum_{j=1}^{k}\Big[|C^n_j| - |C^n_{j-1}|\Big]\\ &\leq |C^n_0| + \sum_{j=1}^{k}\frac{C_o}{N^{3/2}} + \frac{C_o}{N^2}\sum_{j=1}^{k}\sum_{n=1}^{N}|C^n_{j-1}|\\ &\leq |C^n_0| + \frac{C_o}{\sqrt{N}} + \frac{C_o}{N^2}\sum_{j=1}^{k}\sum_{n=1}^{N}|C^n_{j-1}|. \qquad (19.6)\end{aligned}$$

This implies that

$$\frac{1}{N}\sum_{n=1}^{N}|C^n_k| \leq \frac{1}{N}\sum_{n=1}^{N}|C^n_0| + \frac{C_o}{\sqrt{N}} + \frac{C_o}{N^2}\sum_{j=1}^{k}\sum_{n=1}^{N}|C^n_{j-1}|.$$

Let us now define $\gamma^N_k = \frac{1}{N}\sum_{n=1}^{N}|C^n_k|$. Since the random variables C^i_0 take values in a compact set, we have that $\frac{1}{N}\sum_{n=1}^{N}|C^n_0| + \frac{C_o}{\sqrt{N}} < C_o < \infty$. Then,

$$\gamma^N_k \leq C_o + \frac{C_o}{N}\sum_{j=1}^{k}\gamma^N_{j-1}.$$

By the discrete Gronwall lemma and using $k/N \leq T$,

$$\gamma^N_k \leq C_o \exp\left(\frac{C_o k}{N}\right) \leq C_o. \qquad (19.7)$$

We can now combine the bounds (19.7) and (19.6) to yield, for any $0 \leq k \leq TN$,

$$\begin{aligned} |C_k^n| &\leq |C_0^n| + \frac{C_o}{\sqrt{N}} + \frac{C_o}{N}\sum_{j=1}^{k} \gamma_{j-1}^N \\ &\leq |C_0^n| + \frac{C_o}{\sqrt{N}} + \frac{C_o}{N}\sum_{j=1}^{k} C_o \\ &\leq |C_0^n| + \frac{C_o}{\sqrt{N}} + C_o \\ &\leq C_o, \end{aligned} \tag{19.8}$$

where the last inequality follows from the random variables C_0^n taking values in a compact set. Note that the constant $C_o < \infty$ may depend on T and it changes from line to line.

Now, we turn to the bound for $\| W_k^n \|$. We start with the bound (using Young's inequality),

$$\begin{aligned} \| W_{k+1}^n \| &\leq \| W_k^n \| + \frac{C_o}{N^{3/2}}\left(|y_k| + \frac{1}{\sqrt{N}}\sum_{n=1}^{N} |C_k^n|\right)|C_k^n||\sigma'(W_k^n \cdot x_k)| \, \| x_k \| \\ &\leq \| W_k^n \| + C_o\left(\frac{1}{N}|y_k|^2 + \frac{1}{N^2}\sum_{n=1}^{N} |C_k^n|^2 + \frac{1}{N}|C_k^n|^2 \, \| x_k \|^2\right) \\ &\leq \| W_k^n \| + C_o\left(\frac{1}{N}|y_k|^2 + \frac{1}{N^2}\sum_{n=1}^{N} |C_k^n|^2 + \frac{1}{N}|C_k^n|^4 + \frac{1}{N} \| x_k \|^4\right), \end{aligned}$$

for a constant $C_o < \infty$ that may change from line to line. Taking an expectation, using Assumption 19.2, the bound (19.8), and using the fact that $k/N \leq T$, we obtain

$$\mathbb{E} \| W_k^n \| \leq C_o < \infty,$$

for all $i \in \mathbb{N}$ and all k such that $k/N \leq T$, which concludes the proof of the lemma. □

Note that for any given $T < \infty$, the bounds of Lemma 19.9 are uniform in $k/N \leq T$ and $N \in \mathbb{N}$. Using the bounds from Lemma 19.9, we can now establish a bound for $\mathfrak{m}_k^N(x)$ for $x \in \mathcal{D}$.

Lemma 19.10. *Let $T < \infty$ be given. There is a universal constant $C_o < \infty$, such that for all $k \in \mathbb{N}$ such that $k/N \leq T$, and any $x \in \mathcal{D}$,*

$$\mathbb{E}\left[|\mathfrak{m}_k^N(x)|^2\right] < C_o < \infty.$$

Proof. The first step is to represent the evolution of the network output $\mathfrak{m}_k^N(x)$. In particular, we notice that

$$
\begin{aligned}
\mathfrak{m}_{k+1}^N(x) &= \mathfrak{m}_k^N(x) + \frac{1}{\sqrt{N}}\sum_{n=1}^N C_{k+1}^n \sigma(W_{k+1}^n \cdot x) - \frac{1}{\sqrt{N}}\sum_{n=1}^N C_k^n \sigma(W_k^n \cdot x) \\
&= \mathfrak{m}_k^N(x) + \frac{1}{\sqrt{N}}\sum_{n=1}^N \left(C_{k+1}^n \sigma(W_{k+1}^n \cdot x) - C_k^n \sigma(W_k^n \cdot x)\right) \\
&= \mathfrak{m}_k^N(x) \\
&\quad + \frac{1}{\sqrt{N}}\sum_{n=1}^N \left((C_{k+1}^n - C_k^n)\sigma(W_{k+1}^n \cdot x) + (\sigma(W_{k+1}^n \cdot x) - \sigma(W_k^n \cdot x))C_k^n\right) \\
&= \mathfrak{m}_k^N(x) \\
&\quad + \frac{1}{\sqrt{N}}\sum_{n=1}^N \left((C_{k+1}^n - C_k^n)\left[\sigma(W_k^n \cdot x) + \sigma'(W_k^{n,*} \cdot x_k)x \cdot (W_{k+1}^n - W_k^n)\right]\right. \\
&\quad \left. + \left[\sigma'(W_k^n \cdot x)x \cdot (W_{k+1}^n - W_k^n) + \frac{1}{2}\sigma''(W_{k+1}^{n,**} \cdot x)\left((W_{k+1}^n - W_k^n)\cdot x\right)^2\right]C_k^n\right)
\end{aligned}
\tag{19.9}
$$

for points $W_k^{n,*}$ and $W_k^{n,*,*}$ in the line segment connecting the points W_k^n and W_{k+1}^n. Recall that $\eta_k^N = \frac{\eta}{N}$. Substituting (19.3) into (19.9) yields

$$
\begin{aligned}
\mathfrak{m}_{k+1}^N(x) &= \mathfrak{m}_k^N(x) + \frac{\eta}{N^2}\sum_{n=1}^N (y_k - \mathfrak{m}_k^N(x_k))\sigma(W_k^n \cdot x_k)\sigma(W_k^n \cdot x) \\
&\quad + \frac{\eta}{N^2}\sum_{n=1}^N \sigma'(W_k^n \cdot x)(y_k - \mathfrak{m}_k^N(x_k))\sigma'(W_k^n \cdot x_k)x_k \cdot x\left(C_k^n\right)^2 + \mathcal{O}(N^{-3/2}),
\end{aligned}
\tag{19.10}
$$

where we recall Definition 19.8 for the remainder term $\mathcal{O}(N^{-3/2})$.

This leads to the bound

$$
\begin{aligned}
|\mathfrak{m}_{k+1}^N(x)| &\le |\mathfrak{m}_k^N(x)| + \frac{\eta}{N^2}\sum_{n=1}^N |y_k - \mathfrak{m}_k^N(x_k)| + \frac{\eta}{N^2}\sum_{n=1}^N |y_k - \mathfrak{m}_k^N(x_k)|\left(C_k^n\right)^2 \\
&\quad + \frac{C_o}{N^{3/2}} \\
&\le |\mathfrak{m}_k^N(x)| + \frac{C_o}{N}|\mathfrak{m}_k^N(x_k)| + \frac{C_o}{N}.
\end{aligned}
$$

We now square both sides of the above inequality.

$$\begin{aligned}|\mathfrak{m}_{k+1}^N(x)|^2 &\le \Big(|\mathfrak{m}_k^N(x)|+\frac{C_o}{N}|\mathfrak{m}_k^N(x_k)|+\frac{C_o}{N}\Big)^2\\ &\le |\mathfrak{m}_k^N(x)|^2+2|\mathfrak{m}_k^N(x)|(\frac{C_o}{N}|\mathfrak{m}_k^N(x_k)|+\frac{C_o}{N})+(\frac{C_o}{N}|\mathfrak{m}_k^N(x_k)|+\frac{C_o}{N})^2\\ &\le |\mathfrak{m}_k^N(x)|^2+\frac{C_o}{N}|\mathfrak{m}_k^N(x)|^2+\frac{C_o}{N},\end{aligned}$$

where the last line used Young's inequality. Therefore, we have

$$|\mathfrak{m}_{k+1}^N(x)|^2-|\mathfrak{m}_k^N(x)|^2 \le \frac{C_o}{N}|\mathfrak{m}_k^N(x_k)|^2+\frac{C_o}{N}.$$

Then, using a telescoping series,

$$\begin{aligned}|\mathfrak{m}_k^N(x)|^2 &= |\mathfrak{m}_0^N(x)|^2+\sum_{j=1}^k\Big(|\mathfrak{m}_j^N(x)|^2-|\mathfrak{m}_{j-1}^N(x)|^2\Big)\\ &\le |\mathfrak{m}_0^N(x)|^2+\sum_{j=1}^k\Big(\frac{C_o}{N}|\mathfrak{m}_{j-1}^N(x_{j-1})|^2+\frac{C}{N}\Big)\\ &\le |\mathfrak{m}_0^N(x)|^2+C_o+\frac{C_o}{N}\sum_{j=1}^k|\mathfrak{m}_{j-1}^N(x_{j-1})|^2.\end{aligned}$$

Taking expectations,

$$\mathbb{E}\Big[|\mathfrak{m}_k^N(x)|^2\Big]\le \mathbb{E}\Big[|\mathfrak{m}_0^N(x)|^2\Big]+C_o+\frac{C_o}{N}\sum_{j=1}^k\mathbb{E}\Big[|\mathfrak{m}_{j-1}^N(x_{j-1})|^2\Big].$$

Taking advantage of the fact that x_j is sampled from a fixed dataset $\mathcal{D}$ of M data samples,

$$\mathbb{E}\Big[|\mathfrak{m}_k^N(x)|^2\Big]\le \mathbb{E}\Big[|\mathfrak{m}_0^N(x)|^2\Big]+C_o+\frac{C_o}{N}\sum_{j=1}^k\sum_{x'\in\mathcal{D}}\mathbb{E}\Big[|\mathfrak{m}_{j-1}^N(x')|^2\Big], \tag{19.11}$$

and therefore

$$\begin{aligned}\sum_{x\in\mathcal{D}}\mathbb{E}\Big[|\mathfrak{m}_k^N(x)|^2\Big] &\le \sum_{x\in\mathcal{D}}\mathbb{E}\Big[|\mathfrak{m}_0^N(x)|^2\Big]+MC_o+\frac{C_oM}{N}\sum_{j=1}^k\sum_{x'\in\mathcal{D}}\mathbb{E}\Big[|\mathfrak{m}_{j-1}^N(x')|^2\Big]\\ &\le \sum_{x\in\mathcal{D}}\mathbb{E}\Big[|\mathfrak{m}_0^N(x)|^2\Big]+C_o+\frac{C_o}{N}\sum_{j=1}^k\sum_{x\in\mathcal{D}}\mathbb{E}\Big[|\mathfrak{m}_{j-1}^N(x)|^2\Big]. \qquad (19.12)\end{aligned}$$

Recall that

$$\mathfrak{m}_0^N(x)=\frac{1}{\sqrt{N}}\sum_{n=1}^N C_0^n\sigma(W_0^n\cdot x),$$

where (C_0^n, W_0^n) are i.i.d., mean-zero random variables. Then,

$$\mathbb{E}\Big[|\mathfrak{m}_0^N(x)|^2\Big] \leq \mathbb{E}\Big[\Big(\frac{1}{\sqrt{N}}\sum_{n=1}^{N} C_0^n\sigma(W_0^n\cdot x)\Big)^2\Big]$$
$$\leq \frac{C_o}{N}\sum_{n=1}^{N}\mathbb{E}\Big[(C_0^n)^2\Big]$$
$$\leq C_o.$$

Combining this bound with the bound (19.12) and using the discrete Gronwall lemma yields, for any $0 \leq k \leq TN$,

$$\sum_{x\in\mathcal{D}}\mathbb{E}\Big[|\mathfrak{m}_k^N(x)|\Big] \leq C_o.$$

Substituting this bound into equation (19.11) produces the desired bound

$$\mathbb{E}\Big[|\mathfrak{m}_k^N(x)|^2\Big] \leq C_o$$

for any $0 \leq k \leq TN$. □

19.5.2. Tightness of the Scaled Empirical Measure. The first step into establishing that the family $\{(\mu_t^N, h_t^N), t \in [0,T]\}_{N\in\mathbb{N}}$ has a limit as N grows to infinity is to prove a form of compact containment, which in our case translates into showing that there is a compact set that contains (μ_t^N, h_t^N) for all $N \in \mathbb{N}$ and $t \in [0,T]$. Recall that $(\mu_t^N, h_t^N) \in D_E([0,T])$, where $D_E([0,T])$ is the set of maps from $[0,T]$ into E which are right-continuous and which have left-hand limits, $E = \mathcal{M}(\mathbb{R}^{1+d}) \times \mathbb{R}^M$ and $\mathcal{M}(\mathbb{R}^{1+d})$ is the space of probability measures in $\mathbb{R}^{1+d}$ (see also Section A.4).

Lemma 19.11. *For each $\delta > 0$, there is a compact subset $\mathcal{K}$ of E such that*

$$\sup_{N\in\mathbb{N},0\leq t\leq T}\mathbb{P}[(\mu_t^N, h_t^N)\notin\mathcal{K}] < \delta.$$

Proof. For each $L > 0$, define $K_L = [-L,L]^{1+d}$. Then, we have that K_L is a compact subset of $\mathbb{R}^{1+d}$, and for each $t \geq 0$ and $N \in \mathbb{N}$,

$$\mathbb{E}\left[\mu_t^N(\mathbb{R}^{1+d}\setminus K_L)\right] = \frac{1}{N}\sum_{n=1}^{N}\mathbb{P}\left[|C_{\lfloor Nt\rfloor}^n| + \| W_{\lfloor Nt\rfloor}^n \| \geq L\right] \leq \frac{C_o}{L},$$

where we have used Markov's inequality and the bounds from Lemma 19.9. We define the compact subsets of $\mathcal{M}(\mathbb{R}^{1+d})$

$$\hat{K}_L = \overline{\left\{\nu : \nu(\mathbb{R}^{1+d}\setminus K_{(L+j)^2}) < \frac{1}{\sqrt{L+j}} \text{ for all } j\in\mathbb{N}\right\}}$$

and we observe that

$$\begin{aligned}\mathbb{P}\{\mu_t^N \notin \hat{K}_L\} &\leq \sum_{j=1}^{\infty} \mathbb{P}\left[\mu_t^N(\mathbb{R}^{1+d} \setminus K_{(L+j)^2}) > \frac{1}{\sqrt{L+j}}\right] \\ &\leq \sum_{j=1}^{\infty} \frac{\mathbb{E}[\mu_t^N(\mathbb{R}^{1+d} \setminus K_{(L+j)^2})]}{1/\sqrt{L+j}} \\ &\leq \sum_{j=1}^{\infty} \frac{C_o}{(L+j)^2/\sqrt{L+j}} \leq \sum_{j=1}^{\infty} \frac{C_o}{(L+j)^{3/2}}.\end{aligned}$$

Given that $\lim_{L\to\infty} \sum_{j=1}^{\infty} \frac{C_o}{(L+j)^{3/2}} = 0$, we have that, for each $\delta > 0$, there exists a compact set $\hat{K}_L$ such that

$$\sup_{N\in\mathbb{N}, 0\leq t\leq T} \mathbb{P}[\mu_t^N \notin \hat{K}_L] < \frac{\delta}{2}.$$

Due to Lemma 19.10 and Markov's inequality, we also know that, for each $\delta > 0$, there exists a compact set $U = [-B, B]^M$ such that

$$\sup_{N\in\mathbb{N}, 0\leq t\leq T} \mathbb{P}[h_t^N \notin U] < \frac{\delta}{2}.$$

Therefore, for each $\delta > 0$, there exists a compact set $\hat{K}_L \times [-B, B]^M \subset E$ such that

$$\sup_{N\in\mathbb{N}, 0\leq t\leq T} \mathbb{P}[(\mu_t^N, h_t^N) \notin \hat{K}_L \times [-B, B]^M] < \delta. \qquad \square$$

The next step is to establish regularity of the process μ^N in $D_{\mathcal{M}(\mathbb{R}^{1+d})}([0,T])$. Define the function $q(z_1, z_2) = \min\{|z_1 - z_2|, 1\}$, where $z_1, z_2 \in \mathbb{R}$. Let $\mathcal{F}_t^N$ be the σ-algebra generated by $\{(C_0^i, W_0^i)\}_{i=1}^N$ and $\{x_j\}_{j=0}^{\lfloor Nt\rfloor -1}$, i.e., $\mathcal{F}_t^N$ contains the information generated by $\{(C_0^i, W_0^i)\}_{i=1}^N$ and $\{x_j\}_{j=0}^{\lfloor Nt\rfloor -1}$.

Lemma 19.12. *Let $f \in C_b^2(\mathbb{R}^{1+d})$. For any $\delta \in (0,1)$, there is a constant $C_o < \infty$ such that for $0 \leq u \leq \delta$, $0 \leq v \leq \delta \wedge t$, $t \in [0,T]$,*

$$\mathbb{E}\left[q(\langle f, \mu_{t+u}^N\rangle, \langle f, \mu_t^N\rangle) q(\langle f, \mu_t^N\rangle, \langle f, \mu_{t-v}^N\rangle) | \mathcal{F}_t^N\right] \leq C_o\delta + \frac{C_o}{N^{3/2}}.$$

Proof. We start by noticing that a Taylor expansion gives for $0 \le s \le t \le T$,

$$
\begin{aligned}
|\langle f,\mu_t^N\rangle - \langle f,\mu_s^N\rangle| &= |\langle f,\nu_{\lfloor Nt\rfloor}^N\rangle - \langle f,\nu_{\lfloor Ns\rfloor}^N\rangle| \\
&\le \frac{1}{N}\sum_{n=1}^{N} |f(C_{\lfloor Nt\rfloor}^n, W_{\lfloor Nt\rfloor}^n) - f(C_{\lfloor Ns\rfloor}^n, W_{\lfloor Ns\rfloor}^n)| \\
&\le \frac{1}{N}\sum_{n=1}^{N} |\partial_c f(\bar{C}_{\lfloor Nt\rfloor}^n, \bar{W}_{\lfloor Nt\rfloor}^n)||C_{\lfloor Nt\rfloor}^n - C_{\lfloor Ns\rfloor}^n| \\
&\quad + \frac{1}{N}\sum_{n=1}^{N} \| \nabla_w f(\hat{C}_{\lfloor Nt\rfloor}^n, \hat{W}_{\lfloor Nt\rfloor}^n) \| \| W_{\lfloor Nt\rfloor}^n - W_{\lfloor Ns\rfloor}^n \| \qquad (19.13)
\end{aligned}
$$

for points $(\bar{C}^n, \bar{W}^n)$ and $(\hat{C}^n, \hat{W}^n)$ in the segments connecting $C_{\lfloor Ns\rfloor}^n$ with $C_{\lfloor Nt\rfloor}^n$ and $W_{\lfloor Ns\rfloor}^n$ with $W_{\lfloor Nt\rfloor}^n$, respectively.

Let's now establish a bound on $|C_{\lfloor Nt\rfloor}^n - C_{\lfloor Ns\rfloor}^n|$ for $s < t \le T$ with $0 < t-s \le \delta < 1$.

$$
\begin{aligned}
\mathbb{E}\left[|C_{\lfloor Nt\rfloor}^n - C_{\lfloor Ns\rfloor}^n|\Big|\mathcal{F}_s^N\right] &= \mathbb{E}\left[|\sum_{k=\lfloor Ns\rfloor}^{\lfloor Nt\rfloor-1} (C_{k+1}^n - C_k^n)|\Big|\mathcal{F}_s^N\right] \\
&\le \mathbb{E}\left[\sum_{k=\lfloor Ns\rfloor}^{\lfloor Nt\rfloor-1} |\eta(y_k - \mathfrak{m}_k^N(x_k))\frac{1}{N^{3/2}}\sigma(W_k^n\cdot x_k)|\Big|\mathcal{F}_s^N\right] \\
&\le \frac{1}{N^{3/2}}\sum_{k=\lfloor Ns\rfloor}^{\lfloor Nt\rfloor-1} C_o \\
&\le \frac{C_o}{\sqrt{N}}(t-s) + \frac{C_o}{N^{3/2}} \\
&\le \frac{C_o}{\sqrt{N}}\delta + \frac{C_o}{N^{3/2}}, \qquad (19.14)
\end{aligned}
$$

where Assumption 19.2 was used as well as the bounds from Lemmas 19.9 and 19.10.

Let's now establish a bound on $\| W^n_{\lfloor Nt \rfloor} - W^n_{\lfloor Ns \rfloor} \|$ for $s < t \leq T$ with $0 < t - s \leq \delta < 1$. We obtain

$$
\begin{aligned}
\mathbb{E}\Big[\| W^n_{\lfloor Nt \rfloor} - W^n_{\lfloor Ns \rfloor} \| \Big| \mathcal{F}^N_s \Big] &= \mathbb{E}\Big[\| \sum_{k=\lfloor Ns \rfloor}^{\lfloor Nt \rfloor - 1} (W^n_{k+1} - W^n_k) \| \Big| \mathcal{F}^N_s \Big] \\
&\leq \mathbb{E}\Big[\sum_{k=\lfloor Ns \rfloor}^{\lfloor Nt \rfloor - 1} \| \eta(y_k - \mathfrak{m}^N_k(x_k)) \frac{1}{N^{3/2}} C^n_k \sigma'(W^n_k \cdot x_k) x_k \| \Big| \mathcal{F}^N_s \Big] \\
&\leq \frac{1}{N^{3/2}} \sum_{k=\lfloor Ns \rfloor}^{\lfloor Nt \rfloor - 1} C_o \\
&\leq \frac{C_o}{\sqrt{N}}(t-s) + \frac{C_o}{N^{3/2}} \\
&\leq \frac{C_o}{\sqrt{N}}\delta + \frac{C_o}{N^{3/2}},
\end{aligned} \tag{19.15}
$$

where we have again used the bounds from Lemmas 19.9 and 19.10.

Now, we return to equation (19.13). Due to Lemma 19.9, the quantities $(\bar{C}^n_{\lfloor Nt \rfloor}, \bar{W}^n_{\lfloor Nt \rfloor})$ are bounded in expectation for $0 < s < t \leq T$. Therefore, for $0 < s < t \leq T$ with $0 < t - s \leq \delta < 1$,

$$
\mathbb{E}\left[| \langle f, \mu^N_t \rangle - \langle f, \mu^N_s \rangle | | \mathcal{F}^N_s \right] \leq C_o \delta + \frac{C_o}{N^{3/2}},
$$

where $C_o < \infty$ is some unimportant finite constant that may depend on the magnitude of the first partial derivatives of f. Then, the statement of the lemma follows. □

We next establish regularity of the process h^N_t in $D_{\mathbb{R}^M}([0,T])$. For the purposes of the following lemma, let the function $q(z_1, z_2) = \min\{\|z_1 - z_2\|, 1\}$, where $z_1, z_2 \in \mathbb{R}^M$ and $\|z\| = |z_1| + \cdots + |z_M|$.

Lemma 19.13. *For any $\delta \in (0,1)$, there is a constant $C_o < \infty$ such that for $0 \leq u \leq \delta < 1$, $0 \leq v \leq \delta \wedge t$, $t \in [0,T]$,*

$$
\mathbb{E}\left[q(h^N_{t+u}, h^N_t) q(h^N_t, h^N_{t-v}) | \mathcal{F}^N_t \right] \leq C_o \delta + \frac{C_o}{N}.
$$

Proof. Recall that

$$
\begin{aligned}
\mathfrak{m}^N_{k+1}(x) &= \mathfrak{m}^N_k(x) \\
&+ \frac{1}{\sqrt{N}} \sum_{n=1}^{N} \Big(C^n_{k+1} - C^n_k)\sigma(W^n_{k+1} \cdot x) + \sigma'(W^{i,*}_k \cdot x) x \cdot (W^n_{k+1} - W^n_k) C^n_k \Big).
\end{aligned}
$$

Therefore,

$$
\begin{aligned}
h_t^N(x) - h_s^N(x) &= \mathfrak{m}_{\lfloor Nt \rfloor}(x) - \mathfrak{m}_{\lfloor Ns \rfloor}(x) \\
&= \sum_{k=\lfloor Ns \rfloor}^{\lfloor Nt \rfloor} (\mathfrak{m}_{k+1}^N(x) - \mathfrak{m}_k^N(x)) \\
&= \sum_{k=\lfloor Ns \rfloor}^{\lfloor Nt \rfloor} \frac{1}{\sqrt{N}} \sum_{n=1}^{N} \Big((C_{k+1}^n - C_k^n)\sigma(W_{k+1}^n \cdot x) + \sigma'(W_k^{n,*} \cdot x) x \cdot (W_{k+1}^n - W_k^n) C_k^n \Big).
\end{aligned}
$$

This yields the bound

$$
\begin{aligned}
|h_t^N(x) - h_s^N(x)| &\leq \sum_{k=\lfloor Ns \rfloor}^{\lfloor Nt \rfloor} |\mathfrak{m}_{k+1}^N(x) - \mathfrak{m}_k^N(x)| \\
&\leq \sum_{k=\lfloor Ns \rfloor}^{\lfloor Nt \rfloor} \frac{1}{\sqrt{N}} \sum_{n=1}^{N} \Big(|C_{k+1}^n - C_k^n| + \|W_{k+1}^n - W_k^n\| \Big),
\end{aligned}
$$

where we have used the boundedness of $\sigma'(\cdot)$ (from Assumption 19.2) and the bounds from Lemma 19.9.

Taking expectations,

$$
\begin{aligned}
&\mathbb{E}\Big[|h_t^N(x) - h_s^N(x)| \Big| \mathcal{F}_s^N \Big] \\
&\quad \leq \frac{1}{\sqrt{N}} \sum_{n=1}^{N} \sum_{k=\lfloor Ns \rfloor}^{\lfloor Nt \rfloor} \mathbb{E}\Big[|C_{k+1}^n - C_k^n| + \|W_{k+1}^n - W_k^n\| \Big| \mathcal{F}_s^N \Big].
\end{aligned}
$$

Using the bounds (19.14) and (19.15),

$$
\begin{aligned}
\mathbb{E}\Big[|h_t^N(x) - h_s^N(x)| \Big| \mathcal{F}_s^N \Big] &\leq \frac{1}{\sqrt{N}} \sum_{n=1}^{N} \left(\frac{C_o}{\sqrt{N}}(t-s) + \frac{C_o}{N^{3/2}} \right) \\
&= C_o(t-s) + \frac{C_o}{N}. \qquad (19.16)
\end{aligned}
$$

The bound (19.16) holds for each $x \in \mathcal{D}$. Therefore,

$$
\mathbb{E}\Big[\|h_t^N - h_s^N\| \Big| \mathcal{F}_s^N \Big] \leq C_o(t-s) + \frac{C_o}{N}.
$$

The statement of the lemma then follows. □

Last step is to combine the compact containment and regularity results in order to claim that the family of processes $\{(\mu_t^N, h_t^N), t \in [0,T]\}$ has a limit as $N \to \infty$. Indeed, we have the following lemma.

Lemma 19.14. *The family of processes $\{\mu^N, h^N\}_{N \in \mathbb{N}}$ is relatively compact in $D_E([0,T])$.*

Proof. Combining Lemmas 19.11 and 19.12 and the results of Section A.4 proves that $\{\mu^N\}_{N\in\mathbb{N}}$ is relatively compact in $D_{\mathcal{M}(\mathbb{R}^{1+d})}([0,T])$ (see also Theorem 8.6, Remark 8.7 B, and Theorem 9.1 of Chapter 3 of [**EK86**] as well as Theorem 4.6 in [**Jak86**] and Section 3 of [**Led16**]).

Similarly, combining Lemmas 19.11 and 19.13 proves that $\{h^N\}_{N\in\mathbb{N}}$ is relatively compact in $D_{\mathbb{R}^M}([0,T])$.

Since relative compactness is equivalent to tightness, we have that the probability measures of the family of processes $\{\mu^N\}_{N\in\mathbb{N}}$ are tight. Similarly, we have that the probability measures of the family of process $\{h^N\}_{N\in\mathbb{N}}$ are tight. Therefore, $\{\mu^N, h^N\}_{N\in\mathbb{N}}$ is tight. Then, $\{\mu^N, h^N\}_{N\in\mathbb{N}}$ is also relatively compact. □

19.6. Alternative Representation of the Prelimit Process

Let us now build towards identifying the limit in Theorem 19.3. Recall that $\eta_k^N = \frac{\eta}{N}$ and equation (19.10), which describes the evolution of $\mathfrak{m}_k^N(x)$,

$$\mathfrak{m}_{k+1}^N(x) = \mathfrak{m}_k^N(x) + \frac{\eta}{N^2}\sum_{n=1}^{N}(y_k - \mathfrak{m}_k^N(x_k))\sigma(W_k^n \cdot x_k)\sigma(W_k^n \cdot x)$$
$$+ \frac{\eta}{N^2}\sum_{n=1}^{N}\sigma'(W_k^n \cdot x)(y_k - \mathfrak{m}_k^N(x_k))\sigma'(W_k^n \cdot x_k)x_k \cdot x\big(C_k^n\big)^2 + \mathcal{O}(N^{-3/2}).$$

We can rewrite the evolution of $\mathfrak{m}_k^N(x)$ in terms of the empirical measure ν_k^N,

$$\mathfrak{m}_{k+1}^N(x) = \mathfrak{m}_k^N(x) + \frac{\eta}{N}(y_k - \mathfrak{m}_k^N(x_k))\left\langle\sigma(w \cdot x_k)\sigma(w \cdot x), \nu_k^N\right\rangle$$
$$+ \frac{\eta}{N}(y_k - \mathfrak{m}_k^N(x_k))x_k \cdot x\left\langle\sigma'(w \cdot x)\sigma'(w \cdot x_k)c^2, \nu_k^N\right\rangle + \mathcal{O}(N^{-3/2}). \tag{19.17}$$

Using (19.17), we can write the evolution of h_t^N for $t \in [0,T]$ as

$$\begin{aligned} h_t^N &= h_0^N + \sum_{k=0}^{\lfloor Nt\rfloor-1}(\mathfrak{m}_{k+1}^N - \mathfrak{m}_k^N) \\ &= h_0^N + \frac{\eta}{N}\sum_{k=0}^{\lfloor Nt\rfloor-1}(y_k - \mathfrak{m}_k^N(x_k))\left\langle\sigma(w \cdot x_k)\sigma(w \cdot x), \nu_k^N\right\rangle \\ &\quad + \frac{\eta}{N}\sum_{k=0}^{\lfloor Nt\rfloor-1}(y_k - \mathfrak{m}_k^N(x_k))x_k \cdot x\left\langle\sigma'(w \cdot x)\sigma'(w \cdot x_k)c^2, \nu_k^N\right\rangle \\ &\quad + \mathcal{O}(N^{-1/2}). \end{aligned} \tag{19.18}$$

Next, we decompose the summations into a drift and martingale component:

$$
\begin{aligned}
h_t^N = h_0^N &+ \frac{\eta}{N}\sum_{k=0}^{\lfloor Nt\rfloor-1}\int_{\mathcal{X}\times\mathcal{Y}}(y-\mathfrak{m}_k^N(x'))\left\langle\sigma(w\cdot x')\sigma(w\cdot x),\nu_k^N\right\rangle\pi(dx',dy)\\
&+\frac{\eta}{N}\sum_{k=0}^{\lfloor Nt\rfloor-1}\int_{\mathcal{X}\times\mathcal{Y}}(y-\mathfrak{m}_k^N(x'))x\cdot x'\left\langle\sigma'(w\cdot x)\sigma'(w\cdot x')c^2,\nu_k^N\right\rangle\pi(dx',dy)\\
&+\frac{\eta}{N}\sum_{k=0}^{\lfloor Nt\rfloor-1}\Big((y_k-\mathfrak{m}_k^N(x_k))\left\langle\sigma(w\cdot x_k)\sigma(w\cdot x),\nu_k^N\right\rangle\\
&\qquad-\int_{\mathcal{X}\times\mathcal{Y}}(y-\mathfrak{m}_k^N(x'))\left\langle\sigma(w\cdot x')\sigma(w\cdot x),\nu_k^N\right\rangle\pi(dx',dy)\Big)\\
&+\frac{\eta}{N}\sum_{k=0}^{\lfloor Nt\rfloor-1}\Big((y_k-\mathfrak{m}_k^N(x_k))x_k\cdot x\left\langle\sigma'(w\cdot x)\sigma'(w\cdot x_k)c^2,\nu_k^N\right\rangle\\
&\qquad-\int_{\mathcal{X}\times\mathcal{Y}}(y-\mathfrak{m}_k^N(x'))x\cdot x'\left\langle\sigma'(w\cdot x)\sigma'(w\cdot x')c^2,\nu_k^N\right\rangle\pi(dx',dy)\Big)\\
&+\mathcal{O}(N^{-1/2}).
\end{aligned}
$$

For convenience, we define the martingale terms (the third and fourth terms in the equation above) as $M_t^{N,1}$ and $M_t^{N,2}$, respectively. The equation for h_t^N can be rewritten in terms of a Riemann integral and the scaled measure μ_t^N, yielding

$$
\begin{aligned}
h_t^N = h_0^N &+ \eta\int_0^t\int_{\mathcal{X}\times\mathcal{Y}}(y-h_s^N(x'))\left\langle\sigma(w\cdot x')\sigma(w\cdot x),\mu_s^N\right\rangle\pi(dx',dy)ds\\
&+\eta\int_0^t\int_{\mathcal{X}\times\mathcal{Y}}(y-h_s^N(x'))x\cdot x'\left\langle\sigma'(w\cdot x)\sigma'(w\cdot x')c^2,\mu_s^N\right\rangle\pi(dx',dy)ds\\
&+M_t^{N,1}+M_t^{N,2}+\mathcal{O}(N^{-1/2}).
\end{aligned}
\tag{19.19}
$$

In addition, using conditional independence of the terms in the series for $M_t^{N,1}$ and $M_t^{N,2}$ as well as the bounds from Lemmas 19.10 and 19.9, we have for a finite constant $C_o<\infty$ that

$$
\begin{aligned}
\mathbb{E}\left[\left(M_t^{N,1}\right)^2\right] &\le \frac{C_o}{N},\\
\mathbb{E}\left[\left(M_t^{N,2}\right)^2\right] &\le \frac{C_o}{N}.
\end{aligned}
$$

We can also analyze the evolution of the empirical measure ν_k^N in terms of test functions $f \in C_b^2(\mathbb{R}^{1+d})$. Using a Taylor expansion, we find that

$$\langle f, \nu_{k+1}^N \rangle - \langle f, \nu_k^N \rangle = \frac{1}{N} \sum_{n=1}^{N} \Big(f(C_{k+1}^n, W_{k+1}^n) - f(C_k^n, W_k^n) \Big)$$

$$= \frac{1}{N} \sum_{n=1}^{N} \partial_c f(C_k^n, W_k^n)(C_{k+1}^n - C_k^n) + \frac{1}{N} \sum_{n=1}^{N} \nabla_w f(C_k^n, W_k^n)^\top (W_{k+1}^n - W_k^n)$$

$$+ \frac{1}{N} \sum_{n=1}^{N} \partial_c^2 f(\bar{C}_k^n, \bar{W}_k^n)(C_{k+1}^n - C_k^n)^2$$

$$+ \frac{1}{N} \sum_{n=1}^{N} (C_{k+1}^n - C_k^n) \nabla_{cw} f(\hat{C}_k^n, \hat{W}_k^n) \cdot (W_{k+1}^n - W_k^n)$$

$$+ \frac{1}{N} \sum_{n=1}^{N} (W_{k+1}^n - W_k^n) \cdot \nabla_w^2 f(\tilde{C}_k^n, \tilde{W}_k^n)(W_{k+1}^n - W_k^n) \tag{19.20}$$

for points $(\bar{C}_k^n, \bar{W}_k^n)$, $(\hat{C}_k^n, \hat{W}_k^n)$, and $(\tilde{C}_k^n, \tilde{W}_k^n)$ in the segments connecting C_{k+1}^n with C_k^n and W_{k+1}^n with W_k^n, respectively.

Substituting (19.3) into (19.20) yields

$$\langle f, \nu_{k+1}^N \rangle - \langle f, \nu_k^N \rangle = N^{-5/2} \sum_{n=1}^{N} \partial_c f(C_k^n, W_k^n) \eta (y_k - \mathfrak{m}_k^N(x_k)) \sigma(W_k^n \cdot x_k)$$

$$+ N^{-5/2} \sum_{n=1}^{N} \eta(y_k - \mathfrak{m}_k^N(x_k)) C_k^n \sigma'(W_k^n \cdot x_k) \nabla_w f(C_k^n, W_k^n) \cdot x_k + \mathcal{O}_p\left(N^{-2}\right)$$

$$= N^{-3/2} \eta(y_k - \mathfrak{m}_k^N(x_k)) \left\langle \partial_c f(c,w) \sigma(w \cdot x_k), \nu_k^N \right\rangle$$
$$+ N^{-3/2} \eta(y_k - \mathfrak{m}_k^N(x_k)) \left\langle c\sigma'(w \cdot x_k) \nabla_w f(c,w) \cdot x_k, \nu_k^N \right\rangle + \mathcal{O}_p\left(N^{-2}\right),$$

where we recall Definition 19.8 for the definition of $\mathcal{O}_p\left(N^{-2}\right)$. Therefore, we have

$$\langle f, \mu_t^N \rangle = \langle f, \mu_0^N \rangle + \sum_{k=0}^{\lfloor Nt \rfloor - 1} \Big(\langle f, \nu_{k+1}^N \rangle - \langle f, \nu_k^N \rangle \Big)$$

$$= \langle f, \mu_0^N \rangle + N^{-3/2} \sum_{k=0}^{\lfloor Nt \rfloor - 1} \eta(y_k - \mathfrak{m}_k^N(x_k)) \left\langle \partial_c f(c,w) \sigma(w \cdot x_k), \nu_k^N \right\rangle$$

$$+ N^{-3/2} \sum_{k=0}^{\lfloor Nt \rfloor - 1} \eta(y_k - \mathfrak{m}_k^N(x_k)) \left\langle c\sigma'(w \cdot x_k) \nabla_w f(c,w) \cdot x_k, \nu_k^N \right\rangle$$

$$+ \mathcal{O}_p\left(N^{-1}\right). \tag{19.21}$$

19.7. Proof of Main Convergence Results

Intuitively, the limit then can be seen to be the claimed one by taking $N \to \infty$ to (19.19) and (19.21). Let us now study how to rigorously claim the convergence result.

Let π^N be the probability measure of a convergent subsequence of the sequence $\left\{(\mu^N, h^N)_{0\le t\le T}\right\}$. Each π^N takes values in the set of probability measures $\mathcal{M}\left(D_E([0,T])\right)$. The established relative compactness implies that there is a subsequence π^{N_k} which weakly converges. We must prove that any limit point π of a convergent subsequence π^{N_k} will satisfy the evolution equation (19.4).

Lemma 19.15. *Let π^{N_k} be a convergent subsequence with a limit point π. Then, π is a Dirac measure concentrated on $(\mu, h) \in D_E([0,T])$ and (μ, h) satisfies equation* (19.4).

Proof. We define a map $F(\mu, h) : D_E([0,T]) \to \mathbb{R}_+$ for each $t \in [0,T]$, $f \in C_b^2(\mathbb{R}^{1+d})$, $g_1, \ldots, g_p \in C_b(\mathbb{R}^{1+d})$, $q_1, \ldots, q_p \in C_b(\mathbb{R}^M)$, and $0 \le s_1 < \cdots < s_p \le t$.

$$
\begin{aligned}
F(\mu, h) = &\left| (\langle f, \mu_t\rangle - \langle f, \mu_0\rangle) \times \langle g_1, \mu_{s_1}\rangle \times \cdots \times \langle g_p, \mu_{s_p}\rangle \right| \\
&+ \sum_{x\in\mathcal{D}} \Bigg| \Bigg(h_t(x) - h_0(x) \\
&- \eta \int_0^t \int_{\mathcal{X}\times\mathcal{Y}} (y - h_s(x')) \langle \sigma(w\cdot x)\sigma(w\cdot x'), \mu_s\rangle \, \pi(dx', dy) ds \\
&- \eta \int_0^t \int_{\mathcal{X}\times\mathcal{Y}} (y - h_s(x')) \langle c^2\sigma'(w\cdot x')\sigma'(w\cdot x) x\cdot x', \mu_s\rangle \, \pi(dx, dy) ds \Bigg) \\
&\qquad \times q_1(h_{s_1}) \times \cdots \times q_p(h_{s_p}) \Bigg|.
\end{aligned}
$$

Then, using equations (19.19) and (19.21), we obtain

$$
\begin{aligned}
\mathbb{E}_{\pi^N}[F(\mu,h)] &= \mathbb{E}[F(\mu^N,h^N)]\\
&= \mathbb{E}\left|\mathcal{O}_p(N^{-1/2})\times\prod_{i=1}^{p}\langle g_i,\mu^N_{s_i}\rangle\right|\\
&\quad+\mathbb{E}\left|(M_t^{N,1}+M_t^{N,2}+\mathcal{O}_p(N^{-1/2}))\times\prod_{i=1}^{p}q_i(h^N_{s_i})\right|\\
&\le C_o\left(\mathbb{E}\Big[|M^{1,N}(t)|^2\Big]^{\frac{1}{2}}+\mathbb{E}\Big[|M^{2,N}(t)|^2\Big]^{\frac{1}{2}}\right)+\mathcal{O}(N^{-1/2})\\
&\le C_o\left(\frac{1}{\sqrt{N}}+\frac{1}{N}\right),
\end{aligned}
$$

where we have used the Cauchy-Schwarz inequality, see Appendix B.

Therefore,

$$
\lim_{N\to\infty}\mathbb{E}_{\pi^N}[F(\mu,h)]=0.
$$

Since $F(\cdot)$ is continuous and $F(\mu^N)$ is uniformly bounded (due to the uniform boundedness results established earlier),

$$
\mathbb{E}_{\pi}[F(\mu,h)]=0.
$$

Since this holds for each $t\in[0,T]$, all $f\in C_b^2(\mathbb{R}^{1+d})$, and for all functions $g_1,\dots,g_p,q_1,\dots,q_p\in C_b(\mathbb{R}^{1+d})$, we obtain that (μ,h) satisfies the evolution equation (19.4). □

Proof of Theorem 19.3. We now combine the previous results, tightness and identification results to prove Theorem 19.3. Let π^N be the probability measure corresponding to (μ^N,h^N). Each π^N takes values in the set of probability measures $\mathcal{M}\big(D_E([0,T])\big)$. Relative compactness implies that every subsequence π^{N_k} has a further subsequence $\pi^{N_{k_m}}$ which weakly converges. By the identification results any limit point π of $\pi^{N_{k_m}}$ will satisfy the evolution equation (19.4). Equation (19.4) is a finite-dimensional linear equation and therefore has a unique solution. Therefore, by Prokhorov's theorem, π^N weakly converges to π, where π is the distribution of (μ,h), the unique solution of (19.4). That is, (μ^N,h^N) converges in distribution to (μ,h). □

Let us now prove Corollary 19.6, which shows that under reasonable hyperparameter choices, the matrix A in the limit equation will be positive definite.

Proof of Corollary 19.6 . We first show that A is equivalent to the covariance matrix of the random variables $U = \Big(U(x_1), \dots, U(x_M)\Big)$, which are defined as

$$U(x) = \sqrt{\frac{\eta}{M}}\sigma(W \cdot x) + \sqrt{\frac{\eta}{M}}C\sigma'(W \cdot x)x,$$

where $(W, C) \sim \mu_0$ and $x \in \mathcal{D}$. Due to the fact that C is a mean zero random variable and independent of W, we have

$$\begin{aligned}\mathbb{E}\Big[U(x)U(x')\Big] &= \mathbb{E}\Big[\frac{\eta}{M}\sigma(W \cdot x)\sigma(W \cdot x') + \frac{\eta}{M}C^2\sigma'(W \cdot x)\sigma'(W \cdot x')x \cdot x'\Big] \\ &= A(x, x').\end{aligned}$$

To prove that A is positive definite, we need to show that $z^\top Az > 0$ for every non-zero $z \in \mathbb{R}^M$.

$$\begin{aligned} z^\top Az &= z^\top \mathbb{E}\Big[UU^\top\Big]z \\ &= \mathbb{E}\Big[(z^\top U)^2\Big] \\ &= \frac{\eta}{M}\mathbb{E}\Big[\Big(\sum_{m=1}^{M} z_m\left(\sigma(x_m \cdot W) + C\sigma'(W \cdot x_m)x_m\right)\Big)^2\Big]. \end{aligned}$$

The functions $\sigma(x_m \cdot W)$ are linearly independent since the x_m are in district directions (see Remark 3.1 of [**Ito96**]). Therefore, for each non-zero z, there exists a point w^* such that

$$\sum_{m=1}^{M} z_m\sigma(x_m \cdot w^*) \neq 0.$$

Consequently, there exists an $\epsilon > 0$ such that

$$\Big(\sum_{m=1}^{M} z_m\sigma(x_m \cdot w^*)\Big)^2 > \epsilon.$$

Since $\sigma(w \cdot x) + c\sigma'(w \cdot x)x$ is a continuous function, there exists a set $B = \{(c, w) : \|w - w^*\| + \|c\| < \kappa\}$ for some $\kappa > 0$ such that for $(c, w) \in B$,

$$\Big(\sum_{m=1}^{M} z_m\left(\sigma(x_m \cdot w) + C\sigma'(W \cdot x_m)x_m\right)\Big)^2 > \frac{\epsilon}{2}.$$

Then,

$$
\begin{aligned}
&\mathbb{E}\Big[\Big(\sum_{m=1}^{M} z_m\left(\sigma(x_m \cdot W) + C\sigma'(W \cdot x_m)x_m\right)\Big)^2\Big] \\
&\geq \mathbb{E}\Big[\Big(\sum_{m=1}^{M} z_m\left(\sigma(x_m \cdot W) + C\sigma'(W \cdot x_m)x_m\right)\Big)^2 \mathbf{1}_{W \in B}\Big] \\
&\geq \mathbb{E}\Big[\frac{\epsilon}{2}\mathbf{1}_{(C,W)\in B}\Big] \\
&= \frac{\epsilon}{2}K,
\end{aligned}
$$

for some constant $K > 0$.

Therefore, for every non-zero $z \in \mathbb{R}^M$,

$$z^\top A z > 0,$$

and A is positive definite, which concludes the proof of the corollary. □

19.8. Brief Concluding Remarks

Some results on the shape of the energy landscape associated with neural networks can be found in [**CHM$^+$15, PDGB14, PB17**]. He initialization [**HZRS15**] and Xavier initialization [**GB10**] have both been very influential initialization schemes. Their practical success led to the NTK development (the linear regime) that was first developed in [**JGH18**] and shows the convergence of trained neural networks to the ground truth. Related results on the NTK can also be found in [**ADH$^+$19, DLL$^+$19**].

The linear regime is further explored in [**MM23, MZ22, BMR21**] where generalization bounds are also established. In those works one can also find bounds addressing how wide a neural network should be so that the test error is well approximated by the infinite-width limit.

The presentation of the linear regime we followed in this chapter as well as the proofs of the results in Sections 19.5, 19.6, and 19.7 are based on the articles [**SS19, SS22**].

19.9. Exercises

Exercise 19.1. Consider a one-layer neural network with sigmoid activation function. Assume that the input is $X \sim N(0, 1)$ and the output is

$$Y = \sum_{n=1}^{N} c^n \sigma(w^n X + b^n).$$

Find a formula for the variance of Y in terms of $\{c^n, w^n, b^n\}_{n=1}^N$.

Exercise 19.2. Consider a shallow neural network with the bias term present. Namely, consider the neural network model

$$\mathfrak{m}^N(x;\theta) = \frac{1}{\sqrt{N}} \sum_{n=1}^{N} C^n \sigma(W^n \cdot x + b^n).$$

(1) Prove that for all $i \in \mathbb{N}$ and all k such that $k/N \leq T$, we have that $\mathbb{E}|b_k^i| < C$.

(2) Rework the arguments to state the limit problem analogously to what is in Theorem 19.3 without the bias term.

(3) Is Corollary 19.6 now true? Justify your answer.

Exercise 19.3. Let now $\gamma \in (1/2, 1)$ be a given parameter and consider the model

$$\mathfrak{m}^N(x;\theta) = \frac{1}{N^\gamma} \sum_{n=1}^{N} C^n \sigma(W^n \cdot x),$$

where $\theta = (C^1, W^1, \ldots, C^N, W^N)$ and with loss function

$$\Lambda(\theta) = \frac{1}{2M} \sum_{m=1}^{M} (y_m - \mathfrak{m}(x_m;\theta))^2.$$

(1) Write down the SGD updating equations for this model that are analogous to (19.3).

(2) Choose the learning rate to be $\eta_k^N = \eta/N^{2(1-\gamma)}$ with $\eta \in (0, \infty)$. Derive the evolution equation for the analogous h_t^N of equation (19.18) but for this model.

(3) Prove that if we choose the learning rate to be $\eta_k^N = \eta/N^{2(1-\gamma)}$ with $\eta \in (0, \infty)$ some constant, then the statement of Theorem 19.3 remains true for any given value of $\gamma \in [1/2, 1)$.

Exercise 19.4. Consider the single layer neural network model

$$\mathfrak{m}^N(x;\theta) = \frac{1}{\sqrt{N}} \sum_{n=1}^{N} C^n \sigma(W^n \cdot x),$$

where $\theta = (C^1, W^1, \ldots, C^N, W^N)$ and with loss function

$$\Lambda(\theta) = \frac{1}{2M} \sum_{m=1}^{M} (y_m - \mathfrak{m}(x_m;\theta))^2.$$

We train the model parameters θ using continuous-time gradient descent, i.e.,

$$\frac{d\theta_t}{dt} = -\eta \nabla \Lambda(\theta_t), \tag{19.22}$$

where $\eta > 0$ is the learning rate. At time $t = 0$, θ_0 are initialized as i.i.d. Gaussian random variables.

(1) Show that the Euler discretization of the continuous-time algorithm (19.22) is the standard gradient-descent algorithm.

(2) Prove that $t \mapsto \Lambda(\theta_t)$ is monotonically decreasing.

(3) What is the limit of $\mathfrak{m}(x; \theta_0)$ as $N \to \infty$?

(4) Using the chain rule derive a system of ODEs for the vector

$$h^N(t)) = (\mathfrak{m}(x_1; \theta_t), \ldots, \mathfrak{m}(x_M; \theta_t)).$$

(5) Show that $h^N(t)$ converges as $N \to \infty$ to the solution $h(t)$ of a system of linear ODEs. Compare your answer to (19.4).

(6) Let $\hat{\Lambda}(h) = \frac{1}{2M} \sum_{m=1}^{M} |y_m - h_m|^2$ where h_m is the mth element of the vector h. Show that $\lim_{t \to \infty} \hat{\Lambda}(h(t)) = 0$.

(7) Explain how the analysis above shows that the neural network *convexifies* as the number of hidden units tend to infinity.

Chapter 20

Optimization in the Feature Learning Regime: Mean Field Scaling

20.1. Introduction

In Chapter 19, we studied the so-called linear asymptotic regime, i.e., the limit as $N \to \infty$ of the single layer neural network

$$\mathfrak{m}^N(x;\theta) = \frac{1}{\sqrt{N}} \sum_{n=1}^{N} C^n \sigma(W^n \cdot x).$$

As we saw in Chapter 19, this limit gives rise to the neural tangent kernel and can be thought of as the linear regime given the linear nature of the limit (19.5).

The goal of this chapter is to study the so-called nonlinear regime which rises when we scale the neural network by $1/N$ instead of $1/\sqrt{N}$. In particular, let us consider

$$\mathfrak{m}^N(x;\theta) = \frac{1}{N} \sum_{n=1}^{N} C^n \sigma(W^n \cdot x).$$

Interestingly enough, as we shall see below, this scaling regime exhibits different behavior than what we have seen in Chapter 19, leading to a genuinely nonlinear limiting behavior with good generalization properties.

20.2. Preliminary Thoughts

Generally speaking, we can think of the model

$$\mathfrak{m}^N(x;\theta) = \frac{1}{N}\sum_{n=1}^{N} \psi(x;\theta^n),$$

where $\psi(x;\theta)$ could for example be

(1) a radial basis function $\psi(x;\theta) = e^{-\frac{k}{2}\|x-\theta\|^2}$.
(2) a shallow neural network $\psi(x;\theta) = c\sigma(w\cdot x+b)$, where $\theta = (c,w,b)$.
(3) a deep neural network $\psi(x;\theta) = c\sigma(w_1\sigma(w_2\cdot x+b_2)+b_1)$, where $\theta = (c,w_1,w_2,b_1,b_2)$.

For the sake of concreteness, let us consider the (population) loss function to be the quadratic error loss

$$\Lambda^N_{\text{pop}}(\theta) = \frac{1}{2}\mathbb{E}\Big[(\bar{\mathfrak{m}}(X) - \mathfrak{m}^N(X;\theta))^2\Big],$$

where $\bar{\mathfrak{m}}(x)$ is the target data. Notice that we deliberately expressed the loss function in terms of the expectation operator $\mathbb{E}$ associated with the underlying probability measure $\mathbb{P}$. This is the so-called population loss function, as opposed to the empirical loss function that has been the object of study in the vast majority of the book so far. We study the population loss function here as it makes some of the subsequent argument easier to present. However, as we shall see in Section 20.3, where we analyze the mean field scaling using the actual empirical loss function (which is what is used in practice), the intuition developed in this section working with the population loss function is consistent with what is seen when working with the empirical loss function.

For concreteness, let us focus on the shallow neural network case. Notice that we can write

$$\Lambda^N_{\text{pop}}(\theta) = D - \frac{1}{N}\sum_{i=1}^{N} h(\theta^i) - \frac{1}{2N^2}\sum_{i,j=1}^{N} K(\theta^i,\theta^j),$$

where $D = \frac{1}{2}\mathbb{E}|\bar{\mathfrak{m}}(X)|^2$, $h(\theta) = C\mathbb{E}\left[\bar{\mathfrak{m}}(X)\sigma(W\cdot X)\right]$, and

$$K(\theta^i,\theta^j) = C^iC^j\mathbb{E}\left[\sigma(W^i\cdot X)\sigma(W^j\cdot X)\right].$$

Gradient descent in continuous time takes the form

$$\dot{\theta}^i_t = \nabla h(\theta^i_t) - \frac{1}{N}\sum_{j=1}^{N} \nabla_{\theta^i_t} K(\theta^i_t,\theta^j_t).$$

As it turns out if we set

$$\mathfrak{m}^N(x,t) = \frac{1}{N}\sum_{n=1}^{N} \psi(x;\theta_t^n),$$

then $\lim_{N\to\infty} \mathfrak{m}^N(x,t) = \mathfrak{m}(x,t)$, where $\mathfrak{m}(x,t)$ satisfies

$$\dot{\mathfrak{m}}(x,t) = -\int_{\mathcal{X}} A_t(x,x')(\mathfrak{m}(x',t) - \bar{\mathfrak{m}}(x'))\pi(dx'). \tag{20.1}$$

Here $\bar{\mathfrak{m}}(x)$ is the target data and $A_t(x,x')$ is a positive semidefinite kernel, explicitly defined in (20.3). Some remarks are now in order.

Remark 20.1. It is interesting to contrast this with the $1/\sqrt{N}$ normalization of Chapter 19 where the limit is effectively as above but with the kernel $A_t(x,x') = A(x,x')$ being constant in time. In that case we have the linear regime and $A(x,x')$ is called the NTK. However, in the $1/N$ scaling case, the kernel $A_t(x,x')$ truly depends on time t and it corresponds to the nonlinear regime.

Define the empirical measure

$$\mu_t^N = \frac{1}{N}\sum_{n=1}^{N} \delta_{\theta_t^n}.$$

Note that the neural network output can be written as the inner-product

$$\mathfrak{m}^N(x;\theta_t) = \frac{1}{N}\sum_{n=1}^{N} C_t^n \sigma(W_t^n \cdot x) = \left\langle c\sigma(w\cdot x), \mu_t^N \right\rangle,$$

which is an affine function of the empirical measure μ_t^N. It also turns out that μ_t^N converges in the appropriate sense to a measure $\bar{\mu}_t$ whose density, say $q_t(\theta)$ with $\theta = (c,w)$, will satisfy the partial differential equation

$$\partial_t q_t(\theta) = \nabla\cdot\left(\nabla_\theta \tilde{K}(\theta,q_t)q_t(\theta)\right), \tag{20.2}$$

where $\tilde{K}(\theta,q_t) = -h(\theta) + \int_\Theta K(\theta,\theta')q_t(\theta')d\theta'$.

What about the loss function? Notice that $\Lambda^N(\theta)$ can also be seen as a function of the empirical measure μ_t^N. Indeed, we see that

$$\begin{aligned}\Lambda_{\text{pop}}^N(\theta) &= \frac{1}{2}\mathbb{E}\left[(\bar{\mathfrak{m}}(X) - \mathfrak{m}^N(X;\theta))^2\right]\\ &= \frac{1}{2}\mathbb{E}\left[(\bar{\mathfrak{m}}(X) - \left\langle c\sigma(w\cdot X),\mu_t^N\right\rangle)^2\right].\end{aligned}$$

So, let us write $\Lambda_{\text{pop}}^N(\theta) = \Lambda_{\text{pop}}(\mu_t^N)$ to emphasize the dependence on the empirical measure μ_t^N. Now, the convergence $\mu_t^N \to \bar{\mu}_t$ implies

$$\Lambda_{\text{pop}}(\mu_t^N) \to \bar{\Lambda}(\bar{\mu}_t), \text{ as } N\to\infty,$$

where

$$\bar{\Lambda}(\bar{\mu}_t) = \frac{1}{2}\int_{\mathcal{X}} (\bar{\mathrm{m}}(X) - \langle c\sigma(w\cdot x), \bar{\mu}_t\rangle)^2 \pi(dx).$$

A simple calculation shows that

$$\frac{d}{dt}\bar{\Lambda}(\bar{\mu}_t) = -\int_{\Theta} \|\nabla_\theta \tilde{K}(\theta, \bar{\mu}_t)\|_2^2 \bar{\mu}_t(d\theta) \leq 0.$$

Remark 20.2. The latter shows that the function $t \mapsto \bar{\Lambda}(\bar{\mu}_t)$ is non-increasing. While this does not guarantee convergence of $\bar{\Lambda}(\bar{\mu}_t)$ to zero, it is at least in the right direction! In fact, as we shall see in Theorem 20.21 for a deep neural network with mean field scaling, if it converges during training, it must converge to the global minimum (and not a local minimum). A similar result is true for shallow neural networks.

In order now to see, at least heuristically, that (20.1) holds, we will use (20.2). Consider the function $\rho(\theta, x) = c\sigma(w\cdot x)$ and recall that $\theta = (c, w)$. We have

$$\begin{aligned}
\partial_t \mathrm{m}(x,t) &= \partial_t \langle \rho(\cdot, x), \bar{\mu}_t\rangle \\
&= -\int_{\Theta} \nabla_\theta \rho(\theta, x)\left(\int_{\mathcal{X}} \nabla_\theta \rho(\theta, x')\,(\mathrm{m}(x',t) - \bar{\mathrm{m}}(x'))\,\pi(dx')\right)\bar{\mu}_t(d\theta) \\
&= -\int_{\mathcal{X}} A_t(x,x')\,(\mathrm{m}(x',t) - \bar{\mathrm{m}}(x'))\,\pi(dx'),
\end{aligned}$$

where in the last part we integrated by parts and defined

$$A_t(x,x') = \int_{\Theta} \nabla_\theta \rho(\theta, x) \nabla_\theta \rho(\theta, x') \bar{\mu}_t(d\theta), \tag{20.3}$$

which is a symmetric, positive semidefinite kernel. Hence, the key is to establish (20.2). In fact, under the proper assumptions, it can be shown that the flow converges to the target, i.e.,

$$\lim_{N,t\to\infty} \mathrm{m}^N(x,t) = \bar{\mathrm{m}}(x),$$

namely in the limit as $N \to \infty$ and $t \to \infty$ the algorithm recovers the target data. These points are made rigorous in the sections that follow. Section 20.6 compares the scaling of Chapter 19 and of this chapter.

20.3. Mean Field Limit for Shallow Neural Networks

Let us consider the simplest possible setting where we set

$$\mathrm{m}^N(x;\theta) = \frac{1}{N}\sum_{n=1}^{N} C^n \sigma(W^n \cdot x),$$

where $C^n \in \mathbb{R}$, $W^n \in \mathbb{R}^d$, $x \in \mathbb{R}^d$, and $\sigma(\cdot) : \mathbb{R} \to \mathbb{R}$. The number of hidden units is N and the output is scaled by a factor $\frac{1}{N}$; the so-called mean field scaling.

We turn to the empirical objective function, and we have

$$\Lambda^N(\theta) = \frac{1}{2}\frac{1}{M}\sum_{m=1}^{M}\left(y_m - \mathfrak{m}^N(x_m;\theta)\right)^2,$$

$y_m \in \mathbb{R}$, $x_m \in \mathbb{R}^d$ and the parameters $\theta = (C^1,\dots,C^N,W^1,\dots,W^N) \in \mathbb{R}^{N\times(1+d)}$. For notational convenience, we may refer to $\mathfrak{m}^N(x;\theta)$ as $\mathfrak{m}^N(x)$ in our analysis below.

The model parameters θ are trained using stochastic gradient descent. The parameter updates are given by

$$\begin{aligned} C_{k+1}^n &= C_k^n + \frac{\eta_k^N}{N}(y_k - \mathfrak{m}_k^N(x_k))\sigma(W_k^n \cdot x_k), \\ W_{k+1}^n &= W_k^n + \frac{\eta_k^N}{N}(y_k - \mathfrak{m}_k^N(x_k))C_k^n\sigma'(W_k^n \cdot x_k)x_k, \\ \mathfrak{m}_k^N(x) &= \frac{1}{N}\sum_{n=1}^{N} C_k^n\sigma(W_k^n \cdot x), \end{aligned} \tag{20.4}$$

for $k = 0,1,\dots,\lfloor TN\rfloor$ where $T > 0$. $\eta_k^N = \eta$ is the learning rate (chosen to be constant). The data samples (x_k, y_k) are i.i.d. samples from a distribution $\pi(dx,dy)$.

We impose the following assumption.

Assumption 20.3. We have that:

- The activation function $\sigma \in C_b^2(\mathbb{R})$, i.e., σ is two times continuously differentiable and bounded. Additionally, we shall assume that σ has two bounded derivatives.
- The data $(X,Y) \in \mathcal{X} \times \mathcal{Y}$ is compactly supported.
- The sequence of data samples (x_k, y_k) samples are i.i.d. from a distribution $\pi(dx,dy)$
- The randomly initialized parameters (C_0^n, W_0^n) are i.i.d., mean-zero random variables with a distribution $\mu_0(dc,dw)$ that has compact support.

As we shall see in Remark 20.7, Assumption 20.3 can be weakened considerably, but we will present the rigorous arguments under the stronger Assumption 20.3. We study the limiting behavior of the network output $\mathfrak{m}_k^N(x)$ for as the number of hidden units N and stochastic gradient descent steps $k = \lfloor TN\rfloor$

simultaneously become large. Let us now make this precise. Define the empirical measure

$$\nu_k^N = \frac{1}{N}\sum_{n=1}^{N}\delta_{C_k^n, W_k^n}.$$

Note that the neural network output can be written as the inner-product

$$\mathfrak{m}_k^N(x) = \left\langle c\sigma(w\cdot x), \nu_k^N\right\rangle,$$

i.e., it is linear in the empirical measure ν_k^N. Define the scaled processes

$$\mu_t^N = \nu_{\lfloor Nt\rfloor}^N.$$

Note that under Assumption 20.3, the initial empirical measure satisfies $\mu_0^N \xrightarrow{d} \bar{\mu}_0$ as $N\to\infty$. In addition, due to our assumption on the distribution of the (x_k, y_k) data and of the initialization $(C_0^n, W_0^n)_{n=1}^N$, the joint distribution of $(C_k^n, W_k^n)_{i=1}^N \in (\mathbb{R}^{1+d})^{\otimes N}$ is exchangeable and, consequently, ν_k^N is a Markov chain in the space of probability measures on E.

We are now ready to discuss the main result of this chapter.

Theorem 20.4. *Assume that Assumption 20.3 holds, and let the learning rate be given by $\eta_k^N = \eta$ for $0<\eta<\infty$, a fixed constant. The scaled empirical measure μ_t^N converges in probability to a limit measure $\bar{\mu}_t$ with values in $D_E([0,T])$ as $N\to\infty$ (where $E=\mathcal{M}(\mathbb{R}^{1+d})$). For every $f\in C_b^2(\mathbb{R}^{1+d})$, $\bar{\mu}$ is the unique deterministic solution of the measure evolution equation*

$$\begin{aligned}
\langle f,\bar{\mu}_t\rangle &= \langle f,\bar{\mu}_0\rangle \\
&\quad + \int_0^t\left(\int_{\mathcal{X}\times\mathcal{Y}}\eta\big(y-\langle c'\sigma(w'\cdot x),\bar{\mu}_s\rangle\big)\langle\sigma(w\cdot x)\partial_c f,\bar{\mu}_s\rangle\,\pi(dx,dy)\right)ds \\
&\quad + \int_0^t\left(\int_{\mathcal{X}\times\mathcal{Y}}\eta\big(y-\langle c'\sigma(w'\cdot x),\bar{\mu}_s\rangle\big)\langle c\sigma'(w\cdot x)x\cdot\nabla_w f,\bar{\mu}_s\rangle\,\pi(dx,dy)\right)ds \\
&= \langle f,\bar{\mu}_0\rangle \\
&\quad + \int_0^t\left(\int_{\mathcal{X}\times\mathcal{Y}}\eta\big(y-\langle c'\sigma(w'\cdot x),\bar{\mu}_s\rangle\big)\langle\nabla(c\sigma(w\cdot x))\cdot\nabla f,\bar{\mu}_s\rangle\,\pi(dx,dy)\right)ds,
\end{aligned}\tag{20.5}$$

where $\nabla f = (\partial_c f, \nabla_w f)$.

Corollary 20.5. *Assume Assumption 20.3 holds, and for a given $q(t,c,w)$ set*

$$\begin{aligned}
&v(\theta, q(t,\cdot)) \\
&\quad = \int_{\mathcal{X}\times\mathcal{Y}}\left(\left(y-\int_{\mathbb{R}^{1+d}}c'\sigma(w'\cdot x)q(t,c',w')dc'dw'\right)c\sigma(w\cdot x)\right)\pi(dx,dy).
\end{aligned}$$

Suppose that $\bar{\mu}_0$ admits a density $q_0(c, w)$ and there exists a unique solution to the nonlinear partial differential equation

$$\begin{aligned}\frac{\partial q(t,\theta)}{\partial t} &= -\eta \operatorname{div}_\theta \left(q(t,\theta)\nabla_\theta v(\theta, q(t,\cdot))\right)\\ q(0,\theta) &= q_0(\theta),\end{aligned} \tag{20.6}$$

where $\operatorname{div}_\theta$ is the divergence operator with respect to the variable $\theta = (c, w)$ and $q(t, c, w)$ vanishes as $|c|, \| w \| \to \infty$. Then, we have that the solution to the measure evolution equation (20.5) *is such that*

$$\bar{\mu}_t(d\theta) = q(t,\theta)d\theta.$$

Remark 20.6. Theorem 20.4 and Corollary 20.5 imply that the objective function $\Lambda^N(\theta)$ satisfies

(20.7)

$$\lim_{N\to\infty} \Lambda^N_{\text{pop}}(\theta_{\lfloor Nt \rfloor}) = \bar{\Lambda}(q(t,\cdot)) = \frac{1}{2}\int_{\mathcal{X}\times\mathcal{Y}} \left(y - \bar{\mathfrak{m}}(x; q(t,\cdot))\right)^2 \pi(dx, dy), \quad \text{where}$$

$$\bar{\mathfrak{m}}(x; q(t,\cdot)) = \int_{\mathbb{R}^{1+d}} c\sigma(w\cdot x) q(t, c, w) dc dw.$$

Classical results in the literature (see, e.g., [**AGS08, CMV03, JKO98**]) show that the PDE (20.6) is a gradient flow for the limiting objective function (20.7) in the space of probability measures on $\mathbb{R}^{1+d}$ endowed with the Wasserstein metric. This means that the trajectory's $t \mapsto q(t,\cdot)$ goal is to minimize the limit objective function $\Lambda(q)$ as defined by (20.7). More details on the optimal transportation theory as related to the problems of interest here can be found in [**AGS08, CB18**].

Remark 20.7. We note that, as [**SS20b**] shows, Theorem 20.4 holds with the considerable weaker Assumption 20.8 instead of Assumption 20.3. In particular, the compact support assumption of the distributions under which the data samples and parameters at initialization are generated is not needed and can be replaced by appropriate moment conditions.

Assumption 20.8. We have that:

- The activation function $\sigma \in C_b^2(\mathbb{R})$, i.e., σ is bounded and twice continuously differentiable. We further assume that it has two bounded derivatives.
- The randomly initialized parameters (C_0^n, W_0^n) are i.i.d., mean-zero random variables with a distribution $\mu_0(dc, dw)$, such that for some $0 < q < \infty$, we have $\mathbb{E}e^{qC_0^n} < \infty$ and $\mathbb{E}\left\|W_0^n\right\|^2 < \infty$.
- The sequence of data samples (x_m, y_m) is i.i.d. from the probability distribution $\pi(dx, dy)$ such that $\mathbb{E}x_m^4 + \mathbb{E}y_m^4 < \infty$.

We present the proof under the stronger Assumption 20.3 in order to focus on the main ideas and main intuition. We refer the interested reader to **[SS20b]** for the more involved technical details in the case of the weaker Assumption 20.8 instead of Assumption 20.3.

Proof of Theorem 20.4. Let us assume that we can indeed show relative compactness of the family $\{\mu^N\}_{N\in\mathbb{N}}$ in $D_E([0,T])$ where $E = \mathcal{M}(\mathbb{R}^{1+d})$ (this follows exactly along the lines of the calculations in Section 19.5.2 and the calculations are included in Section 20.3.1). This will guarantee that the family $\{\mu^N\}_{N\in\mathbb{N}}$ indeed has a limit as $N \to \infty$. Consider a test function $f \in C_b^2(\mathbb{R}^{1+d})$. By Taylor expansion, we shall have that

$$
\begin{aligned}
\langle f, \nu_{k+1}^N \rangle - \langle f, \nu_k^N \rangle &= \frac{1}{N}\sum_{n=1}^{N}\Big(f(c_{k+1}^n, w_{k+1}^n) - f(c_k^n, w_k^n)\Big) \\
&= \frac{1}{N}\sum_{n=1}^{N} \partial_c f(c_k^n, w_k^n)(c_{k+1}^n - c_k^n) + \frac{1}{N}\sum_{n=1}^{N} \nabla_w f(c_k^n, w_k^n)(w_{k+1}^n - w_k^n) \\
&\quad + \frac{1}{N}\sum_{n=1}^{N} \partial_c^2 f(\bar{c}_k^n, \bar{w}_k^n)(c_{k+1}^n - c_k^n)^2 \\
&\quad + \frac{1}{N}\sum_{n=1}^{N} (c_{k+1}^n - c_k^n)\nabla_{cw} f(\bar{c}_k^n, \bar{w}_k^n)(w_{k+1}^n - w_k^n) \\
&\quad + \frac{1}{N}\sum_{n=1}^{N} (w_{k+1}^n - w_k^n)^\top \nabla_w^2 f(\bar{c}_k^n, \bar{w}_k^n)(w_{k+1}^n - w_k^n),
\end{aligned}
$$

for points $\bar{c}_k^n, \bar{w}_k^n$ in the segments connecting c_{k+1}^n with c_k^n and w_{k+1}^n with w_k^n, respectively. Notice now that the uniform bounds of Exercise 20.1 and the relation (20.4) imply that as N gets large

$$
\begin{aligned}
\langle f, \nu_{k+1}^N \rangle - \langle f, \nu_k^N \rangle &= \frac{1}{N^2}\sum_{n=1}^{N} \partial_c f(c_k^n, w_k^n)\eta(y_k - \mathfrak{m}^N(x_k;\theta_k))\sigma(w_k^n \cdot x_k) \\
&\quad + \frac{1}{N^2}\sum_{n=1}^{N} \eta(y_k - \mathfrak{m}^N(x_k;\theta_k))c_k^n \sigma'(w_k^n \cdot x_k)\nabla_w f(c_k^n, w_k^n)\cdot x_k + \mathcal{O}_p\left(N^{-2}\right),
\end{aligned}
$$

where we recall Definition 19.8 for the notation $\mathcal{O}_p\left(N^{-2}\right)$. The term $\mathcal{O}_p\left(N^{-2}\right)$ is a result of $f \in C_b^2$, the bounds from of Exercise 20.1, as well as the moment bounds on (x_k, y_k) from Assumption 19.2. We next define the components

$$D_k^{1,N} = \frac{1}{N}\int_{\mathcal{X}\times\mathcal{Y}} \eta\big(y - \langle c\sigma(w\cdot x), \nu_k^N\rangle\big)\langle \sigma(w\cdot x)\partial_c f, \nu_k^N\rangle\,\pi(dx,dy),$$
$$D_k^{2,N} = \frac{1}{N}\int_{\mathcal{X}\times\mathcal{Y}} \eta\big(y - \langle c\sigma(w\cdot x), \nu_k^N\rangle\big)\langle c\sigma'(w\cdot x)x\cdot\nabla_w f, \nu_k^N\rangle\,\pi(dx,dy),$$
$$M_k^{1,N} = \frac{1}{N}\eta\big(y_k - \langle c\sigma(w\cdot x_k), \nu_k^N\rangle\big)\langle \sigma(w\cdot x_k)\nabla_c f, \nu_k^N\rangle - D_k^{1,N},$$
$$M_k^{2,N} = \frac{1}{N}\eta\big(y_k - \langle c\sigma(w\cdot x_k), \nu_k^N\rangle\big)\langle c\sigma'(w\cdot x_k)x\cdot\nabla_w f, \nu_k^N\rangle - D_k^{2,N},$$

which leads to the expression

$$\langle f, \nu_{k+1}^N\rangle - \langle f, \nu_k^N\rangle = D_k^{1,N} + D_k^{2,N} + M_k^{1,N} + M_k^{2,N} + \mathcal{O}_p\left(N^{-2}\right).$$

Next, we define $D^{1,N}, D^{2,N}, M^{1,N}$, and $M^{2,N}$ as sums over indexes $k \in \{0,\dots,\lfloor Nt\rfloor - 1\}$ of $D_k^{1,N}, D_k^{2,N}, M_k^{1,N}$, and $M_k^{2,N}$, respectively

$$D^{1,N}(t) = \sum_{k=0}^{\lfloor Nt\rfloor-1} D_k^{1,N}, \qquad D^{2,N}(t) = \sum_{k=0}^{\lfloor Nt\rfloor-1} D_k^{2,N},$$
$$M^{1,N}(t) = \sum_{k=0}^{\lfloor Nt\rfloor-1} M_k^{1,N}, \qquad M^{2,N}(t) = \sum_{k=0}^{\lfloor Nt\rfloor-1} M_k^{2,N}. \tag{20.8}$$

The scaled empirical measure can be written as the telescoping sum

$$\begin{aligned}\langle f, \mu_t^N\rangle - \langle f, \mu_0^N\rangle &= \langle f, \nu_{\lfloor Nt\rfloor}^N\rangle - \langle f, \nu_0^N\rangle \\ &= \sum_{k=0}^{\lfloor Nt\rfloor-1}\Big(\langle f, \nu_{k+1}^N\rangle - \langle f, \nu_k^N\rangle\Big).\end{aligned}$$

Therefore, the scaled empirical measure satisfies, as N grows,

$$\begin{aligned}\langle f, \mu_t^N\rangle - \langle f, \mu_0^N\rangle &= \sum_{k=0}^{\lfloor Nt\rfloor-1}\Big(\langle f, \nu_{k+1}^N\rangle - \langle f, \nu_k^N\rangle\Big)\\
&= \sum_{k=0}^{\lfloor Nt\rfloor-1}\Big(D_k^{1,N} + D_k^{2,N} + M_k^{1,N} + M_k^{2,N}\Big) + \mathcal{O}_p\left(N^{-1}\right)\\
&= \int_0^t\Big(\int_{\mathcal{X}\times\mathcal{Y}} \eta\big(y - \langle c\sigma(w\cdot x), \mu_s^N\rangle\big)\langle \sigma(w\cdot x)\nabla_c f, \mu_s^N\rangle\,\pi(dx,dy)\Big)ds\\
&\quad + \int_0^t\Big(\int_{\mathcal{X}\times\mathcal{Y}} \eta\big(y - \langle c\sigma(w\cdot x), \mu_s^N\rangle\big)\langle c\sigma'(w\cdot x)x\cdot\nabla_w f, \mu_s^N\rangle\,\pi(dx,dy)\Big)ds\\
&\quad + M^{1,N}(t) + M^{2,N}(t) + \mathcal{O}_p\left(N^{-1}\right).\end{aligned} \tag{20.9}$$

In fact Exercise 20.2 claims $M^{1,N}(t)$ and $M^{2,N}(t)$ converge to 0 in L^2 as $N \to \infty$. Tightness together with relation (20.9) shows that (20.5) is the limit equation, similar to the proof of Theorem 19.3.

In particular, let π^N be the probability measure of a convergent subsequence of $\{\mu^N\}_{0 \le t \le T}$. Each π^N takes values in the set of probability measures $\mathcal{M}(D_E([0,T]))$. The established relative compactness implies that there is a subsequence π^{N_k} which weakly converges. Then, as in Lemma 19.15, we get that if π^{N_k} is a convergent subsequence with a limit point π, then, π is a Dirac measure concentrated on $\bar{\mu} \in D_E([0,T])$ and $\bar{\mu}$ satisfies equation (20.5).

Lastly, it remains to show uniqueness of the solution to (20.5). This is proven via a contraction argument, as detailed in Section 20.3.2. This completes the proof of Theorem 20.4. □

An important consequence of Theorem 20.4 is that the neural network has the *propagation of chaos* property. This is the content of Theorem 20.9, presented here without proof.

Theorem 20.9 ([**SS20b**]). *Assume that Assumption* 19.2 *holds. Consider* $T < \infty$ *and let* $t \in (0,T]$. *Define the probability measure* $\rho_t^N \in \mathcal{M}(\mathbb{R}^{(1+d)N})$ *where*

$$\rho_t^N(dx^1, \dots, dx^N) = \mathbb{P}[(c^1_{\lfloor Nt \rfloor}, w^1_{\lfloor Nt \rfloor}) \in dx^1, \dots, (c^N_{\lfloor Nt \rfloor}, w^N_{\lfloor Nt \rfloor}) \in dx^N].$$

Then, the sequence of probability measures $\rho_\cdot^N$ *is* $\bar{\mu}_\cdot$*-chaotic. That is, for* $k \in \mathbb{N}$

$$\lim_{N\to\infty} \left\langle f_1(x^1) \times \cdots \times f_k(x^k), \rho_\cdot^N(dx^1, \dots, dx^N) \right\rangle = \prod_{i=1}^{k} \langle f_i, \bar{\mu}_\cdot \rangle, \tag{20.10}$$

for all $f_1, \dots, f_k \in C_b^2(\mathbb{R}^{1+d})$.

Theorem 20.9 implies asymptotic independence of the particles as $N \to \infty$. Indeed, by (20.10), as $N \to \infty$, the dynamics of the weights (c_k^i, w_k^i) will become independent of the dynamics of the weights (c_k^j, w_k^j) for any $i \neq j$ in the limit as $N \to \infty$. It is perhaps interesting to remark here that the dynamics (c_k^i, w_k^i) are still random due to the random initialization. Let us finally discuss insights from the mean field limit of Theorem 20.4.

Remark 20.10. As $N \to \infty$, the neural network converges in probability to a deterministic model. This is despite the fact that the neural network is randomly initialized and it is trained on a random sequence of data samples via stochastic gradient descent.

Remark 20.11. For finite N, η must decay with the number of iterations in order for stochastic gradient descent to converge. Despite this, the noise disappears and the neural network's parameter distribution converges to a deterministic evolution equation. This is due to the normalization of $\frac{1}{N}$ in the hidden layer replacing the role of the learning rate decay.

Remark 20.12. As it also discussed in Remark 20.6, the partial differential equation (20.6) is a gradient flow for the limiting objective function (20.7) in the space of probability measures on $\mathbb{R}^{1+d}$ endowed with the Wasserstein metric. Hence, the target of the limiting law of large numbers result is to minimize the limit objective function $\bar{\Lambda}(q)$ as defined by (20.7).

20.3.1. Tightness in mean field scaling.

The process for establishing tightness here is parallel to the process we followed in Chapter 19. The first step into establishing that the family $\{\mu_t^N, t \in [0,T]\}_{N\in\mathbb{N}}$ has a limit as N grows to infinity is to prove an appropriate form of compact containment, and, in our case, it is enough to show that there is a compact set that contains $\{\mu_t^N\}$ for all $N \in \mathbb{N}$ and $t \in [0,T]$. Recall that $\mu_t^N \in D_E([0,T])$, where $D_E([0,T])$ is the set of maps from $[0,T]$ into E which are right-continuous and which have left-hand limits, $E = \mathcal{M}(\mathbb{R}^{1+d})$, and $\mathcal{M}(\mathbb{R}^{1+d})$ is the space of probability measures in $\mathbb{R}^{1+d}$ (see also Section A.4).

Lemma 20.13. *For each $\delta > 0$, there is a compact subset $\mathcal{K}$ of E such that*

$$\sup_{N\in\mathbb{N},0\le t\le T} \mathbb{P}[\mu_t^N \notin \mathcal{K}] < \delta.$$

Proof. Given the a priori bounds established in Exercise 20.1, the proof is completely analogous to that of Lemma 19.11. □

Next we establish regularity of the family of measures $\{\mu_t^N : t \in [0,T]\}_{N\in\mathbb{N}}$. As in Chapter 19 consider the function $q(z_1,z_2) = \min\{|z_1-z_2|,1\}$ with $z_1,z_2 \in \mathbb{R}$. Recall that $\mathcal{F}_t^N$ is the σ-algebra generated by $\{(C_0^i, W_0^i)\}_{i=1}^N$ and $\{x_j\}_{j=0}^{\lfloor Nt\rfloor-1}$, i.e., $\mathcal{F}_t^N$ contains the information generated by $\{(C_0^i, W_0^i)\}_{i=1}^N$ and $\{x_j\}_{j=0}^{\lfloor Nt\rfloor-1}$.

Lemma 20.14. *Let $f \in C_b^2(\mathbb{R}^{1+d})$. For any $p \in (0,1)$ and $\delta \in (0,1)$, there is a constant $C_o < \infty$ such that for $0 \le u \le \delta$, $0 \le v \le \delta \wedge t$, $t \in [0,T]$,*

$$\mathbb{E}\left[q(\langle f,\mu_{t+u}^N\rangle,\langle f,\mu_t^N\rangle)q(\langle f,\mu_t^N\rangle,\langle f,\mu_{t-v}^N\rangle)|\mathcal{F}_t^N\right] \le C_o\delta^p + \frac{C_o}{N}.$$

Proof. The proof is parallel to that of Lemma 19.12 of Chapter 19. A Taylor expansion gives for $0 \leq s \leq t \leq T$

$$
\begin{aligned}
|\langle f, \mu_t^N \rangle - \langle f, \mu_s^N \rangle| &= |\langle f, \nu_{\lfloor Nt \rfloor}^N \rangle - \langle f, \nu_{\lfloor Ns \rfloor}^N \rangle| \\
&\leq \frac{1}{N} \sum_{n=1}^{N} |f(C_{\lfloor Nt \rfloor}^n, W_{\lfloor Nt \rfloor}^n) - f(C_{\lfloor Ns \rfloor}^n, W_{\lfloor Ns \rfloor}^n)| \\
&\leq \frac{1}{N} \sum_{n=1}^{N} |\partial_c f(\bar{C}_{\lfloor Nt \rfloor}^n, \bar{W}_{\lfloor Nt \rfloor}^n)| |C_{\lfloor Nt \rfloor}^n - C_{\lfloor Ns \rfloor}^n| \\
&\quad + \frac{1}{N} \sum_{n=1}^{N} \| \nabla_w f(\hat{C}_{\lfloor Nt \rfloor}^n, \hat{W}_{\lfloor Nt \rfloor}^n) \| \| W_{\lfloor Nt \rfloor}^n - W_{\lfloor Ns \rfloor}^n \|, \qquad (20.11)
\end{aligned}
$$

for points $(\bar{C}^n, \bar{W}^n)$ and $(\hat{C}^n, \hat{W}^n)$ in the segments connecting $C_{\lfloor Ns \rfloor}^n$ with $C_{\lfloor Nt \rfloor}^n$ and $W_{\lfloor Ns \rfloor}^n$ with $W_{\lfloor Nt \rfloor}^n$, respectively.

The next step is to establish a bound on $|C_{\lfloor Nt \rfloor}^n - C_{\lfloor Ns \rfloor}^n|$ for $s < t \leq T$ with $0 < t - s \leq \delta < 1$. For $p \in (0, 1)$ we have

$$
\begin{aligned}
\mathbb{E}\left[|C_{\lfloor Nt \rfloor}^n - C_{\lfloor Ns \rfloor}^n| \middle| \mathcal{F}_s^N \right] &= \mathbb{E}\left[\left| \sum_{k=\lfloor Ns \rfloor}^{\lfloor Nt \rfloor - 1} (C_{k+1}^n - C_k^n) \right| \middle| \mathcal{F}_s^N \right] \\
&\leq \mathbb{E}\left[\sum_{k=\lfloor Ns \rfloor}^{\lfloor Nt \rfloor - 1} \left| \eta (y_k - \mathfrak{m}_k^N(x_k)) \frac{1}{N} \sigma(W_k^n \cdot x_k) \right| \middle| \mathcal{F}_s^N \right] \\
&\leq \frac{1}{N} \sum_{k=\lfloor Ns \rfloor}^{\lfloor Nt \rfloor - 1} C_o \\
&\leq C_o(t-s) + \frac{C_o}{N} \\
&\leq C_o(t-s)^p \mathbf{1}_{t-s<1} + C_o(t-s)^p T^{1/p} \mathbf{1}_{t-s \geq 1} + \frac{C_o}{N} \\
&\leq C_o \delta^p + \frac{C_o}{N}, \qquad (20.12)
\end{aligned}
$$

where Assumption 20.3 and the bounds from Exercise 20.1 were used. Also, $0 < C_o < \infty$ is an unimportant finite constant that may change from line to line.

Let's now establish a bound on $\| W^n_{\lfloor Nt\rfloor} - W^n_{\lfloor Ns\rfloor} \|$ for $s < t \le T$ with $0 < t - s \le \delta < 1$. We obtain

$$
\begin{aligned}
\mathbb{E}\Big[\| W^n_{\lfloor Nt\rfloor} - W^n_{\lfloor Ns\rfloor} \| \Big| \mathcal{F}^N_s \Big] &= \mathbb{E}\Big[\| \sum_{k=\lfloor Ns\rfloor}^{\lfloor Nt\rfloor -1} (W^n_{k+1} - W^n_k) \| \Big| \mathcal{F}^N_s \Big] \\
&\le \mathbb{E}\Big[\sum_{k=\lfloor Ns\rfloor}^{\lfloor Nt\rfloor -1} \| \eta(y_k - \mathfrak{m}^N_k(x_k)) \frac{1}{N} C^n_k \sigma'(W^n_k \cdot x_k) x_k \| \Big| \mathcal{F}^N_s \Big] \\
&\le \frac{1}{N} \sum_{k=\lfloor Ns\rfloor}^{\lfloor Nt\rfloor -1} C_o \\
&\le C_o \delta^p + \frac{C_o}{N},
\end{aligned}
$$

where we have again used the bounds from Exercise 20.1.

Thus, going back to (20.11), due to the a priori bounds from Exercise 20.1, the quantities $(\bar{C}^n_{\lfloor Nt\rfloor}, \bar{W}^n_{\lfloor Nt\rfloor})$ are bounded in expectation for $0 < s < t \le T$. Therefore, for $0 < s < t \le T$ with $0 < t - s \le \delta < 1$

$$
\mathbb{E}\left[| \langle f, \mu^N_t \rangle - \langle f, \mu^N_s \rangle | | \mathcal{F}^N_s \right] \le C_o \delta^p + \frac{C_o}{N},
$$

where $C_o < \infty$ is some unimportant finite constant which may depend on the magnitude of the first partial derivatives of f. This concludes the proof of the lemma. □

We can now establish that the family of processes $\{\mu^N_t, t \in [0, T]\}$ has a limit as $N \to \infty$. Indeed, we have the following lemma.

Lemma 20.15. *The family of processes $\{\mu^N\}_{N\in\mathbb{N}}$ is relatively compact in $D_E([0, T])$.*

Proof. Combining Lemmas 20.13 and 20.14 and the results of Section A.4, proves that $\{\mu^N\}_{N\in\mathbb{N}}$ is relatively compact in $D_{\mathcal{M}(\mathbb{R}^{1+d})}([0, T])$ (see also Theorem 8.6, Remark 8.7 B and Theorem 9.1 of Chapter 3 of [**EK86**], as well as Theorem 4.6 in [**Jak86**] and Section 3 of [**Led16**]). □

20.3.2. Uniqueness in mean field scaling. The goal of this section is to prove uniqueness of the evolution equation (20.5). The strategy is to set up a Picard type of iteration and prove that it has a unique fixed point in the appropriate space through a contraction mapping. To this end, notice that

$$\langle f, \bar{\mu}_t \rangle = \langle f, \bar{\mu}_0 \rangle$$
$$+ \int_0^t \left(\int_{\mathcal{X}\times\mathcal{Y}} \eta(y - \langle c'\sigma(w' \cdot x), \bar{\mu}_s \rangle) \langle \sigma(w \cdot x) \partial_c f, \bar{\mu}_s \rangle \pi(dx, dy) \right) ds$$
$$+ \int_0^t \left(\int_{\mathcal{X}\times\mathcal{Y}} \eta(y - \langle c'\sigma(w' \cdot x), \bar{\mu}_s \rangle) \langle c\sigma'(w \cdot x) x \nabla_w f, \bar{\mu}_s \rangle \pi(dx, dy) \right) ds.$$

(20.13)

$$= \langle f, \bar{\mu}_0 \rangle + \int_0^t \langle G(z, Q(\bar{\mu}_s, \cdot)) \cdot \nabla f, \bar{\mu}_s \rangle \, ds,$$

where for $z = (c, w_1, \dots, w_d) \in \mathbb{R}^{1+d}$, $Q(\bar{\mu}, x) = \langle c\sigma(w \cdot x), \bar{\mu} \rangle$ we have

$$G(z, Q(\bar{\mu}, \cdot)) = (G_1(z, Q(\bar{\mu}, \cdot)), G_2(z, Q(\bar{\mu}, \cdot))) \in \mathbb{R}^{1+d},$$

with

$$G_1(z, Q(\bar{\mu}, \cdot)) = \int_{\mathcal{X}\times\mathcal{Y}} \eta(y - Q(\bar{\mu}, x)) \sigma(w \cdot x) \pi(dx, dy) \in \mathbb{R},$$
$$G_2(z, Q(\bar{\mu}, \cdot)) = \int_{\mathcal{X}\times\mathcal{Y}} \eta(y - Q(\bar{\mu}, x)) c\sigma'(w \cdot x) x \pi(dx, dy) \in \mathbb{R}^d.$$

A solution to (20.13), $\bar{\mu}_\cdot$, is associated to the nonlinear random process Z_t (see for example [**Kol14**]) satisfying the random ordinary differential equation

$$Z_t = Z_0 + \int_0^t G(Z_s, Q(\bar{\mu}_s, \cdot)) ds,$$
$$Z_0 \sim \bar{\mu}(0, c, w),$$
$$\bar{\mu}_t = \mathrm{Law}(Z_t). \tag{20.14}$$

This ODE is random due to the random initial data. Let us define the mapping $F : D_{\mathbb{R}}([0,T]) \mapsto D_{\mathcal{M}(\mathbb{R}^{1+d})}([0,T])$ such that for a path $(R_t)_{t\in[0,T]} \in D_{\mathbb{R}}([0,T])$, we have $F(R_\cdot) = \mathrm{Law}(Y_\cdot)$ where $Y_\cdot$ is given by

$$Y_t = Y_0 + \int_0^t G(Y_s, R_s) ds,$$
$$Y_0 \sim \bar{\mu}(0, c, w).$$

Next, define the map $L : D_{\mathcal{M}(\mathbb{R}^{1+d})}([0,T]) \mapsto D_{\mathbb{R}}([0,T])$ taking a measure valued process μ_t and mapping it to $Q(\mu_t, x) = L(\mu)$, where

$$Q(\mu_t, x) = \langle c\sigma(w \cdot x), \mu_t \rangle,$$

and the map $H : D_{\mathcal{M}(\mathbb{R}^{1+d})}([0,T]) \mapsto D_{\mathcal{M}(\mathbb{R}^{1+d})}([0,T])$ via the composition of the mappings F and L, i.e., we set $H = F \circ L$. Oftentimes, if we want to emphasize the dependence on T, we may write H_T for H.

The discussion above show that if $(\mu_t)_{t\in[0,T]}$ is a fixed point of H, then $\text{Law}(Z_t) = H_t(\mu_\cdot)$ is a solution to (20.13). In the reverse direction, if $(Z_t)_{t\in[0,T]}$ is a solution to (20.14), then its law will be a fixed point of H, implying that $\text{Law}(Z_t) = H_t(\mu)$. It is also a fact that if μ is a weak measure valued solution to (20.13), then it must be a fixed point of H, satisfying (20.14) and consequently proving our result.

We next show that H is a contraction mapping for $t \in [0,T]$. For this purpose, we first show that in order to study the fixed point of H, it is enough to consider $H : C([0,T]; M(\mathbb{R}^{1+d})) \mapsto C([0,T]; M(\mathbb{R}^{1+d}))$. Then this will allow us to work in $C([0,T]; M(\mathbb{R}^{1+d}))$ instead of working in the larger space $D_{\mathcal{M}(\mathbb{R}^{1+d})}([0,T])$, which in turn simplifies some of the arguments.

Therefore, we derive in Lemma 20.16 appropriate a priori bounds for the parameters c_t and w_t and study their regularity in time. If we denote by $\mathbb{E}$ the expectation operator taken with respect to the measure governing the evolution of parameters (notice that here (x,y) are considered to be integration variables) we have the following system of random ODEs.

$$
\begin{aligned}
c_t &= c_0 + \int_0^t \alpha \int_{\mathcal{X}\times\mathcal{Y}} (y - \mathbb{E}[c_s\sigma(w_s\cdot x)])\sigma(w_s\cdot x)\pi(dx,dy)ds,\\
w_t &= w_0 + \int_0^t \alpha \int_{\mathcal{X}\times\mathcal{Y}} (y - \mathbb{E}[c_s\sigma(w_s\cdot x)])c_s\sigma'(w_s\cdot x)x\pi(dx,dy)ds.
\end{aligned}
\tag{20.15}
$$
$$(c_0, w_0) \sim \bar{\mu}(0,c,w).$$

Lemma 20.16 provides us with the necessary a priori uniform bounds on the parameters and also shows that there is regularity in time.

Lemma 20.16. *There is a constant $C_o < \infty$, depending on T, such that*

$$\sup_{t\in[0,T]} (|c_t| + \| w_t \|) \le C_o,$$

and for every $0 \le s \le t \le T$ we have that

$$|c_t - c_s| + \| w_t - w_s \| \le C_o(t-s).$$

Proof. Let's examine c_t first and establish a bound on its growth. The finite constant $C_o < \infty$ may change from line to line, and it may also depend upon the final time T.

$$c_t = c_0 + \int_0^t \eta \int_{\mathcal{X}\times\mathcal{Y}} (y - \mathbb{E}[c_s\sigma(w_s \cdot x)])\sigma(w_s \cdot x)\pi(dx, dy)ds.$$

$$\begin{aligned} c_t\sigma(w_t \cdot x) &= \sigma(w_t \cdot x)c_0 \\ &\quad + \sigma(w_t \cdot x)\int_0^t \eta \int_{\mathcal{X}\times\mathcal{Y}} (y - \mathbb{E}[c_s\sigma(w_s \cdot x)])\sigma(w_s \cdot x)\pi(dx, dy)ds. \end{aligned}$$

(20.16)

$$|c_t\sigma(w_t \cdot x)| \leq C_o|c_0| + C_o \int_0^t \int_{\mathcal{X}\times\mathcal{Y}} |(y - \mathbb{E}[c_s\sigma(w_s \cdot x)])\sigma(w_s \cdot x)|\pi(dx, dy)ds.$$

We have used the fact that $\sigma(\cdot)$ is bounded. Now, we will use the facts that c_0, X, and Y have compact support.

$$|c_t\sigma(w_t \cdot x)| \leq C_o + C_o \int_0^t \int_{\mathcal{X}\times\mathcal{Y}} \mathbb{E}[|c_s\sigma(w_s \cdot x)|]\pi(dx, dy)ds.$$

$$|c_t\sigma(w_t \cdot x)| \leq C_o + C_o \int_0^t \int_{\mathcal{X}\times\mathcal{Y}} \sup_{x'\in\mathcal{X}} \mathbb{E}[|c_s\sigma(w_s \cdot x')|]\pi(dx, dy)ds.$$

$$\sup_{x\in\mathcal{X}} \mathbb{E}[|c_t\sigma(w_t \cdot x)|] \leq C_o + C_o \int_0^t \sup_{x'\in\mathcal{X}} \mathbb{E}[|c_s\sigma(w_s \cdot x')|]ds.$$

Therefore, by Gronwall's inequality,

$$\sup_{x\in\mathcal{X}} \mathbb{E}[|c_t\sigma(w_t \cdot x)|] \leq C_o$$

for $0 \leq t \leq T$. Therefore, going back to (20.16) and recalling Assumption 19.2, we get that uniformly in $t \in [0, T]$,

$$|c_t| \leq C_o.$$

Similarly, now from (20.15) we also obtain that there is a constant $C_o < \infty$, uniform in $t \in [0, T]$ such that

$$\| w_t \| \leq C_o.$$

The latter statements imply the first statement of the lemma. Let us now prove the second statement of the lemma. Similarly to the calculations above and using the uniform bounds on c_t and w_t together with Assumption 19.2, we have

$$\begin{aligned} |c_t - c_s| &= \left| \int_s^t \eta \int_{\mathcal{X}\times\mathcal{Y}} (y - \mathbb{E}[c_u\sigma(w_u \cdot x)])\sigma(w_u \cdot x)\pi(dx, dy)du \right| \\ &\leq C_o(t - s). \end{aligned}$$

The corresponding statement $\| w_t - w_s \| \leq C_o(t-s)$ follows along the same lines, concluding the proof of the lemma. □

As a consequence of the regularity result in Lemma 20.16, (20.15) is a continuous process. Therefore, we can prove a contraction in $C([0,T];M(\mathbb{R}^{1+d}))$ (instead of studying the process in the larger space $D_{\mathcal{M}(\mathbb{R}^{1+d})}([0,T])$).

Now that we have established this a priori boundedness and regularity result, let us go back to the proof of uniqueness. Notice that Lemma 20.16 shows that c_t and w_t are bounded on $[0,T]$. Motivated by this fact, let us define the *bump* function $b(z) \in C^\infty$, which is one for $|z| \leq B$ and zero for $|z| \geq 2B$. If, for example, $\sup_{t\in[0,T]} |c_t| \leq C_o$, then we set $B = 2C_o$. Lemma 20.16 allows us to do so.

Let us define for notational convenience $C_T = C([0,T], \mathbb{R}^{1+d})$ and let M_T be the set of probability measures on C_T. Consider an element $\kappa \in M_T$. Motivated by the discussion before Lemma 20.16, let us set $\text{Law}(Y) = H(\kappa_\cdot)$, where, slightly abusing notation, $Y = (c, w)$ with

$$
\begin{aligned}
c_t &= c_0 + \int_0^t \int_{\mathcal{X}\times\mathcal{Y}} \eta\Big(y - \langle G_{s,x}, \kappa\rangle\Big)\sigma(w_s \cdot x)\pi(dx,dy)ds,\\
w_t &= w_0 + \int_0^t \int_{\mathcal{X}\times\mathcal{Y}} \eta\Big(y - \langle G_{s,x}, \kappa\rangle\Big)c_s\sigma'(w_s \cdot x)x\pi(dx,dy)ds,\\
G_{s,x} &= c_s'\sigma(w_s' \cdot x)b(c_s'),
\end{aligned}
$$

$$(c_0, w_0) \sim \bar{\mu}(0, c, w). \tag{20.17}$$

We next show existence and uniqueness of a fixed point $\text{Law}(c_t, w_t)$ for the mapping H, as defined via (20.17). For $\kappa, \kappa' \in M_T$ and $p \geq 1$ define the metric

$$D_{T,p}(\kappa,\kappa') = \inf\left\{\left(\int_{C_T\times C_T} \sup_{s\leq T} \|x_s - y_s\|_p^p \wedge 1 d\nu(x,y)\right)^{1/p}, \nu \in P(\kappa,\kappa')\right\},$$

where $P(\kappa,\kappa')$ is the set of probability measures on $C_T \times C_T$ such that the marginal distributions are κ and κ', respectively. The space M_T endowed with the metric D_T is a complete metric space.

If a solution to (20.14) exists, then it must be a fixed point of H (defined via equation (20.17)). This is an immediate consequence of Lemma 20.16. Therefore, if H has a unique solution, there can be at most one solution to (20.14). If (20.14) has at most one solution, (20.13) has at most one solution. Therefore, if H has a unique fixed point, this proves uniqueness for (20.13).

Now, for two elements $\kappa^1, \kappa^2 \in M_T$, let us set $\text{Law}(Y_\cdot^i) = \text{Law}((c_\cdot^i, w_\cdot^i)) = H(\kappa_\cdot^i)$ for $t \in [0,T]$ with $i = 1,2$. So, let (c_t^1, w_t^1) satisfy (20.17) with $\kappa = \kappa^1$, and let (c_t^2, w_t^2) satisfy (20.17) with $\kappa = \kappa^2$. The processes (c_t^1, w_t^1) and (c_t^2, w_t^2) have the same initial conditions. That is,

$$(c_0^1, w_0^1) = (c_0^2, w_0^2) = (c_0, w_0),$$
$$(c_0, w_0) \sim \bar{\mu}(0, c, w).$$

We now prove a contraction for the mapping H for some $0 < T_0 < T$. By definition, (c_t^1, w_t^1) and (c_t^2, w_t^2) have marginal distributions $H(\kappa^1)$ and $H(\kappa^2)$, respectively, on the time interval $[0, T_0]$. Once this is proven, we can extend this to the entire interval $[0, T]$ since T_0 is not affected by the input measures κ^1, κ^2 or by which subinterval of $[0, T]$ we are considering. The following lemma is going into this direction.

Lemma 20.17. *Let $\kappa^1, \kappa^2 \in M_T$ and $T < \infty$. Then there exists a finite constant $C_o < \infty$ that may depend on T such that*

$$D_{t,1}(H(\kappa^1), H(\kappa^2)) \leq C_o \int_0^t D_{u,1}(\kappa^1, \kappa^2) du$$

for any $0 < t < T$.

Proof. The formula (20.17) yields

$$\begin{aligned}
c_t^1 - c_t^2 &= \int_0^t \int_{\mathcal{X}\times\mathcal{Y}} \eta\Big(y - \langle G_{s,x}, \kappa^1\rangle\Big)\sigma(w_s^1 \cdot x)\pi(dx, dy)ds \\
&\quad - \int_0^t \int_{\mathcal{X}\times\mathcal{Y}} \eta\Big(y - \langle G_{s,x}, \kappa^2\rangle\Big)\sigma(w_s^2 \cdot x)\pi(dx, dy)ds \\
&= \int_0^t \int_{\mathcal{X}\times\mathcal{Y}} \eta y\Big(\sigma(w_s^1 \cdot x) - \sigma(w_s^2 \cdot x)\Big)\pi(dx, dy)ds \\
&\quad + \int_0^t \int_{\mathcal{X}\times\mathcal{Y}} \eta \langle G_{s,x}, \kappa^2\rangle \sigma(w_s^2 \cdot x)\pi(dx, dy)ds \\
&\quad - \int_0^t \int_{\mathcal{X}\times\mathcal{Y}} \eta \langle G_{s,x}, \kappa^1\rangle \sigma(w_s^1 \cdot x)\pi(dx, dy)ds \\
&= \int_0^t \int_{\mathcal{X}\times\mathcal{Y}} \eta y\Big(\sigma(w_s^1 \cdot x) - \sigma(w_s^2 \cdot x)\Big)\pi(dx, dy)ds \\
&\quad + \int_0^t \int_{\mathcal{X}\times\mathcal{Y}} \eta \langle G_{s,x}, \kappa^2\rangle \left(\sigma(w_s^2 \cdot x) - \sigma(w_s^1 \cdot x)\right) \pi(dx, dy)ds \\
&\quad + \int_0^t \int_{\mathcal{X}\times\mathcal{Y}} \eta \langle G_{s,x}, \kappa^2 - \kappa^1\rangle \sigma(w_s^1 \cdot x)\pi(dx, dy)ds.
\end{aligned}$$

In order to address the mean-field term, we recall that $c'_s\sigma(w'_s x)b(c'_s)$ and $\sigma'(\cdot)$ are bounded and that X, Y have compact support. Therefore, we get that

$$\Big|\int_0^t\int_{\mathcal{X}\times\mathcal{Y}}\langle c'_s\sigma(w'_s x)b(c'_s),\kappa^2\rangle\Big(\sigma(w^2_s\cdot x)-\sigma(w^1_s\cdot x)\Big)\pi(dx,dy)ds\Big|$$
$$\leq C_o\int_0^t \parallel w^2_s - w^1_s \parallel ds.$$

We next bound the term

$$\Big|\int_0^t\int\Big(\langle c'_s\sigma(w'_s x)b(c'_s),\kappa^2-\kappa^1\rangle\,\sigma(w^1_s\cdot x)\pi(dx,dy)ds\Big|.$$

Since the map $(c,w)\mapsto c\sigma(w\cdot x)b(c)$ is globally Lipschitz, we have that

$$|c^2\sigma(w^2\cdot x)b(c^2)-c^1\sigma(w^1\cdot x)b(c^1)|\leq K(|c^2-c^1|+\|w^2-w^1\|),$$

where the constant $K<\infty$ does not depend upon x (since X has compact support). Then, for $0\leq s\leq T$,

$$\Big|\int_0^t\int_{\mathcal{X}\times\mathcal{Y}}\Big(\langle c'_s\sigma(w'_s x)b(c'_s),\kappa^2-\kappa^1\rangle\Big)\sigma(w^1_s\cdot x)\pi(dx,dy)ds\Big|$$
$$\leq K\int_0^t D_{s,1}(\kappa^1,\kappa^2)ds.$$

Similar calculations also give the necessary bound for the difference $w^1_t - w^2_t$. Hence, for $0\leq s\leq T$, we eventually have the bound

$$\sup_{u\leq s}\big[|c^1_u-c^2_u|+\parallel w^1_u-w^2_u\parallel\big]\leq C_1\int_0^s\Big(|c^2_u-c^1_u|+\parallel w^2_u-w^1_u\parallel\Big)du$$
$$+C_2\int_0^s D_{u,1}(\kappa^1,\kappa^2)du,$$

for finite constants $C_1, C_2<\infty$. We then also have that

$$\mathbb{E}\Big[\sup_{u\leq s}\big[|c^1_u-c^2_u|+\parallel w^1_u-w^2_u\parallel\big]\Big]$$
$$\leq C_1\int_0^s\mathbb{E}\Big[\sup_{\tau\leq u}\big[|c^2_\tau-c^1_\tau|+\parallel w^2_\tau-w^1_\tau\parallel\big]\Big]du$$
$$+C_2\int_0^s D_{u,1}(\kappa^1,\kappa^2)du.$$

By Gronwall's inequality, we then get for $s\leq T$,

$$\mathbb{E}\Big[\sup_{u\leq s}\big[|c^1_u-c^2_u|+\parallel w^1_u-w^2_u\parallel\big]\Big]\leq C_2\exp(C_1 s)\int_0^s D_{u,1}(\kappa^1,\kappa^2)du.$$

The latter display immediately implies the statement of the lemma. □

Lemma 20.17 immediately proves there is a contraction on the interval $[0, T_0]$.

$$\begin{aligned}D_{t,1}(H(\kappa^1), H(\kappa^2)) &\leq C_o \int_0^t D_{u,1}(\kappa^1, \kappa^2) du \\ &\leq C_o \int_0^t D_{t,1}(\kappa^1, \kappa^2) du \\ &\leq C_o t D_{t,1}(\kappa^1, \kappa^2).\end{aligned}$$

Then, choose T_0 such that $C_o T_0 < 1$. In fact we have Lemma 20.18.

Lemma 20.18. *Let $T < \infty$. The mapping $H_T = (F \circ F)_T$ has a unique fixed point.*

Proof. By Lemma 20.17 and the Banach fixed-point theorem we obtain that there is $0 < T_0 < \infty$ such that $H_{T_0}(m)$ will be a contraction map. This then implies that (20.17) has a unique solution on $[0, T_0]$. We can then extend this construction to the whole interval $[0, T]$ by dividing the interval $[0, T]$ into subintervals $[0, T_0], [T_0, 2T_0], \ldots, [T-T_0, T]$. In each subinterval, it can be shown that the solution is unique by proving a contraction as was done in Lemma 20.17, which can be done as T_0 can be always taken to be of the same magnitude, i.e., it does not depend on which subinterval is being examined. This concludes the proof of the lemma. □

20.4. Central Limit Theorem Behavior for Shallow Neural Networks

In this subsection we show that shallow neural networks satisfy a central limit type of theorem as the size of the network and the number of training steps become large. The *central limit theorem* (CLT) quantifies the speed of convergence of the finite neural network to its mean-field limit as well as how the finite neural network fluctuates around the mean-field limit for large N. Instead of presenting the full details, we shall only present the main elements that will allow us to guess what the limit would be and refer the interested reader to [**SS20a**] for the proof details.

We start by defining the fluctuation process

$$\alpha_t^N = \sqrt{N}(\mu_t^N - \bar{\mu}_t).$$

Then $\alpha^N \xrightarrow{d} \bar{\alpha}$, where $\bar{\alpha}$ satisfies a stochastic partial differential equation. This result characterizes the fluctuations of the finite empirical measure μ^N around its mean-field limit $\bar{\mu}$ for large N. Interestingly, the limit $\bar{\alpha}$ has a Gaussian distribution.

In order to motivate the result, let us recall the formula for $\langle f, \mu_t^N \rangle$ by (20.9) and the formula for $\langle f, \bar{\mu}_t \rangle$ by (20.5). Let us take the difference of the two formulas, scaling the result by $\sqrt{N}$, in order to get a formula for $\langle f, \alpha_t^N \rangle$:

$$\begin{aligned}
&\langle f, \alpha_t^N \rangle - \langle f, \alpha_0^N \rangle \\
&= \int_0^t \left(\int_{\mathcal{X}\times\mathcal{Y}} \eta(y - \langle c\sigma(w \cdot x), \bar{\mu}_s \rangle) \langle \sigma(w \cdot x)\partial_c f, \alpha_s^N \rangle \, \pi(dx, dy) \right) ds \\
&- \int_0^t \left(\int_{\mathcal{X}\times\mathcal{Y}} \eta \langle c\sigma(w \cdot x), \alpha_s^N \rangle \langle \sigma(w \cdot x)\partial_c f, \bar{\mu}_s \rangle \, \pi(dx, dy) \right) ds \\
&+ \int_0^t \left(\int_{\mathcal{X}\times\mathcal{Y}} \eta(y - \langle c\sigma(w \cdot x), \bar{\mu}_s \rangle) \langle c\sigma'(w \cdot x)x \cdot \nabla_w f, \alpha_s^N \rangle \, \pi(dx, dy) \right) ds \\
&- \int_0^t \left(\int_{\mathcal{X}\times\mathcal{Y}} \eta \langle c\sigma(w \cdot x), \alpha_s^N \rangle) \langle c\sigma'(w \cdot x)x \cdot \nabla_w f, \bar{\mu}_s \rangle \, \pi(dx, dy) \right) ds \\
&+ \sqrt{N} \left(M_t^{1,N} + M_t^{2,N} \right) + R_t^N,
\end{aligned}$$

where it can be shown that the remainder term R_t^N goes to zero as $N \to \infty$ uniformly in $t \in [0, T]$, while $\sqrt{N}\left(M_t^{1,N} + M_t^{2,N}\right)$ behaves asymptotically as $N \to \infty$ as a Gaussian martingale.

In particular, for test functions in the appropriate space we have that the following stochastic partial differential equation characterizes the Gaussian evolution of the limit α_t,

$$\begin{aligned}
\langle f, \bar{\alpha}_t \rangle &= \langle f, \bar{\alpha}_0 \rangle \\
&+ \int_0^t \int_{\mathcal{X}\times\mathcal{Y}} \eta(y - \langle c\sigma(w \cdot x), \bar{\mu}_s \rangle) \langle \nabla(c\sigma(w \cdot x)) \cdot \nabla f, \bar{\alpha}_s \rangle \, \pi(dx, dy) ds \\
&- \int_0^t \int_{\mathcal{X}\times\mathcal{Y}} \eta \langle c\sigma(w \cdot x), \bar{\alpha}_s \rangle \langle \nabla(c\sigma(w \cdot x)) \cdot \nabla f, \bar{\mu}_s \rangle \, \pi(dx, dy) ds + \langle f, \bar{M}_t \rangle .
\end{aligned} \tag{20.18}$$

$\bar{M}_t$ is a mean-zero Gaussian process with variance-covariance structure given explicitly as follows. For $\mu \in \mathcal{M}(\mathbb{R}^{1+d})$ and $h \in \mathcal{C}_0^1(\mathbb{R}^{1+d})$ define the operator

$$\mathcal{R}_{x,y,\mu}[h] = (y - \langle c\sigma(w \cdot x), \mu \rangle) \langle \nabla(c\sigma(w \cdot x)) \cdot \nabla h, \mu \rangle .$$

Then, we shall have that $(\sqrt{N}\langle f, M_t^N \rangle, \sqrt{N}\langle g, M_t^N \rangle) \in D_{\mathbb{R}^2}([0, T])$ converges to a distribution valued mean-zero Gaussian martingale with covariance function

$$
\begin{aligned}
&\operatorname{Cov}\Big[\langle f, \bar{M}_t\rangle, \langle g, \bar{M}_t\rangle\Big] \\
&= \eta^2 \int_0^t \Big[\int_{\mathcal{X}\times\mathcal{Y}} \Big(\mathcal{R}_{x,y,\bar{\mu}_s}[f] - \int_{\mathcal{X}\times\mathcal{Y}} \mathcal{R}_{x,y,\bar{\mu}_s}[f]\pi(dx,dy)\Big) \\
&\qquad \times \Big(\mathcal{R}_{x,y,\bar{\mu}_s}[g] - \int_{\mathcal{X}\times\mathcal{Y}} \mathcal{R}_{x,y,\bar{\mu}_s}[g]\pi(dx,dy)\Big)\pi(dx,dy)\Big] ds.
\end{aligned}
$$

Finally, the stochastic evolution equation (20.18) has a unique solution, which implies that $\bar{\alpha}$ is unique.

The CLT stochastic evolution equation (20.18) is coupled with the mean-field limit PDE. The stochastic evolution equation (20.18) is linear in $\bar{\alpha}$ and driven by a Gaussian process; therefore, the limit $\bar{\alpha}_t$ itself is a Gaussian process.

The convergence of the fluctuation process α_t^N indicates that for large N the empirical distribution of the neural network's parameters behaves as

$$
\nu^N_{\lfloor N\cdot\rfloor} = \mu^N_{\cdot} \approx \bar{\mu}_{\cdot} + \frac{1}{\sqrt{N}}\bar{\alpha}_{\cdot},
$$

where $\bar{\alpha}$ has a Gaussian distribution. Combined, the fluctuations result and the law of large numbers results show that the relation between the number of particles (*hidden units*, in the language of neural networks) and the number of stochastic gradient steps should be of the same order to have convergence and statistically good behavior. Under this scaling, as a measure valued process, the empirical distribution of the parameters behaves as a Gaussian distribution.

20.5. Deep Neural Networks in Mean Field Scaling

In this section we shall briefly consider the ideas behind the mean field limits for deep neural networks. For illustration purposes, let us consider a multilayer neural network with two hidden layers. The extension to even deeper neural networks with more layers is analogous.

$$
\mathfrak{m}^{N_1,N_2}(x;\theta) = \frac{1}{N_2}\sum_{i=1}^{N_2} C^i \sigma\left(\frac{1}{N_1}\sum_{j=1}^{N_1} W^{2,i,j}\sigma\left(W^{1,j}\cdot x\right)\right). \tag{20.19}
$$

Notice now that (20.19) can be also written as

$$
\begin{aligned}
H^{1,j}(x) &= \sigma(W^{1,j} \cdot x), \quad j = 1, \dots, N_1, \\
Z^{2,i}(x) &= \frac{1}{N_1} \sum_{j=1}^{N_1} W^{2,i,j} H^{1,j}(x), \quad i = 1, \dots, N_2, \\
H^{2,i}(x) &= \sigma\Big(Z^{2,i}(x)\Big), \\
\mathfrak{m}_\theta^{N_1,N_2}(x) &= \frac{1}{N_2} \sum_{i=1}^{N_2} C^i H^{2,i}(x), \qquad (20.20)
\end{aligned}
$$

where $C^i, W^{2,i,j} \in \mathbb{R}$ and $x, W^{1,j} \in \mathbb{R}^d$. The neural network model has parameters

$$\theta = (C^1, \dots, C^{N_2}, W^{2,1,1}, \dots, W^{2,N_1,N_2}, W^{1,1}, \dots, W^{1,N_1}),$$

which must be estimated from data. The number of hidden units in the first layer is N_1 and the number of hidden units in the second layer is N_2. The multilayer neural network (20.20) includes a normalization factor of $\frac{1}{N_1}$ in the first hidden layer and $\frac{1}{N_2}$ in the second hidden layer.

Consider the mean square error loss again given by

$$\Lambda_{\text{pop}}^{N_1,N_2}(\theta) = \frac{1}{2}\mathbb{E}_{X,Y}\Big[(Y - \mathfrak{m}^{N_1,N_2}(X;\theta))^2\Big]$$

for the population loss function, where the data $(X, Y) \sim \pi(dx, dy)$, and

$$\Lambda^{N_1,N_2}(\theta) = \frac{1}{2}\frac{1}{|\mathcal{D}|} \sum_{(x,y)\in\mathcal{D}} (y - \mathfrak{m}^{N_1,N_2}(x;\theta))^2 \qquad (20.21)$$

for the empirical loss function used in practice. The goal is to estimate a set of parameters θ which minimizes the objective function (20.21).

The stochastic gradient descent algorithm for estimating the parameters θ is, for $k \in \mathbb{N}$,

$$
\begin{aligned}
C_{k+1}^{i} &= C_{k}^{i} + \frac{\eta_C^{N_1,N_2}}{N_2}(y_k - \mathfrak{m}^{N_1,N_2}(x_k;\theta_k))H_k^{2,i}(x_k),\\
W_{k+1}^{1,j} &= W_k^{1,j} + \frac{\eta_{W,1}^{N_1,N_2}}{N_1}(y_k - \mathfrak{m}^{N_1,N_2}(x_k;\theta_k))\\
&\quad \times \left(\frac{1}{N_2}\sum_{i=1}^{N_2} C_k^i \sigma'(Z_k^{2,i}(x_k))W_k^{2,i,j}\right)\sigma'(W_k^{1,j}\cdot x_k)x_k,\\
W_{k+1}^{2,i,j} &= W_k^{2,i,j} + \frac{\eta_{W,2}^{N_1,N_2}}{N_1N_2}(y_k - \mathfrak{m}^{N_1,N_2}(x_k;\theta_k))C_k^i\sigma'(Z_k^{2,i}(x_k))H_k^{1,j}(x_k),\\
H_k^{1,i}(x_k) &= \sigma(W_k^{1,i}\cdot x_k),\\
Z_k^{2,i}(x_k) &= \frac{1}{N_1}\sum_{j=1}^{N_1} W_k^{2,i,j}H_k^{1,j}(x_k),\\
H_k^{2,i}(x_k) &= \sigma(Z_k^{2,i}(x_k)),\\
\mathfrak{m}^{N_1,N_2}(x_k;\theta_k) &= \frac{1}{N_2}\sum_{i=1}^{N_2} C_k^i H_k^{2,i}(x_k),
\end{aligned}
$$

where $\eta_C^{N_1,N_2}$, $\eta_{W,1}^{N_1,N_2}$, and $\eta_{W,2}^{N_1,N_2}$ are the learning rates. The learning rates may depend upon N_1 and N_2. The parameters at step k are

$$\theta_k = (C_k^1,\dots,C_k^{N_2}, W_k^{2,1,1},\dots,W_k^{2,N_1,N_2}, W_k^{1,1},\dots,W_k^{1,N_1}).$$

(x_k, y_k) are samples of the random variables (X, Y).

Assumption 20.19. We assume the following conditions.

- $\sigma(\cdot) \in C_b^2$, i.e., it is twice continuously differentiable and bounded. Additionally, we shall assume that σ has two bounded derivatives.
- The distribution $\pi(dx, dy)$ has compact support, i.e., the data (x_k, y_k) takes values in the compact set $\mathcal{X} \times \mathcal{Y}$.
- The random initialization of the parameters, i.e., $\{C_\circ^i\}_i$, $\{W_\circ^{2,i,j}\}_{i,j}$ and $\{W_\circ^{1,j}\}_j$, are i.i.d. and take values in compact sets $\mathcal{C}$, $\mathcal{W}^1$, and $\mathcal{W}^2$.
- The probability distribution of initial parameters $(C_\circ^i, W_\circ^{2,i,j}, W_\circ^{1,j})_{i,j}$ admits continuous probability density functions.

We denote by $\mu_c(dc)$, $\mu_{W^2}(du)$, and $\mu_{W^1}(dw)$ the probability distributions of $\{C_\circ^i\}_i$, $\{W_\circ^{2,i,j}\}_{i,j}$, and $\{W_\circ^{1,j}\}_j$, respectively.

Theorem 20.20. *Let $T > 0$ be given, let Assumption 20.19 hold, and choose the learning rates to be*

$$\eta_C^{N_1,N_2} = \frac{N_2}{N_1}, \quad \eta_{W,1}^{N_1,N_2} = 1, \quad \text{and} \quad \eta_{W,2}^{N_1,N_2} = N_2.$$

Then, for any $t \in [0,T]$ and $x \in \mathcal{X}$,

$$\lim_{N_2\to\infty}\lim_{N_1\to\infty} \mathfrak{m}^{N_1,N_2}(x;\theta_{\lfloor N_1 t\rfloor}) = \mathfrak{m}_t(x),$$

in probability, where we have that

$$\mathfrak{m}_t(x) = \int_{\mathcal{C}} \tilde{C}_t^c \tilde{H}_t^{2,c}(x)\mu_c(dc),$$

with

$$\begin{aligned}
d\tilde{C}_t^c &= \int_{\mathcal{X}\times\mathcal{Y}} (y-\mathfrak{m}_t(x))\tilde{H}_t^{2,c}(x)\pi(dx,dy)dt, \quad \tilde{C}_0^c = c,\\
d\tilde{W}_t^{1,w} &= \int_{\mathcal{X}\times\mathcal{Y}} (y-\mathfrak{m}_t(x))V_t^w(x)\sigma'(\tilde{W}_t^{1,w}\cdot x)x\pi(dx,dy)dt, \tilde{W}_0^{1,w} = w,\\
d\tilde{W}_t^{2,c,w,u} &= \int_{\mathcal{X}\times\mathcal{Y}} (y-\mathfrak{m}_t(x))\tilde{C}_t^c\sigma'(\tilde{Z}_t^c(x))\tilde{H}_t^{1,w}(x)\pi(dx,dy)dt, \tilde{W}_0^{2,c,w,u} = u,\\
\tilde{H}_t^{1,w}(x) &= \sigma(\tilde{W}_t^{1,w}\cdot x),\\
\tilde{Z}_t^c(x) &= \int_{\mathcal{W}^1}\int_{\mathcal{W}^2} \tilde{W}_t^{2,c,w,u}\tilde{H}_t^{1,w}(x)\mu_{W^2}(du)\mu_{W^1}(dw),\\
\tilde{H}_t^{2,c}(x) &= \sigma(\tilde{Z}_t^c(x)),
\end{aligned}$$

(20.22)

$$V_t^w(x) = \int_{\mathcal{C}} \tilde{C}_t^c\sigma'(\tilde{Z}_t^c(x))\left(\int_{\mathcal{W}^2}\tilde{W}_t^{2,c,w,u}\mu_{W^2}(du)\right)\mu_c(dc).$$

The system in (20.22) *has a unique solution.*

Notice that we can also write that $\mathfrak{m}_t(x)$ satisfies

(20.23)

$$\mathfrak{m}_t(x) = \int_{\mathcal{C}} \tilde{C}_t^c\sigma\left(\int_{\mathcal{W}^1}\int_{\mathcal{W}^2}\tilde{W}_t^{2,c,w,u}\sigma(\tilde{W}_t^{1,w}\cdot x)\mu_{W^2}(du)\mu_{W^1}(dw)\right)\mu_c(dc).$$

Note that the learning rates in the second layer are trained faster than the other parameters. This choice of learning rates is necessary for convergence to a non-trivial limit as $N_1, N_2 \to \infty$. If the parameters in all the layers are trained with the same learning rate, it can be shown that the network will not train as N_1, N_2 become large. The proof of Theorem 20.20 can be found in [**SS21**]; see also the papers [**AOY19, Ngu19, NP23**] for related results.

20.5.1. Convergence Properties of the Limit as Time Grows. Let us now conclude this section by discussing the convergence properties of $\mathfrak{m}_t(x)$ as $t \to \infty$.

Let us denote $\Theta_t(c, w, u) = (\tilde{C}_t^c, \tilde{W}_t^{1,w}, \tilde{W}_t^{2,c,w,u})$ for the components of the ODE in (20.22). For notational convenience, we shall often write $\theta = (c, w, u)$. Then, we obviously have that $\mathfrak{m}_t(x)$ depends on t only through $[\Theta_t] = \{\Theta_t(\theta)\}_{\theta \in \mathcal{C} \times \mathcal{W}^1 \times \mathcal{W}^2}$. In order to emphasize that, we shall write $\mathfrak{m}_t(x) = g(x; [\Theta_t])$.

Analogously, we will denote

$$[\Theta_t(c, \cdot)] = \{\Theta_t(c, w, u)\}_{(w,u) \in \mathcal{W}^1 \times \mathcal{W}^2} \text{ and } [\Theta_t(\cdot, w, \cdot)] = \{\Theta_t(c, w, u)\}_{(c,u) \in \mathcal{C} \times \mathcal{W}^2},$$

which then leads to the notation

$$\tilde{Z}_t^c(x) = \tilde{Z}(x; [\Theta_t(c, \cdot, \cdot)]) \text{ and } V_t^w(x) = V(x; [\Theta_t(\cdot, w, \cdot)]).$$

For notational convenience let us also denote $h(x; [\Theta_t]) = (\bar{\mathfrak{m}}(x) - \mathfrak{m}_t(x))$.

Then, for $(c, w) \in \mathcal{C} \times \mathcal{W}^1$, we also define

$$R_1([\Theta_t], c) = \int_{\mathcal{X}} h(x; [\Theta_t]) \sigma(\tilde{Z}(x; [\Theta_t(c, \cdot)])) \pi(dx),$$

$$R_2([\Theta_t], \tilde{W}_t^{1,w}, w) = \int_{\mathcal{X}} h(x; [\Theta_t]) V(x; [\Theta_t(\cdot, w, \cdot)]) \sigma'(\tilde{W}_t^{1,w} \cdot x) x \pi(dx),$$

$$R_3([\Theta_t], \tilde{C}_t^c, \tilde{W}_t^{1,w}, c) = \int_{\mathcal{X}} h(x; [\Theta_t]) \tilde{C}_t^c \sigma'(\tilde{Z}(x; [\Theta_t(c, \cdot)]))) \sigma(\tilde{W}_t^{1,w} \cdot x) \pi(dx),$$

and we set

$$H([\Theta_t], \Theta_t(\theta), \theta) = (R_1([\Theta_t], c), R_2([\Theta_t], \tilde{W}_t^{1,w}, w), R_3([\Theta_t], \tilde{C}_t^c, \tilde{W}_t^{1,w}, c)).$$

The notation used above makes it clear that the functions $H(\cdot)$ depend on $[\Theta_t]$, on $\Theta_t(\theta)$, and on θ separately. The ODE system in (20.22) can be written in the form

$$\dot{\Theta}_t(\Theta_0) = H([\Theta_t], \Theta_t(\Theta_0), \Theta_0), \text{ such that } \Theta_0 = (c, w, u). \tag{20.24}$$

The limiting objective function can be written as

$$\begin{aligned} \bar{\Lambda}(\Theta_t) &= \lim_{N_2 \to \infty} \lim_{N_1 \to \infty} \Lambda_{\text{pop}}^{N_1, N_2}(\theta_{\lfloor N_1 t \rfloor}) \\ &= \lim_{N_2 \to \infty} \lim_{N_1 \to \infty} \frac{1}{2} \mathbb{E}_X \left[(\bar{\mathfrak{m}}(X) - \mathfrak{m}^{N_1, N_2}(X; \theta_{\lfloor N_1 t \rfloor}))^2 \right] \\ &= \frac{1}{2} \int_{\mathcal{X}} \left[(\bar{\mathfrak{m}}(x) - \mathfrak{m}(x; [\Theta_t]))^2 \right] \pi(dx), \end{aligned}$$

where $\mathfrak{m}(x; [\Theta_t]) = \mathfrak{m}_t(x)$ is given by (20.23) and $\bar{\mathfrak{m}}(x)$ is the target function of x.

By inspection of the previous formula, we see that the function $\Theta \mapsto \bar{\Lambda}(\Theta)$ is non-negative and becomes zero only at the global minimum.

In fact, as we shall now see, under the appropriate conditions and as $t \to \infty$, the global minimum is achieved. Namely, we have that

$$\lim_{t\to\infty} g(x; [\Theta_t]) = \bar{\mathfrak{m}}(x) \quad \text{for almost all } x \in \mathcal{X}.$$

The limiting loss function $\bar{\Lambda}$ acts as a Lyapunov function for the dynamical system (20.24) (see Exercise 20.4)

$$\frac{d}{dt}\bar{\Lambda}(\Theta_t) = \nabla_\Theta \bar{\Lambda}(\Theta_t) \cdot \dot{\Theta}_t = \nabla_\Theta \bar{L}(\Theta_t) \cdot H(\Theta_t) \leq 0. \tag{20.25}$$

The fact that $\frac{d}{dt}\bar{\Lambda}(\Theta_t) \leq 0$ means that $\bar{\Lambda}(\Theta_t)$ is at least decreasing (albeit not strictly) in the gradient direction of the paths governing the limiting behavior of the weights.

Let us define ζ_0 to be the joint measure for the random initialization of the parameters, i.e., for $\{C_\circ^i\}_i, \{W_\circ^{1,j}\}_j, \{W_\circ^{2,i,j}\}_{i,j}$. By Assumption 20.19 we have that this is the product measure $\zeta_0 = \mu_c \times \mu_{W^1} \times \mu_{W^2}$. By analogy, let us now define ζ_t to be the probability measure at time t of the random vector $\Theta_t = (\tilde{C}_t, \tilde{W}_t^1, \tilde{W}_t^2)$ as governed by the solution to the random ODE system (20.22). Then, ζ_t is the pushforward of ζ_0 under Θ_t given by (20.24), i.e.,

$$\zeta_t = (\Theta_t)_\sharp \zeta_0;$$

see for example Chapter 8 in [**AGS08**]. Then, we have the following result.

Theorem 20.21. *Let us assume that* $\mathrm{support}(\zeta_0) = \mathcal{C} \times \mathcal{W}^1 \times \mathcal{W}^2$ *and that the activation function* $\sigma(\cdot)$ *is real analytic, bounded, and* $\sigma'(\cdot) > 0$. *If* $\zeta_t \to \zeta^*$ *weakly, where* ζ^* *is a non-degenerate measure that admits a density with finite first moments, then we have that* ζ^* *is a global minimum with zero loss.*

The proof of Theorem 20.21 can be found in [**SS21**].

Remark 20.22. We mention here that perhaps what is important is not so much the exact form of the limit formula for $\mathfrak{m}_t(x)$, but rather the fact that for the right choice of the learning rates, such a limit exists.

20.6. In Between the Linear and the Nonlinear Regime

Let us consider now the same setup as in Section 20.3 but instead of mean-field scaling, consider the model

$$\mathfrak{m}^N(x;\theta) = \frac{1}{N^\gamma}\sum_{n=1}^{N} C^n \sigma(W^n \cdot x), \tag{20.26}$$

where now $\gamma \in (1/2, 1)$ and as a before $C^n \in \mathbb{R}$, $W^n \in \mathbb{R}^d$, $x \in \mathbb{R}^d$, and $\sigma(\cdot) : \mathbb{R} \to \mathbb{R}$.

In Chapter 19 we studied the behavior of (20.26) under stochastic gradient descent when $\gamma = 1/2$ whereas in this chapter we studied it in the case of $\gamma = 1$. Hence, natural questions arise:

- What happens in between (i.e., for $\gamma \in (1/2, 1)$)?
- How do the different scalings behave?
- How does that behavior extend to multi-layer neural networks?

Let us first investigate the situation in the case of a shallow neural network. We continue working with the mean square error loss function

$$\Lambda^N(\theta) = \frac{1}{2}\frac{1}{M}\sum_{m=1}^{M}\left(y_m - \mathfrak{m}^N(x_m;\theta)\right)^2,$$

and the model parameters θ are trained by stochastic gradient descent:

$$\begin{aligned}
C^n_{k+1} &= C^n_k + \frac{\eta^N_k}{N^\gamma}\left(y_k - \mathfrak{m}^N_k(x_k)\right)\sigma(W^n_k x_k),\\
W^n_{k+1} &= W^n_k + \frac{\eta^N_k}{N^\gamma}\left(y_k - \mathfrak{m}^N_k(x_k)\right)C^n_k\sigma'(W^n_k x_k)x_k,
\end{aligned}$$

for $k \in \{0, 1, 2, \dots\}$, η^N_k is the learning rate. As usual, we set

$$\mathfrak{m}^N_k(x) = \frac{1}{N^\gamma}\sum_{n=1}^{N} C^i_k\sigma(W^n_k x),$$

define the empirical measure

$$\nu^N_k = \frac{1}{N}\sum_{n=1}^{N}\delta_{C^n_k, W^n_k},$$

and define the usual scaled processes

$$\mu^N_t = \nu^N_{\lfloor Nt\rfloor}, \quad h^N_t = \mathfrak{m}^N_{\lfloor Nt\rfloor},$$

where $\mathfrak{m}^N_k = \left(\mathfrak{m}^N_k(x_1), \dots, \mathfrak{m}^N_k(x_M)\right)$, $h^N_t = \left(h^N_t(x_1), \dots, h^N_t(x_M)\right)$, and we have defined $h^N_t(x) = \mathfrak{m}^N_{\lfloor Nt\rfloor}(x)$.

Under Assumption 19.2, as $N \to \infty$ and for $x \in \mathcal{D}$,

$$\mathfrak{m}^N_0(x) \xrightarrow{d} \mathcal{G}(x), \tag{20.27}$$

where $\mathcal{G} \in \mathbb{R}^M$ is the Gaussian random variable such that

$$\mathcal{G}(x) \sim N(0, \langle |c\sigma(w\cdot x)|^2, \mu_0\rangle).$$

We also of course have that

$$\nu_0^N \xrightarrow{p} \nu_0 \equiv \mu_0.$$

Following the procedure developed in Chapter 19 (see also Exercise 19.3 for the derivation), we get

Theorem 20.23. *Let $T < \infty$ be given, and assume that Assumption 19.2 holds. Fix some $\gamma \in [1/2, 1)$ and let the learning rate be $\eta^N = \eta/N^{2(1-\gamma)}$ with constant $0 < \eta < \infty$. Then, as $N \to \infty$, the process (μ_t^N, h_t^N) converges in probability in the space $D_E([0,T])$ to (μ_t, h_t), which for $t \in [0,T]$, satisfies (19.4) in Theorem 19.3.*

Essentially, Theorem 20.23 says that no matter the value of $\gamma \in [1/2, 1)$, for γ in that range, the limit of (μ_t^N, h_t^N) as $N \to \infty$ will be the same.

Hence, one cannot really infer anything useful in terms of comparing the different γ scalings from Theorem 20.23. The idea that we explore here is to view h_t^N as a stochastic process, and as such it has randomness. For any value of γ the limit of h_t^N as $N \to \infty$ is the same. Natural questions arise:

- What about the error in the convergence?
- What about the variance of h_t^N for large, but fixed N?
- What about the behavior as t grows?

Clearly, smaller variance would imply less error in the approximation and perhaps also lead to better generalization properties.

That is the point of view taken in [**SY21**], which goes one step further from Theorem 20.23. In [**SY21**], the fluctuation corrections to the limit for any $\gamma \in (1/2, 1)$ are being derived, and in the end one obtains that in distribution an asymptotic expansion of h_t^N in N as $N \to \infty$ holds. In particular, for a given but fixed $\nu \in \{1, 2, 3, \dots\}$ and for any $\gamma \in \left(\frac{2\nu-1}{2\nu}, \frac{2\nu+1}{2\nu+2}\right) \subset \left(\frac{1}{2}, 1\right)$, we have, in distribution as $N \to \infty$,

(20.28)

$$h_t^{N,\gamma} \approx h_t + \sum_{j=1}^{\nu-1} N^{-j(1-\gamma)} Q_t^j + N^{-(\gamma-1/2)} e^{-At} \mathcal{G} + \text{ lower order terms in } N,$$

where h_t is the limit of $h_t^{N,\gamma}$ as $N \to \infty$ per Theorem 20.23, Q_t^j are deterministic quantities defined recursively, A is a positive definite matrix (same as in the NTK case of Chapter 19) and $\mathcal{G}$ is a Gaussian vector of mean zero and known variance-covariance structure (composed of the elements $\mathcal{G}(x)$ of (20.27)). In addition, the quantities h_t, Q_t^j, A, and $\mathcal{G}$ are independent of $N < \infty$ and $\gamma > 0$. For fixed $j \in \mathbb{N}$, one can also show that $Q_t^j \to 0$ exponentially fast as $t \to \infty$, see [**SY21**] for proofs.

Notice that the asymptotic expansion (20.28) leads to an important conclusion. In particular, for fixed (but large) $N < \infty$ and $t < \infty$, the magnitude of the variance of the neural network output to leading order in N is $N^{-2(\gamma-1/2)}\|e^{-At}\operatorname{Var}(\mathcal{G})e^{-A^\top t}\|$. This is monotonically decreasing as $\gamma \to 1$. In addition, in the case of $\gamma = 1/2$, the variance to the leading order in N, is of order $\|e^{-At}\operatorname{Var}(\mathcal{G})e^{-A^\top t}\|$, i.e., it is independent of N.

This suggests that even though for any $\gamma \in [1/2, 1)$ the limit of the neural network model as $N \to \infty$ is the same, the variance to the leading order in N is *not* the same. In particular, the variance decays as $\gamma \to 1$.

The next natural question is how this variance reduction translates to performance and generalization, i.e., to out-of-sample accuracy. To answer the question, shallow neural networks of the form (20.26) were trained via both cross-entropy loss and via mean-square error loss for the MNIST dataset [**LBBH98**]. We recall that the MNIST dataset includes 70,000 images of handwritten integers from 0 to 9. In the MNIST dataset, each image has 784 pixels, 60,000 images are used as training images and 10,000 images are testing images. The learning rate is taken to be $\alpha^N = 1/N^{2-2\gamma}$, as suggested by the theoretical results. The neural networks are trained to identify the handwritten numbers using the image pixels as an input. As this is about a categorical problem (image recognition), cross-entropy is a more appropriate loss function, but given that the theory has been developed for mean square error loss, we present both.

We observe that test accuracy for each network increases as $\gamma \in [1/2, 1]$ increases (see Figure 20.1).

So in conclusion, in the case of shallow neural networks we have observed,

- The variance of the stochastic process governing the behavior of the neural network (to leading order in N) is monotonically decreasing in $\gamma \in (1/2, 1)$.
- Generalization properties and out-of-sample performance increases monotonically in $\gamma \in (1/2, 1)$.

Hence, there is evidence to suggest that the mean-field scaling has certain advantages in regards to the generalization performance of shallow neural networks trained with stochastic gradient descent for the regression problem, even though its mathematical analysis is more complicated.

Finally, we discuss what happens in the case of deep neural networks. In particular, let us consider the following neural network with two hidden layers:

$$\mathfrak{m}^{N_1,N_2}(x;\theta) = \frac{1}{N_2^{\gamma_2}} \sum_{i=1}^{N_2} C^i \sigma\left(\frac{1}{N_1^{\gamma_1}} \sum_{j=1}^{N_1} W^{2,j,i}\sigma(W^{1,j}x) \right),$$

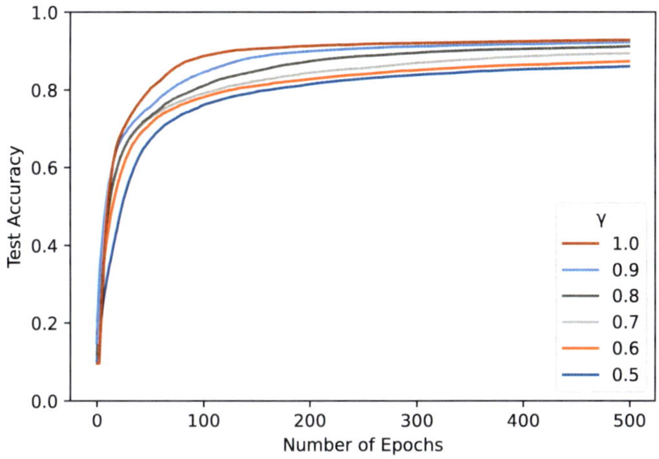

(a) $N = 1,000$ hidden units, cross entropy loss

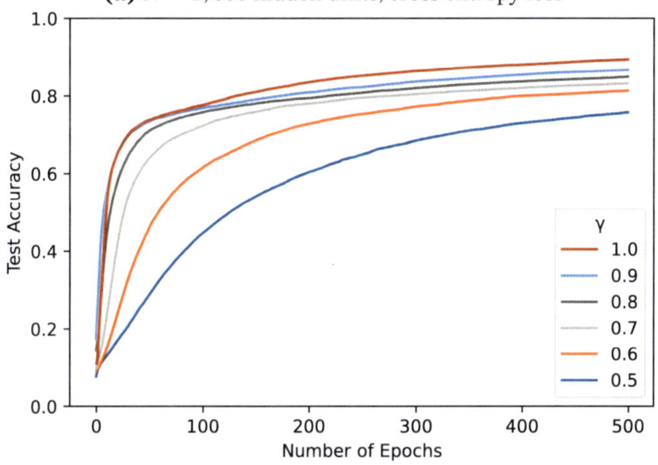

(b) $N = 1,000$ hidden units, mean square loss

Figure 20.1. Performance of scaled neural networks on MNIST test dataset; see [**SY21**]. The top figure is for cross-entropy loss, and the bottom figure is for mean-square loss. Accuracy increases monotonically in γ in both figures.

where $C^i, W^{2,j,i} \in \mathbb{R}$, $x, W^{1,j} \in \mathbb{R}^d$, and $\gamma_1, \gamma_2 \in [1/2, 1)$ are fixed scaling parameters. The neural network model has parameters

$$\theta = \left(C^1, \dots, C^{N_2}, W^{2,1,1}, \dots, W^{2,N_1,N_2}, W^{1,1}, \dots, W^{1,N_1}\right),$$

which are to be estimated from data $(X, Y) \sim \pi(dx, dy)$.

This problem has been recently studied in [**YS23**] for deep neural networks of arbitrary depth. Albeit more complicated analysis and notation, it is demonstrated there that an asymptotic expansion in the spirit of (20.28), as N_2 grows to infinity, holds with the same conclusion for variance reduction. Namely, variance of $\mathfrak{m}^{N_1,N_2}_{\lfloor N_2 t\rfloor}$ to leading order in N_2 (and N_1) is smaller when $\gamma_1, \gamma_2 \to 1$, i.e., when the scalings tend to the mean field scaling.

To make the discussion below simpler and more intuitive, we will set $N_1 = N_2 = N$, which is what is typically done in practice. For the standard mean-square error loss, the standard SGD yields the update equations

$$
\begin{aligned}
C^i_{k+1} &= C^i_k + \frac{\eta^N_C}{N^{\gamma_2}}\left(y_k - \mathfrak{m}^N_k(x_k)\right) H^{2,i}_k(x_k),\\
W^{1,j}_{k+1} &= W^{1,j}_k + \frac{\eta^N_{W,1}}{N^{\gamma_1}}\left(y_k - \mathfrak{m}^N_k(x_k)\right)\left(\frac{1}{N^{\gamma_2}}\sum_{i=1}^N C^i_k \sigma'(Z^{2,i}_k(x_k)) W^{2,j,i}_k\right)\\
&\qquad \times \sigma'(W^{1,j}_k x_k) x_k,\\
W^{2,j,i}_{k+1} &= W^{2,j,i}_k + \frac{\eta^N_{W,2}}{N^{\gamma_1}N^{\gamma_2}}\left(y_k - \mathfrak{m}^N_k(x_k)\right) C^i_k \sigma'(Z^{2,i}_k(x_k)) H^{1,j}_k(x_k),
\end{aligned}
$$

where

$$
H^{1,j}_k(x) = \sigma(W^{1,j}_k x), Z^{2,i}_k(x) = \frac{1}{N^{\gamma_1}}\sum_{j=1}^N W^{2,j,i}_k H^{1,j}_k(x), H^{2,i}_k(x) = \sigma(Z^{2,i}_k(x)).
$$

and $\eta^N_c, \eta^N_{W,1}, \eta^N_{W,2}$ are the learning rates.

As demonstrated in [**YS23**], the asymptotic analysis goes through if the learning rates are chosen to be of specific order with respect to the number of hidden units per layer and the γ_1, γ_2 scalings. In particular, in the usual case in practice where $N_1 = N_2 = N$, one would pick

$$
\eta^N_C = \frac{\eta_C}{N^{2-2\gamma_2}}, \quad \eta^N_{W,1} = \frac{\eta_{W,1}}{N^{4-2(\gamma_1+\gamma_2)}}, \quad \eta^N_{W,2} = \frac{\eta_{W,2}}{N^{3-2(\gamma_1+\gamma_2)}}, \tag{20.29}
$$

where the coefficients $\eta_C, \eta_{W,1}, \eta_{W,2} \in (0,\infty)$ are chosen to be of order 1 with respect to N.

The numerical studies of [**YS23**] for deep neural networks also demonstrate improved out-of-sample performance when $\gamma_1 = \gamma_2 = 1$ in a monotonic way in $\gamma_1, \gamma_2 \in (1/2, 1)$. In addition, the same conclusions hold for deep neural networks of arbitrary (but fixed) depth with the appropriate choice for the learning rates.

The results presented here were derived for feed forward neural networks trained with standard stochastic gradient descent. The papers [**SY21, YS23**] also contain numerical studies for the CIFAR10 dataset [**KH09**], another image recognition dataset, which contains 60,000 color images in 10 classes (airplane, automobile, bird, cat, deer, dog, frog, horse, ship, and truck). For that dataset, convolutional neural networks (see Chapter 14) were applied (as opposed to feed forward neural networks) and the conclusions in regards to the effect of the γ-scalings were the same.

Also, we note that no attempt was made in these numerical studies to further optimize the out-of-sample accuracy. This is an apples-to-apples comparison of the effect of scaling, which was the only parameter varied in these studies. This means that mean-field scaling gives a good initial architecture and then out-of-sample accuracy could be further optimized by tweaking the learning rate (e.g., the constant coefficients $\eta_C, \eta_{W,1}, \eta_{W,2} \in (0, \infty)$ in (20.29)) for instance.

Therefore, the conclusion is that in deep neural networks of arbitrary (but fixed) depth trained with stochastic gradient descent for the regression problem, there is both mathematical (variance reduction) and numerical (improved out-of-sample accuracy, generalization performance) evidence that the mean-field scaling $\gamma_i = 1$ has certain advantages compared to the scalings with $\gamma_i < 1$. However, the mathematical analysis for mean field scaling is certainly more involved.

We highlight that one important practical conclusion of the mathematical analysis is that it suggests a closed form formula for the choice of the learning rates hyperparameters; see in this section for neural networks of depth 1 and 2 and Section 4 of [**YS23**] for neural networks of arbitrary depth.

20.7. Elements of Generalization Performance

Based on the discussion of the previous section, the specific chosen neural network architecture can have a profound effect on how the model behaves on unseen data, stated otherwise on its generalization performance properties. In fact characterizing the generalization performance of models is a very active area of research in deep learning. Below we comment on some of its main elements.

For the purposes of this section, it is instructive to view the loss function Λ as a function of the model $\mathfrak{m} = \mathfrak{m}(x; \theta)$ instead of its parameters θ. So, we shall write $\Lambda(\mathfrak{m})$ and $\Lambda_{\text{pop}}(\mathfrak{m})$ for the empirical and population loss functions, respectively. We are minimizing $\Lambda(\mathfrak{m})$ within a class of models, say for $\mathfrak{m} \in \bar{\mathcal{M}}$. Of course, we would like to be able to solve the problem

$$\mathfrak{m}^\star_{\text{pop}} = \text{argmin}_{\mathfrak{m}\in\bar{\mathcal{M}}} \Lambda_{\text{pop}}(\mathfrak{m}).$$

It is important to mention the set $\bar{\mathcal{M}}$ may not contain the unconstrained minimizer of the population loss function.

For comparison purposes, let us also denote

$$\mathfrak{m}^\star = \text{argmin}_{\mathfrak{m}\in\bar{\mathcal{M}}} \Lambda(\mathfrak{m}).$$

In a generalization bound, the question we ask is: given a model $\mathfrak{m}^\star$ that has been chosen because it performs well on training data, is it also true that

$\Lambda_{\text{pop}}(\mathfrak{m}^\star)$ is small? A positive answer to this question would then indicate that the performance of the model $\mathfrak{m}^\star$ is good on the entire distribution.

A partial answer to this question can be given by the Chernoff type bounds, see Lemma A.30. Consider the dataset $\mathcal{D} = \{(x_m, y_m)\}_{m=1}^M$. Since the data is assumed to be independent and identically distributed, the Chernoff bound gives for all $\mathfrak{m} \in \bar{\mathcal{M}}$ and all $\delta > 0$

$$\mathbb{P}\left(|\Lambda(\mathfrak{m}) - \Lambda_{\text{pop}}(\mathfrak{m})| > \delta\right) \leq 2e^{-2M\delta^2}.$$

Recall now that for any countable set $\{A_1, A_2, A_3, \ldots\}$ one has the union bound

$$\mathbb{P}\left(\bigcup_{j\geq 1} A_j\right) \leq \sum_{j\geq 1} \mathbb{P}(A_j).$$

Assume that the class of available models $\bar{\mathcal{M}}$ is finite. Applying first the union bound and then the Chernoff bound yields

$$\begin{aligned}
&\mathbb{P}\left(\text{There exists } \mathfrak{m} \in \bar{\mathcal{M}} \text{ such that } |\Lambda(\mathfrak{m}) - \Lambda_{\text{pop}}(\mathfrak{m})| > \delta\right)\\
&\leq \sum_{\mathfrak{m}\in\bar{\mathcal{M}}} \mathbb{P}\left(|\Lambda(\mathfrak{m}) - \Lambda_{\text{pop}}(\mathfrak{m})| > \delta\right)\\
&\leq |\bar{\mathcal{M}}| \left[2e^{-2M\delta^2}\right],
\end{aligned}$$

where $|\bar{\mathcal{M}}|$ is the size of the set of allowable models. This bound shows that if we want the upper bound for this probability to be bounded by some $\epsilon > 0$, i.e., if we want $|\bar{\mathcal{M}}|\left[2e^{-2M\delta^2}\right] \leq \epsilon$, then we would need to have

$$M \geq \frac{1}{2\delta^2} \log \frac{2|\bar{\mathcal{M}}|}{\epsilon},$$

as the size of the training dataset. The Chenroff bound then yields that with probability at least $1 - \epsilon$,

$$\Lambda_{\text{pop}}(\mathfrak{m}^\star) - \Lambda(\mathfrak{m}^\star) \leq \delta$$

and that

$$\Lambda(\mathfrak{m}^\star_{\text{pop}}) - \Lambda_{\text{pop}}(\mathfrak{m}^\star_{\text{pop}}) \leq \delta.$$

Combining these two facts with the estimate $\Lambda(\mathfrak{m}^\star) \leq \Lambda(\mathfrak{m}^\star_{\text{pop}})$ (true by definition) yields the estimate

$$\Lambda_{\text{pop}}(\mathfrak{m}^\star) - \Lambda_{\text{pop}}(\mathfrak{m}^\star_{\text{pop}}) \leq 2\delta.$$

Combining this further with the estimate

$$\left[M \geq \frac{1}{2\delta^2} \log \frac{2|\bar{\mathcal{M}}|}{\epsilon}\right] \Rightarrow \left[\delta \leq \sqrt{\frac{1}{2M} \log \frac{2|\bar{\mathcal{M}}|}{\epsilon}}\right],$$

yields that with probability $1-\epsilon$ the bound holds

$$\Lambda_{\text{pop}}(\mathfrak{m}^\star) - \Lambda_{\text{pop}}(\mathfrak{m}^\star_{\text{pop}}) \leq 2\sqrt{\frac{1}{2M}\log\frac{2|\bar{\mathcal{M}}|}{\epsilon}}. \tag{20.30}$$

Despite the attractiveness of this conclusion, this bound is not that useful for deep learning. This is because the set of neural networks has infinite size, i.e., $|\bar{\mathcal{M}}| = \infty$ (thus the union bound is not meaningful). Hence, this line of approach does not yield a useful result in the context of infinite possible models to choose from, as it suggests that we would need an infinite amount of data. So the question is whether we can still get good bounds with a finite set of training data in the case where $|\bar{\mathcal{M}}| = \infty$.

The celebrated VC theory introduced in [**VC71**, **Vap99**] was developed to answer the latter question. The VC dimension of a given class $\bar{\mathcal{M}}$ is a measure of the expressive power of the set $\bar{\mathcal{M}}$. In order to define what VC dimension is, we first need to introduce the notion of *shattering*. We say that the class of models $\bar{\mathcal{M}}$ shatters a set of points $\{x_1, \dots, x_M\}$ if for every possible training set $\mathcal{D} = \{(x_m, y_m)\}_{m=1}^M$ there exists a model $\mathfrak{m} \in \bar{\mathcal{M}}$ that results in zero training error. The VC dimension of $\bar{\mathcal{M}}$ is then the maximum amount of data samples M such that $\bar{\mathcal{M}}$ shatters the set $\{x_1, \dots, x_M\}$. If no such maximal values exist, then the VC dimension is defined to be infinity.

The VC dimension can be used to give a probabilistic upper bound on the test error of a classification model. In particular, for $VC \ll M$, as it is shown in [**Vap99**]

$$\Lambda_{\text{pop}}(\mathfrak{m}^\star) - \Lambda_{\text{pop}}(\mathfrak{m}^\star_{\text{pop}}) \leq 2\sqrt{\frac{VC\left(1+\log\frac{2M}{VC}\right) - \log\frac{\epsilon}{4}}{M}}, \tag{20.31}$$

with probability at least $1-\epsilon$. It is clear that (20.31) is a considerable improvement over (20.30), especially when it comes to deep learning where $|\bar{\mathcal{M}}| = \infty$.

The next question is whether one can understand generalization of deep neural networks via the VC theory. The first thing to do then would be to compute the VC dimension of neural networks, which however turns out to not be a trivial task. Typically, one can get lower and upper bounds for the VC dimension in the spirit of [**BHLM19**]. In [**BHLM19**] lower and upper bounds for the VC dimension of deep neural networks with ReLU activation functions are computed. In particular, it is shown in that paper that if there are W many weights and L many layers, then the VC dimension of such a neural network will be of the order of VC$= O(WL\log(W))$. Unfortunately, such bounds are not necessarily useful in the context of VC theory. For instance if $L=3$, $W=10^3$, $M=10^5$, and $\epsilon = 0.01$, then one has $VC \approx 2 * 10^4$ and (20.31) gives

$$\Lambda_{\text{pop}}(\mathfrak{m}^\star) - \Lambda_{\text{pop}}(\mathfrak{m}^\star_{\text{pop}}) \leq 1.63,$$

which is not necessarily that informative. However, as we show in Section 20.6 for example and as it is indisputably the case in the empirical literature, deep neural networks have very good out-of-sample accuracy. Closing this gap between theory and empirical evidence is currently an active research area. Beyond the VC-dimension approach (that we briefly discussed in this section) to understanding generalization, there are other methods too. One of those is the Rademacher complexity approach (see for example [**SSBD14**]), which has also been well developed. A nice review book chapter summarizing the main methods is [**JGR19**]. A nice related literature review that goes over some of the main strategies for obtaining bounds for the statistical risk can also be found in [**SH17**]. The paper [**LFK$^+$22**] describes a PAC-Bayes (Probably Approximately Correct Bayes) framework approach for obtaining generalization bounds for deep learning models.

20.8. Brief Concluding Remarks

The mean field scaling for shallow neural networks was studied by various authors around the same time, each one using a slightly different set of tools, see [**CB18, MMN18, RVE18, SS20b**]. In the exposition that we followed here, we largely adopted the presentation of [**SS20b**] which is based mainly on stochastic analysis and weak convergence types of arguments. Modulo Remark 20.7, the proofs of Section 20.3 are based on [**SS20b**]. The mean field scaling for deep neural networks was analyzed in [**AOY19, Ngu19, NP23, SS21**], and the presentation that we followed was based on [**SS21**].[1] The investigation of the regimes between the linear and the nonlinear regime was studied in [**SY21**] for the shallow case and in [**YS23**] for the deep neural network case.

In [**MMM19**] bounds are established quantifying the accuracy of mean field scaling in terms of regularity properties of the data. In addition, in [**MMM19**] it is shown that the mean field scaling recovers the kernel ridge regression as a special limit case.

In this chapter we also argued that there is theoretical and empirical evidence to support the hypothesis that neural network architectures with mean field scaling generalize better than neural network architectures with square-root scaling for example. Affirmatively answering this question is part of the research on generalization theory, which is an active area of research, see [**VC71, Vap99, BHLM19, JGR19, Yar17, SH17, SSBD14, LFK$^+$22**] for a nonexhaustive list of earlier and more recent related works.

[1] Copyrighted to INFORMS and republished with permission.

20.9. Exercises

Exercise 20.1. Consider the system

$$C^n_{k+1} = C^n_k + \frac{\eta^N_k}{N}(y_k - \mathfrak{m}^N_k(x_k))\sigma(W^n_k \cdot x_k),$$

$$W^n_{k+1} = W^n_k + \frac{\eta^N_k}{N}(y_k - \mathfrak{m}^N_k(x_k))C^n_k\sigma'(W^n_k \cdot x_k)x_k,$$

$$\mathfrak{m}^N_k(x) = \frac{1}{N}\sum_{n=1}^{N} C^n_k\sigma(W^n_k \cdot x),$$

for $k = 0, 1, \dots, \lfloor TN \rfloor$, where $T > 0$. $\eta^N_k = \eta$ is the learning rate with $0 < \eta < \infty$ a fixed constant. Prove that, for $k \le \lfloor TN \rfloor$ and uniformly in $k, N \in \mathbb{N}$, there exists a constant $C_o < \infty$ such that

$$\sup_{N\in\mathbb{N}, k/N \le T} \frac{1}{N}\sum_{n=1}^{N} \mathbb{E}\left[|C^n_k| + \| W^n_k \|\right] \le C_o.$$

Exercise 20.2. Let $M^{1,N}(t)$ and $M^{2,N}(t)$ be defined by (20.8) in Section 20.3. Prove that

$$\lim_{N\to\infty}\left[\mathbb{E}\left(\left(M^{1,N}(t)\right)^2\right) + \mathbb{E}\left(\left(M^{2,N}(t)\right)^2\right)\right] = 0.$$

Exercise 20.3. Let $\alpha^N_t = \sqrt{N}(\mu^N_t - \bar{\mu}_t)$ be the fluctuation process. For a test function $f \in C^2_b(\mathbb{R}^{1+d})$ derive exactly the formula for $\langle f, \alpha^N_t\rangle$ as indicated in Section 20.4 characterizing the remainder term R^N_t.

Exercise 20.4. Show that (20.25) holds, i.e., show that $\frac{d}{dt}\bar{\Lambda}(\Theta_t) \le 0$ holds.

Exercise 20.5. Consider the multilayer feed forward neural network in the mean field scaling

$$\mathfrak{m}^{N_1,N_2,N_3}(x;\theta) = \frac{1}{N_3}\sum_{i=1}^{N_3} C^i\sigma\left(\frac{1}{N_2}\sum_{j=1}^{N_2} W^{3,i,j}\sigma\left(\frac{1}{N_1}\sum_{\nu=1}^{N_1} W^{2,j,\nu}\sigma\left(W^{1,\nu}\cdot x\right)\right)\right),$$

where $C^i, W^{2,j,\nu}, W^{3,i,j} \in \mathbb{R}$, and $x, W^{1,\nu} \in \mathbb{R}^d$. For the quadratic cost

$$\Lambda^{N_1,N_2,N_3}(\theta) = \frac{1}{2}\frac{1}{|\mathcal{D}|}\sum_{(x,y)\in\mathcal{D}} (y - \mathfrak{m}^{N_1,N_2,N_3}(x;\theta))^2,$$

derive the SGD updating equations.

Chapter 21

Reinforcement Learning

21.1. Introduction

Reinforcement learning is a subfield of artificial intelligence that has enjoyed a lot of success in recent years, ranging from mastering the game of Go, to robotics, to video games and self-driving cars; see [**SSS⁺17**, **KP12**, **MKS⁺15**, **MKS⁺13**] for a non-comprehensive list of application examples.

Our goal in this chapter is not to exhaust this very rich and deep topic but rather to lay down the main framework and discuss convergence properties of deep reinforcement learning where a neural network is trained to learn the optimal action given the current state. We start with a motivating example in Section 21.2 where we build from scratch and step by step the basic Q-learning formulation. Deep reinforcement learning is studied in Section 21.3, Q-learning in Section 21.4, and the convergence analysis of Q-learning is presented in Section 21.5.

21.2. Motivating Reinforcement Learning Through an Example

Suppose we are driving on a hill with profile

$$U(q) = 0.45\sin(3q) + 0.55, \quad q \in \mathbb{R},$$

in a car with mass m. Our goal is to drive the car from a given location, say $q = -0.5$, to another location, say $q_+ = 0.45$; see Figure 21.1.

Our car has three control settings:

- moving forward; a unit force to the right (control is set to $+1$).

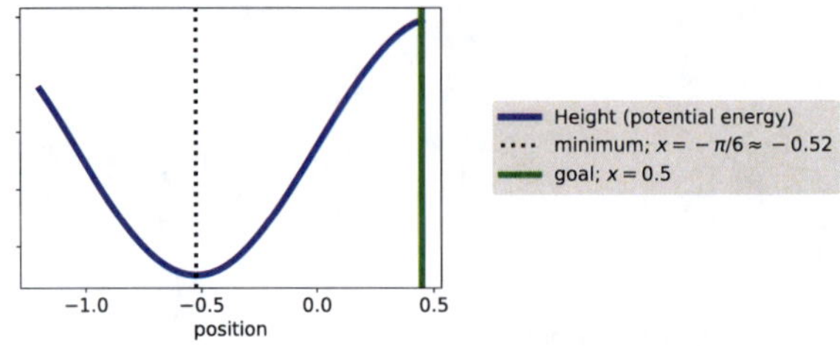

Figure 21.1. Profile of the hill and target

- moving backward; a unit force to the left (control is set to -1).
- neutral; zero force (control is set to 0).

How can we learn to drive the car in order to best reach the goal? To answer the question, let us first mathematically formalize the previous word description.

Let us set the acceleration at time t to be $\alpha(t) \in \mathsf{A} = \{-1, 0, 1\}$. By the second law of Newton, the position and velocity of the car will be given by the system of equations

$$\begin{aligned} \dot{q}(t) &= v(t) \quad \text{(horizontal position and velocity)}, \\ m\dot{v}(t) &= \text{horizontal component of acceleration and gravity}. \end{aligned}$$

Let us now compute the horizontal component of acceleration and gravity in this case. First we observe that the component of gravitational force that is tangent to the curve $y = U'(q)$ is

$$(F_g)_T(q) = \frac{-gU'(q)}{\sqrt{(U'(q))^2 + 1}},$$

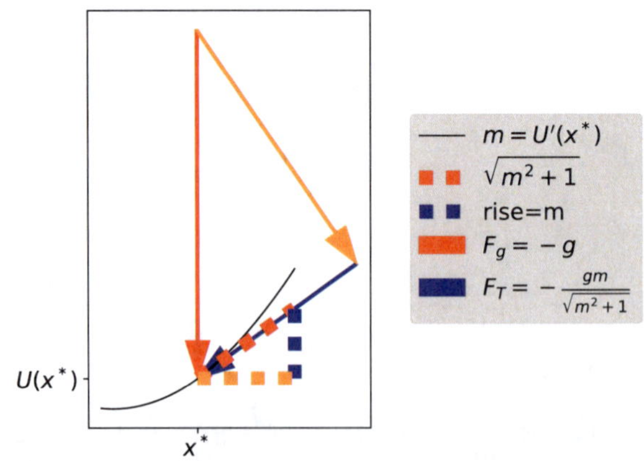

Figure 21.2. Gravitational force tangent to the curve $y = U'(q)$

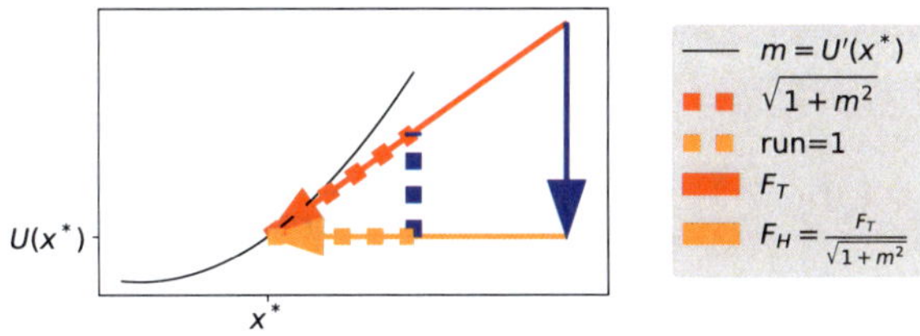

Figure 21.3. Tangential to horizontal force

where $g = 9.81\text{m}/\sec^2$ or $g = 32\text{ft}/\sec^2$ is the gravitational constant (see Figure 21.2).

The force tangential to U at time t takes the form

$$F_T(t) = \alpha(t) + (F_g)_T(q(t))$$
$$= \alpha(t) - \frac{gU'(q(t))}{\sqrt{(U'(q(t)))^2 + 1}}.$$

By projection, the horizontal force to U becomes, Figure 21.3,

$$F_H(q(t)) = \frac{F_T(t)}{\sqrt{(U'(q(t)))^2 + 1}}$$
$$= \frac{\alpha(t)}{\sqrt{(U'(q(t)))^2 + 1}} - g\frac{U'(q(t))}{\sqrt{(U'(q(t)))^2 + 1}}.$$

Note now that if $\sup_q |U'(q)| \ll 1$, then we can approximate

$$F_H(q(t)) \approx \alpha(t) - gU'(q(t)).$$

So, we have arrived at the following set of equations

$$\dot{q}(t) = v(t),$$
$$m\dot{v}(t) = \frac{\alpha(t)}{\sqrt{(U'(q(t)))^2 + 1}} - g\frac{U'(q(t))}{\sqrt{(U'(q(t)))^2 + 1}},$$

the former being the horizontal position and velocity and the latter being the horizontal component of acceleration and gravity.

If the condition $|U'| \ll 1$ is valid in the region of interest, then we can simplify the previous equations by

(21.1)
$$\dot{q}(t) = v(t) \quad \text{(horizontal position and velocity)},$$
$$m\dot{v}(t) = \alpha(t) - gU'(q(t)) \quad \text{(horizontal component of acceleration and gravity)}.$$

Introducing a timestep parameter $\delta > 0$, we can discretize the previous system of equations as

$$(21.2)\qquad \begin{aligned} q(t+\delta) &= q(t) + v(t)\delta + \frac{\alpha(t)}{2m}\delta^2, \\ v(t+\delta) &= v(t) + \frac{\delta}{m}\left(\alpha(t) - gU'(q(t))\right). \end{aligned}$$

(we have included the second-order δ^2 term in the discretization of (21.1) to ensure that the control acts tranversally the boundary of B).

In our specific example $U(q) = 0.45\sin(3q) + 0.55$, which gives $U'(q) = 1.35\cos(3q)$. As we discussed $g = 9.81\text{m/sec}^2$ or $g = 32\text{ft/sec}^2$ is the gravitational constant, m is the mass of the object, and δ is the descritization step.

How do we choose $\alpha(t)$ to best achieve our goal? Let us define the state vector $x = (q, v) \in \mathsf{X} = \mathbb{R}^2$ and an action $\alpha \in \mathsf{A}$. We set

$$\Phi_\alpha(x) = \left(q + v\delta + \frac{\alpha}{2m}\delta^2, v + \frac{\delta}{m}\left(\alpha - gU'(q)\right)\right),$$

which describes the one-step dynamics if we use action $\alpha \in \mathsf{A}$.

Having this description in mind as motivation and a concrete example, let us start now building towards a more abstract setup.

A control policy $\lambda : \mathbb{R}^2 \mapsto \mathsf{A}$ amounts to taking a certain action based on the state we are currently in. Let us denote by $\mathcal{P}$ the collection of all policies. For $\lambda \in \mathcal{P}$ and initial state $x = (q, v) \in \mathbb{R}^2$, let us denote by $\mathcal{R}_n(x;\lambda)$ the dynamics of the vehicle when we use policy λ at the current state x during time-iteration $n \in \mathbb{N}$. In particular, we have

$$\begin{aligned} \mathcal{R}_{n+1}(x;\lambda) &= \Phi_{\lambda(\mathcal{R}_n(x;\lambda))}\left(\mathcal{R}_n(x;\lambda)\right), \\ \mathcal{R}_n(x;\lambda) &= x. \end{aligned}$$

The best policy is the one achieving the quickest time to reach the target set $\mathsf{B} = (q_+, \infty) \times \mathbb{R}$. Hence, for $x \in \mathsf{X}$ and $\lambda \in \mathcal{P}$ we define the map

$$T_\lambda(x) = \min\{n\delta \geq 0, \mathcal{R}_n(x;\lambda) \in \mathsf{B}\} \quad (\min \emptyset = \infty)$$

(recall that δ is our timestep), which gives rise to the value function

$$V(x) = \inf_{\lambda \in \mathcal{P}} \{T_\lambda(x)\}.$$

Hence, the goal is to find $\lambda^* \in \mathcal{P}$ so that $V(x) = T_{\lambda^*}(x)$. In order to demonstrate that there are different ways to reach the goal, let us consider the following three scenarios.

Strategy 1. Applying positive force when the velocity is already positive and to the left of the minimum, see Figure 21.4.

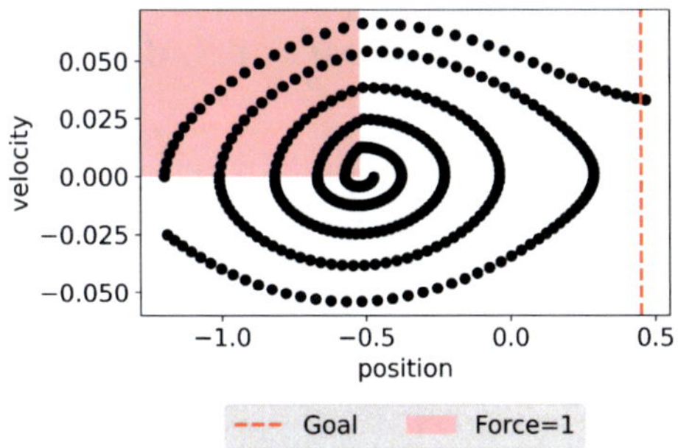

Figure 21.4. Strategy 1: Force is positive when the velocity is already positive and to the left of the minimum. Goal is reached in 374 steps.

Strategy 2. Applying positive force when the velocity is already positive, see Figure 21.5.

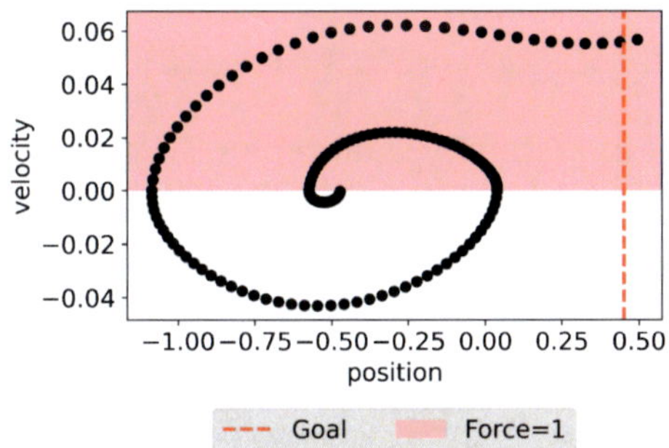

Figure 21.5. Strategy 2: Force is positive when the velocity is already positive. Goal is reached in 161 steps.

Strategy 3. Applying a force which agrees with the sign of the velocity, see Figure 21.6.

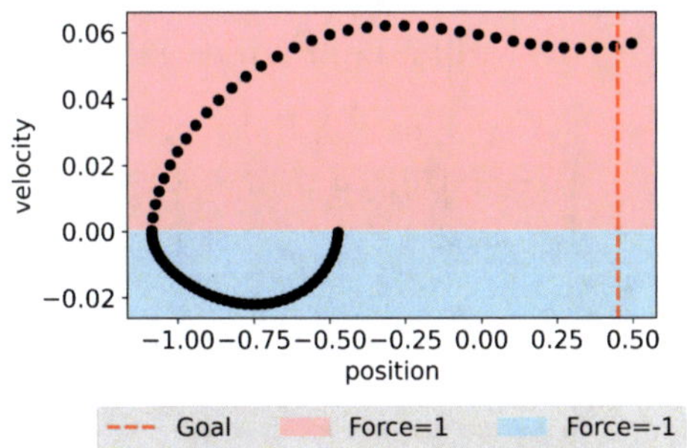

Figure 21.6. Strategy 3: Force agrees with the sign of the velocity. Goal is reached in81 steps.

In this demonstration, the conclusion is that Strategy 3 (see Figure 21.6) reaches the goal in fewer steps compared to the other two strategies.

Now treating $\mathcal{R}$ as a dynamical system, we can do things in an iterative manner by restarting the system. In particular, if we fix momentarily $\lambda \in \mathcal{P}$ and $x \in \mathsf{X}$, we have

$$\mathcal{R}_1(x;\lambda) = \Phi_{\lambda(x)}(x) = \mathcal{R}_0\left(\Phi_{\lambda(x)}(x);\lambda\right).$$

Proceeding now iteratively, if for some $n \in \mathbb{N}$, we have

$$\mathcal{R}_n(x;\lambda) = \mathcal{R}_{n-1}\left(\Phi_{\lambda(x)}(x);\lambda\right) = \mathcal{R}_{n-1}\left(\mathcal{R}_1(x;\lambda);\lambda\right).$$

Hence, we subsequently have the following restart dynamics

$$\begin{aligned}\mathcal{R}_{n+1}(x;\lambda) &= \Phi_{\lambda(\mathcal{R}_n(x;\lambda))}\left(\mathcal{R}_n(x;\lambda)\right)\\ &= \Phi_{\lambda(\mathcal{R}_{n-1}(\Phi_{\lambda(x)}(x);\lambda)}\left(\mathcal{R}_{n-1}\left(\Phi_{\lambda(x)}(x);\lambda\right)\right)\\ &= \mathcal{R}_n\left(\Phi_{\lambda(x)}(x);\lambda\right).\end{aligned}$$

Thus in general, for $n \in \mathbb{N}$, we can write

$$\mathcal{R}_n(x;\lambda) = \mathcal{R}_{n-1}\left(\Phi_{\lambda(x)}(x);\lambda\right).$$

Let us next derive an iterative equation for $T_\lambda(x)$. Fix $\lambda \in \mathcal{P}$ and $x \notin \mathsf{B}$. We certainly have that $J(x;\lambda) \geq 1$. In addition, we also have

$$\begin{aligned}T_\lambda(x) &= \min\{n\delta : n \geq 0, \mathcal{R}_n(x;\lambda) \in \mathsf{B}\} \quad \text{(by definition of } T_\lambda(x))\\ &= \min\left\{n\delta : n \geq 1, \mathcal{R}_{n-1}(\Phi_{\lambda(x)}(x);\lambda) \in \mathsf{B}\right\} \quad \text{(because } x \notin \mathsf{B})\\ &= \delta + \min\left\{(n-1)\delta : n-1 \geq 0, \mathcal{R}_{n-1}(\Phi_{\lambda(x)}(x);\lambda) \in \mathsf{B}\right\} \quad (n = 1 + n - 1)\\ &= \delta + T_\lambda(\Phi_\lambda(x)) \quad \text{(by definition of } T_\lambda(\Phi_\lambda(x))).\end{aligned}$$

Hence, collecting the calculations above we have for $\lambda \in \mathcal{P}$

$$T_\lambda(x) = \begin{cases}\delta + T_\lambda(\Phi_\lambda(x)) & \text{if } x \notin \mathsf{B} \text{ (restart dynamics)}\\ 0 & \text{if } x \in \mathsf{B} \text{ (boundary condition).}\end{cases}$$

Now, we are ready to derive the Bellman equation for this setting. Notice that we can write

$$\begin{aligned}T_\lambda(x) &= \begin{cases}\delta + T_\lambda(\Phi_{\lambda(z)}(x)) & \text{if } x \notin \mathsf{B}\\ 0 & \text{if } x \in \mathsf{B}\end{cases}\\ &\geq \begin{cases}\delta + V(\Phi_{\lambda(x)}(x)) & \text{if } x \notin \mathsf{B}\\ 0 & \text{if } x \in \mathsf{B}\end{cases}\\ &\geq \begin{cases}\min_{\alpha \in \mathsf{A}}\{\delta + V(\Phi_\alpha(x))\} & \text{if } x \notin \mathsf{B}\\ 0 & \text{if } x \in \mathsf{B}.\end{cases}\end{aligned}$$

Minimizing over $\lambda \in \mathcal{P}$ on the left-hand side, we obtain

$$V(x) \geq \begin{cases} \min_{\alpha \in \mathsf{A}} \{\delta + V(\Phi_\alpha(x))\} & \text{if } x \notin \mathsf{B} \\ 0 & \text{if } x \in \mathsf{B}. \end{cases} \tag{21.3}$$

This finally suggests the *Bellman equation* for the value function

$$V(x) = \begin{cases} \min_{\alpha \in \mathsf{A}} \{\delta + V(\Phi_\alpha(x))\} & \text{if } x \notin \mathsf{B} \\ 0 & \text{if } x \in \mathsf{B}. \end{cases} \tag{21.4}$$

The best policy then is

$$\lambda^*(x) = \operatorname*{argmin}_{\alpha \in \mathsf{A}} \{\delta + V(\Phi_\alpha(x))\}. \tag{21.5}$$

21.2.1. Basic formulation of Q-learning. Now that we have formulated the Bellman equation (21.4) and the associated optimal policy (21.5), we can turn this into a learning problem.

Let us define the so-called Q function

$$Q(x, \alpha) = \delta + V(\Phi_\alpha(x)) \tag{21.6}$$

for $x \notin \mathsf{B}$. If the Bellman equation is satisfied, then

$$V(x) = \min_{\alpha \in \mathsf{A}} Q(x, \alpha)$$

for $x \notin \mathsf{B}$. Using this fact in the right-hand side of (21.6), we have that

$$V(\Phi_\alpha(x)) = \inf_{\alpha' \in \mathsf{A}} Q(\Phi_\alpha(x), \alpha')$$

if $\Phi_\alpha(x) \notin \mathsf{B}$. On the other hand, $V(\Phi_\alpha(x)) = 0$ if $\Phi_\alpha(x) \in \mathsf{B}$, so if (and only if) V satisfies the Bellman equation,

$$\begin{aligned} Q(x, \alpha) &= \begin{cases} \delta + \min_{\alpha' \in \mathsf{A}} Q(\Phi_\alpha(x), \alpha') & \text{if } \Phi_\alpha(x) \notin \mathsf{B} \\ \delta & \text{if } \Phi_\alpha(x) \in \mathsf{B} \end{cases} \\ &= \delta + \mathbf{1}_{\{\Phi_\alpha(x) \notin \mathsf{B}\}} \min_{\alpha' \in \mathsf{A}} Q(\Phi_\alpha(x), \alpha'). \end{aligned} \tag{21.7}$$

If we have solved (21.7), the equation (21.5) for our best policy will then be

$$\lambda^*(x) = \operatorname*{argmin}_{\alpha \in \mathsf{A}} Q(x, \alpha). \tag{21.8}$$

Let's convert (21.7) to a deep learning problem; let's try to find a function $Q : \mathsf{X} \times \mathsf{A} \mapsto \mathbb{R}$ which minimizes

$$\left| Q(x, \alpha) - \left\{ \delta + \mathbf{1}_{\{\Phi_\alpha(x) \notin \mathsf{B}\}} \inf_{\alpha' \in \mathsf{A}} Q(\Phi_\alpha(x), \alpha') \right\} \right|^2 \tag{21.9}$$

over (in some appropriate sense) all $(x, \alpha) \in \mathsf{X} \times \mathsf{A}$.

Let's think through a computational framework. Let's assume that we have

- A history $\mathcal{H}$ of state-action-next state triples. Namely, $\mathcal{H}$ is a (multi)set of points in $\mathsf{X} \times \mathsf{A} \times \mathsf{X}$. Each (x, α, x') in $\mathcal{H}$ is of the form $\Phi_\alpha(x) = x'$. We want to understand how to get to B, so none of the x's themselves are in B. We may have to initially observe the system to build $\mathcal{H}$.
- A parametrized map $Q : \mathsf{X} \times \mathsf{A} \times \mathcal{P} \to \mathbb{R}$, where $\mathcal{P}$ is some Euclidean parameter space.

We would like to find a $\theta \in \mathcal{P}$ such that $Q(\cdot, \cdot, \theta^*)$ satisfies (21.7) (as much as possible).

Let's make some definitions based on $\mathcal{H}$. Define

$$\Pi\mathcal{H} \overset{\text{def}}{=} \{(x, \alpha) : (x, \alpha, x') \in \mathcal{H}\} \tag{21.10}$$

as the (multi)set of state-action points. In our deterministic case, the state-action-next state triples also implies that

$$\Phi_\alpha^{\mathcal{H}}(x) = x' \tag{21.11}$$

is well defined for $(x, \alpha) \in \Pi\mathcal{H}$ (and we know $\Phi^{\mathcal{H}}$ only for $(x, \alpha) \in \Pi\mathcal{H}$).

Let's next assume that we have some current value θ_n at iteration n (perhaps θ_0 is random). Let's construct the cost function

$$\begin{aligned}\Lambda_n(\theta) \overset{\text{def}}{=} \frac{1}{|\mathcal{H}|}\Bigg\{ \sum_{(x,\alpha)\in\Pi\mathcal{H}} &\Big|Q(x, \alpha, \theta) \\ &- \Big\{\delta + \mathbf{1}_{\{\Phi_\alpha^{\mathcal{H}}(x)\notin \mathsf{B}\}} \inf_{\alpha' \in \mathsf{A}} Q(\Phi_\alpha^{\mathcal{H}}(x), \alpha', \theta_n)\Big\}\Big|^2\Bigg\}\end{aligned} \tag{21.12}$$

(reflecting the minimization problem (21.9)). We can then compute $\nabla\Lambda_n(\theta_n)$ and, given some learning rate η, define

$$\theta_{n+1} \overset{\text{def}}{=} \theta_n - \eta\nabla\Lambda_n(\theta_n). \tag{21.13}$$

After N steps (e.g., when we want to stop our iteration), our approximation of the optimal policy will be

$$\lambda_N(x) \overset{\text{def}}{=} \operatorname*{argmin}_{\alpha\in\mathsf{A}}(x, \alpha', \theta_N).$$

Several broad comments are in order.

- The size of $\mathcal{H}$ and the complexity of Q are related; a more complex collection Q of parametrized maps will in general require more data-points to optimize (21.9).
- We may of course increase or modify $\mathcal{H}$ as our iteration proceeds. At each step, we might randomly choose an action (i.e., *explore*) to build a more comprehensive history. Conversely, at some point we may fix our history and focus more on minimizing (21.9) (i.e., *exploiting*).

- We are ultimately interested in (21.8). Assuming at A is finite, we can divide X up into regions where we would use the different elements of A. Informally, we would like $\mathcal{H}$ to give us information about *changes* in the optimal action, i.e., in the boundaries of the different regions.

21.2.2. Introducing Randomness in the Basic Framework. Let us go back to our basic car driving example earlier in Section 21.2 and introduce randomness. In particular, let $\zeta_n, n \in \mathbb{N}$ be a sequence of i.i.d. random variables taking values in some countable subset Z of $\mathbb{R}$ and with probability mass function p. Let us consider the following modification to the dynamics (21.2):

$$\begin{aligned} q(n+\delta) &= q(n) + v(n)\delta + \frac{\alpha(n)}{2m}\delta^2, \\ v(n+\delta) &= v(n) + \frac{\delta}{m}\left(\alpha(n) - gU'(q(n))\right) + \zeta_{n+1}. \end{aligned}$$

This leads to the mapping

$$\Phi_\alpha(x,\zeta) = \left(q + v\delta + \frac{\alpha}{2m}\delta^2, v + \frac{\delta}{m}\left(\alpha - gU'(q)\right) + \zeta\right),$$

which describes the one-step dynamics with action $\alpha \in \mathsf{A}$. The random recursion reads as follows

$$\begin{aligned} \mathcal{R}_{n+1}(x;\lambda) &= \Phi_{\lambda(\mathcal{R}_n(x;\lambda))}\left(\mathcal{R}_n(x;\lambda), \zeta_{n+1}\right), \\ \mathcal{R}_n(x;\lambda) &= x. \end{aligned}$$

The time to reach B is now random (due to the randomness via $\{\zeta_n\}_{n\in\mathbb{N}}$):

$$T_\lambda(x) = \min\{n\delta \geq 0, \mathcal{R}_n(x;\lambda) \in \mathsf{B}\} \quad (\min\emptyset = \infty)$$

and, thus, it makes sense to define the cost function as the expected time to reach the target set B:

$$W_\lambda(x) = \mathbb{E}\left[T_\lambda(x)\right].$$

Let us now follow the procedure of Section 21.2 to derive the Bellman equation in this case. Fix $\lambda \in \mathcal{P}$ and $x \in \mathsf{X}$. For a given vector $(x_1, \ldots, x_n) \in \mathsf{X}^n$,

$$\begin{aligned} &\mathbb{P}\{\mathcal{R}_1(x;\lambda) = x_1, \mathcal{R}_2(x;\lambda) = x_2, \ldots, \mathcal{R}_n(x;\lambda) = x_n\} \\ &\quad = \mathbb{P}\left\{\Phi_{\lambda(x)}(x,\zeta_1) = x_1, \Phi_{\lambda(x_1)}(x_1,\zeta_2) = x_2, \ldots, \Phi_{\lambda(x_{n-1})}(x_{n-1},\zeta_n) = x_n\right\} \\ &\quad = \sum_{\zeta\in\mathsf{Z}} \mathbb{P}\left\{\Phi_{\lambda(x)}(x,\zeta) = x_1, \Phi_{\lambda(x_1)}(x_1,\zeta_1) = x_2, \ldots,\right. \\ &\qquad\qquad\qquad \left.\Phi_{\lambda(x_{n-1})}(x_{n-1},\zeta_{n-1}) = x_n\right\} p(\zeta) \\ &\quad = \sum_{\zeta\in\mathsf{Z}} \mathbb{P}\left\{\Phi_{\lambda(x)}(x,\zeta) = x_1, \mathcal{R}_1(x_1;\lambda) = x_2, \ldots, \mathcal{R}_{n-1}(x_{n-1};\lambda) = x_n\right\} p(\zeta), \end{aligned}$$

where we used the fact that ζ_1 has probability mass function p and the vectors $(\zeta_1, \ldots, \zeta_{n-1})$ and $(\zeta_2, \ldots, \zeta_n)$ have the same distribution.

Let's now generalize the previously derived equation. If $x \notin \mathsf{B}$, we can write

$$
\begin{aligned}
&\mathbb{P}\left\{\mathcal{R}_1(x;\lambda) \in A_1, \{\mathcal{R}_{n'}(x;\lambda)\}_{n'=2}^{n} \in A\right\} \\
&= \sum_{\zeta\in\mathsf{Z}} \mathbb{P}\left\{\Phi_{\lambda(x)}(x,\zeta) \in A_1, \{\mathcal{R}_{n'}\left(\Phi_{\lambda(x)}(x,\zeta);\lambda\right)\}_{n'=1}^{n-1} \in A\right\} p(\zeta),
\end{aligned}
\tag{21.14}
$$

for some $A_1 \in \mathsf{X}$ and $A \in \mathsf{X}^{n-1}$.

If $x \notin \mathsf{B}$, then $T_\lambda(x) \geq \delta$, and for any $n \geq 1$, we get

$$
\begin{aligned}
\mathbb{P}\{T_\lambda(x) > n\delta\} &= \mathbb{P}\left\{\bigcap_{n'=1}^{n} \{\mathcal{R}_{n'}(x;\lambda) \notin \mathsf{B}\}\right\} \\
&= \sum_{\zeta\in\mathsf{Z}} \mathbb{P}\left\{\bigcap_{n'=0}^{n-1} \{\mathcal{R}_{n'}(\Phi_{\lambda(x)}(x,\zeta);\lambda) \notin \mathsf{B}\}\right\} \\
&= \sum_{\zeta\in\mathsf{Z}} \mathbb{P}\left\{T_\lambda(\Phi_{\lambda(x)}(x,\zeta)) > (n-1)\delta\right\} p(\zeta) \\
&= \sum_{\zeta\in\mathsf{Z}} \mathbb{P}\left\{T_\lambda(\Phi_{\lambda(x)}(x,\zeta)) + \delta > n\delta\right\} p(\zeta).
\end{aligned}
$$

Thus, we will generally have

$$
\mathbb{P}\{T_\lambda(x) \in A\} = \sum_{\zeta\in\mathsf{Z}} \mathbb{P}\left\{T_\lambda(\Phi_{\lambda(x)}(x,\zeta)) + \delta \in A\right\} p(\zeta).
$$

The latter relation implies

$$
\mathbb{E}\left[T_\lambda(x)\right] = \sum_{\zeta\in\mathsf{Z}} \mathbb{E}\left[T_\lambda(\Phi_{\lambda(x)}(x,\zeta)) + \delta\right] p(\zeta).
$$

Therefore, we may now define

$$
\begin{aligned}
W_\lambda(x) &= \mathbb{E}\left[T_\lambda(x)\right] \\
&= \begin{cases} \sum_{\zeta\in\mathsf{Z}} \{\delta + W_\lambda(\Phi_{\lambda(x)}(x,\zeta))\} p(\zeta) & \text{if } x \notin \mathsf{B} \\ 0 & \text{if } x \in \mathsf{B}. \end{cases}
\end{aligned}
$$

For the value function $G(x) = \inf_{\lambda\in\mathcal{P}} W_\lambda(x)$ and for $x \in \mathsf{X}$, we shall have

$$
W_\lambda(x) \geq \begin{cases} \min_{\alpha\in\mathsf{A}} \sum_{\zeta\in\mathsf{Z}} \{\delta + G(\Phi_\alpha(x,\zeta))\} p(\zeta) & \text{if } x \notin \mathsf{B} \\ 0 & \text{if } x \in \mathsf{B}. \end{cases}
$$

This gives the corresponding Bellman equation for the value function

$$
V(x) = \begin{cases} \min_{\alpha\in\mathsf{A}} \sum_{\zeta\in\mathsf{Z}} \{\delta + V(\Phi_\alpha(x,\zeta))\} p(\zeta) & \text{if } x \notin \mathsf{B} \\ 0 & \text{if } x \in \mathsf{B}. \end{cases}
$$

The best policy then is

$$\lambda^*(z) = \operatorname*{argmin}_{\alpha\in\mathsf{A}} \left\{ \sum_{\zeta\in\mathsf{Z}} \{\delta + V(\Phi_\alpha(x,\zeta))\}\, p(\zeta) \right\}.$$

Mimicking now the discussion in Section 21.2.1, the basic Q-learning framework is as follows. For $x \notin \mathsf{B}$, define the so-called Q function

$$Q(x,\alpha) = \sum_{\zeta\in\mathsf{Z}} \{\delta + V(\Phi_\alpha(x,\zeta))\}\, p(\zeta).$$

Then, we get

$$Q(x,\alpha) = \delta + \sum_{\zeta\in\mathsf{Z}:\Phi_\alpha(x,\zeta)\notin\mathsf{B}} \inf_{\alpha'\in\mathsf{A}} Q(\Phi_\alpha(x,\zeta),\alpha')p(\zeta), \tag{21.15}$$

with the last equality following from the observation that

$$V\left(\Phi_\alpha(x,\zeta)\right) = \min_{\alpha'\in\mathsf{A}} Q\left(\Phi_\alpha(x,\zeta),\alpha'\right)$$

if $\Phi_\alpha(x,\zeta) \notin \mathsf{B}$. We can then try to solve (21.15) via a sequence of loss functions analogous to (21.12).

Given the history $\mathcal{H}$, we again project the history to $\Pi\mathcal{H}$ as in (21.10). Here, however, the next state is random, and we would like to construct its *distribution*, as a function of the state-action pair (x,α); i.e.,

$$\mu^{\mathcal{H}}_{(x,\alpha)}(S) \stackrel{\text{def}}{=} \frac{1}{|\mathcal{H}|} \sum_{x'\in\mathsf{X}:(x,\alpha,x')\in\mathcal{H}} \mathbf{1}_S(x') \tag{21.16}$$

for all measurable subsets S of X and all $(x,\alpha)\in\Pi\mathcal{H}$. If $\mathcal{H}$ is rich enough,

$$\mu^{\mathcal{H}}_{(x,\alpha)}(S) \approx \sum_{\zeta\in\mathsf{Z}} \mathbf{1}_S(\Phi_\alpha(x,\zeta))p(\zeta) \tag{21.17}$$

for all measurable subsets S of X and all $(x,\alpha)\in\Pi\mathcal{H}$. In fact, (21.17) is only approximate, as variations in (x,α) will change the statistics of $\Phi_\alpha(x,\zeta)$.

Namely, if we have a current $\theta_n \in \mathcal{P}$ such that $Q(\cdot,\cdot,\theta_n)$ is our best current estimate of the solution of (21.15), then let's consider the cost function

$$\begin{aligned}\Lambda_n(\theta) \stackrel{\text{def}}{=} \frac{1}{|\mathcal{H}|}\Bigg\{ \sum_{(x,\alpha)\in\Pi\mathcal{H}} &\Bigg|Q(x,\alpha,\theta_n) \\ &- \left\{\delta + \int_{x'\in\mathsf{X}\setminus\mathsf{B}} \inf_{\alpha'\in\mathsf{A}} Q(x',a',\theta_n)\mu^{\mathcal{H}}_{(x,\alpha)}(dx')\right\}\Bigg|^2 \Bigg\}\end{aligned} \tag{21.18}$$

and carry out a gradient descent step as in (21.13) to improve θ_n. Of course, we are here assuming that $\mathcal{H}$ is rich enough that

$$\sum_{\zeta:\Phi_\alpha(x,\zeta)\notin\mathsf{B}} \inf_{\alpha'\in\mathsf{A}} Q(\Phi_\alpha(x,\zeta),\alpha')p(\zeta)$$

is approximated by the empirical average over $\mathcal{H}$.

21.2.3. Generalizing the Cost Functional. In the preceding discussion of this section we introduced $T_\lambda(x)$ as the time to the target set B, and we were interested in choosing the policy λ that would minimize either $T_\lambda(x)$ or its expected value (depending on whether the time T was deterministic or random). Let us now modify our objective, and let us replace this goal with something more complex. In particular, let us define

- $r(x, x', \alpha)\delta \geq 0$ to be the cost for using action $\alpha \in \mathsf{A}$ if the current state is $x \in \mathsf{X}$ and the state transitions to $x' \in \mathsf{X}$. To emphasize similarity with our previous two cases (Sections 21.2.1 and 21.2.2), we have scaled the cost function by the time step δ.
- $\beta > 0$ is a discount factor (future cost may have different value than present cost).

Our goal now is to minimize

$$
\begin{aligned}
W_\lambda(x) &= \mathbb{E}\left[\sum_{n: n\delta < T_\lambda(x)} \beta^n r\left(\mathcal{R}_n(x;\lambda), \mathcal{R}_{n+1}(x;\lambda), \lambda(\mathcal{R}_n(x;\lambda))\right)\delta\right] \\
&= \mathbb{E}\left[\sum_{n=0}^{\infty} \beta^n r\left(\mathcal{R}_n(x;\lambda), \mathcal{R}_{n+1}(x;\lambda), \lambda(\mathcal{R}_n(x;\lambda))\right)\mathbf{1}_{\{T_\lambda(x) > n\delta\}}\right]\delta.
\end{aligned}
$$

We notice that if we set $r = 1$ and $\beta = 1$, we recover $\mathbb{E}[T_\lambda(x)]$, which is what we studied in Section 21.2.2.

Of course, we shall have that if $x \in \mathsf{B}$, then $W_\lambda(x) = 0$. Therefore, let us now suppose that $x \neq 0$ and notice that we can write

$$
\begin{aligned}
W_\lambda(x) &= \mathbb{E}\left[r\left(x, \mathcal{R}_1(x;\lambda), \lambda(x)\right)\right]\delta \\
&\quad + \mathbb{E}\left[\sum_{n=1}^{\infty} \beta^n r\left(\mathcal{R}_n(x;\lambda), \mathcal{R}_{n+1}(x;\lambda), \lambda(\mathcal{R}_n(x;\lambda))\right)\mathbf{1}_{\{T_\lambda(x) > n\delta\}}\right]\delta \\
&= \mathbb{E}\left[r\left(x, \mathcal{R}_1(x;\lambda), \lambda(x)\right)\right]\delta \\
&\quad + \sum_{n=1}^{\infty} \beta^n \mathbb{E}\left[r\left(\mathcal{R}_n(x;\lambda), \mathcal{R}_{n+1}(x;\lambda), \lambda(\mathcal{R}_n(x;\lambda))\right)\mathbf{1}_{\{T_\lambda(x) > n\delta\}}\right]\delta. \qquad (21.19)
\end{aligned}
$$

Note that we can further write

$$
\mathbb{E}\left[r\left(x, \mathcal{R}_1(x;\lambda), \lambda(x)\right)\right] = \sum_{\zeta \in \mathsf{Z}} r\left(x, \Phi_{\lambda(x)}(x, \zeta), \lambda(x)\right) p(\zeta).
$$

Recalling relation (21.14) we next see that for $x \notin \mathsf{B}$ and $n \geq 1$, we have

$$
\begin{aligned}
&\mathbb{E}\left[r\left(\mathcal{R}_n(x;\lambda), \mathcal{R}_{n+1}(x;\lambda), \lambda(\mathcal{R}_n(x;\lambda))\right)\mathbf{1}_{\{T_\lambda(x)>n\delta\}}\right] \\
&= \mathbb{E}\left[r\left(\mathcal{R}_n(x;\lambda), \mathcal{R}_{n+1}(x;\lambda), \lambda(\mathcal{R}_n(x;\lambda))\right)\left\{\prod_{n'=1}^{n}\mathbf{1}_{\{\mathcal{R}_{n'}(x;\lambda)\notin \mathsf{B}\}}\right\}\right] \\
&= \sum_{\zeta\in \mathsf{Z}}\Bigg[r\left(\mathcal{R}_{n-1}(\Phi_{\lambda(x)}(x,\zeta);\lambda), \mathcal{R}_n(\Phi_{\lambda(z)}(x,\zeta);\lambda), \lambda(\mathcal{R}_{n-1}(\Phi_{\lambda(x)}(x,\zeta);\lambda))\right) \\
&\qquad \times\left\{\prod_{n'=1}^{n}\mathbf{1}_{\{\mathcal{R}_{n'-1}(\Phi_{\lambda(x)}(x,\zeta);\lambda)\notin \mathsf{B}\}}\right\}\Bigg]p(\zeta) \\
&= \sum_{\zeta\in \mathsf{Z}}\Big[r\left(\mathcal{R}_{n-1}(\Phi_{\lambda(x)}(x,\zeta);\lambda), \mathcal{R}_n(\Phi_{\lambda(x)}(x,\zeta);\lambda), \lambda(\mathcal{R}_{n-1}(\Phi_{\lambda(x)}(x,\zeta);\lambda))\right) \\
&\qquad \times\left\{\mathbf{1}_{T_\lambda(\Phi_{\lambda(x)}(x,\zeta))>(n-1)\delta}\right\}\Big]\,p(\zeta).
\end{aligned}
$$

Going back to (21.19), we have

$$
\begin{aligned}
&\sum_{n=1}^{\infty}\beta^n\mathbb{E}\left[r\left(\mathcal{R}_n(x;\lambda), \mathcal{R}_{n+1}(x;\lambda), \lambda(\mathcal{R}_n(x;\lambda))\right)\mathbf{1}_{\{T_\lambda(x)>n\delta\}}\right]\delta \\
&= \sum_{n=1}^{\infty}\beta^n\sum_{\zeta\in \mathsf{Z}} \\
&\quad \times\Big[r\left(\mathcal{R}_{n-1}(\Phi_{\lambda(x)}(x,\zeta);\lambda), \mathcal{R}_n(\Phi_{\lambda(x)}(x,\zeta);\lambda), \lambda(\mathcal{R}_{n-1}(\Phi_{\lambda(x)}(x,\zeta);\lambda))\right) \\
&\qquad \times\left\{\mathbf{1}_{T_\lambda(\Phi_{\lambda(x)}(x,\zeta))>(n-1)\delta}\right\}\Big]\,\delta p(\zeta) \\
&= \beta\sum_{\zeta\in \mathsf{Z}}\sum_{n=0}^{\infty}\beta^n \\
&\quad \times\Big[r\left(\mathcal{R}_n(\Phi_{\lambda(x)}(z,\zeta);\lambda), \mathcal{R}_{n+1}(\Phi_{\lambda(x)}(x,\zeta);\lambda), \lambda(\mathcal{R}_{n-1}(\Phi_{\lambda(x)}(x,\zeta);\lambda))\right) \\
&\qquad \times\left\{\mathbf{1}_{T_\lambda(\Phi_{\lambda(x)}(x,\zeta))>n\delta}\right\}\Big]\,\delta p(\zeta) \\
&= \beta\sum_{\zeta\in \mathsf{Z}}\ \sum_{n:\,n\delta<T_\lambda(\Phi_{\lambda(x)}(x,\zeta))}\beta^n \\
&\quad \times\left[r\left(\mathcal{R}_n(\Phi_{\lambda(x)}(z,\zeta);\lambda), \mathcal{R}_{n+1}(\Phi_{\lambda(x)}(x,\zeta);\lambda), \lambda(\mathcal{R}_n(\Phi_{\lambda(x)}(x,\zeta);\lambda))\right)\right]\delta p(\zeta) \\
&= \beta\sum_{\zeta\in \mathsf{Z}} W_\lambda(\Phi_{\lambda(x)}(x,\zeta))p(\zeta).
\end{aligned}
$$

For $\lambda \in \mathcal{P}$ we thus have

$$
W_\lambda(x) \geq \begin{cases}\sum_{\zeta\in \mathsf{Z}}\left\{r(x, \Phi_{\lambda(x)}(x,\zeta)); \lambda(x))\delta + \beta W_\lambda\left(\Phi_{\lambda(x)}(x,\zeta)\right)\right\}p(\zeta) & \text{if } x \notin \mathsf{B} \\ 0 & \text{if } x \in \mathsf{B}.\end{cases}
$$

Proceeding now in parallel with Section 21.2.2, we have the corresponding Bellman equation for the value function

(21.20)

$$V(x) = \begin{cases} \min_{\alpha \in \mathsf{A}} \sum_{\zeta \in \mathsf{Z}} \{r(x, \Phi_\alpha(x,\zeta); \alpha)\delta + \beta V(\Phi_\alpha(x,\zeta))\}\, p(\zeta) & \text{if } x \notin \mathsf{B} \\ 0 & \text{if } x \in \mathsf{B}. \end{cases}$$

The best policy then is

$$\lambda^*(x) = \underset{\alpha \in \mathsf{A}}{\operatorname{argmin}} \left\{ \sum_{\zeta \in \mathsf{Z}} \{r(x, \Phi_\alpha(x,\zeta); \alpha)\delta + \beta V(\Phi_\alpha(x,\zeta))\}\, p(\zeta) \right\}.$$

Mimicking now the discussion in Section 21.2.1, the basic Q-learning framework is as follows. Let us define the so-called Q function

$$\begin{aligned} Q(x,\alpha) &= \sum_{\zeta \in \mathsf{Z}} \{r(x, \Phi_\alpha(x,\zeta); \alpha)\delta + \beta V(\Phi_\alpha(x,\zeta))\}\, p(\zeta) \\ &= \sum_{\zeta \in \mathsf{Z}} \left\{ r(x, \Phi_\alpha(x,\zeta); \alpha)\delta + \beta \mathbf{1}_{\{\Phi_\alpha(x,\zeta) \notin \mathsf{B}\}} \inf_{\alpha' \in \mathsf{A}} Q(\Phi_\alpha(x,\zeta), \alpha') \right\} p(\zeta), \end{aligned}$$

for $x \notin \mathsf{B}$, the last equality following from the observation $V(x) = \min_{\alpha \in \mathsf{A}} Q(x,\alpha)$ if $x \notin \mathsf{B}$. Here the analogue of (21.18) is

$$\begin{aligned} \Lambda_n(\theta) \overset{\text{def}}{=} \frac{1}{|\mathcal{H}|} \Bigg\{ \sum_{(x,\alpha) \in \Pi \mathcal{H}} \Bigg| Q(x,\alpha,\theta_n) - \Bigg\{ \delta \int_{x' \in \mathsf{X}} r(x,x',\alpha) \mu^{\mathcal{H}}_{(x,\alpha)}(dx') \\ + \int_{x' \in \mathsf{X} \setminus \mathsf{B}} \inf_{\alpha' \in \mathsf{A}} Q(x', a', \theta_n) \mu^{\mathcal{H}}_{(x,\alpha)}(dx') \Bigg\} \Bigg|^2 \Bigg\}. \end{aligned}$$

21.3. Deep Reinforcement Learning

In Section 21.2 we introduced the basic reinforcement learning framework through the example of driving a car on a hill. We formulated the basic optimization problem where the function $Q(x,a)$, that approximates the solution to the Bellman equation, minimizes an appropriate loss function. As mentioned there, in typical applications, the Q-function is modeled as a neural network and the optimization problem is usually being solved with some variant of the stochastic gradient algorithm. The subfield of reinforcement learning that is using neural networks to learn the optimal control given the current state of the system is called *deep reinforcement learning* (DRL). The goal of this section is to lay down the generic formulation in its general case. In addition, for presentation purposes, we ignore the effect of the timestep parameter δ that was introduced in the concrete example of Section 21.2. In Section 21.4 we will discuss convergence properties of this algorithm in a mathematically rigorous way.

Consider a Markov decision problem defined on a finite state space $\mathsf{X} \subset \mathbb{R}^{d_x}$. For every $x \in \mathsf{X}$, there is a finite set $\mathsf{A} \subset \mathbb{R}^{d_a}$ of actions that can be taken such that:

- $p(z|x,a) = \mathbb{P}(X_{j+1} = z|X_j = x, a_j = a)$ is the transition probability function.
- given a state x and an action a, a reward/cost $r(x,a)$ is collected.
- λ is a control policy that depends on the history up to the present.
- $\beta \in (0,1]$ denotes a discount factor.

Then, typically, two types of problems are being considered:

Infinite time horizon reward, which is defined to be

$$W_\lambda(x) = \mathbb{E}_\lambda\left[\sum_{j=0}^{\infty} \beta^j r(x_j, a_j)|X_0 = x\right]. \tag{21.21}$$

Letting $V(x,a)$ be the reward, given that we start at position $x \in \mathsf{X}$, action $a \in \mathsf{A}$ is taken and the optimal value is then being used. Optimal control theory (see for example [**KY03**]),

$$\max_{a\in\mathsf{A}} V(x,a) = \sup_\lambda W_\lambda(x)$$

and the dynamic programming principle gives the Bellman equation

$$V(x,a) = r(x,a) + \beta \sum_{z\in\mathsf{X}} \max_{a'\in\mathsf{A}} V(z,a')p(z|x,a),$$

and the optimal policy is $a^*(x) = \operatorname{argmax}_{a\in\mathsf{A}} V(x,a)$.

Finite time horizon reward, which is defined to be

$$W_\lambda(J,x) = \mathbb{E}_\lambda\left[\sum_{j=0}^{J} \beta^j r_j|X_0 = x\right], \tag{21.22}$$

where $r_j = r(j, x_j, a_j)$ for $j = 0, 1, \ldots, J-1$ and $r_J = r(J, x_J)$. Here J is a deterministic time horizon.

The optimal control $a^*(J,x)$ is given by the solution to the Bellman equation

$$V(j,x,a) = r(j,x,a) + \beta \sum_{z} \max_{a'\in\mathsf{A}} V(j+1|z,a')p(z|x,a),$$

$$V(J,x,a) = r(J,x).$$

Optimal policy is $a^*(j,x) = \operatorname{argmax}_{a\in\mathsf{A}} V(j,x,a)$ and the principle of optimality dictates that

$$\max_{a\in\mathsf{A}} V(0,x,a) = \sup_\lambda W_\lambda(J,x).$$

In principle, we may be able to solve the Bellman equation, but

- $p(z|x, a)$ may not be known.
- state space may be too high dimensional.

In this case, $V(x, a)$ would need to be approximated. To do so, one can indeed use neural networks. Exploring this idea is the content of Sections 21.4 and 21.5.

21.4. Q-learning

Reinforcement learning approximates the solution to the Bellman equation using a function approximator. Typically, a neural network is being used, $Q(x, a, \theta)$, where θ is the parameter to be learned in training. We will focus the discussion on the infinite time horizon reward (21.21). The case of the finite time horizon reward problem (21.22) is similar.

The goal of the Q-learning algorithm is to minimize the objective function

$$\Lambda(\theta) = \frac{1}{2} \sum_{(x,a) \in \mathsf{X} \times \mathsf{A}} [Y(x, a) - Q(x, a, \theta)]^2 \pi(x, a), \tag{21.23}$$

where $\pi(x, a)$ is a probability function to be specified and the target function (for example in the infinite horizon setting) becomes

$$Y(x, a) = r(x, a) + \beta \sum_{x' \in \mathsf{X}} \max_{a' \in \mathsf{A}} Q(x', a', \theta) p(x'|x, a).$$

Notice that if $\Lambda(\theta) = 0$, then $Q(x, a, \theta)$ is a solution to the Bellman equation. So, it makes sense to try to find values for θ so that $\Lambda(\theta)$ is as close to zero as possible. To do so, one may use stochastic gradient descent, which, in this case, becomes

$$\theta_{k+1} = \theta_k + \eta_k g_k,$$

where η_k is the learning rate and g_k is

$$g_k = \left(r(x_k, a_k) + \beta \max_{a' \in \mathsf{A}} Q(x_{k+1}, a', \theta_k) - Q(x_k, a_k, \theta_k) \right) \nabla_\theta Q(x_k, a_k, \theta_k)$$

with (x_k, a_k) being an ergodic Markov chain with $\pi(x, a)$ as its stationary distribution.

In its simplest form, one may define

$$Q(x, a, \theta) = \frac{1}{N^\gamma} \sum_{n=1}^{N} C^n \sigma(W^n \cdot (x, a) + b^n), \tag{21.24}$$

where $\gamma \in [1/2, 1]$, $\theta = (C^1, \dots, C^N, W^1, \dots, W^N, b_1, \dots, b^N) \in \mathbb{R}^{(2+d)N}$, $d = d_x + d_a$. In the infinite horizon setting, the SGD algorithm reads as follows

$$C^n_{k+1} = C^n_k + \frac{\eta^N_k}{N^\gamma}\Big(r(x_k, a_k) + \beta \max_{a' \in \mathsf{A}} Q(x_{k+1}, a', \theta_k) - Q(x_k, a_k, \theta_k)\Big) \times \sigma(W^n_k \cdot (x_k, a_k) + b^n_k), \tag{21.25}$$

$$\begin{aligned} W^n_{k+1} &= W^n_k + \frac{\eta^N_k}{N^\gamma}\Big(r(x_k, a_k) + \beta \max_{a' \in \mathsf{A}} Q(x_{k+1}, a', \theta_k) - Q(x_k, a_k, \theta_k)\Big) \\ &\quad \times C^n_k \sigma'(W^n_k \cdot (x_k, a_k) + b^n_k)(x_k, a_k), \\ b^n_{k+1} &= b^n_k + \frac{\eta^N_k}{N^\gamma}\Big(r(x_k, a_k) + \beta \max_{a' \in \mathsf{A}} Q(x_{k+1}, a', \theta_k) - Q(x_k, a_k, \theta_k)\Big) \\ &\quad \times C^n_k \sigma'(W^n_k \cdot (x_k, a_k) + b^n_k), \end{aligned} \tag{21.26}$$

where the learning rate η^N_k may depend on both k and N. A few remarks are in order.

Remark 21.1 (On the choice of the next action). One of the most common choices for the distribution π is that of pure exploration, which, at the very beginning of training without prior knowledge, typically amounts to sampling uniformly at random from all possible $(x_k, a_k) \in \mathsf{X} \times \mathsf{A}$. Even though many typical applications use pure exploration as the default choice for choosing the action taken at time k, other choices do exist. In particular, some of the most used ones (besides pure exploration) are

- Greedy action: $a_k = \operatorname{argmax}_{a \in \mathsf{A}} Q(x_k, a, \theta)$.
- ϵ-greedy algorithm, where

$$a_k = \begin{cases} \text{Uniform}\{a : a \in \mathsf{A}\}, & \text{with probability } \epsilon \\ \operatorname{argmax}_{a \in \mathsf{A}} Q(x_k, a, \theta), & \text{with probability } 1 - \epsilon. \end{cases}$$

We note that the ϵ-greedy algorithm typically has $\epsilon \downarrow 0$ as the number of training epochs m increases. As the model $Q(x_k, a, \theta)$ becomes more accurate, we would like to more frequently take the greedy action.

For the purposes of the presentation in this chapter, it is sufficient to have in mind the choice of pure exploration.

Remark 21.2 (Actor-critic setting). A more advanced setup is the actor-critic setting (especially useful when the action space is continuous), where both the action and the value function are modeled as neural networks. We will not analyze the actor-critic setting in this chapter, but let us briefly discuss the setting for completeness. In the action-critic setting both the value function and the optimal policy are being learned. In particular,

- *Actor model*: a neural network based model $p((x,a);\theta^A) : \mathsf{X}\times\mathsf{A} \mapsto \mathbb{R}^K$ that gives an approximation to the optimal policy. It can be thought of as the probability of selecting action $a \in \mathsf{A}$ at state $x \in \mathsf{X}$. The parameter θ^A is to be estimated during the learning process.
- *Critic model*: a neural network $Q(x,a;\theta^Q)$ that gives the state-action value function for a pair $(x,a) \in \mathbb{R}^d \times \mathsf{A}$.
- The critic minimizes the objective function

$$\Lambda(\theta^Q) = \left(Y_k - Q(X_k, p((x_k,a_k);\theta^A);\theta^Q)\right)^2,$$
$$Y_k = r(X_k,a_k) + \beta Q(x_{k+1}, p((x_{k+1},a_{k+1});\theta^A);\theta^Q),$$

  where Y_k is treated as constant.
- The actor maximizes the objective function

$$\max_\lambda \mathbb{E}_\lambda\left[\sum_{j=0}^{\infty} \beta^j r(x_j,a_j)\right] = \sum_{x\in\mathsf{X}} W_\lambda(x)\rho_0(x),$$

  where $W_\lambda(x)$ is defined in (21.21) and $\rho_0(x)$ is the initial distribution of states. In practice, when doing stochastic gradient descent, the state-action value function is being replaced by its estimate, leading to maximizing the objective function

$$G(\theta^A) = Q(x_k, p((x_k,a_k);\theta^A);\theta^Q).$$

  Here $p((x,a);\theta^A)$ is the actor model, which could for example be $S_{\text{softmax}}(P((x,a);\theta_A))$ (i.e., a probability distribution) with $P((x,a);\theta_A)$ being the neural network

$$P((x,a);\theta_A) = \frac{1}{N^\gamma}\sum_{n=1}^{N} B^n\sigma(U^n\cdot(x,a) + d^n),$$

  where $\{B^n, U^n, d^n\}_{n=1}^N$ are parameters to be estimated via the learning process.

We do not analyze further the action-critic setting in the chapter. The analysis in the subsequent sections of this chapter focuses on the case of pure-exploration for the actor or more generally on the case of any (general) fixed policy for which the Markov chain (x_k,a_k) is ergodic bounded away from zero ergodic distribution.

Remark 21.3. The Q-learning algorithm calculates its updates by taking the derivative of $\Lambda(\theta)$ while treating the function Y as constant. But Y is not constant and does depend on θ, so there is bias in the sense that

$$\mathbb{E}[g_k|\theta_k, x_k, a_k] \neq \frac{1}{2}\nabla_\theta \left(Y(x_k, a_k) - G(x_k, a_k, \theta_k)\right)^2.$$

This fact yields certain difficulties in the proofs and it also forces the proofs to require β to be small in order for global convergence to be realized, see Theorem 21.7. Essentially, the Q-learning algorithm calculates the update by treating the target Y as a constant. Consequently, this implies that the asymptotic dynamics of the corresponding neural network as N and k increase may not move in the descent direction of the limiting objective function for an arbitrary value of β. As we prove in Section 21.5 however, taking β to be small does guarantee that the algorithm is moving in the descent direction of the objective function and thus in that case convergence can be realized. At this point, we mention that a related phenomenon has also been observed in [**CYLW19**].

21.5. Convergence Properties of the Q-learning Algorithm

Let us now investigate how the Q-learning algorithm behaves when the number of hidden units N and number of stochastic gradient descent iterates k grow to infinity. We shall study the behavior of single layer Q-networks in the case of the infinite time horizon reward problem (21.21). We prove that the Q-network (which models the value function for the related optimal control problem) converges to the solution of a random ordinary differential equation. We characterize the limiting random ODE. We also study the behavior of the solution to the limiting random ODE as time $t \to \infty$.

We start with an assumption that will be assumed throughout this section. Without loss of generality we may and will assume that the bias terms $b^n = 0$ for all $n = 1, \dots, N$.

Assumption 21.4. We are assuming the following:

(1) $(C_0^n, W_0^n)_{n=1}^N$ are independent and identically distributed random variables with mean zero and joint distribution denoted by $\mu_0(dc, dw)$.
(2) C_0^n are bounded and $\int \|w\|\mu_0(dcdw) < \infty$.
(3) The cost r is uniformly bounded.
(4) $\lim_{N\to\infty} \frac{1}{N}\sum_{n=1}^N \mathbf{1}_{X_n=x|X_0=z} = \pi(x) > 0$ almost surely.
(5) The spaces X and A are finite discrete spaces.
(6) We take $\gamma = 1/2$ in (21.24).

To have a concrete example in mind, we assume that the action $a \in \mathsf{A}$ is sampled uniformly at random from the set of all possible actions. That is, we consider $\pi(x,a) = \frac{1}{|\mathsf{A}|}\pi(x)$. However, what we shall discuss below is more general and the proofs go through if (x_k, a_k) is assumed to be ergodic with distribution $\pi(x,a) > 0$ for all $(x,a) \in \mathsf{X} \times \mathsf{A}$.

Next, let us define the quantities

$$\begin{aligned} \nu_k^N &= \frac{1}{N}\sum_{n=1}^{N} \delta_{C_k^n, W_k^n}, \\ \mu_t^N &= \nu_{\lfloor Nt \rfloor}^N, \\ h_t^N(x,a) &= Q^N(x,a,\theta_{\lfloor Nt \rfloor}). \end{aligned}$$

Note that Assumption 21.4 guarantees that $\mu_0^N \to \mu_0$ and

$$h_0^N \to N(0, \langle |c\sigma(w \cdot (x,a))|^2, \mu_0 \rangle),$$

i.e., h_0^N converges to a Gaussian distribution.

Assumption 21.5. We assume that

(1) σ is non-polynomial and slowly increasing, i.e., $\frac{\sigma(x)}{|x|^a} \to 0$ for all $a > 0$.

(2) $\mu_0(\Gamma) > 0$ if Γ has positive Lebesgue measure.

Let $\xi = (x,a) \in \mathsf{X} \times \mathsf{A}$ and consider the matrix A with elements

$$A_{\xi,\xi'} = \left\langle \sigma(w\cdot\xi)\sigma(w\cdot\xi') + c^2\sigma'(w\cdot\xi)\sigma'(w\cdot\xi')\xi\cdot\xi', \mu_0 \right\rangle. \tag{21.27}$$

Then, we have the following results.

Theorem 21.6. *Assume that Assumption* 21.4 *holds, and choose the learning rate to be* $\eta_k^N = \frac{\eta}{N}$ *for* $0 < \eta < \infty$ *a fixed constant. Then the process* h_t^N *converges, as* $N \to \infty$*, in distribution in the space* $D_{\mathbb{R}^{|\mathsf{X}\times\mathsf{A}|}}([0,T])$ *to* h_t*, the solution of the ordinary differential equation*

$$\begin{aligned} h_t(x,a) &= h_0(x,a) + \eta \int_0^t \sum_{(x',a')} \pi(x',a') A_{(x,a),(x',a')} \\ &\quad \times \left(r(x',a') + \beta \sum_{z\in\mathsf{X}} \max_{a'\in\mathsf{A}} h_s(z,a'')p(z|x',a') - h_s(x',a') \right) ds, \end{aligned}$$

(21.28)

$$h_0(x,a) = G(x,a) \sim N(0, \langle |c\sigma(w\cdot(x,a))|^2, \mu_0 \rangle).$$

As a matter of fact, if β is small enough, then $h_t(x,a)$ is guaranteed to converge to the value function $V(x,a)$ as $t \to \infty$. This is the content of Theorem 21.7.

Theorem 21.7. *Assume that $\beta < \frac{2}{1+|\mathsf{A}|}$, and let Assumption* 21.5 *hold. Then we have that*

$$\lim_{t\to\infty} \sup_{(x,a)\in \mathsf{X}\times\mathsf{A}} |h_t(x,a) - V(x,a)| = 0.$$

21.5.1. Proof of Theorem 21.6. The proof of Theorem 21.6 will be split into three steps.

21.5.1.1. *Step 1: Characterization of the limit.* Let us start by deriving an equivalent representation for $Q_k^N(x,a) = Q^N(x,a,\theta_k)$, and let us set $\xi = (x,a)$. Consider the neural network model (21.24) with the bias term $b^n = 0$ (for simplicity of exposition) and $\gamma = 1/2$. Employing a Taylor expansion, we notice that

$$\begin{aligned}
Q_{k+1}^N(\xi) &= Q_k^N(\xi) + \frac{1}{\sqrt{N}} \sum_{n=1}^N \left(C_{k+1}^n \sigma(W_{k+1}^n \cdot \xi) - C_k^n \sigma(W_k^n \cdot \xi)\right) \\
&= Q_k^N(\xi) + \frac{1}{\sqrt{N}} \sum_{n=1}^N \left((C_{k+1}^n - C_k^n)\sigma(W_k^n \cdot \xi)(W_{k+1}^n - W_k^n)\right) \\
&\quad + \frac{1}{\sqrt{N}} \sum_{n=1}^N \sum_{n=1}^N C_k^n \Big(\sigma'(W_k^{n,*} \cdot \xi)\xi \cdot (W_{k+1}^n - W_k^n) \\
&\quad + \frac{1}{2}\sigma''(\tilde{W}_k^{n,*} \cdot \xi)\left(\xi \cdot (W_{k+1}^n - W_k^n)\right)^2\Big),
\end{aligned}$$

where $W_k^{n,*}, \tilde{W}_k^{n,*}$ are points in the line segments that connect W_k^n and W_{k+1}^n. Let us now recall the SGD update (21.26) and use $\eta_k^N = \frac{\eta}{N}$ for the learning rate where $0 < \eta < \infty$ is a fixed constant. Plugging that into the previous display, we subsequently obtain (with $\xi = (x,a)$ and $\xi_k = (x_k, a_k)$)

$$\begin{aligned}
Q_{k+1}^N(\xi) &= Q_k^N(\xi) \\
&\quad + \frac{\eta}{N^2}\Big(r(\xi_k) + \beta \max_{a'\in\mathsf{A}} Q_k(x_{k+1}, a') - Q_k(\xi_k)\Big) \sum_{n=1}^N \sigma(W_k^n \cdot \xi_k)\sigma(W_k^n \cdot \xi) \\
&\quad + \frac{\eta}{N^2}\Big(r(\xi_k) + \beta \max_{a'\in\mathsf{A}} Q_k(x_{k+1}, a') - Q_k(\xi_k)\Big) \\
&\qquad \times \sum_{n=1}^N (C_k^n)^2 \sigma'(W_k^n \cdot \xi_k)\sigma'(W_k^n \cdot \xi)\xi_k \cdot \xi + \mathcal{O}_p\left(N^{-3/2}\right),
\end{aligned}$$

where we recall Definition 19.8 for the notation $\mathcal{O}_p\left(N^{-3/2}\right)$.

Writing this in terms of the empirical measure ν_k^N, we then get

$$\begin{aligned}
Q_{k+1}^N(\xi) &= Q_k^N(\xi) \\
&\quad + \frac{\eta}{N}\left(r(\xi_k) + \beta \max_{a' \in \mathsf{A}} Q_k(x_{k+1}, a') - Q_k(\xi_k)\right)\langle \sigma(w \cdot \xi_k)\sigma(w \cdot \xi), \nu_k^N\rangle \\
&\quad + \frac{\eta}{N}\left(r(\xi_k) + \beta \max_{a' \in \mathsf{A}} Q_k(x_{k+1}, a') - Q_k(\xi_k)\right) \\
&\quad \times \langle (c)^2 \sigma'(w \cdot \xi_k)\sigma'(w \cdot \xi)\xi_k \cdot \xi, \nu_k^N\rangle + \mathcal{O}_p(N^{-3/2}),
\end{aligned} \tag{21.29}$$

Recalling now the definition of $h_t^N = Q_{\lfloor Nt \rfloor}^N$, we subsequently obtain

$$\begin{aligned}
h_t^N(\xi) &= h_0^N(\xi) + \sum_{k=0}^{\lfloor Nt \rfloor - 1} (Q_{k+1}^N(\xi) - Q_k^N(\xi)) \\
&= h_0^N(\xi) + \frac{\eta}{N} \sum_{k=0}^{\lfloor Nt \rfloor - 1} \left(r(\xi_k) + \beta \max_{a' \in \mathsf{A}} Q_k(x_{k+1}, a') - Q_k(\xi_k)\right) \\
&\quad \times \langle \sigma(w \cdot \xi_k)\sigma(w \cdot \xi), \nu_k^N\rangle \\
&\quad + \frac{\eta}{N} \sum_{k=0}^{\lfloor Nt \rfloor - 1} \left(r(\xi_k) + \beta \max_{a' \in \mathsf{A}} Q_k(x_{k+1}, a') - Q_k(\xi_k)\right) \\
&\quad \times \langle (c)^2 \sigma'(w \cdot \xi_k)\sigma'(w \cdot \xi)\xi_k \cdot \xi, \nu_k^N\rangle + \mathcal{O}_p(N^{-1/2}) \\
&= h_0^N(\xi) + \eta \int_0^t \sum_{\xi' \in \mathsf{X} \times \mathsf{A}, x'' \in \mathsf{X}} \left(r(\xi') + \beta \max_{a' \in \mathsf{A}} h_s^N(x'', a'') - h_s^N(\xi')\right) \\
&\quad \times \langle \sigma(w \cdot \xi')\sigma(w \cdot \xi), \mu_s^N\rangle \pi(dx'', \xi') ds \\
&\quad + \eta \int_0^t \sum_{\xi' \in \mathsf{X} \times \mathsf{A}, x'' \in \mathsf{X}} \left(r(\xi') + \beta \max_{a' \in \mathsf{A}} h_s^N(x'', a'') - h_s^N(\xi')\right) \\
&\quad \times \langle (c)^2 \sigma'(w \cdot \xi')\sigma'(w \cdot \xi)\xi' \cdot \xi, \mu_s^N\rangle \pi(dx'', \xi') ds \\
&\quad + R_t^N + \mathcal{O}_p(N^{-1/2}),
\end{aligned} \tag{21.30}$$

where R_t^N is the error being made by replacing the sums by the integrals above.

By Lemma 5.9 of [**SS22**] (we omit the proof here due to its technical nature) we have that

$$\lim_{N \to \infty} \sup_{t \in [0,T]} \mathbb{E}|R_t^N| = 0.$$

Similarly for a test function $f \in \mathcal{C}_b^2(\mathbb{R}^{1+d})$ we have

$$\begin{aligned}\langle f, \mu_t^N\rangle &= \langle f, \mu_0^N\rangle + \sum_{k=0}^{\lfloor Nt\rfloor -1} \left(\langle f, \nu_{k+1}^N\rangle - \langle f, \nu_k^N\rangle\right) \\ &= \langle f, \mu_0^N\rangle + \frac{\eta}{N^{3/2}} \sum_{k=0}^{\lfloor Nt\rfloor -1} \left(r(\xi_k) + \beta \max_{a'\in \mathsf{A}} Q_k(x_{k+1}, a') - Q_k(\xi_k)\right) \\ &\quad \times \left\langle \operatorname{div} f \cdot (\sigma(w\cdot \xi_k) + c\sigma'(w\cdot\xi_k)\xi_k), \nu_k^N\right\rangle \\ &\quad + \mathcal{O}_p(N^{-1}). \end{aligned} \tag{21.31}$$

21.5.1.2. *Step 2: Existence and identification of the limit.* By Exercises 21.1 and 21.2, respectively, we have that the following two a priori bounds hold

$$\max_{n\in\mathbb{N}} \max_{k\le \lfloor NT\rfloor} \left(|C_k^n| + \mathbb{E}\|W_k^n\|\right) \le C_o < \infty$$

and

$$\max_{N\in\mathbb{N}} \max_{k\le \lfloor NT\rfloor} \mathbb{E} \sup_{(x,a)\in \mathsf{X}\times\mathsf{A}} \left|Q_k^N(x,a)\right| \le C_o < \infty,$$

for an appropriate finite constant $C_o < \infty$.

These a priori bounds lead to the following bounds in Lemmas 21.8 and 21.9, whose proof are given at the end of this subsection. Lemma 21.8 is about appropriate compact containment of the involved processes.

Lemma 21.8. *Let $\zeta > 0$. Then there exists a compact subset $\mathcal{K}$ of E such that*

$$\sup_{N\in\mathbb{N}, 0\le t\le T} \mathbb{P}[(\mu_t^N, h_t^N) \notin \mathcal{K}] < \zeta.$$

Lemma 21.9 is about regularity of the involved processes. As in Chapter 19 consider the function $q(z_1, z_2) = \min\{|z_1 - z_2|, 1\}$ with $z_1, z_2 \in \mathbb{R}$. Recall that $\mathcal{F}_t^N$ is the σ-algebra generated by $\{(C_0^i, W_0^i)\}_{i=1}^N$ and $\{x_j\}_{j=0}^{\lfloor Nt\rfloor-1}$, i.e., $\mathcal{F}_t^N$ contains the information generated by $\{(C_0^i, W_0^i)\}_{i=1}^N$ and $\{x_j\}_{j=0}^{\lfloor Nt\rfloor-1}$.

Lemma 21.9. *Let $f \in C_b^2(\mathbb{R}^{1+d})$. For any $\delta_o \in (0,1)$, there is a constant $C_o < \infty$ such that for $0 \le u \le \delta_o$, $0 \le v \le \delta_o \wedge t$, $t \in [0,T]$,*

$$\mathbb{E}\left[q(\langle f, \mu_{t+u}^N\rangle, \langle f, \mu_t^N\rangle) q(\langle f, \mu_t^N\rangle, \langle f, \mu_{t-v}^N\rangle) | \mathcal{F}_t^N\right] \le C_o\delta_o + \frac{C_o}{N^{3/2}}$$

and

$$\mathbb{E}\left[q(h^N_{t+u}, h^N_t)q(h^N_t, h^N_{t-v})|\mathcal{F}^N_t\right] \leq C_o\delta_o + \frac{C_o}{N}.$$

By Lemmas 21.8 and 21.9 and the convergence results of Section A.4, we have that $\{\mu^N, h^N\}_{N\in\mathbb{N}}$ is relatively compact as a $D_E([0,T])$- valued random variable where $E = \mathcal{M}(\mathbb{R}^{1+d}) \times \mathbb{R}^M$.

Subsequently, any subsequence of $\{\mu^N, h^N\}_{N\in\mathbb{N}}$ will have a convergent sub-subsequence. Any such convergent subsequence can be identified via relations (21.30) and (21.31).

In particular, (21.31) shows that the limit as $N \to \infty$, $\left\langle f, \mu^N_t\right\rangle$ will not be changing in t. Therefore $\left\langle f, \mu^N_t\right\rangle \to \langle f, \mu_0\rangle$ as $N \to \infty$ for all $t \in [0,T]$. The latter and (21.30) would then yield the limit of h^N_t as stated in Theorem 21.6. It remains to show that such a limit is unique. This is being taken care of by step 3 of the proof. But before getting into step 3 of the proof let us present the proofs of Lemmas 21.8 and 21.9.

Proof of Lemma 21.8. Given the a priori bounds established in Exercises 21.1 and 21.2, the proof is completely analogous to that of Lemma 19.11. □

Proof of Lemma 21.9. First, we prove the first statement of the lemma, i.e., the regularity of $\{\mu^N\}_{N\in\mathbb{N}}$. By the standard Taylor expansion, we have for $0 \leq s \leq t \leq T$:

$$\begin{aligned}
|\left\langle f, \mu^N_t\right\rangle - \left\langle f, \mu^N_s\right\rangle| &= |\left\langle f, \nu^N_{\lfloor Nt\rfloor}\right\rangle - \left\langle f, \nu^N_{\lfloor Ns\rfloor}\right\rangle| \\
&\leq \frac{1}{N}\sum_{n=1}^N |f(C^n_{\lfloor Nt\rfloor}, W^n_{\lfloor Nt\rfloor}) - f(C^n_{\lfloor Ns\rfloor}, W^n_{\lfloor Ns\rfloor})| \\
&\leq \frac{1}{N}\sum_{n=1}^N |\partial_c f(\bar{C}^n_{\lfloor Nt\rfloor}, \bar{W}^n_{\lfloor Nt\rfloor})||C^n_{\lfloor Nt\rfloor} - C^n_{\lfloor Ns\rfloor}| \\
&\quad + \frac{1}{N}\sum_{n=1}^N \parallel \nabla_w f(\bar{C}^n_{\lfloor Nt\rfloor}, \bar{W}^n_{\lfloor Nt\rfloor}) \parallel\parallel W^n_{\lfloor Nt\rfloor} - W^n_{\lfloor Ns\rfloor} \parallel
\end{aligned} \tag{21.32}$$

for points $\bar{C}^n, \bar{W}^n$ in the segments connecting $C^n_{\lfloor Ns\rfloor}$ with $C^n_{\lfloor Nt\rfloor}$ and $W^n_{\lfloor Ns\rfloor}$ with $W^n_{\lfloor Nt\rfloor}$, respectively.

Let's now establish a bound on $|C^n_{\lfloor Nt \rfloor} - C^n_{\lfloor Ns \rfloor}|$ for $s < t \le T$ with $0 < t - s \le \delta_o < 1$. By Assumption 21.4 and Exercises 21.1 and 21.2 we have

$$\begin{aligned}
\mathbb{E}\Big[|C^n_{\lfloor Nt\rfloor} - C^n_{\lfloor Ns\rfloor}|\Big|\mathcal{F}^N_s\Big] &= \mathbb{E}\Big[|\sum_{k=\lfloor Ns\rfloor}^{\lfloor Nt\rfloor-1}(C^n_{k+1} - C^n_k)|\Big|\mathcal{F}^N_s\Big] \\
&\le \mathbb{E}\Big[\sum_{k=\lfloor Ns\rfloor}^{\lfloor Nt\rfloor-1} |\eta(r_k + \gamma \max_{a'\in A} Q^N_k(x_{k+1}, a') \\
&\quad - Q^N_k(x_k, a_k))\frac{1}{N^{3/2}}\sigma(W^n_k \cdot x_k)|\Big|\mathcal{F}^N_s\Big] \\
&\le \frac{1}{N^{3/2}} \sum_{k=\lfloor Ns\rfloor}^{\lfloor Nt\rfloor-1} C_o \\
&\le \frac{C_o}{\sqrt{N}}(t-s) + \frac{C_o}{N^{3/2}} \\
&\le \frac{C_o}{\sqrt{N}}\delta_o + \frac{C_o}{N^{3/2}}, \qquad (21.33)
\end{aligned}$$

for an appropriate constant $C_o < \infty$ that may change from line to line. Similarly, we have

$$\begin{aligned}
\mathbb{E}\Big[\parallel W^n_{\lfloor Nt\rfloor} - W^n_{\lfloor Ns\rfloor} \parallel \Big|\mathcal{F}^N_s\Big] &= \mathbb{E}\Big[\parallel \sum_{k=\lfloor Ns\rfloor}^{\lfloor Nt\rfloor-1}(W^n_{k+1} - W^n_k) \parallel \Big|\mathcal{F}^N_s\Big] \\
&\le \mathbb{E}\Big[\sum_{k=\lfloor Ns\rfloor}^{\lfloor Nt\rfloor-1} \parallel \eta(r_k + \gamma \max_{a'\in A} Q^N_k(x_{k+1}, a') - Q^N_k(x_k, a_k)) \\
&\qquad \times \frac{1}{N^{3/2}} C^i_k \sigma'(W^n_k \cdot x_k) x_k \parallel \Big|\mathcal{F}^N_s\Big] \\
&\le \frac{1}{N^{3/2}} \sum_{k=\lfloor Ns\rfloor}^{\lfloor Nt\rfloor-1} C_o \\
&\le \frac{C_o}{\sqrt{N}}(t-s) + \frac{C_o}{N^{3/2}} \\
&\le \frac{C_o}{\sqrt{N}}\delta_o + \frac{C_o}{N^{3/2}}. \qquad (21.34)
\end{aligned}$$

Returning to equation (21.32), the previous bounds together with Exercise 21.1 yield for $0 < s < t \le T$ with $0 < t - s \le \delta_o < 1$

$$\mathbb{E}\left[|\langle f, \mu^N_t\rangle - \langle f, \mu^N_s\rangle||\mathcal{F}^N_s\right] \le C_o\delta_o + \frac{C_o}{N^{3/2}},$$

where $C_o < \infty$ is an unimportant constant. Then, the first statement of the lemma follows.

Next, we establish the second statement of the lemma, i.e., the regularity of $\{h^N\}_{N\in\mathbb{N}}$. Recalling

$$Q^N_{k+1}(\xi) = Q^N_k(\xi)$$
$$+\frac{1}{\sqrt{N}}\sum_{n=1}^{N}\Big((C^n_{k+1} - C^n_k)\sigma\big(W^n_{k+1}\cdot\xi\big) + \sigma'(W^{n,*}_k\cdot\xi)\xi^\top(W^n_{k+1} - W^n_k)C^i_k\Big),$$

we obtain

$$h^N_t(\xi) - h^N_s(\xi) = \sum_{k=\lfloor Ns\rfloor}^{\lfloor Nt\rfloor}(Q^N_{k+1}(\xi) - Q^N_k(\xi))$$
$$= \sum_{k=\lfloor Ns\rfloor}^{\lfloor Nt\rfloor}\frac{1}{\sqrt{N}}\sum_{n=1}^{N}\Big((C^n_{k+1} - C^n_k)\sigma(W^n_{k+1}\cdot\xi) + \sigma'(W^{n,*}_k\cdot\xi)\xi^\top(W^n_{k+1} - W^n_k)C^n_k\Big).$$

The latter together with the boundedness of $\sigma'(\cdot)$ and the bound of Exercise 21.1 yield

$$|h^N_t(\xi) - h^N_s(\xi)| \le \sum_{k=\lfloor Ns\rfloor}^{\lfloor Nt\rfloor}|Q^N_{k+1}(\xi) - Q^N_k(\xi)|$$
$$\le \sum_{k=\lfloor Ns\rfloor}^{\lfloor Nt\rfloor}\frac{1}{\sqrt{N}}\sum_{n=1}^{N}\Big(|C^n_{k+1} - C^n_k| + \|W^n_{k+1} - W^n_k\|\Big).$$

Taking expectations, we have

$$\mathbb{E}\Big[\sup_\xi|h^N_t(\xi) - h^N_s(\xi)|\Big|\mathcal{F}^N_s\Big]$$
$$\le \frac{1}{\sqrt{N}}\sum_{n=1}^{N}\sum_{k=\lfloor Ns\rfloor}^{\lfloor Nt\rfloor}\mathbb{E}\Big[|C^n_{k+1} - C^n_k| + \|W^n_{k+1} - W^n_k\|\Big|\mathcal{F}^N_s\Big].$$

Next, we use the bounds (21.33) and (21.34),

$$\mathbb{E}\Big[\sup_\xi|h^N_t(\xi) - h^N_s(\xi)|\Big|\mathcal{F}^N_s\Big] \le \frac{1}{\sqrt{N}}\sum_{i=1}^{N}\Big(\frac{C_o}{\sqrt{N}}(t-s) + \frac{C_o}{N^{3/2}}\Big)$$
$$= C_o(t-s) + \frac{C_o}{N},$$

to conclude

$$\mathbb{E}\Big[\|h^N_t - h^N_s\|\Big|\mathcal{F}^N_s\Big] \le C_o(t-s) + \frac{C_o}{N}.$$

This then yields the second statement of the lemma, concluding the proof. □

21.5.1.3. *Step 3: Uniqueness of the limit.* Let us assume that there are two possible solutions $h_t^{(1)}$ and $h_t^{(2)}$ of (21.28). Recall now the matrix A with elements $A_{\xi,\xi'}$ as defined by (21.27). By Lemma 21.10 we have that the matrix A is positive definite.

Next, we set

$$\mathcal{G}_{\xi,t} = \beta \sum_{x''\in\mathsf{X}} \left[\max_{a''\in\mathsf{A}} h_t^{(1)}(x'',a'') - \max_{a''\in\mathsf{A}} h_t^{(2)}(x'',a'')\right] \mathbb{P}[x''|\xi].$$

If we set now $\psi_t(x,a) = h_t^{(1)}(x,a) - h_t^{(2)}(x,a)$, we find that for some constant $C_o < \infty$,

$$\begin{aligned}|\mathcal{G}_{\xi,t}| &\leq \beta \sum_{x''\in\mathsf{X}} \max_{a''\in\mathsf{A}} |\psi_t(x'',a'')|\, \mathbb{P}[x''|\xi] \\ &\leq C_o \sum_{\xi\in\mathsf{X}\times\mathsf{A}} |\psi_t(\xi)|\,.\end{aligned}$$

At the same time, by definition, we have

$$\begin{aligned}\psi_t(\xi) &= \int_0^t \sum_{\xi'\in\mathsf{X}\times\mathsf{A}} (\mathcal{G}_{\xi',s} - \psi_s(\xi'))A_{\xi,\xi'}\pi(\xi')ds, \\ \psi_0(\xi) &= 0.\end{aligned}$$

Using the bound for the term $|\mathcal{G}_{\xi,t}|$ and the boundedness of the elements $A_{\xi,\xi'}$ would then give,

$$\begin{aligned}|\psi_t(\xi)|^2 &= 2\int_0^t \psi_s(\xi)d\psi_s(\xi) \\ &= 2\int_0^t \psi_s(\xi) \sum_{\xi'\in\mathsf{X}\times\mathsf{A}} (\mathcal{G}_{\xi',s} - \psi_s(\xi'))A_{\xi,\xi'}\pi(\xi')ds \\ &\leq C_o \int_0^t \psi_s(\xi) \sum_{\xi\in\mathsf{X}\times\mathsf{A}} |\psi_s(\xi)|ds,\end{aligned}$$

for some finite constant $C_o < \infty$. The next step is to sum over all possible $\xi \in \mathsf{X}\times\mathsf{A}$. Doing so and using the finiteness of the state space gives

$$\begin{aligned}\sum_{\xi\in\mathsf{X}\times\mathsf{A}} |\psi_t(\xi)|^2 &\leq C_o \int_0^t \Big|\sum_{\xi\in\mathsf{X}\times\mathsf{A}} |\psi_s(\xi)|\Big|^2 ds \\ &\leq C_o \int_0^t \sum_{\xi\in\mathsf{X}\times\mathsf{A}} |\psi_s(\xi)|^2 ds.\end{aligned}$$

The derivation of uniqueness is concluded by applying the Gronwall lemma (see Section B.1) which then yields that $\psi_t(\xi) = 0$ for all $0 \leq t \leq T$ and for all $\xi \in \mathsf{X} \times \mathsf{A}$. Therefore, the solution h_t is indeed unique.

21.5.2. Proof of Theorem 21.7. In order to prove Theorem 21.7 we first need a preliminary lemma.

Lemma 21.10. *Set* $\xi = (x, a) \in \mathsf{X} \times \mathsf{A}$. *Let Assumption* 21.5 *hold. Then, the matrix A with elements*

$$A_{\xi,\xi'} = \left\langle \sigma(w \cdot \xi)\sigma(w \cdot \xi') + c^2\sigma'(w \cdot \xi)\sigma'(w \cdot \xi')\xi \cdot \xi', \mu_0 \right\rangle,$$

is positive definite, as long as ξ, ξ' are in district directions.[1]

Proof of Lemma 21.10. Let us first define $\Sigma(\xi) = \sigma(w \cdot \xi) + c\sigma'(w \cdot \xi)$ and set the vector $\Sigma = (\Sigma(\xi_1, \ldots, \xi_M))$ where $M = |\mathsf{X} \times \mathsf{A}|$. Given that we have assumed that C is mean zero and independent of W, we obtain that

$$\begin{aligned} A_{\xi,\xi'} &= \mathbb{E}_{\mu_0}\left[\sigma(w \cdot \xi)\sigma(w \cdot \xi') + c^2\sigma'(w \cdot \xi)\sigma'(w \cdot \xi')\xi \cdot \xi'\right] \\ &= \mathbb{E}_{\mu_0}\left[\Sigma(\xi)\Sigma(\xi')\right]. \end{aligned}$$

The next step is to show that for all $z \in \mathbb{R}^M$ we have that $z^\top Az > 0$. Indeed, notice that

$$\begin{aligned} z^\top Az = z^\top \mathbb{E}_{\mu_0}\left[\Sigma^\top\Sigma\right] z &= \mathbb{E}_{\mu_0}\left[(\Sigma z)^\top(\Sigma z)\right] \\ &= \mathbb{E}_{\mu_0}\left[\left(\sum_{m=1}^{M} (z_m\sigma(w \cdot \xi_m) + c\sigma'(w \cdot \xi_m)\xi_m)\right)^2\right]. \end{aligned}$$

Due to the fact that ξ_m are assumed to be in district directions, by [**Ito96**] we have that $\sigma(w \cdot \xi_m)$ are linearly independent. This means that for non-zero z, there exists some $\hat{w}$ such that $\sum_{m=1}^{M} z_m\sigma(\hat{w} \cdot \xi_m) \neq 0$. But this means that there exists $\epsilon > 0$ so that $\left(\sum_{m=1}^{M} z_m\sigma(\hat{w} \cdot \xi_m)\right)^2 > \epsilon > 0$. By continuity now, there exists a set $\Gamma = \{(c, w) : \|c\| + \|w - \hat{w}\| < \eta\}$ for some $\eta > 0$ so that for $(c, w) \in \Gamma$

$$\left(\sum_{m=1}^{M} z_m\sigma(\hat{w} \cdot \xi_m) + c\sigma'(w \cdot \xi_m)\xi_m\right)^2 > \frac{\epsilon}{2} > 0.$$

[1] Recall Definition 19.5. For a given vector ξ define the line $L_\xi = \{y \in \mathbb{R}^d : y = t\xi, t \in \mathbb{R}\}$. Two vectors ξ and ξ' are said to be in distinct directions if they are nonzero and the lines $L_\xi, L_{\xi'}$ meet only at the origin; see [**Ito96**].

Therefore, we can conclude that

$$
\begin{aligned}
&\mathbb{E}_{\mu_0}\left[\left(\sum_{m=1}^{M}(z_m\sigma(w\cdot\xi_m)+c\sigma'(w\cdot\xi_m)\xi_m)\right)^2\right] \\
&\geq \mathbb{E}_{\mu_0}\left[\left(\sum_{m=1}^{M}(z_m\sigma(w\cdot\xi_m)+c\sigma'(w\cdot\xi_m)\xi_m)\right)^2\mathbf{1}_{(c,w)\in\Gamma}\right] \\
&\geq \frac{\epsilon}{2}\mathbb{P}\left[(c,w)\in\Gamma\right] \\
&> 0.
\end{aligned}
$$

Hence, we indeed have that for all $z \in \mathbb{R}^M$, $z^\top Az > 0$, which proves that the matrix A is positive definite. □

Let us now proceed with the proof of Theorem 21.7. We want to prove convergence of h_t to V when $t \to \infty$ at least when β is small. We start by defining

$$
\mathcal{H}_{\xi,t} = \sum_{x'',x\in\mathsf{X}}\left[\max_{a''\in\mathsf{A}} h_t(x'',a'') - \max_{a''\in\mathsf{A}} V(x'',a'')\right]\mathbb{P}[x''|\xi].
$$

If we set now $\psi_t(x,a) = h_t(x,a) - V(x,a)$, we find that

$$
\frac{d\psi_t}{dt} = -A\left(\pi\odot(\psi_t - \beta\mathcal{H}_t)\right).
$$

Next, let us define $\Delta_t = \beta\psi_t\cdot(\pi\odot\mathcal{H}_t)$. Then, we have the bound

$$
\begin{aligned}
|\Delta_t| &\leq \beta\sum_{\xi\in\mathsf{X}\times\mathsf{A}}\|\pi(\xi)\psi_t(\xi)\mathcal{H}_{\xi,t}\| \\
&\leq \frac{\beta}{2}\sum_{\xi\in\mathsf{X}\times\mathsf{A}}\pi(\xi)|\psi_t(\xi)|^2 + \frac{\beta}{2}\sum_{\xi\in\mathsf{X}\times\mathsf{A}}\pi(\xi)|\mathcal{H}_{\xi,t}|^2 \\
&\leq \frac{\beta}{2}\sum_{\xi\in\mathsf{X}\times\mathsf{A}}\pi(\xi)|\psi_t(\xi)|^2 + \frac{\beta}{2}\sum_{\xi\in\mathsf{X}\times\mathsf{A}}\pi(\xi)\sum_{(x'',a'')\in\mathsf{X}\times\mathsf{A}}|\psi_t(x'',a'')|^2\mathbb{P}(x''|\xi) \\
&= \frac{\beta}{2}\pi\cdot\psi_t\odot\psi_t + \frac{\beta}{2}|\mathsf{A}|\pi\cdot\psi_t\odot\psi_t \\
&= \frac{|\mathsf{A}|+1}{2}\beta\pi\cdot\psi_t\odot\psi_t.
\end{aligned}
$$

By Lemma 21.10 we have that the matrix $A > 0$ is positive definite. This means that A^{-1} exists and that A^{-1} is also positive definite. Letting now $Z_t = \frac{1}{2}\psi_t \cdot A^{-1}\psi_t$, we obtain

$$\begin{aligned}\frac{dZ_t}{dt} &= -\psi_t \cdot A^{-1}A(\pi \odot (\psi_t - \beta\mathcal{H}_t)) \\ &= -\pi\psi_t \odot \psi_t + \beta\psi_t \cdot (\pi \odot \mathcal{H}_t)) \\ &\leq -\pi\psi_t \odot \psi_t + |\Delta_t| \\ &\leq -\pi\psi_t \odot \psi_t + \frac{|\mathsf{A}| + 1}{2}\beta\pi \cdot \psi_t \odot \psi_t \\ &= \left(-1 + \frac{|\mathsf{A}| + 1}{2}\beta\right)\pi\psi_t \odot \psi_t.\end{aligned}$$

Thus, if $\beta < \frac{2}{1+|\mathsf{A}|}$, we indeed then obtain that there is some $\delta^* > 0$ so that

$$\frac{dZ_t}{dt} \leq -\delta^*\pi\psi_t \odot \psi_t.$$

At the same time, we also have that there is some $C_o < \infty$ such that

$$Z_t \leq C_o\psi_t \odot \psi_t,$$

showing that $\psi_t \odot \psi_t \geq \frac{Z_t}{C_o}$. Thus we then obtain for some unimportant constant $0 < C_o' < \infty$ that

$$\begin{aligned}\frac{dZ_t}{dt} &\leq -\delta^*\pi\frac{Z_t}{C_o} \\ &\leq -C_o'Z_t,\end{aligned}$$

which yields

$$Z_t \leq Z_0e^{-C_o't}$$

proving that

$$\lim_{t\to\infty} Z_t = 0.$$

This also concludes the proof of the theorem.

21.6. Brief Concluding Remarks

Many excellent texts have been devoted to reinforcement learning itself, see for example [**SB18**, **GK20**], and the survey by [**ADBB17**]. The books [**BPM90**, **KY03**] are classical resources on stochastic approximation and recursive algorithms.

Most of the known reinforcement learning algorithms are based on some variation of the Q-learning or policy gradient methods [**SMSM00**]. The idea of Q-learning finds its origins in [**Wat89**]. Later on, proofs of convergence were developed in [**WD92, Tsi94**]. Neural networks, as function approximators, in reinforcement learning (i.e., using Q-networks) were introduced later in [**MKS$^+$15**]. Since then the field has exploded, and recent developments include deep recurrent Q-networks [**HS15**], dueling architectures for deep reinforcement learning [**WSH$^+$16**], double Q-learning [**HGS16**], bootstrapped deep Q-networks [**OBPR16**], and asynchronous methods for deep reinforcement learning [**MBM$^+$16**] to name just a few.

The convergence proof of Q-learning that we presented in this chapter in Section 21.5 is based on the article [**SS22**].

21.7. Exercises

Exercise 21.1. In the context of the SGD algorithm (21.26) (set the bias $b = 0$ for convenience),

$$\begin{aligned}
C^n_{k+1} &= C^n_k + \frac{\eta^N_k}{N^\gamma}\left(r(x_k, a_k) + \beta \max_{a' \in \mathsf{A}} Q(x_{k+1}, a', \theta_k) - Q(x_k, a_k, \theta_k)\right) \\
&\quad \times \sigma(W^n_k \cdot (x_k, a_k)) \\
W^n_{k+1} &= W^n_k + \frac{\eta^N_k}{N^\gamma}\left(r(x_k, a_k) + \beta \max_{a' \in \mathsf{A}} Q(x_{k+1}, a', \theta_k) - Q(x_k, a_k, \theta_k)\right) \\
&\quad \times C^n_k \sigma'(W^n_k \cdot (x_k, a_k))(x_k, a_k),
\end{aligned}$$

and after choosing the learning rate to be $\eta^N_k = \frac{\eta}{N}$ for some $0 < \eta < N$ constant, prove that

$$\max_{n \in \mathbb{N}} \max_{k \leq \lfloor NT \rfloor} \left(|C^n_k| + \mathbb{E}\|W^n_k\|\right) \leq C_o < \infty,$$

for some constant $C_o < \infty$.

Exercise 21.2. In the context of the SGD algorithm (21.26) (set the bias $b = 0$ for convenience),

$$\begin{aligned}
C^n_{k+1} &= C^n_k + \frac{\eta^N_k}{N^\gamma}\left(r(x_k, a_k) + \beta \max_{a' \in \mathsf{A}} Q(x_{k+1}, a', \theta_k) - Q(x_k, a_k, \theta_k)\right) \\
&\quad \times \sigma(W^n_k \cdot (x_k, a_k)) \\
W^n_{k+1} &= W^n_k + \frac{\eta^N_k}{N^\gamma}\left(r(x_k, a_k) + \beta \max_{a' \in \mathsf{A}} Q(x_{k+1}, a', \theta_k) - Q(x_k, a_k, \theta_k)\right) \\
&\quad \times C^n_k \sigma'(W^n_k \cdot (x_k, a_k))(x_k, a_k),
\end{aligned}$$

and after choosing the learning rate to be $\eta^N_k = \frac{\eta}{N}$ for some $0 < \eta < N$ constant, prove that

$$\max_{N \in \mathbb{N}} \max_{k \leq \lfloor NT \rfloor} \mathbb{E} \sup_{(x,a) \in \mathsf{X} \times \mathsf{A}} \left|Q^N_k(x, a)\right|^2 \leq C_o < \infty,$$

for some constant $C_o < \infty$. Recall that $Q^N_k(x, a) = Q^N(x, a; \theta_k)$ for $k \leq \lfloor Nt \rfloor$.

Exercise 21.3. In the context of the finite time horizon problem (21.22):

(1) Write down the equivalent to the infinite time horizon problem SGD algorithm (21.26).

(2) Write down the expansion for Q^N_{k+1} that is the equivalent to the corresponding expansion for the infinite time horizon problem (21.29).

(3) Define $h^N_t = Q^N_{\lfloor Nt \rfloor}$. What is the prelimit expression for h^N_t that is equivalent to the one in the infinite time horizon problem (21.30)?

Exercise 21.4. Consider the regression problem where the objective function is (21.23),

$$\Lambda(\theta) = \frac{1}{2} \sum_{(x,a)\in \mathsf{X}\times\mathsf{A}} [Y(x,a) - Q(x,a,\theta)]^2\, \pi(x,a),$$

but with y_k now being independent samples from a fixed distribution. Then, (21.23) is simply the mean-squared error objective function for regression. In particular, the corresponding (population) loss function is

$$\Lambda_{\text{pop}}(\theta) = \frac{1}{2}\mathbb{E}\left[\left(Y - \mathfrak{m}^N(X;\theta)\right)^2\right],$$

where the data $(X,Y) \sim \pi(dx,dy)$, the model $Y \in \mathbb{R}$,

$$\mathfrak{m}^N(x;\theta) = \frac{1}{\sqrt{N}} \sum_{n=1}^{N} C^n \sigma(W^n \cdot x),$$

and the parameters $\theta = (C^1,\ldots,C^N,W^1,\ldots,W^N) \in \mathbb{R}^{N\times(1+d)}$. This is the setup studied in the optimization in the feature learning regime; see Chapter 19. Show that this is a special case of the reinforcement learning setup studied in this chapter.

Exercise 21.5. The *tabular* Q-learning algorithm directly estimates a value for every state-action pair (x,a) via the learning algorithm:

$$Q_{t+1}(X_t,A_t) = Q_t(X_t,A_t) + \eta_t\Big(r(X_t,A_t) + \beta \max_{a'} Q_t(X_{t+1},a') - Q_t(X_t,A_t)\Big),$$

where $Q_t(x,a)$ is the estimate of the value for action a in state x at time step t, $r(x,a)$ is the reward for action a in state x, $A_t \in \mathsf{A}$ is the action at time step t, and $X_t \in \mathsf{X}$ is a sample from the Markov chain.

What are the advantages of the deep Q-learning algorithm in comparison to the tabular Q-learning algorithm?

Exercise 21.6. Suppose we select actions A_t uniformly at random. Let $\pi(x,x',a)$ be the stationary distribution of the Markov chain (X_t, X_{t+1}, A_t) and furthermore suppose we can directly generate i.i.d. samples (x_i, x'_i, a_i) from

$\pi(x, x', a)$. A tabular Q-learning algorithm could then be constructed using these i.i.d. samples:

(21.35)

$$Q_{i+1}(x_i, a_i) = Q_i(x_i, a_i) + \eta_i\Big(r(x_i, a_i) + \beta \max_{a'} Q_i(x_i', a') - Q_t(x_i, a_i)\Big).$$

(1) Rewrite (21.35) as the stochastic gradient descent algorithm for an appropriate objective function $\Lambda(Q)$, where the parameters that are being optimized over are $Q = \{Q_{x,a}\}_{x,a \in \mathsf{X} \times \mathsf{A}}$.

(2) Suppose that $\Lambda(Q) = 0$ and $\pi(x, a) > 0 \;\forall\; x, a \in \mathsf{X} \times \mathsf{A}$. Prove that Q is a solution to the Bellman equation and $a(x) = \arg\max_a Q(x, a)$ is the optimal policy.

Exercise 21.7. Consider a linear model $Q(x, a; \theta) = \theta \cdot (x, a)$ for the value of selecting action a in state x. What would be the *linear* Q-learning algorithm?

Exercise 21.8. Consider the deep Q-learning algorithm, and suppose that we use the policy $A_t = \arg\max_a Q(X_t, a; \theta_t)$ to select actions. Construct a simple example to show that the model may not train (i.e., it may not converge to a suboptimal action). What would be a better policy for selecting actions?

Chapter 22

Neural Differential Equations

22.1. Introduction

A common problem in applied mathematics and more generally in science is that we are given a target profile and we want to built a dynamical system whose behavior matches that given profile. This chapter introduces the idea of accomplishing this goal by using neural networks to model the underlying dynamical system. Due to the universal approximation properties of neural networks (see Chapter 16) and the power of stochastic gradient descent based methods (see Chapters 7, 8, and 18) such methods have proven to be very efficient in practice and are typically called neural differential equation methods. In this chapter we focus on ordinary and stochastic differential equations but such ideas have been applied to partial differential equations as well (see the brief concluding remarks in Section 22.7 for some related literature).

22.2. Ordinary Differential Equations with Neural Network Dynamics

Consider the ordinary differential equation (ODE) whose dynamics is given by a neural network $\mathfrak{m}(u;\theta)$ with parameters $\theta \in \Theta$,

$$\frac{du(t)}{dt} = \mathfrak{m}(u(t);\theta), \tag{22.1}$$

with the initial condition $u(0) = u_0$. The ODE solution is d-dimensional, i.e., $u(t) = (u_1(t), u_2(t), \dots, u_d(t))$. Let $h(t)$ be the target function that we would like to predict. A reasonable formulation is to define the objective function

$$\Lambda(\theta) = \frac{1}{2}\int_0^T |h(t) - u(t)|^2 dt + \frac{1}{2}|h(T) - u(T)|^2,$$

and the goal is to select θ such that the solution of the neural ODE (22.1) minimizes $\Lambda(\theta)$. Minimizing the objective function $\Lambda(\theta)$ requires optimizing over the ODE (22.1).

As we shall see, a direct minimization of $\Lambda(\theta)$ leads to solving a system of equations which is as big as the dimension of the parameter space $\theta \in \Theta$ (which can become pretty large for deep learning models). To go around this issue, we first demonstrate how to derive an adjoint ODE for (22.1) which allows for a computationally efficient evaluation of the gradient $\Lambda(\theta)$. Once we can evaluate $\nabla_\theta \Lambda(\theta)$, it is easy to optimize over $\Lambda(\theta)$ via a (stochastic) gradient descent type of method.

Define the gradient of the ODE solution with respect to the parameters θ as

$$\tilde{u}(t) = \nabla_\theta u(t).$$

Then, $\tilde{u}$ satisfies the ODE

$$\frac{d\tilde{u}(t)}{dt} = \sum_{i=1}^{d} \frac{\partial \mathfrak{m}}{\partial u_i}(u(t);\theta)\tilde{u}_i(t) + \nabla_\theta \mathfrak{m}(u(t);\theta). \tag{22.2}$$

Similarly,

$$\nabla_\theta \Lambda(\theta) = \int_0^T \tilde{u}(t)^\top \Big(h(t) - u(t)\Big) dt + \tilde{u}(T)^\top \Big(h(T) - u(T)\Big). \tag{22.3}$$

In principle, we could solve (22.2) and then evaluate $\nabla_\theta \Lambda(\theta)$ via the formula (22.3). However, the parameters θ are very high dimensional in deep learning (e.g., 10^6 or 10^7 parameters). The dimension of the ODE (22.2) is the same as the parameter dimension, which makes (22.2) computationally challenging to solve in practice. For example, if the ODE is d-dimensional and the number of parameters is $P = \dim(\Theta)$, the ODE system (22.2) has $d \times P$ ODEs. Instead, we will derive the *adjoint* ODE for (22.1) which will allow for the computationally efficient evaluation of the gradient $\nabla_\theta \Lambda(\theta)$.

Let $\hat{u}$ satisfy the ODE,

$$-\frac{d\hat{u}(t)}{dt} = \frac{\partial \mathfrak{m}}{\partial u}(u(t);\theta)^\top \hat{u}(t) + \Big(h(t) - u(t)\Big), \tag{22.4}$$

with final condition $\hat{u}(T) = h(T) - u(T)$. Then, multiply (22.2) by $\hat{u}$, which yields

$$\hat{u}(t)\frac{d\tilde{u}(t)}{dt} = \hat{u}(t)\sum_{i=1}^{d}\frac{\partial \mathfrak{m}}{\partial u_i}(u(t);\theta)\tilde{u}_i(t) + \hat{u}(t)\nabla_\theta \mathfrak{m}(u(t);\theta).$$

Integrating over the time interval $[0, T]$ produces

$$\int_0^T \hat{u}(t)\frac{d\tilde{u}(t)}{dt}dt = \int_0^T \Big(\hat{u}(t)\sum_{i=1}^{d}\frac{\partial \mathfrak{m}}{\partial u_i}(u(t);\theta)\tilde{u}_i + \hat{u}(t)\nabla_\theta \mathfrak{m}(u(t);\theta)\Big)dt.$$

Integration by parts yields

$$-\int_0^T \frac{d\hat{u}(t)}{dt}\tilde{u}(t)dt + \Big(\hat{u}(T)\tilde{u}(T) - \hat{u}(0)\tilde{u}(0)\Big)$$
$$= \int_0^T \Big(\hat{u}(t)\sum_{i=1}^{d}\frac{\partial \mathfrak{m}}{\partial u_i}(u(t);\theta)\tilde{u}_i(t) + \hat{u}(t)\nabla_\theta \mathfrak{m}(u(t);\theta)\Big)dt.$$

Since $u(0)$ is a constant, we have that $\tilde{u}(0) = 0$. Collecting terms yields

$$\int_0^T \tilde{u}^\top(t)\Big(-\frac{d\hat{u}}{dt} - \frac{\partial \mathfrak{m}}{\partial u}(u(t);\theta)^\top \hat{u}(t)\Big)dt + \tilde{u}(T)^\top \hat{u}(T)$$
$$= \int_0^T \nabla_\theta \mathfrak{m}(u(t);\theta)^\top \hat{u}(t)dt.$$

Substituting (22.4) into the above equation yields

$$\int_0^T \tilde{u}^\top(t)\Big(h(t) - u(t)\Big)dt + \tilde{u}(T)\Big(h(T) - u(T)\Big) = \int_0^T \nabla_\theta g(u(t);\theta)^\top \hat{u}(t)dt.$$

Therefore, using equation (22.3), we have a formula for the gradient of the objective function

$$\nabla_\theta \Lambda(\theta) = \int_0^T \nabla_\theta \mathfrak{m}(u(t);\theta)^\top \hat{u}(t)dt. \tag{22.5}$$

A key feature is that the dimension of the adjoint ODE (22.4) is d no matter how large the dimension $P = \dim(\Theta)$ of the parameters $\theta \in \Theta$ is. This stands in contrast to the forward ODE (22.2), whose dimension is proportional to the dimension of the parameters. Therefore, the adjoint method provides a highly computationally efficient method for optimizing over neural ODEs with high-dimensional parameters.

The gradient descent algorithm for optimizing the neural ODE model therefore is as follows:

- For $k = 0, 1, 2, \dots,$
  - Solve the forward ODE on $[0, T]$,
  $$\frac{du(t)}{dt} = \mathfrak{m}(u(t); \theta_k),$$
  with the initial condition $u(0) = u_0$.
  - Solve the adjoint ODE on $[0, T]$,
  $$-\frac{d\hat{u}(t)}{dt} = \frac{\partial \mathfrak{m}}{\partial u}(u(t); \theta_k)^\top \hat{u}(t) + \Big(h(t) - u(t)\Big),$$
  with final condition $\hat{u}(T) = h(T) - u(T)$.
  - Calculate the gradient,
  $$\nabla_\theta \Lambda(\theta_k) = \int_0^T \nabla_\theta \mathfrak{m}(u(t); \theta_k)^\top \hat{u}(t) dt.$$
  - Update the parameters with a gradient descent step,
  $$\theta_{k+1} = \theta_k - \eta_k \nabla_\theta \Lambda(\theta_k),$$
  where η_k is the learning rate.

22.3. Backpropagation Formula from the Euler Discretization

There is an alternative derivation of the adjoint ODEs for training neural ODEs which directly connects to the backpropagation algorithm for neural networks that we visited in Chapters 6 and 7. First, we discretize the neural ODE using an Euler discretization, which yields the discrete equations

$$u_{i+1} = u_i + \mathfrak{m}(u_i; \theta)\Delta, \tag{22.6}$$

where u_i is an approximation for the solution $u(i\Delta)$ to the ODE (22.1), Δ is the time step size, and $u_0 = u(0)$. As $\Delta \to 0$, the Euler approximation (22.6) will converge to the solution of the ODE (22.1).

The objective function becomes the sum

$$\Lambda(\theta) = \sum_{i=1}^{N} J_i + \frac{1}{2}|h(N\Delta) - u_N|^2,$$

where $N\Delta = T$ and $J_i = \frac{1}{2}|h(i\Delta) - u_i|^2\Delta$. The (discretized) neural ODE model can be trained with gradient descent

$$\theta_{k+1} = \theta_k - \eta_k \nabla_\theta \Lambda(\theta_k).$$

We will now derive the gradient of the objective function using the chain rule (i.e., the backpropagation algorithm!). Define

$$\hat{u}_i = \frac{\partial \Lambda}{\partial u_i}.$$

Then, for the final time N,

$$\hat{u}_N = \big(h(N\Delta) - u_N\big)\Delta + \big(h(N\Delta) - u_N\big).$$

Using the chain rule, we can derive for $i < N$,

$$\begin{aligned}\hat{u}_i &= \frac{\partial u_{i+1}}{\partial u_i}^{\top} \frac{\partial \Lambda}{\partial u_{i+1}} + \frac{\partial J_i}{\partial u_i} \\ &= \frac{\partial u_{i+1}}{\partial u_i}^{\top} \hat{u}_{i+1} + \big(h(i\Delta) - u_i\big)\Delta \\ &= \hat{u}_{i+1} + \frac{\partial \mathfrak{m}}{\partial u}(u_i;\theta)^{\top} \hat{u}_{i+1}\Delta + \big(h(i\Delta) - u_i\big)\Delta. \qquad (22.7)\end{aligned}$$

The gradients of the loss with respect to the discretized ODE solution u_i can therefore be solved—similarly to the backward step of the backpropagation algorithm—by calculating the update formula (22.7) backwards in time from $i = N \to 0$. Equation (22.7) can therefore be viewed as a backpropagation algorithm for the discrete ODE (22.6).

We can also derive a formula for the gradient of the objective function with respect to the model parameters,

$$\nabla_\theta \Lambda(\theta) = \sum_{i=0}^{N-1} \nabla_\theta \mathfrak{m}(u_i;\theta)^{\top} \hat{u}_{i+1}\Delta. \qquad (22.8)$$

Equations (22.7) and (22.8) are often referred to as *discrete adjoint equations*. These equations can be used to optimize over the discretized neural ODE (22.6). However, it is interesting to investigate the connection between the discrete adjoint equations and the continuous adjoint equations derived in the previous section.

Rearranging (22.7) yields

$$-\frac{\hat{u}_{i+1} - \hat{u}_i}{\Delta} = \frac{\partial \mathfrak{m}}{\partial u}(u_i;\theta)^{\top} \hat{u}_{i+1} + \big(h(i\Delta) - u_i\big),$$

with the final condition $\hat{u}_N = \big(h(N\Delta) - u_N\big)\Delta + \big(h(N\Delta) - u_N\big)$. Recall that $N = \frac{T}{\Delta}$ and let $\Delta \to 0$. Then, we can clearly see that the discrete adjoint equations converge to the continuous adjoint equations in the previous section.

22.4. Training Neural ODEs with Minibatch Datasets

The neural ODE framework can be easily extended to training on multiple data samples. For example, consider again the ODE model with neural network dynamics

$$\frac{du(t)}{dt} = \mathfrak{m}(u(t), x; \theta) \tag{22.9}$$

with initial condition $u(t = 0) = u_0(x)$, where $x \in \mathbb{R}^d$ is a data feature. The (unknown) target solution is a function $y(t)$. However, a dataset $(x^{(i)}, y^{(i)})_{i=1}^N$ of observations for the input data features x and target solution y is available. This dataset can be used to train the neural network parameters θ using a generalization of the adjoint formula described previously.

Define $u(t; \theta, x)$ as the solution of (22.9) with parameters θ and data feature x. The objective function becomes

$$\Lambda(\theta) = \frac{1}{N}\sum_{i=1}^{N}\left(\frac{1}{2}\int_0^T |y^{(i)}(t) - u(t;\theta, x^{(i)})|^2 dt + \frac{1}{2}|y^{(i)}(T) - u(T;\theta, x^{(i)})|^2\right).$$

Define the adjoint ODEs for the data samples $i = 1, 2, \ldots, N$ as

$$-\frac{d\hat{u}^{(i)}(t)}{dt} = \frac{\partial \mathfrak{m}}{\partial u}(u^{(i)}(t);\theta)^\top \hat{u}^{(i)}(t) + \left(y^{(i)}(t) - u^{(i)}(t)\right) \tag{22.10}$$

with final condition $\hat{u}^{(i)}(T) = y^{(i)}(T) - u^{(i)}(T)$ and where $u^{(i)}$ is the solution to (22.1) with data feature $x = x^{(i)}$.

In a completely analogous way to how we derived (22.5), we obtain that the gradient of the objective function can be evaluated via

$$\nabla_\theta \Lambda(\theta) = \frac{1}{N}\sum_{i=1}^{N}\int_0^T \nabla_\theta \mathfrak{m}(u^{(i)}(t);\theta)^\top \hat{u}^{(i)}(t)dt.$$

Each gradient descent step therefore requires the solution of N adjoint PDEs (22.10), where N is the number of data samples. This can be computationally expensive for large datasets. Training can be accelerated using stochastic gradient descent where at each iteration a minibatch of data samples is randomly selected and the corresponding adjoint PDEs are calculated. The stochastic gradient descent algorithm is outlined below.

- For $k = 0, 1, 2, \ldots$:
  - Select uniformly at random $(x^{(j)}, y^{(j)})_{j=1}^M$ from the dataset $(x^{(i)}, y^{(i)})_{i=1}^N$ where $M < N$.
  - Solve the forward ODEs on $[0, T]$ for $j = 1, \ldots, M$,

    $$\frac{du^{(j)}(t)}{dt} = \mathfrak{m}(u^{(j)}(t);\theta_k, x^{(j)}),$$

    with the initial conditions $u(0) = u_0(x^{(j)})$.

- Solve the adjoint ODEs on $[0, T]$ for $j = 1, \dots, M$,

$$-\frac{d\hat{u}^{(j)}(t)}{dt} = \frac{\partial \mathfrak{m}}{\partial u}(u^{(j)}(t); \theta_k, x^{(j)})^\top \hat{u}^{(j)}(t) + \left(y^{(j)}(t) - u^{(j)}(t)\right),$$

  with final conditions $\hat{u}^{(j)}(T) = y^{(j)}(T) - u^{(j)}(T)$.
- Calculate an unbiased estimate for the gradient $\nabla_\theta \Lambda(\theta_k)$,

$$G_k = \frac{1}{M}\sum_{j=1}^{M}\int_0^T \nabla_\theta \mathfrak{m}(u^{(j)}(t); \theta_k, x^{(j)})^\top \hat{u}^{(j)}(t)dt,$$

  where $\mathbb{E}[G_k|\theta_k] = \nabla_\theta \Lambda(\theta_k)$.
- Update the parameters with a minibatch stochastic gradient descent step,

$$\theta_{k+1} = \theta_k - \eta_k G_k,$$

  where $\eta_k > 0$ is the learning rate.

It is important to recognize that the j for loop in the above algorithm can be easily vectorized. If computed sequentially, the for loop $j = 1, \dots, M$ has a computational cost of $\mathcal{O}(M)$ for both the forward solution of the neural ODE and the evaluation its adjoint equations. However, using parallelization (on either GPUs or CPUs), the minibatch can be perfectly parallelized with a computational cost of $\mathcal{O}(1)$ no matter how large the minibatch is.

22.5. Neural Stochastic Differential Equations

Neural networks can also be used to model the dynamics of SDEs. Neural SDEs can use a neural network to model to the drift and diffusion (sometimes also referred to as volatility) coefficient in the SDE,

$$dX_t^\theta = \mu(X_t^\theta; \theta)dt + \sigma(X_t^\theta; \theta)dW_t, \tag{22.11}$$

where W_t is a standard Brownian motion (see Appendix A for an introductory discussion to Brownian motion and SDEs). The drift and diffusion functions $\mu(x; \theta)$ and $\sigma(x; \theta)$ are neural networks with parameters θ. The stochastic process X_t^θ depends upon the parameters θ, which is indicated with the corresponding superscript.

The objective function for training the parameters θ will depend upon the specific application. Typically, optimization will require discretizing equation (22.11) on a time grid $t_i = i\Delta$,

$$X_{i+1}^\theta = X_i^\theta + \mu(X_i^\theta; \theta)\Delta + \sigma(X_i^\theta; \theta)\sqrt{\Delta}W_i, \tag{22.12}$$

where W_i are standard normal random variables and X_i^θ is an approximation to $X_{i\Delta}^\theta$. Under appropriate technical conditions, as $\Delta \to 0$, the Euler approximation (22.12) will converge to the solution of the SDE (22.11), see for example [**KP92**].

Suppose we are trying to calibrate θ such that the process X_t^θ matches a target process Y_t as closely as possible. Furthermore, the path of Y_t is completely observed at all points in time for the time interval $[0, T]$. Then, we can calibrate θ by maximizing the log-likelihood objective function for (22.12). The objective function is

$$\Lambda(\theta) = \frac{1}{N}\sum_{i=1}^{N} \log\left[\phi\Big(Y_i - X_i^\theta, X_{i-1}^\theta + \mu(X_{i-1}^\theta;\theta)\Delta, \sigma(X_{i-1}^\theta;\theta)\sqrt{\Delta}\Big)\right], \tag{22.13}$$

where $Y_i = Y_{i\Delta}$ and $\phi(z, \mu_0, \sigma_0)$ is the probability of z for a Gaussian distribution with mean μ_0 and standard deviation σ_0. The number of datapoints N is selected to be $N = \lfloor \frac{T}{\Delta} \rfloor$. Equation (22.13) can be directly maximized using the backpropagation algorithm and gradient descent.

Alternatively, in other applications we may wish to select the parameters such that the expectation of a function of X_t^θ, say for example $g(x)$, matches the target data. Specifically, the objective function might be

$$\Lambda(\theta) = \frac{1}{2N}\sum_{i=1}^{N}\Big(\mathbb{E}[g^{(i)}(X_T^\theta)] - Y^{(i)}\Big)^2. \tag{22.14}$$

We wish to select the neural network parameters θ such that $\mathbb{E}\big[g^{(i)}(X_T^\theta)\big]$ is as close as possible to the target data $Y^{(i)}$ for $i = 1, 2, \ldots, N$. This objective function would for example be used in option pricing in finance where the functions $g^{(i)}$ are different payoff functions (e.g., depending upon the strike price) and $Y^{(i)}$ are the observed option prices in the market.

Since an expectation appears in (22.14), minimizing the objective function (22.14) requires calculating the gradient of the distribution of the process X_t^θ with respect to θ. The gradient of $\Lambda(\theta)$ with respect to the parameters θ is

$$\nabla_\theta \Lambda(\theta) = \frac{1}{N}\sum_{i=1}^{N}\Big(\mathbb{E}[g^{(i)}(X_T^\theta)] - Y^{(i)}\Big)\nabla_\theta \mathbb{E}\big[g^{(i)}(X_T^\theta)\big].$$

Calculating $\nabla_\theta \mathbb{E}\big[g^{(i)}(X_T^\theta)\big]$ is typically not computationally tractable. Standard automatic differentiation cannot be directly applied to (22.14) to calculate the gradient due to the expectation being inside the squared error. A naive implementation of stochastic gradient descent would be

- Simulate a Monte Carlo path $X_t^{\theta_k,\ell}$ of $X_t^{\theta_k}$.

- Calculate $\nabla_\theta X_T^{\theta_k,\ell}$ via the backpropagation algorithm (or automatic differentiation).
- Update the parameters via

$$\theta_{k+1} = \theta_k - \eta_k G_k,$$

$$G_k = \frac{1}{N}\sum_{i=1}^{N}\Big(g^{(i)}(X_T^{\theta_k,\ell}) - Y^{(i)}\Big)\frac{\partial g^{(i)}}{\partial x}(X_T^{\theta_k,\ell})\nabla_\theta X_T^{\theta_k,\ell}.$$

The above algorithm can be implemented very easily by applying automatic differentiation to (22.12). In particular, one could simply apply automatic differentiation to

$$\tilde{\Lambda}(\theta_k) = \frac{1}{N2}\sum_{i=1}^{N}\Big(g^{(i)}(X_T^{\theta_k,\ell}) - Y^{(i)}\Big)^2.$$

Although $\tilde{\Lambda}(\theta_k)$ is an unbiased estimate of the objective function $\Lambda(\theta_k)$, the gradient of $\tilde{\Lambda}(\theta_k)$ is not an unbiased estimate of the gradient of $\Lambda(\theta_k)$. Specifically, G_k is not an unbiased estimate of the gradient of the objective function (22.14):

$$\mathbb{E}\Big[G_k|\theta_k\Big] \neq \nabla_\theta\Lambda(\theta_k).$$

The reason that G_k is not an unbiased estimate of $\nabla_\theta\Lambda(\theta_k)$ is due to the term $(g^{(i)}(X_T^{\theta_k,\ell}) - Y^{(i)})$ *not being independent* of the term $\nabla_\theta g^{(i)}(X_T^{\theta_k,\ell}) = \frac{\partial g^{(i)}}{\partial x}(X_T^{\theta_k,\ell})\nabla_\theta X_T^{\theta_k,\ell}$. Due to this lack of independence,

$$\mathbb{E}\Big[\Big(g^{(i)}(X_T^{\theta_k,\ell}) - Y^{(i)}\Big)\nabla_\theta g^{(i)}(X_T^{\theta_k,\ell})\Big|\theta_k\Big]$$
$$\neq \mathbb{E}\Big[\Big(g^{(i)}(X_T^{\theta_k,\ell}) - Y^{(i)}\Big)\Big|\theta_k\Big]\mathbb{E}\Big[\nabla_\theta g^{(i)}(X_T^{\theta_k,\ell})\Big|\theta_k\Big].$$

However, an unbiased stochastic estimate can be computed via an alternative method which simulates two independent paths of the SDE X_t^θ:

- Simulate independent Monte Carlo paths $X_t^{\theta,\ell}$ and $\bar{X}_t^{\theta,\ell}$ for X_t^θ.
- Calculate $\nabla_\theta g^{(i)}(X_T^{\theta,\ell})$ via the backpropagation algorithm (or automatic differentiation).
- Update the parameters via a stochastic gradient descent step,

$$\theta_{k+1} = \theta_k - \eta_k V_k, \text{ where}$$

$$V_k = \frac{1}{N}\sum_{i=1}^{N}\Big(g^{(i)}(\bar{X}_T^{\theta_k,\ell}) - Y^{(i)}\Big)\frac{\partial g^{(i)}}{\partial x}(X_T^{\theta_k,\ell})\nabla_\theta X_T^{\theta_k,\ell}. \tag{22.15}$$

Due to the independence of the paths $X_t^{\theta,\ell}$ and $\bar{X}_t^{\theta,\ell}$, V_k is an unbiased estimate of the gradient of the objective function $\Lambda(\theta)$:

$$\begin{aligned}\mathbb{E}\Big[V_k\Big|\theta_k\Big] &= \frac{1}{N}\sum_{i=1}^{N}\Big(\mathbb{E}\Big[g^{(i)}(\bar{X}_T^{\theta_k,\ell})\Big|\theta_k\Big] - Y^{(i)}\Big) \times \mathbb{E}\Big[\nabla_\theta g^{(i)}(X_T^{\theta_k,\ell})\Big|\theta_k\Big] \\ &= \frac{1}{N}\sum_{i=1}^{N}\Big(\mathbb{E}\Big[g^{(i)}(\bar{X}_T^{\theta_k})\Big|\theta_k\Big] - Y^{(i)}\Big) \times \mathbb{E}\Big[\nabla_\theta g^{(i)}(X_T^{\theta_k})\Big|\theta_k\Big] \\ &= \nabla_\theta \Lambda(\theta_k).\end{aligned}$$

This algorithm, using two independent Monte Carlo paths, therefore provides a computationally efficient method for implementing stochastic gradient descent to optimize neural SDE models. At each optimization iteration, two Monte Carlo paths of the SDE are simulated and automatic differentiation is applied to the second path. This can be easily implemented in a deep learning library such as PyTorch.

22.6. Examples in PyTorch

22.6.1. Example 1: ODE. As our first example, we will train a neural ODE using automatic differentiation in PyTorch to match a target ODE,

$$\frac{du}{dt} = -u - u^2,$$

with initial condition $u(t = 0) = 1$. We will present the code in several blocks below.

First, we import the relevant Python and PyTorch modules and simulate the target ODE, which we will train the neural ODE to match. The ODE is numerically simulated using the Euler scheme.

```
import torch
import numpy as np
import torch.nn as nn
import torch.nn.functional as F
import torch.optim as optim
from torch.autograd import Variable

#Simulate the target ODE which we will train the neural ODE to match.

#Time step size
dt = 0.01

#Number of time steps
L = 100

u = torch.cuda.FloatTensor( np.ones(L+1) )

for i in range(L):
    u[i+1] = u[i] + dt*( -u[i] - u[i]**2 )
```

Then, we initialize a neural network for the ODE dynamics and specify several relevant hyperparameters. The neural network has two hidden layers with ReLU activation functions. The RMSProp optimization algorithm will be used to train the neural network parameters.

```
#Initial Learning rate
LR = 0.01

#Number of hidden units
H = 200

num_inputs = 1

num_outputs = 1

class NeuralNetworkModel(nn.Module):
    def __init__(self):
        super(NeuralNetworkModel, self).__init__()

        self.fc1 = nn.Linear(num_inputs, H).type(torch.FloatTensor)
        self.fc2 = nn.Linear(H, H).type(torch.FloatTensor)
        self.fc3 = nn.Linear(H, num_outputs).type(torch.FloatTensor)

    def forward(self, x):

        L1 =  self.fc1( x )

        H1 = torch.relu( L1 )

        L2 = self.fc2( H1 )

        H2 = torch.relu(L2)

        f_out = self.fc3(H2)

        return f_out

model = NeuralNetworkModel()
model.cuda()
optimizer = optim.RMSprop(model.parameters(), lr= LR, momentum=0.0)
```

Using automatic differentiation, we train the neural ODE model to match the target ODE. A piecewise constant learning rate schedule is used for the training. The loss function that will be minimized is

$$\Lambda(\theta) = \frac{1}{2}\int_0^1 \Big(u(t) - v(t)\Big)^2 dt,$$

where $u(t)$ is the target data and $v(t) = v(t;\theta)$ is the neural ODE model.

```
#Number of optimization iterations
K = 20000

v_np = np.ones(L+1)

for k in range(K):

    if (k < 1000):
        LR_k = 0.01
    elif ( (k>= 1000) & (k < 2000) ):
        LR_k = 0.005
    elif ( (k>= 2000) & (k < 3000) ):
        LR_k = 0.0025
    elif ( (k>= 3000) & (k < 4000) ):
        LR_k = 0.001
    elif ( (k>= 4000) & (k < 5000) ):
        LR_k = 0.0005
    else:
        LR_k = 0.0001

    for param_group in optimizer.param_groups:
        param_group['lr'] = LR_k

    v = torch.cuda.FloatTensor( np.ones(1) )
    Loss = torch.cuda.FloatTensor( np.zeros(1) )

    optimizer.zero_grad()

    for i in range(L):

        v = v + dt*model(v)

        Loss = Loss + dt*0.5*(v - u[i])**2

        v_np[i+1] = v.detach().cpu().numpy()

    Loss = Loss/float(L)

    Loss.backward()

    optimizer.step()

    if (k print(k, 'Current Loss', Loss.detach().cpu().numpy() )
```

In the end of training, the reported loss in one of our experiments was $2.93e - 05$.

22.6.2. Example 2: SDE. In our second example, we train a neural SDE model to minimize the objective function (22.14). As a simple demonstration of the methods from Section 22.5, we will use $N = 1$. The target datapoint will be produced using the price of a call option from the Black-Scholes model. In particular, we have the SDE

$$dS_t = rS_t dt + \sigma \bar{\mathfrak{m}}(S) dW_t,$$

where W_t is a standard Wiener process and r is the interest rate. In the classical Black-Scholes model $\bar{\mathfrak{m}}(S) = S$, the price of the call option with time horizon T and strike price K takes the form

$$C = e^{-rT} \max\{S - K, 0\}.$$

In the application below, data are assumed to be from the classical Black-Scholes model, i.e., when $\bar{\mathfrak{m}}(S) = S$. Then, the goal is to train a model with $\bar{\mathfrak{m}}(S)$ being a neural network $\mathfrak{m}(S;\theta)$ to match such data.

More specifically, the neural SDE will be trained such that it generates the same price for this call option. Although the method is demonstrated for a single datapoint, the code can be easily generalized to the case with $N > 1$ (e.g., market prices are observed for multiple options and the neural SDE is trained to match these market prices).

First, the training datapoint is generated from the Black-Scholes model.

```
#Generate training data

L = 10000000

S = torch.cuda.FloatTensor(np.ones( (L,1) ) )

dt = 0.01
N = 100
r = 0.05

K = 1.05

sigma = 0.5

for i in range(N):
    Z = torch.randn( (L,1), dtype = torch.float32 ).cuda(device='cuda:0' )
    S = S + r*S*dt + sigma*S*Z*np.sqrt(dt)

Payoff = S - K
Payoff[ Payoff <= 0.0] = 0.0
Price = torch.mean( np.exp(-r*1)*Payoff )
print(Price)
```

The returned price from the algorithm here is $C = 0.1981$. Let's now see how well the neural SDE is doing in recovering this price. We train the neural SDE using the method from Section 22.5.

```
LR = 0.001

model = NeuralNetworkModel()
model.cuda()
optimizer = optim.RMSprop(model.parameters(), lr= LR, momentum=0.0)
```

```
L = 10000

dt = 0.01
N = 100
r = 0.05

K = 1.05

Number_of_iterations = 100

Price_eval_list = []

for j in range(Number_of_iterations):

    optimizer.zero_grad()

    S1 = torch.cuda.FloatTensor(np.ones( (L,1) ) )
    S2 = torch.cuda.FloatTensor(np.ones( (L,1) ) )

    for i in range(N):

        Z1 = torch.randn( (L,1), dtype = torch.float32 ).cuda()
        Z2 = torch.randn( (L,1), dtype = torch.float32 ).cuda()

        vol1 = model(S1)
        vol2 = model(S2)

        S1 = S1 + r*S1*dt + vol1*Z1*np.sqrt(dt)
        S2 = S2 + r*S2*dt + vol2*Z2*np.sqrt(dt)

    Payoff1 = S1.detach() - K
    Payoff1[ Payoff1 <= 0.0] = 0.0
    Model_Price1 = torch.mean( np.exp(-r*1)*Payoff1 )

    Payoff2 = S2 - K
    Payoff2[ Payoff2 <= 0.0] = 0.0
    Model_Price2 = torch.mean( np.exp(-r*1)*Payoff2 )

    G = (Model_Price1 - Price.detach())*Model_Price2
    G.backward()
    optimizer.step()

    #Evaluate trained neural network
    L_eval = 100000
    with torch.no_grad():

      S = torch.cuda.FloatTensor(np.ones( (L_eval,1) ) )

      for i in range(N):

          Z = torch.randn( (L_eval,1), dtype = torch.float32 ).cuda()

          vol = model(S)

          S = S + r*S*dt + vol*Z*np.sqrt(dt)
```

```
Payoff = S.detach() - K
Payoff[ Payoff <= 0.0] = 0.0
Model_Price = torch.mean( np.exp(-r*1)*Payoff )
Price_eval_list.append( Model_Price.cpu().numpy())

Model_Price_Ave = np.mean(Price_eval_list[-10:])

print(j, Model_Price_Ave )
```

At the end, the algorithm based on the neural SDE returned the estimated price $\hat{C} = 0.19809113$, which is indeed very close to the target value of $C = 0.1981$.

The key line of the above computational method is the line which calculates the variable G. When automatic differentiation is applied to the variable G, it will produce the gradient V_k from equation (22.15). The detach command truncates the chain rule at the variable where it is applied (i.e., it treats the variable as a constant when automatic differentiation is applied).

22.7. Brief Concluding Remarks

In this chapter we studied neural ODE and SDE and the goal was to demonstrate that one can learn the dynamics of an ODE or an SDE to match given data. This idea has been explored in [**SS17**]-[**SS20c**] to develop the stochastic gradient descent algorithm in continuous time (SGDCT), providing a computationally efficient method for statistical learning of complex models potentially over long time periods.

Even though we did not present any here, the same kinds of questions can be asked and answered for PDEs, see for example [**SMS23**] for the case of linear PDEs. Deep learning has proven to be very successful in approximating the solution to oftentimes high-dimensional partial differential equations. Some of the early works in this field include [**LLF98, LLP00, SS18, BEJ19, RPK19**], but many papers followed thereafter. An exposition to using machine learning for dynamical systems can be found in [**E17**].

Chapter 23

Distributed Training

23.1. Introduction

In practice, both datasets (millions or even billions of data samples) and models (millions to hundreds of millions of parameters) are large. Each data sample itself could be memory-intensive (e.g., a high-resolution image or a video). Due to the large model size and large amount of memory-per-data-samples, it may be challenging to evaluate and calculate the backpropagation step for a large minibatch on a GPU. This forces the training to use a small minibatch size (perhaps even an SGD with only a single data sample), which will slow training. When the minibatch size is small, the noise in each training update will increase. The increased noise will typically slow convergence and require a smaller learning rate. The smaller learning rate will further slow the convergence. Therefore, it is typically optimal to have a larger minibatch size, which reduces the noise in the updates and allows for a larger learning rate magnitude.

If the model parameters are stored as floats (4 bytes per float), the memory cost for a single fully connected layer with H hidden units connected to another layer with H hidden units is

$$\begin{aligned}
\text{Number of parameters} &= H \times H, \\
\text{Memory to store parameters} &= 4 \times H \times H, \\
\text{Number of hidden units in minibatch} &= M \times H, \\
\text{Memory to store hidden units in minibatch} &= 4 \times M \times H,
\end{aligned}$$

where M is the size of the minibatch. Similarly, the computational cost of both the forward and backward step can quickly increase for larger model sizes (e.g., large H) or large minibatch sizes (large M). The number of arithmetic and algebraic operations for the forward step to evaluate a single hidden layer is

$$\text{Number of operations} = M \times H \times H + M \times H,$$

where the second term is due to evaluating the activation function after the linear operation. Similarly, the number of arithmetic and algebraic operations for the backward step for the single hidden layer is

$$\text{Number of operations} = M \times H \times H + M \times H.$$

The memory cost and computational cost quickly grows with the depth of the neural network. For example, for a neural network with L fully connected layers, the costs become

$$\begin{aligned}
&\text{Memory to store hidden units in minibatch} = 4 \times M \times H \times L, \\
&\text{Number of operations in forward step} = L \times (M \times H \times H + M \times H), \\
&\text{Number of operations in backward step} = L \times (M \times H \times H + M \times H).
\end{aligned}$$

Computational problems can quickly be encountered for large (M, H, L). For example, GPUs have a limited memory (often less than the CPUs on the machine). Therefore, the GPU may produce an out-of-memory error for large models and/or large minibatch sizes. Beyond a certain limit, the computational operations may also not be fully parallelized on the GPU, leading to slower computational times and, eventually, out-of-memory errors. Out-of-memory errors can be addressed by reducing the size of the model and/or the minibatch size. However, reducing the minibatch size may reduce the convergence speed (the gradient estimates will be more noisy, which may also require a smaller learning rate). Reducing the model size, although it would address an out-of-memory error, may also reduce the accuracy/performance of the model.

Distributing the training over multiple GPUs is therefore advantageous. The training can be parallelized by dividing the *total* minibatch M into smaller minibatches of size M_0 where $NM_0 = M$ on GPUs $i = 1, \ldots, N$. The GPUs can be on the same machine or on different machines. The minibatch objective function for the model then becomes

$$\begin{aligned}
\Lambda(\theta) &= \frac{1}{M} \sum_{i=1}^{M} \ell_{y^{(i)}}\Big(\mathrm{m}(x^{(i)}; \theta) \Big) \\
&= \underbrace{\frac{1}{M} \sum_{i=1}^{M_0} \ell_{y^{(i)}}\Big(\mathrm{m}(x^{(i)}; \theta) \Big)}_{\text{GPU 1}} + \underbrace{\frac{1}{M} \sum_{i=M_0+1}^{2M_0} \ell_{y^{(i)}}\Big(\mathrm{m}(x^{(i)}; \theta) \Big)}_{\text{GPU 2}} \\
&\quad + \cdots + \underbrace{\frac{1}{M} \sum_{i=(N-1)M_0+1}^{NM_0} \ell_{y^{(i)}}\Big(\mathrm{m}(x^{(i)}; \theta) \Big)}_{\text{GPU N}},
\end{aligned}$$

where $\ell_y(z)$ is the loss function, $y^{(i)}$ is the target data, $\mathfrak{m}(x;\theta)$ is the model, and θ are the model parameters. The computations (both the forward and backpropagation steps) for the data samples $D_i = \{iM_0 + 1, \ldots, (i+1)M_0\}$ are performed *in parallel* on GPUs $i = 1, \ldots, N$. Therefore, in principle, the computational time to calculate the gradient $\nabla_\theta \Lambda(\theta)$ for the total minibatch of size M should be the same computational time as calculating the gradient for an objective function with M_0 data samples. An important consideration is the communication cost of sharing the gradients between the different machines, which may reduce the computational time savings of parallelization. This communication cost will become apparent in the algorithms presented below.

23.2. Synchronous Gradient Descent

We first present the distributed gradient descent algorithm with synchronous parameter updates. In synchronous gradient descent, each GPU calculates a gradient, the gradients are averaged across all GPUs, and then the average gradient is used to update the model. At all update steps, the model copy on each GPU is therefore identical.

- Each GPU holds a copy of the neural network model parameters θ.
- The neural network model parameters on all GPUs $i = 1, \ldots, N$ are initialized to the same initial parameters θ_0.
- For optimization iterations $k = 1, 2, \ldots, K$:
  - A minibatch of size M_0 is randomly selected for each GPU $i = 1, \ldots, N$.
  - On each GPU i, the following objective function is evaluated (forward step):

$$\Lambda^i(\theta) = \frac{1}{M_0} \sum_{j=iM_0}^{(i+1)M_0} \ell_{y^{(j)}}\Big(\mathfrak{m}(x^{(j)}; \theta) \Big).$$

  - Then, the gradient of $\Lambda^i(\theta)$ is calculated (backpropagation step).
  - The gradients from each GPU are averaged:

$$\nabla_\theta \Lambda(\theta) = \frac{1}{N} \sum_{i=1}^{N} \nabla_\theta \Lambda^i(\theta),$$

    which requires communication between the GPUs.
  - The parameters θ are updated using the gradient $\nabla_\theta \Lambda(\theta)$.

The forward and backward steps in the backpropagation algorithm are completed independently on each GPU and do not require communication.

This is referred to as a *perfectly parallel* calculation in parallel computing. However, the parameter update does require communication between the GPUs to average the gradients from all of the GPUs. The communication time depends upon the size of the network, network architecture/connection speeds, and the size of the data that must be transferred. For example, if a large cluster of many machines is used, the communication times will be much larger than a single machine with multiple GPUs. Of course, multiple machines allows for a much larger minibatch size. Communication times will increase as the size of the data is increased. In distributed gradient descent, each machine must communicate the model gradients $\nabla_\theta \Lambda^i(\theta)$. The number of model gradients is equal to the number of model parameters. In deep learning, the standard is to use floats for the model parameters (and their gradients). Therefore, if there are d parameters, the model parameter gradients are $d \times 4$ bytes. For large models, the communication times can potentially be significant, reducing the computational speed of distributed gradient descent.

The synchronous distributed gradient descent algorithm (presented above) has a potential computational disadvantage in that machine i must wait for *all other* machines $j \neq i$ to complete their calculation of their respective model gradients before updating the model. Therefore, the slowest machine's computational time will be a bottleneck for the parameter update. The computational time for each parameter update is restricted by the slowest machine (even if all other machines are significantly faster). This inspires the next distributed gradient descent algorithm that we will discuss: *asynchronous* gradient descent.

23.3. Asynchronous Gradient Descent

Asynchronous gradient descent updates the model parameters on each machine *asynchronously* without waiting for the other machines to complete their gradient calculations. The typical framework includes a machine as the *parameter server* which holds the *master copy* of the model parameters θ. The *worker machines* $i = 1, \ldots, N$ will calculate the gradients of the model. Once a worker has completed its calculation, it will send the model gradients to the parameter server which, immediately upon receiving the parameters and without waiting for the other machines to complete their work, will update the master copy of the model parameters θ. The updated master copy of the model parameters will then be transmitted back to the worker machine i.

The current model parameters on machine i are denoted θ_i. Due to the asynchronous updates, at any single point in time, the parameters $\theta_1, \theta_2, \ldots, \theta_N$ may differ. The asynchronous gradient descent algorithm is summarized below.

- Initialize model parameters θ on the parameter server (machine $i = 1$).

- For $i = 2, \ldots, N$ **(independently in parallel)**:
  - For $k = 1, 2, \ldots, K$:
    * Copy the model parameters θ to machine i. Set $\theta_i = \theta$.
    * Randomly select a minibatch of data and evaluate the objective function,

$$\Lambda^i(\theta) = \frac{1}{M_0} \sum_{j=iM_0}^{(i+1)M_0} \ell_{y^{(j)}}\left(\mathrm{m}(x^{(j)}; \theta) \right).$$

    * Then, the gradient of $\Lambda^i(\theta_i)$ is calculated (backpropagation step).
    * Communicate the gradient $\nabla_\theta \Lambda^i(\theta_i)$ to machine 1.
    * Update θ on machine 1 with stochastic gradient descent.

Asynchronous gradient descent has the advantage that machine i does not have to wait for work to complete on the other machines $j \neq i$. Instead, it is able to complete rapid (noisy) updates to the model. This is particularly advantageous when the cluster consists of multiple machines with different computational capabilities/speeds. A disadvantage is that there is an inconsistency between the model gradients calculated on the different machines. Since updates are performed asynchronously, the model parameters on machine i may not necessarily equal the current model parameters on machines $j \neq i$. Similarly, the model parameters θ_i on machine i may not equal the most recent master copy of the model parameters θ on the parameter server.

Therefore, since the gradients are calculated based upon different parameters than the current master copy of the parameters, the parameter update *is not guaranteed to decrease the objective function*. That is, the gradient calculated on machine i is not necessarily a descent direction for the objective function. This can become more problematic in larger clusters where communication times from worker machines to the parameter server machine may be large. The larger the communication times are, the greater the potential difference between the model parameters θ_i on machine i and the parameter server parameters θ. This increases the error for the gradient calculated on machine i with respect to the objective function $\Lambda(\theta)$.

23.4. Parallel Efficiency

The goal of distributing training across multiple machines is to reduce the computational time required to train a model. The parallel algorithm's performance can be measured via its *parallel efficiency*. Let T be the time required to complete the task using one machine. Let T_N be the time required to complete the same task using N machines. The parallel efficiency is

$$E_N = \frac{T}{N \times T_N} \times 100\%. \tag{23.1}$$

If $T_N = \frac{T}{N}$, the efficiency is $E_N = 100\%$. If $T_N = \frac{2T}{N}$, the efficiency is $E_N = 50\%$. Typically, parallel algorithms do not achieve 100% efficiency since large amounts of data must be communicated between machines. The communication time reduces the performance of the parallel algorithm. Communication costs typically increase as the number of machines N increases and therefore the parallel efficiency will be a function of N. Perfectly parallel tasks requiring no communication have 100% efficiency. An example is Monte Carlo simulation. Monte Carlo simulation can be performed completely independently on each machine with a single communication at the end to average the Monte Carlo samples across all machines. Tasks requiring heavy communication may have significantly less than 100% efficiency.

Formula (23.1) for parallel efficiency is referred to as *strong scaling*. The strong scaling of an algorithm measures how quickly a task can be completed as a function of the number of machines. For example, strong scaling in deep learning would measure how quickly (in computational time) the model training achieves a certain fixed accuracy (e.g., 99% accuracy on the MNIST dataset). *Weak scaling* is given by the formula

$$E_N = \frac{T}{T_N} \times 100\%, \tag{23.2}$$

where T is the computational time to complete X amount of work on a single machine and T_N is the computational time to complete X amount of work on each of N total machines ($N \times X$ total work). For example, here X could be calculating the model parameter gradients on M data samples (i.e., $N \times M$ total data samples).

Distributed gradient descent can typically achieve excellent weak scaling. However, strong scaling may not necessarily perform as well as weak scaling. As the number of total data samples $N \times M \to \infty$, each minibatch stochastic gradient descent update will converge to a deterministic gradient descent update with an L^2-convergence rate $\sim (N \times M)^{-2}$. This may improve the convergence speed to a local minimizer in the total number of parameter update steps (since the gradient estimates will be less noisy and, in addition, the learning rate can be potentially increased since the noise has been reduced). However, the convergence speed will certainly not improve at a rate proportional to N^{-1}. In particular, since it will converge to deterministic gradient descent, the convergence rate will be limited by the convergence rate for deterministic gradient descent. Therefore, the marginal benefits of increasing the number of machines (and the minibatch size) will start to vanish after a certain point.

23.5. MPI Communication

Message passing interface (MPI) is the standard method for communicating data between multiple processes in parallel computing. The multiple processes may be on the same machine or multiple machines. For example, a cluster could have eight machines with 16 processes per machine. In total, there would be 128 processes. Each process performs a task (i.e., calculations) and also communicates with other processes. The completion of a task on process i may require data from the calculations on the other processes $j \neq i$.

Each process $i = 0, 1, \ldots, N-1$ (in the example above, $N = 128$) is assigned a *rank*. Process i's rank is the integer i. Complex calculations—where each process may perform a different task with dependencies on communications between the processes—can be concisely coded as a *single program* by writing the calculation as a function of the rank i. However, in many deep learning applications, the communication is relatively simple. The synchronous gradient descent algorithm (presented in the previous section) assigns an identical calculation to each process with the communication operation being an average of all of the processes' gradients.

PyTorch has MPI capabilities, which can be used via the "torch.distributed" library.

```
import torch.distributed as dist
```

The rank of each process and the total number of processes can be obtained via

```
num_processes = dist.get_world_size()
rank = dist.get_rank()
```

The distributed class is initialized the group of processes via

```
dist.init_process_group(`mpi', rank=rank, world_size=num_nodes)
```

The key MPI operation for synchronous gradient descent is a communication operation to average tensors on all the processes. In particular, if there are tensors $X_0, X_1, \ldots, X_{N-1}$ on the processes $i = 0, 1, \ldots, N-1$, we need a communication operation that calculates

$$\bar{X} = \sum_{i=0}^{N-1} X_i,$$

and returns $\bar{X}$ to all processes $i = 0, 1, \ldots, N-1$. In the context of synchronous gradient descent, the tensors X_i are the minibatch gradients calculated on each process. In MPI, this communication operation is the All Reduce operation and it is implemented in PyTorch as

```
dist.all_reduce(X, op=dist.reduce_op.SUM)
```

X is the tensor on each process. The above command averages the tensors from all of the processes and replaces X with the average. As a concrete example, consider the following command run on N processes.

```
X = torch.FloatTensor( np.ones(1)*rank )
dist.all_reduce(X, op=dist.reduce_op.SUM)
print(rank, X)
```

The printed result will be a scalar value $N-1$ on all processes (ranks) $0, 1, \ldots, N-1$.

Typical clusters will not have GPU-to-GPU communication available between machines. That is, if a rank 1 tensor is on the GPU of machine 1, it cannot be directly communicated to rank 2 on the GPU of machine 2. Instead, the rank 1 tensor must first be moved to the CPU of machine 1, then sent to the CPU of machine 2, and then finally moved to the GPU of machine 2. This is illustrated by the following example.

```
X = torch.cuda.FloatTensor( np.ones(1)*rank )
dist.all_reduce(X.cpu(), op=dist.reduce_op.SUM)
X.cuda()
print(rank, X)
```

In some more recent advanced *high-performance computing* (HPC) architectures, GPU-to-GPU communication between machines is available.

We will now use PyTorch's MPI All Reduce operation to implement synchronous distributed gradient descent to train a neural network. As in Chapter 7 we work with the MNIST dataset [**LBBH98**], available from https://yann.lecun.com/exdb/mnist/. We recall that the original dataset was downloaded and stored in an hdf5 file, with the input data normalized by the maximum value of a pixel (255).

```
import numpy as np
import torch
import torch.nn as nn
import torch.nn.functional as F
import torch.optim as optim
```

```
from torch.autograd import Variable
import torch.distributed as dist

import h5py
import time
import os

import subprocess
from mpi4py import MPI

num_nodes = dist.get_world_size()
rank = dist.get_rank()

backend = 'mpi'
dist.init_process_group(backend, rank=rank, world_size=num_nodes)

dtype = torch.FloatTensor

#Load MNIST dataset
MNIST_data = h5py.File('MNISTdata.hdf5', 'r')
x_train = np.float32(MNIST_data['x_train'][:] )
y_train = np.int32(np.array(MNIST_data['y_train'][:,0]))
x_test = np.float32( MNIST_data['x_test'][:] )
y_test = np.int32( np.array( MNIST_data['y_test'][:,0]  ) )

MNIST_data.close()

#Number of hidden units
H = 100

class MnistModel(nn.Module):
    def __init__(self):
        super(MnistModel, self).__init__()
        # input is 28x28
        # padding=2 for same padding
        self.fc1 = nn.Linear(28*28, H)
        self.fc2 = nn.Linear(H, H)
        self.fc3 = nn.Linear(H, 10)

    def forward(self, x):
        x = F.relu(self.fc1( x ))
        x = F.dropout(x, p = 0.6, training=self.training)
        x = F.relu(self.fc2( x ))
        x = F.dropout(x, p = 0.6, training=self.training)
        x = self.fc3( x )
        return F.log_softmax(x, dim=1)

model = MnistModel()

#All machines should have a copy of the same model
for param in model.parameters():
    tensor0 = param.data
    dist.all_reduce(tensor0, op=dist.reduce_op.SUM)
    param.data = tensor0*np.sqrt(np.float(num_nodes))

model.cuda()
```

```
LR = 0.001

optimizer = optim.Adam(model.parameters(), lr=LR)

batch_size = 1000
L_Y_train = len(y_train)

model.train()

train_loss = []
train_accu = []

for epoch in range(10):
    time1 = time.time()

    I_permutation = np.random.permutation(L_Y_train)
    x_train = x_train[I_permutation,:]
    y_train = y_train[I_permutation]

    for i in range(0, L_Y_train, batch_size):
        #apply .cuda() to move to GPU
        x_train_batch = torch.FloatTensor( x_train[i:i+batch_size,:] )
        y_train_batch = torch.LongTensor( y_train[i:i+batch_size] )
        data, target = Variable(x_train_batch).cuda(), Variable(
                                              y_train_batch).cuda()
        optimizer.zero_grad()
        output = model(data)
        loss = F.nll_loss(output, target)
        loss.backward()    # Calculate gradients
        train_loss.append(loss.data[0])
        #All-Reduce Communication
        for param in model.parameters():
            tensor0 = param.grad.data.cpu()
            dist.all_reduce(tensor0, op=dist.reduce_op.SUM)
            tensor0 /= float(num_nodes)
            param.grad.data = tensor0.cuda()

        optimizer.step()    # Update parameters

        prediction = output.data.max(1)[1]
        accuracy = ( float(prediction.eq(target.data).sum()) / float(
                                              batch_size) )*100
        train_accu.append(accuracy)

    time2 = time.time()
    time_elapsed = time2 - time1
    if (rank == 0):
        print(epoch, accuracy, time_elapsed)
```

A few elements of the above code deserve to be highlighted. First, all processes are initialized with the same model parameters. Second, the parameter gradients are *summed* across all processes using the MPI All Reduce operation. Then, the gradients are normalized by dividing by the number of processes to obtain the average gradient from all of the processes' minibatch gradients. The above code is for the case of N processes (ranks) on N machines each with one GPU (i.e., one process per GPU). MPI-based synchronous gradient descent can also be implemented for the case of a cluster of machines with multiple GPUs per machine.

23.6. Point-to-point MPI Communication

The All Reduce communication operation in the previous section is a form of *collective communication* where all machines (or processes) communicate with each other. This simple communication operation is sufficient for distributed *synchronous* gradient descent and can be used for distributed training of nearly all deep learning models in practice. More complex communication is also possible via *point-to-point* MPI communication operations. These operations can send data from a specific process i to a specific process j.

The following code uses point-to-point communication to move a tensor between a process i and the other processes

```
num_nodes = dist.get_world_size()
rank = dist.get_rank()
i = 0
N = num_nodes
for j in range(N):
        tensor = torch.FloatTensor(k*np.ones(1))
        if (i != j):
                if rank == i:
                        dist.send(tensor=tensor, dst=j)
                if (rank == j):
                        dist.recv(tensor=tensor, src=i)
```

The above code sends a tensor from process i to all of the other ranks. The communication is *blocking* since operations on processes i and j halt until the communication between i and j is completed. The number of communications N can be changed as long as it less than the total number of nodes. That is, it is permissible to use point-to-point communications between a subset of the total number of processes. For example, if $N = 2$, the tensor will be sent from process $i = 0$ to process $j = 1$ only. Point-to-point communication allows for complex distributed computing tasks which, in general, can be a function of all work/operations/outputs across the entire cluster of machines. It should be noted though that constant point-to-point communication of large tensors between large numbers of machines can significantly increase communication costs, slowing down the progress of the overall code.

23.7. Python MPI Communication

MPI communication can also be directly run in Python scripts, which can be used for Python-only (e.g., numpy) operations as well as PyTorch code. The commands are similar to the distributed PyTorch MPI commands described above. Some example scripts are provided below. These examples solve the simple problem of minimizing the objective function,

$$\Lambda(\theta) = \frac{1}{2}\mathbb{E}\Big[(Y-\theta)^2\Big],$$

with stochastic gradient descent. The minibatch stochastic gradient descent algorithm is

$$\theta_{k+1} = \theta_k - \eta_k \frac{1}{M}\sum_{j=1}^{M}(y^{(k,j)} - \theta_k),$$

where $y^{(k,j)}$ are i.i.d. samples from the random variable Y. The distributed (synchronous) stochastic gradient algorithm with N processes becomes

$$G^{(k,i)} = \frac{1}{M_0}\sum_{j=iM_0}^{(i+1)M_0}(y^{(k,j)} - \theta_k),$$

$$\theta_{k+1} = \theta_k - \eta_k \frac{1}{N}\sum_{i=1}^{N} G^{(k,i)},$$

where $G^{(k,i)}$ is the gradient calculated for an independent minibatch of size M_0 on process i. The gradient estimate from process i is therefore $G^{(k,i)}$, and these gradient estimates are averaged together to provide a more accurate gradient estimate to update the parameter θ. The total minibatch size is $N \times M_0$. The distributed training can be implemented via the following code.

```
import numpy as np
from mpi4py import MPI

name = MPI.Get_processor_name()
comm = MPI.COMM_WORLD
rank = comm.Get_rank()
num_nodes = int(comm.Get_size())

N = 100000
Initial_LR = 0.1
M0 = 10000000
theta = np.ones(1)

for k in range(N):

    LR = Initial_LR/(1.0 + k/1000.0)

    Y = np.random.randn(M0)
    G = -1.0*(Y - theta)
    G = np.mean(G)*np.ones(1)

    #if (k #    print('rank', rank,'G on each individual process', G)

    G = comm.allreduce(G, op=MPI.SUM)
```

```
    G = G/float(num_nodes)

    #if (k #    print('rank', rank, 'G after averaging across all processes
                                                            ', G)

    theta = theta - LR*G

    if (k print('k', k, 'rank', rank, 'theta', theta)
```

As before, a rank is assigned to each process and the All Reduce command sums the gradients across all of the different processes. The (commented-out) print statements can be included to confirm that the All Reduce command is summing the gradients from the different processes.

As an example of point-to-point communication, we reimplement the above synchronous distributed gradient descent algorithm using point-to-point communications:

```
import numpy as np
from mpi4py import MPI

name = MPI.Get_processor_name()
comm = MPI.COMM_WORLD
rank = comm.Get_rank()
num_nodes = int(comm.Get_size())

N = 100000
Initial_LR = 0.1
M0 = 1000
theta = np.ones(1)

for k in range(N):

    LR = Initial_LR/(1.0 + k/1000.0)

    if rank > 0:

        Y = np.random.randn(M0)
        G = -1.0*(Y - theta)
        G = np.mean(G)*np.ones(1)

        comm.Send(G,dest=0, tag=13)

    if rank == 0:

        G = np.zeros(1)
        for j in range(1, num_nodes):
            G_temporary = np.ones(1)
            comm.Recv(G_temporary, source = j, tag=13)
            G = G + G_temporary

        G = G/float(num_nodes-1)

        theta = theta - LR*G
```

```
if rank == 0:
    for j in range(1, num_nodes):
        comm.Send(theta , dest=j , tag =13)
if rank > 0:
    comm.Recv(theta , source = 0, tag = 13)

if (k print('k', k, 'rank', rank, 'theta', theta)
```

In the above code, process 0 contains the trained parameter. Processes $1, \ldots, N-1$ calculate gradient estimates and send them to process 0. Using these gradient estimates, process 0 updates the model parameter. Then, the new updated model parameter is communicated back to the other processes $1, \ldots, N-1$. The above point-to-point communications will not be as efficient as the MPI All Reduce command. However, it serves as a simple example demonstrating how to use point-to-point communications.

MPI communication can also be used for model-parallelized training. That is, different parts of the model will be evaluated on different processes/machines. Communication will occur to calculate the overall model output and the model parameter gradients. Model parallelization is useful when the size of the model is so large it may not fit on a single machine. We will demonstrate model parallelization via simple example (which could be extended to more complex models). Consider the model and objective function,

$$\Lambda(\theta) = \frac{1}{2}\mathbb{E}\Big[\ \|Y - \theta X\|^2\ \Big],$$

where Y is an $q \times 1$ vector, θ is a $q \times d$ matrix, and X is a $d \times 1$ vector. If $q \times d$ is very large, the computational cost of the matrix multiplication θX will be large. If the matrix is very large, the matrix θ may not even be able to be stored in memory. The matrix multiplication can be distributed across multiple processes via

$$\begin{aligned} \theta X &= \theta^{(0)} X^{(0)} + \cdots + \theta^{(N-1)} X^{(N-1)} \\ &= \sum_{i=0}^{N-1} \theta^{(i)} X^{(i)}, \end{aligned}$$

where $\theta^{(i)}$ is the matrix $\theta_{:,iM:(i+1)M}$, $X^{(i)}$ is the matrix $X_{iM:(i+1)M,:}$, and $M = \frac{d}{N}$ (assuming for simplicity that d is an integer multiple of N). We will perform the calculation $\theta^{(i)} X^{(i)}$ on process i and then communicate the result to the other processes. Example code is provided below (for simplicity, we let $q = 1$, $Y = \sum_{j=1}^{d} X_i$, and $X_i \sim \mathcal{N}(0, 1)$).

```
import numpy as np
from mpi4py import MPI

name = MPI.Get_processor_name()
comm = MPI.COMM_WORLD
rank = comm.Get_rank()
num_nodes = int(comm.Get_size())

N = 100000
Initial_LR = 0.1
d_per_process = 2
d = d_per_process*num_nodes
theta = np.zeros(d_per_process)

for k in range(N):

    LR = Initial_LR/(1.0 + k/1000.0)

    X = np.random.randn(d_per_process)
    Y = np.sum(X)
    Y = Y*np.ones(1)
    f = np.dot(theta, X)
    f = f*np.ones(1)

    f = comm.allreduce(f, op=MPI.SUM)
    Y = comm.allreduce(Y, op=MPI.SUM)

    G = -1.0*(Y[0] - f[0])*X

    theta = theta - LR*G

    Error = np.mean(np.abs(theta - 1.0 ))

    if (k print('k', k, 'rank', rank, 'Error', Error)
```

In the above code, $\theta^{(i)}X^{(i)}$ is calculated separately on each process. Then, using an allreduce operation, the $\theta^{(i)}X^{(i)}$ are communicated and summed to produce $\sum_{i=0}^{N-1}\theta^{(i)}X^{(i)}$. Finally, the parameter gradients for $\theta^{(i)}$ are calculated separately on each process and updated with stochastic gradient descent.

We now extend this approach for model-parallelized training of a neural network. Recall that we have previously used data-parallelization to parallelize the calculation of gradients across the data samples in a dataset. In data-parallelization, each machine calculates the model parameter gradients on a subset of the overall dataset (or minibatch). Each machine will have a copy of the model, and communication only occurs to sum the gradient estimates from all of the machines. In model-parallelization, each machine only stores part of the model. Communication between machines must therefore occur to evaluate the model output itself even on a single data sample. Furthermore, communication must also necessarily occur to calculate the gradient (i.e., the backpropagation step) for a single data sample.

Consider a single-layer neural network

$$\mathfrak{m}(x;\theta) = \sum_{j=1}^{H} C^j \sigma(W^{j,:}x + b^j),$$

where the parameters are $\theta = \{C, W, b\}$ and H is the number of hidden units. If the number of hidden units H is very large, then $\mathfrak{m}(x;\theta)$ will be computationally expensive to evaluate. The evaluation of the neural network can be easily distributed across multiple machines:

$$\mathfrak{m}(x;\theta) = \sum_{i=0}^{N-1} \mathfrak{m}^i(x;\theta),$$

$$\mathfrak{m}^i(x;\theta) = \sum_{iH_0}^{(i+1)H_0} C^{j,:}\sigma(W^{j,:}x + b^j),$$

where $N \times H_0 = H$. The model $\mathfrak{m}^i(x;\theta)$ is evaluated on machine i and and then the neural network output $\mathfrak{m}(x;\theta)$ is calculated by an All Reduce communication which sums the $\mathfrak{m}^i(x;\theta)$ from all of the machines. Example code is provided below:

```
import numpy as np
from mpi4py import MPI

name = MPI.Get_processor_name()
comm = MPI.COMM_WORLD
rank = comm.Get_rank()
num_nodes = int(comm.Get_size())

H0 = 1000
H = num_nodes*H0
d = 100

#Parameter matrices
W = np.random.randn( H0, d )
b = np.random.randn( H0, 1  )
C = np.random.randn( 1, H0  )

def sigmoid(x):
    sigma = np.exp(x)/(1.0 + np.exp(x))
    return sigma

def Neural_Network(x):
    Hidden_Layer = sigmoid( np.dot(W,x) + b )
    f  = np.dot(C, Hidden_Layer)
    f = comm.allreduce(f, op=MPI.SUM)
    return f[0], Hidden_Layer

#Example
x = np.random.randn(d,1)
f, Hidden_Layer = Neural_Network(x)
if rank == 0:
        print(f)
```

The above code evaluates H_0 hidden units on each process, where the neural network has in total $H = N \times H_0$ with N total processes. Similarly, the backpropagation step can also be distributed across the different machines. In the case of a single-layer network, the only communication required is to calculate the output of the neural network. The model parameter derivatives can then be calculated separately on each process.

```
def Backward(x, y, f, Hidden_Layer):
    e = -(y - f)
    C_deriv = e*Hidden_Layer
    delta = e*C[0,:]
    z_deriv = Hidden_Layer[:,0]*(1.0 - Hidden_Layer[:,0])*delta
    B_deriv = z_deriv
    W_deriv = np.dot(z_deriv[:,None], np.transpose(x) )

    return C_deriv, B_deriv, W_deriv
```

For multi-layer neural networks, the values of the hidden units must also be communicated between the different processes. The implementation of this slightly more complex distributed algorithm is left as an exercise (see Exercise 23.6).

23.8. Brief Concluding Remarks

Training deep learning models can be computationally expensive due to the size of the model (i.e., the large number of parameters) as well as the size of the dataset. In many real-world machine learning training tasks, the computational cost can become a significant obstacle to training models with a single machine. Data-parallelization can address this challenge by distributing subsets of the minibatch data samples across multiple machines in order to calculate their gradients. For large models, the model itself can be distributed across multiple machines, which typically will also require implementation of a distributed version of the backpropagation algorithm. For more examples and additional discussion of parallel computing in deep learning, it is recommended to read PyTorch's documentatation on distributed training of deep learning and the article [**DCM**+**12**].

23.9. Exercises

Exercise 23.1. Using the code provided for the MNIST dataset, measure its strong scaling and weak scaling as a function of the number of machines.

Exercise 23.2. Prove that the distributed minibatch stochastic gradient descent update converges to gradient descent as the number of machines $N \to \infty$.

Exercise 23.3. Implement asynchronous stochastic gradient descent using MPI point-to-point communication operations.

Exercise 23.4. Modify the distributed training code to train a convolution network on the MNIST dataset.

Exercise 23.5. Use the distributed training code to train a convolution network on the CIFAR10 dataset. Measure strong scaling and weak scaling as a function of the number of machines.

Exercise 23.6. Develop a distributed algorithm for training a model-parallelized two-layer neural network and implement it in Python.

Chapter 24

Automatic Differentiation

24.1. Introduction

Deep learning involves large, complex, nonlinear models with large numbers of parameters which must be trained. Examples include multi-layer fully connected networks, deep convolution networks, residual networks, and recurrent networks. There is a highly flexible choice of model architectures within each of these model classes. Developing deep learning models typically involves evaluating a series of different architectures. Each time the model architecture changes, the backpropagation algorithm also changes. Rederiving the backpropagation algorithm via the chain rule for each new model architecture is time-consuming and would substantially limit developing new models.

Automatic differentiation (AD) is a numerical algorithm to evaluate the backpropagation rule (i.e., the gradients with respect to model parameters) for a very general class of functions, including deep learning model architectures. The user only has to define the model architecture (i.e., the function) and then, given the model definition, AD will then evaluate the gradient for the model. This allows for the user to rapidly develop, train, and evaluate a series of deep learning models without having to rederive the backpropagation rule via the chain rule for each new model variation.

Automatic differentiation is especially useful for large-scale models with large numbers of intermediate functions and parameters (e.g., large language models or residual networks with hundreds of layers).

24.2. Reverse-mode versus Forward-mode Differentiation

We will first review reverse-mode and forward-mode differentiation. As an example, consider the sequence of functions

$$z_\ell = \mathfrak{m}_\ell(z_{\ell-1}, x; \theta),$$
$$z_1 = = \mathfrak{m}_1(x; \theta), \tag{24.1}$$

where x is the input data and θ is the parameter to be trained. Let x, θ, and z_ℓ be vectors of length d_x, d_θ, and d_ℓ, respectively. For $\ell > 1$, the function $\mathfrak{m}_\ell$ is therefore a map $\mathfrak{m}_\ell : \mathbb{R}^{z_{\ell-1} \times d_x \times d_\theta} \to \mathbb{R}^{d_\ell}$. For $\ell = 1$, $\mathfrak{m}_1 : \mathbb{R}^{d_x \times d_\theta} \to \mathbb{R}^{d_\ell}$. The final model output is

$$\mathfrak{m}(x; \theta) = z_L,$$

and we are interested in gradients of the following function with respect to the parameters θ,

$$\Lambda(\theta) = \rho(\mathfrak{m}(x; \theta), y),$$

where y is a vector of length d_L and the function $\rho : \mathbb{R}^{d_L \times d_L} \to \mathbb{R}^{d_O}$. In machine learning, $\Lambda(\theta)$ is typically the loss function—which is scalar-valued—and therefore $d_O = 1$.

24.2.1. Reverse-mode Differentiation. We will first review reverse-mode differentiation, which is equivalent to the backpropagation algorithm. Define $\hat{z}_\ell = \frac{\partial \Lambda}{\partial z_\ell}$, which is a $d_O \times d_\ell$ matrix. By the chain rule,

$$\hat{z}_\ell = \hat{z}_{\ell+1} \frac{\partial \mathfrak{m}_\ell}{\partial z}(z_{\ell-1}, x; \theta).$$

Reverse-mode differentiation sequentially calculates $\hat{z}_L \to \hat{z}_{L-1} \to \cdots \to \hat{z}_1$. The parameter gradient is a $d_O \times d_\theta$ matrix which can be evaluated via

$$\frac{\partial \Lambda}{\partial \theta} = \sum_{\ell=1}^{L} \hat{z}_\ell \frac{\partial \mathfrak{m}_\ell}{\partial \theta}(z_{\ell-1}, x; \theta). \tag{24.2}$$

The computational cost of reverse-mode differentiation depends upon the functional form of $\mathfrak{m}_\ell(z, x; \theta)$. Typical deep learning models involve linear matrix multiplications followed by elementwise nonlinearities. Let's consider an example and evaluate its computational cost. Let $\mathfrak{m}_\ell(z, x; \theta) = \sigma(W_z^\ell z + W_x^\ell x)$ where $\sigma(\cdot)$ is an elementwise nonlinear activation function, W_z^ℓ is a parameter matrix with dimensions $d_\ell \times d_{\ell-1}$, and W_x^ℓ is a parameter matrix with dimensions $d_\ell \times d_x$. $\frac{\partial \mathfrak{m}_\ell}{\partial \theta}(z_{\ell-1}, x; \theta) = \sigma'(W_z^\ell z + W_x^\ell x) \odot W_z^\ell$, which is a $d_\ell \times d_{\ell-1}$ matrix. Note that the elementwise multiplication $v = \sigma'(W_z^\ell z + W_x^\ell x) \odot W_z^\ell$ is defined as $v_{ij} = (\sigma'(W_z^\ell z + W_x^\ell x))_i (W_z^\ell)_{ij}$.

The equation $\hat{z}_{\ell+1}\frac{\partial \mathrm{m}_\ell}{\partial z}(z_{\ell-1}, x; \theta) = \hat{z}_{\ell+1}v$ therefore involves a $d_O \times d_\ell$ matrix multiplied by a $d_\ell \times d_{\ell-1}$ matrix, which requires $\mathcal{O}(d_O \times d_\ell \times d_{\ell-1})$ arithmetic operations. Note that $W_z^\ell z + W_x^\ell x)$ has been calculated in the forward evaluation of the function and is therefore not included in the computational cost of the backpropagation algorithm. The evaluation of the elementwise nonlinear activation function will be assumed to be small compared to the matrix multiplications. Consequently, the backpropagation algorithm will approximately require the following number of arithmetic operations to calculate the derivatives $(\hat{z}_\ell)_{\ell=1}^{L-1}$:

$$\sum_{\ell=2}^{L} d_O \times d_\ell \times d_{\ell-1},$$

where we have assumed that the computational cost of calculating $\hat{z}_L$ is relatively small when L is large. If $d_\ell = d$, then the total number of arithmetic operations becomes

$$(L-1) \times (d_O \times d^2).$$

In addition, we have to account for the computational cost of the formula (24.2). $\frac{\partial \mathrm{m}_\ell}{\partial \theta}(z_{\ell-1}, x; \theta)$ involves the calculation of the derivatives

$$\begin{aligned} \frac{\partial \mathrm{m}_\ell}{\partial W_x^\ell}(z_{\ell-1}, x; \theta) &= \sigma'(W_z^\ell z + W_x^\ell x)x^\top, \\ \frac{\partial \mathrm{m}_\ell}{\partial W_z^\ell}(z_{\ell-1}, x; \theta) &= \sigma'(W_z^\ell z + W_x^\ell x)z^\top, \end{aligned} \tag{24.3}$$

which require $\mathcal{O}(d_\ell \times d_x)$ and $\mathcal{O}(d_\ell \times d_{\ell-1})$ arithmetic operations, respectively. Therefore, $\hat{z}_\ell\frac{\partial \mathrm{m}_\ell}{\partial \theta}(z_{\ell-1}, x; \theta)$ requires approximately $\mathcal{O}(d_O \times d_\ell \times (d_x + d_{\ell-1}))$ arithmetic operations. If $d_\ell = d$, the total number of arithmetic operations is approximately

$$(L-1) \times (d_O \times 2d^2) + L \times (d_O \times d \times d_x). \tag{24.4}$$

24.2.2. Forward-mode Differentiation. In forward-mode differentiation, we move from the beginning of the sequence of functions ($\ell = 1$) to the end ($\ell = L$ and $\Lambda(\theta)$) when calculating the derivatives. Forward-mode differentiation tracks the derivative z_ℓ with respect to the parameters θ, in contrast to reverse-mode differentiation which tracks the derivative of $\Lambda(\theta)$ with respect to z_ℓ. Define $\tilde{z}_\ell = \frac{\partial z_\ell}{\partial \theta}$. By the chain rule,

$$\tilde{z}_\ell = \frac{\partial \mathrm{m}_\ell}{\partial \theta}(z_{\ell-1}, x; \theta) + \frac{\partial \mathrm{m}_\ell}{\partial z}(z_{\ell-1}, x; \theta)\tilde{z}_{\ell-1},$$

where $\frac{\partial \mathrm{m}_\ell}{\partial \theta}(z_{\ell-1}, x; \theta)$ is a $d_\ell \times d_\theta$ matrix, $\frac{\partial \mathrm{m}_\ell}{\partial z}(z_{\ell-1}, x; \theta)$ is a $d_\ell \times d_{\ell-1}$ matrix, and $\tilde{z}_{\ell-1}$ is a $d_{\ell-1} \times 1$ matrix. Forward-mode differentiation sequentially tracks

$\tilde{z}_1 \to \tilde{z}_2 \to \cdots \to \tilde{z}_L$. The parameter gradient is

$$\frac{\partial \Lambda}{\partial \theta} = \frac{\partial \rho}{\partial z_L}(z_L, y)\tilde{z}_L.$$

Let us now analyze the computational cost of forward-mode differentiation for the specific function $\mathfrak{m}_\ell(z, x; \theta) = \sigma(W_z^\ell z + W_x^\ell x)$. Since $\frac{\partial \mathfrak{m}_\ell}{\partial z}(z_{\ell-1}, x; \theta)$ is a $d_\ell \times d_{\ell-1}$ matrix and $\tilde{z}_{\ell-1}$ is a $d_{\ell-1} \times d_\Theta$ matrix, the number of arithmetic operations for $\frac{\partial \mathfrak{m}_\ell}{\partial z}(z_{\ell-1}, x; \theta)\tilde{z}_{\ell-1}$ is $\mathcal{O}(d_\ell \times d_{\ell-1} \times d_\Theta)$. $\frac{\partial \mathfrak{m}_\ell}{\partial \theta}(z_{\ell-1}, x; \theta)$ requires $\mathcal{O}(d_\ell \times d_x + d_\ell \times d_{\ell-1})$ operations; see equation (24.3). Combining these estimates (and assuming that the elementwise nonlinear activation function has a small cost compared with the matrix multiplications), the total computational cost is approximately

$$\sum_{\ell=1}^{L} (d_\ell \times d_x + d_\ell \times d_{\ell-1} + d_\ell \times d_{\ell-1} \times d_\Theta), \tag{24.5}$$

where we have assumed that the computational cost of $\frac{\partial \rho}{\partial z_L}(z_L, y)\tilde{z}_L$ is relatively small compared to equation (24.5) when L is large. If $d_\ell = d$, then the total number of arithmetic operations becomes

$$L \times (d \times d_x + d^2 + d^2 \times d_\Theta). \tag{24.6}$$

24.2.3. Comparison of Forward and Reverse Differentiation. Let's now compare the computational costs of reverse-mode and forward-mode differentiation in equations (24.4) and (24.6), respectively. If d_O is large and d_Θ is small, then forward-mode differentiation will have a much lower cost than reverse-mode differentiation. However, in deep learning we typically consider models where $d_O = 1$ (a scalar objective function or loss) and d_Θ (the number of parameters) is very large. Then, it is clear that reverse-mode differentiation has a *much lower* computational cost since (24.6) grows linearly in d_Θ.

24.3. Introduction to PyTorch Automatic Differentiation

Both forward-mode and reverse-mode differentiation can be implemented in computer algorithms for general classes of functions. This is referred to as automatic differentiation.

We will focus on using PyTorch for reverse-mode automatic differentiation. PyTorch is a *define-by-run* framework where the function (which will be differentiated) is determined/defined at runtime and then can be subsequently differentiated. This provides a high degree of flexibility, since the function to be differentiated can dynamically be defined (and change) at runtime. Examples include conditionals (e.g., if-else statements) and for-loops of variable length.

In contrast, *define-and-run* automatic differentiation software would first define a *static* function that cannot be changed during runtime and would then be differentiated. PyTorch is also widely used since its syntax is *Pythonic* (i.e., very similar to standard Python code) and is seamlessly integrated into Python. Function evaluation and differentiation can be easily run on GPUs, which significantly accelerates training of models involving large numbers of parameters and large datasets.

PyTorch provides highly general automatic differentiation for calculating the gradients of scalar functions. Consider the following example:

```
import torch

x = torch.ones(2)
x.requires_grad=True
f = torch.sum( x**3 + x**2 )
f.backward()
```

The first line initializes the tensor which is the input to the function. The second line uses "x.requires_grad=True" to indicate that we would like to take the derivative of the function output with respect to the input variable x. In the fourth line, the ".backward()" operation implements PyTorch's automatic differentiation to differentiate the function output f with respect to the input variable x. The gradients which are calculated by automatic differentiation are stored in the tensor "x.grad.data". Note that PyTorch's standard automatic differentiation with the backward() operation requires that the output of the function to be differentiated be a scalar (and not a vector or tensor). In machine learning we will typically be evaluating the gradient of the loss, which is a scalar, and therefore this covers the vast majority of machine learning applications.

If the tensors are located on the GPU, the automatic differentiation will also be performed on the GPU. For large tensor operations, this typically significantly accelerates the calculation of gradients:

```
import torch

x = torch.ones(2).cuda()
x.requires_grad=True
f = torch.sum( x**3 + x**2 )
f.backward()

print(x.grad.data)
```

If the machine has multiple GPUs, we can place the tensor on the jth GPU using the command "x = torch.ones(2).cuda(device = j)". Then, the gradients would also be calculated on the jth GPU.

The computational graph showing the sequence of operations which PyTorch keeps track of to calculate the chain rule can be displayed. Consider the following sequence of functions:

```
x = torch.ones(2).cuda()
x.requires_grad=True
f = torch.sum( x**3 + x**2 )
f = 2*f
f = torch.sigmoid(f)
f = 5*f
f.backward()
```

PyTorch keeps track of the sequence of operations (and intermediate function outputs) in order to calculate the chain rule. The computational graph (and order in which the chain rule is applied) can be displayed via the following command:

```
g = f.grad_fn
print(g)
g = f.grad_fn.next_functions[0][0]
print(g)

while ( bool( g.next_functions ) ):
  g = g.next_functions[0][0]
  print(g)
```

For the specific example of above, the code above displays the following computational graph:

```
<MulBackward0 object at 0x7811a52425f0>
<SigmoidBackward0 object at 0x7811a5242740>
<MulBackward0 object at 0x7811a52425f0>
<SumBackward0 object at 0x7810e41a53f0>
<AddBackward0 object at 0x7810e41a52a0>
<PowBackward0 object at 0x7811a5242650>
<AccumulateGrad object at 0x7810e41a4f40>
```

PyTorch is able to differentiate a wide class of functions. We next provide a slightly more complex example of a single-layer neural network:

```
#Dimension of input data
d = 2
#Number of hidden units
H = 5
#Input data
x = torch.randn((d,1))

#Weight parameter matrices
W = torch.randn((H,d) )
B = torch.randn((H,1))
C = torch.randn((1,H))

W.requires_grad = True
B.requires_grad = True
C.requires_grad = True

HiddenLayer = torch.sigmoid( torch.matmul(W, x) + B )
f = torch.matmul(C, HiddenLayer)

f.backward()

print(W.grad.data, B.grad.data, C.grad.data)
```

The above example can also be conveniently implemented by creating a neural network module:

```
import torch
import torch.nn as nn
import torch.optim as optim

#Dimension of input data
d = 1
#Number of hidden units
H = 100

class SingleLayerNeuralNetwork(nn.Module):
    def __init__(self):
        super(SingleLayerNeuralNetwork, self).__init__()

        self.fc1 = nn.Linear(d, H).type(torch.FloatTensor)
        self.fc2 = nn.Linear(H, 1, bias=False).type(torch.FloatTensor)

    def forward(self, x ):

        HiddenLayer = torch.sigmoid( self.fc1( x ) )
        f = self.fc2(HiddenLayer)

        return f

model = SingleLayerNeuralNetwork()

#Move model to GPU
model.cuda()
```

```
LR = 0.001
optimizer = optim.RMSprop(model.parameters(), lr= LR, momentum=0.0)

#Minibatch size
M = 10000

#Number of optimization steps
T = 100000

for i in range(T):

    optimizer.zero_grad()

    x = torch.randn(M,1).cuda()
    y = x**2

    f = model(x)

    Loss = torch.mean( (y - f)**2 )

    Loss.backward()

    optimizer.step()

    print(i, Loss.detach().cpu().numpy() )
```

The above example trains a neural network to represent a parabola on a randomly sampled datapoints. The *optimizer* object implements the updates of the parameters. In this case, the RMSProp optimizer is chosen, although other choices exist (e.g., standard gradient descent or ADAM). "Loss.backward()" calculates the gradients via automatic differentiation.

The gradients of the parameters are available in the fields "model.fc1. weight.grad.data", "model.fc1.bias.grad.data", and "model.fc2.weight.grad. data". The parameters are updated with the RMSProp algorithm when "optimizer.step()" is called.

PyTorch, by default, accumulates gradients. This means that every time "optimizer.step()" is called, the calculated gradients are added to the current tensors "model.fc1.weight.grad.data", "model.fc1.bias.grad.data", and "model. fc2.weight.grad.data". That is—by default—these "grad" fields in the neural network model will actually be the sum of all calculated gradients over all time steps. The sum of the gradients would then be used to update the parameters, which is not the correct gradient descent algorithm. Instead, we would like to only use the gradients calculated at the current training iteration i. This can be implemented by including "optimizer.zero _grad()" at the beginning of each training iteration. The command "optimizer.zero _grad()" sets all of the gradient fields in the neural network model to zero.

PyTorch can also differentiate over piecewise continuously differentiable functions which are defined with condition (if-else) statements:

```
x = torch.randn(2)
x.requires_grad=True
if (x[0] < 0):
  f = torch.sum( x**3 + x**2 )
else:
  f = torch.sum(x)
f.backward()

print(x[0], x.grad.data)
```

If we would like to exclude certain parts of the function from being included in the chain rule, we can use the ".detach()" operation:

```
x = torch.ones(2).cuda()
x.requires_grad=True
f = torch.sum( x**3 + (x**2).detach() )
f.backward()

print(x.grad.data)
```

In the example above, the output of x^2 is treated as a constant and is not differentiated. Only the first term x^3 is differentiated. From the perspective of PyTorch's automatic differentiation, the output of x^2 is frozen as a constant whose derivative is zero.

In some cases we may not wish to calculate the gradients of a series of PyTorch operations or functions. An example is when an already trained PyTorch model is being purely used for predictions (inference). Keeping track of the computational graph—including storing data from intermediate function evaluations—in order to calculate the chain rule can have significant memory costs. If the model is only being evaluated for predictions (and it is not necessary to keep track of the computational graph), we can use the following command:

```
with torch.no_grad():
      x = torch.ones(2).cuda()
      x.requires_grad=True
      f = torch.sum( x**3 + (x**2).detach() )
```

The computational graph for the function evaluations within the "with torch.no _grad():" block will not be stored. Automatic differentiation of the function f is, therefore, not possible. In general, (already trained) PyTorch models which are being used for predictions (sometimes referred to as *inference*) should be evaluated using "with torch.no _grad():" or "with torch.inference _mode():" to reduce memory costs.

Automatic differentiation of large models for large minibatches can have both high memory and computational costs. Large numbers of arithmetic operations are required to calculate the chain rule. In the forward evaluation of the model, the intermediate function outputs must be stored to be later used to evaluate the chain rule, which can require large amounts of memory. If an out-of-memory error occurs, the minibatch size can be reduced. The standard choice for data types in deep learning is 32-bit floating point (float32), which has lower memory cost than the default choice in scientific computing of 64-bit floating point (float64). An alternative approach with even lower memory cost is 16-bit floating point (float16). Finally, as discussed in the chapter on distributed model training, Chapter 23, the calculation of the gradient of a minibatch can be parallelized across multiple GPUs/machines.

24.4. Brief Concluding Remarks

A fundamental element of deep learning is the design and evaluation of different model architectures, which include a large number of hyperparameters. Typically, a series of different models will be designed and evaluated on data. The chain rule for the backpropagation algorithm will change each time a new model is designed. If the backpropagation algorithm had to be rederived from scratch for each new model, this would be a significant obstacle to the development of deep learning models. Automatic differentiation addresses this challenge by automatically calculating the chain rule (and gradients with respect to the model parameters). Automatic differentiation therefore facilitates model development and evaluation. For a more detailed discussion of automatic differentiation, we recommend reading [**ea19**], [**Gil08**], and [**GW08**].

Part 3

Appendixes

Appendix A

Background Material in Probability

A.1. Basic Notions in Probability

In this section we review basic things about probability theory and we visit notions that are frequently seen in the book. There are many excellent classical texts for probability theory and convergence topics, see for example [**Bil99**].

Definition A.1. Consider a probability space $(\Omega, \mathcal{F}, \mathbb{P})$ where

- Ω is the sample space.
- $\mathcal{F}$ is the so-called σ-algebra (i.e., a collection of subsets of Ω which is closed under complements and countable unions and which also contains Ω.)
- $\mathbb{P}$ is the probability measure.

Definition A.2. Let $(\Omega_1, \mathcal{F}_1)$ and $(\Omega_2, \mathcal{F}_2)$ be two measurable spaces. A function $X : \Omega_1 \mapsto \Omega_2$ is called a random variable if the event $\{\omega \in \Omega_1 : X(\omega) \in A\} \in \mathcal{F}_1$ for every $A \in \mathcal{F}_2$.

Definition A.3. The expectation of X is defined as

$$\mathbb{E}(X) = \int_\Omega X(\omega) d\mathbb{P}(\omega).$$

More generally, if $X : \Omega_1 \mapsto \Omega_2$ is a random variable per the previous definition and $f : \Omega_2 \mapsto \mathbb{R}$ is an $\mathcal{F}_2$ measurable function, then

$$\mathbb{E}(f(X)) = \int_\Omega f(X(\omega)) d\mathbb{P}(\omega).$$

Note that $\mathbb{P}(X \in A) = \int_A d\mathbb{P}(\omega)$. Hence, $\mathbb{E}(1_{X \in A}) = \mathbb{P}(X \in A)$.

Definition A.4. If we can write $d\mathbb{P}(x) = f(x)dx$, then f is called the probability density of X, or pdf for short.

Definition A.5. Let $p > 0$. We say that $X \in L^p(\mathbb{P})$ if $\|X\|_{L^p} = \left(\mathbb{E}|X|^p\right)^{1/p} < \infty$.

Definition A.6. Let $X \in L^2(\mathbb{P})$. We define the variance of the random variable X as

$$\mathrm{Var}(X) = \mathbb{E}\left[(X - \mathbb{E}(X))^2\right].$$

If X is a multidimensional random variable, then we are talking about the variance-covariance matrix which is defined as

$$\mathrm{Var}(X) = \mathbb{E}\left[(X - \mathbb{E}(X)) \cdot (X - \mathbb{E}(X))^\top\right].$$

Let us see some examples now.

Example A.7. If X is distributed as an $\mathrm{Exp}(\lambda)$ random variable with $\lambda > 0$, then $\mathbb{P}[X > x] = e^{-\lambda x}$. In this case the probability density function takes the form $f(x) = \lambda e^{-\lambda x}$ for $x \geq 0$ and $f(x) = 0$ for $x < 0$. In regards to its expectation and variance we have, respectively,

$$\mathbb{E}X = \int_0^\infty x\lambda e^{-\lambda x} dx = \frac{1}{\lambda},$$

$$\mathrm{Var}\,X = \int_0^\infty \left(x - \frac{1}{\lambda}\right)^2 \lambda e^{-\lambda x} dx = \frac{1}{\lambda^2}.$$

Example A.8. If X is distributed as a $\mathrm{Uniform}(\theta_1, \theta_2)$ random variable with $\theta_1 < \theta_2$, then for $x \in (\theta_1, \theta_2)$, $\mathbb{P}[X > x] = \frac{\theta_2 - x}{\theta_2 - \theta_1}$. In this case the probability density function takes the form $f(x) = \frac{1}{\theta_2 - \theta_1}$ for $x \in (\theta_1, \theta_2)$, and $f(x) = 0$ otherwise. In regards to its expectation and variance we have, respectively,

$$\mathbb{E}X = \frac{\theta_2 + \theta_1}{2},$$

$$\mathrm{Var}\,X = \frac{(\theta_2 - \theta_1)^2}{12}.$$

Example A.9. If X is distributed as a $\mathrm{Normal}(\mu, \sigma^2)$ (or equivalently $X \sim N(\mu, \sigma^2)$) random variable, then the probability density function takes the form $f(x) = \frac{1}{\sqrt{2\pi\sigma^2}} e^{-\frac{1}{2}\frac{(x-\mu)^2}{\sigma^2}}$ for $x \in \mathbb{R}$. In regard to its expectation and variance we have, respectively,

$$\mathbb{E}X = \mu,$$

$$\mathrm{Var}\,X = \sigma^2.$$

Definition A.10. The conditional probability of an event B given another event A is $\mathbb{P}(B|A) = \frac{\mathbb{P}(B\cap A)}{\mathbb{P}(A)}$.

If A, B are independent events, then $\mathbb{P}(B|A) = \mathbb{P}(B)$ since in that case we have that $\mathbb{P}(B \cap A) = \mathbb{P}(B)\mathbb{P}(A)$.

If $f_{X,Y}(x, y)$ is the joint distribution of (X, Y), then for a nice function $g : \mathbb{R}^2 \mapsto \mathbb{R}$,

$$\mathbb{E}\left[g(X, Y)|Y = a\right] = \frac{\int_{-\infty}^{\infty} g(x, a) f_{X,Y}(x, a)dx}{\int_{-\infty}^{\infty} f_{X,Y}(x, a)dx}.$$

Theorem A.11 (Bayes theorem). *Let A, B be two events with $\mathbb{P}(A) \neq 0$ and $\mathbb{P}(B) \neq 0$. Then, we have that*

$$\mathbb{P}(A|B) = \frac{\mathbb{P}(B|A)\mathbb{P}(A)}{\mathbb{P}(B)}.$$

In general, if $\{Z_j\}$ is a countable partition of Ω (i.e., $\bigcup_j Z_j = \Omega$ with $Z_i \cap Z_j = \emptyset$ for $i \neq j$), then

$$\mathbb{P}(Z_j|B) = \frac{\mathbb{P}(B|Z_j)\mathbb{P}(Z_j)}{\sum_j \mathbb{P}(B|Z_j)\mathbb{P}(Z_j)},$$

and $\mathbb{P}(B) = \sum_j \mathbb{P}(B|Z_j)\mathbb{P}(Z_j)$.

A.2. Basics on Stochastic Processes

Definition A.12. A stochastic process X_t for $t \in \mathcal{T}$ is a collection of random variables from $(\Omega_1, \mathcal{F}_1)$ (the sample space) to $(\Omega_2, \mathcal{F}_2)$ (the state space).

We write without distinction $X_t, X(t, \omega), X_t(\omega)$ for a stochastic process with the understanding that all of these notations mean the same thing. In many typical situations $\Omega_2 = \mathbb{R}^d$, and in that case $\mathcal{F}_2$ is the σ-algebra of the subsets of $\mathbb{R}^d$.

Definition A.13. For $k \in \mathbb{N}$ and $(t_1, \dots, t_k) \in \mathcal{T}^k$, the collection of random variables $(X_{t_1}, \dots, X_{t_k})$ is called the finite dimensional distribution of X.

Definition A.14. A filtration on $(\Omega, \mathcal{F})$ is a non-decreasing family $\{\mathcal{F}_t\}_{t\in\mathcal{T}}$ of sub-σ-algebras of $\mathcal{F}$ such that $\mathcal{F}_s \subset \mathcal{F}_t \subset \mathcal{F}$ for $s \leq t$. We set $\mathcal{F}_\infty = \sigma\left(\bigcup_{t\in\mathcal{T}} \mathcal{F}_t\right)$. The filtration generated by X_t is denoted by $\mathcal{F}_t^X = \sigma\left(X_s; s \leq t\right)$. We say that X_t is adapted to $\mathcal{F}_t$ if X_t is $\mathcal{F}_t$-measurable.

Definition A.15. Let X_t be defined on $(\Omega, \mathcal{F}, \mathbb{P})$ and let $\mathcal{F}_t^X = \sigma(X_s, s \leq t)$. We say that $\{X_t\}$ is a Markov process if

$$\mathbb{P}(X_{s+t} \in A|\mathcal{F}_s^X) = \mathbb{P}(X_{s+t} \in A|X_s),$$

for all $s, t \in \mathcal{T}$ and for all A in the state space.

If $\mathcal{F}_t$ is a filtration such that $\mathcal{F}_t^X \subset \mathcal{F}_t$ for every $t \in \mathcal{T}$ and $\mathbb{P}(X_{s+t} \in A|\mathcal{F}_s^X) = \mathbb{P}(X_{s+t} \in A|X_s)$, then we say that X_t is a Markov process with respect to $\mathcal{F}_t$.

Markov property means that the past is irrelevant and only the present matters when it comes to future behaviour!

Definition A.16. We write $\mathbb{P}(X_{s+t} \in A|\mathcal{F}_s^X) = p(s,t,X_s,A)$, and we call the function p the transition function.

The transition function satisfies the Chapman-Kolmogorov equation,

$$\int_E p(s,t,x,dy)p(t,u,y,A) = p(s,u,x,A),$$

for every $s < t < u$ and $A \in \mathcal{B}(E)$.

Definition A.17. Let $\mathcal{F}_t$ be a filtration on $(\Omega, \mathcal{F}, \mathbb{P})$ and let X_t be adapted to $\mathcal{F}_t$. We say that X_t is an $\mathcal{F}_t$-martingale if

(1) $\mathbb{E}|X_t| < \infty$ for all $t \in \mathcal{T}$.

(2) $\mathbb{E}[X_{t+s}|\mathcal{F}_s] = X_s$ for all $s, t \in \mathcal{T}$.

One of the main examples of a Markov stochastic process is the so-called Brownian motion, otherwise called the Wiener process.

Definition A.18. A one-dimensional Brownian motion $W_t : \Omega \times \mathbb{R}^+ \mapsto \mathbb{R}$ is a real-valued stochastic process with the following properties:

(1) $W_0 = 0$ almost surely.

(2) $t \mapsto W_t$ is continuous.

(3) W_t has independent increments. This means that if $t_1 < t_2 < \cdots < t_k$, then $W_{t_{n+1}} - W_{t_n}$ is independent of $W_{t_{m+1}} - W_{t_m}$ for any $n+1 \leq m$.

(4) $W_t - W_s$ is distributed as $N(0, t-s)$.

A d-dimensional Brownian motion is a collection of one-dimensional Brownian motions, i.e., $W_t = (W_t^1, \ldots, W_t^d)$, where $W_t^i, i = 1, \ldots, d$ is a collection of one-dimensional independent Brownian motions.

Some important properties of Brownian motion follow:

(1) $\mathbb{E}W_t = 0$.

(2) $\mathbb{E}W_t W_s = \min(t, s)$.

(3) $\mathbb{E}[W_t|\mathcal{F}_s^W] = \mathbb{E}W_s$ (martingale property).

(4) W_t is a Markov process.

(5) $W_{ct} = \sqrt{c}W_t$ where c is a real-valued positive constant.

(6) $t \mapsto W_t$ is continuous but nowhere-differentiable. In this regard, observe that

$$\begin{aligned}\mathrm{Var}\left[\frac{W_{t+\Delta t} - W_t}{\Delta t}\right] &= \frac{1}{(\Delta t)^2}\mathbb{E}\left[W_{t+\Delta t} - W_t - \mathbb{E}[W_{t+\Delta t} - W_t]\right]^2 \\ &= \frac{1}{(\Delta t)^2}(t + \Delta t - t) \\ &= \frac{1}{\Delta t} \to \infty, \quad \text{as } \Delta t \to 0.\end{aligned}$$

While this calculation is not a proof that $t \mapsto W_t$ is not differentiable, it is strongly suggestive of this fact.

The next result is one of the most used results when it comes to martingales.

Theorem A.19 (Doob, see [**KS98**]). *Let $\{X_t, t \in \mathcal{T}\}$ be a martingale with respect to the filtration $\mathcal{F}_t$ that is also right-continuous. Then, for every $\lambda > 0$, $p \geq 1$, $a > 1$ we have*

(1) $\mathbb{P}\left(\sup_{0\leq s\leq t} |X_s| > \lambda\right) \leq \frac{1}{\lambda^p}\mathbb{E}|X_t|^p$ *for t such that* $\mathbb{E}|X_t|^p < \infty$.

(2) $\mathbb{E}\left(\sup_{0\leq s\leq t} |X_s|^a\right) \leq \left(\frac{a}{a-1}\right)^a \mathbb{E}|X_t|^a$ *for t such that* $\mathbb{E}|X_t|^a < \infty$.

(3) *If* $\sup_t \mathbb{E}|X_t| < \infty$, *then the limit* $Y(\omega) = \lim_{t\to\infty} X_t(\omega)$ *exists almost surely and* $\mathbb{E}|Y| < \infty$.

A.3. Notions of Convergence and Tightness

There are many different ways in which a sequence of random variables $\{X^n\}$ can converge to a random variable X as $n \to \infty$. Below we review the main notions of convergence, discuss their relations and present some of the related key results.

Definition A.20. We say that we have the following.

(1) Convergence in probability ($X^n \overset{p}{\to} X$): for all $\epsilon > 0$,

$$\lim_{n\to\infty} \mathbb{P}(|X^n - X| > \epsilon) = 0.$$

(2) Convergence with probability 1 or almost surely ($X^n \overset{a.s.}{\to} X$):

$$\mathbb{P}(\lim_{n\to\infty} X^n = X) = 1.$$

(3) Convergence in $\mathcal{L}^p$ ($X^n \overset{\mathcal{L}^p}{\to} X$):

$$\lim_{n\to\infty} \mathbb{E}|X^n - X|^p = 0.$$

(4) Weak Convergence ($X^n \stackrel{w}{\to} X$): for every continuous and bounded function $f(x)$ we have that

$$\lim_{n\to\infty} \mathbb{E}f(X^n) = \mathbb{E}f(X).$$

Remark A.21 (Relations between different convergence notions). We have that

- $X^n \stackrel{a.s}{\to} X$ implies $X^n \stackrel{p}{\to} X$.
- $X^n \stackrel{\mathcal{L}^p}{\to} X$ implies $X^n \stackrel{p}{\to} X$.
- $X^n \stackrel{p}{\to} X$ implies $X^n \stackrel{w}{\to} X$

A very useful tool in proving convergence in probability is using Chebychev's inequality: for every $\delta > 0$,

$$\mathbb{P}\left(|X^n - X| > \delta\right) \leq \frac{\mathbb{E}(X^n - X)^2}{\delta^2}.$$

The books [**Bil99**] and [**EK86**] are standard resources on the topic of convergence of probability measures. In particular, very useful tools in proving and characterizing weak convergence are the following theorems.

Theorem A.22. *Let $\{X^n_t\}$ be a sequence of continuous stochastic processes defined on $(\Omega, \mathcal{F}, \mathbb{P})$ satisfying the following conditions*

(1) *There exists $p > 0$ such that $\sup_{n\in\mathbb{N}} \mathbb{E}|X^n_0|^p < \infty$.*

(2) *There exist $a, b > 0$ and $c = c(T) > 0$ such that*

$$\sup_{n\in\mathbb{N}} \mathbb{E}|X^n_t - X^n_s|^a \leq c|t-s|^{1+b}$$

for every T and for every $t, s \in [0, T]$.

Then, if $\mathbb{P}^n$ is the law of X^n, $\{\mathbb{P}^n\}$ is a tight sequence of probability measures.

Theorem A.22(2) shows that X^n_t will have convergent subsequence. In order to uniquely characterize the limit, we need convergence of the finite dimensional distributions. In particular, we have

Theorem A.23. *Let $\{X^n\}$, X be continuous stochastic processes defined on $(\Omega, \mathcal{F}, \mathbb{P})$ satisfying the conditions:*

(1) *$\{X^n\}_{n=1}^{\infty}$ is a tight sequence.*

(2) *The finite dimensional distributions of $\{X^n\}$ converge to those of X in $[0, \infty)$.*

Then, we have that $X^n_\cdot \stackrel{w}{\to} X_\cdot$ in $C([0, \infty))$.

A.4. Convergence in the Skorokhod Space $D_E([0,T])$

Let $0 < T < \infty$, and let E be a given set that will be more precisely characterized below. In this section we briefly review the basics for the Skorokhod space $D_E([0,T])$ (definition follows below) and for some of its convergence properties. This space is very well suited for proving convergence of families of stochastic processes and is heavily used in Chapters 19 and 20. An excellent source on characterization and convergence of stochastic process is the book **[EK86]** to which we refer the interested reader for further details.

In practice, for many stochastic processes we can assume that sample paths are right continuous and have left limits at each time point (the so-called càdlàg processes). This means that for a stochastic process $X_t : [0,T] \mapsto E$, we shall have that for each $t \in [0,T]$, the limits $\lim_{s\to t+} X_s = X_t$ and $\lim_{s\to t-} X_s = X_{t-}$ exist. Such stochastic processes compose the space $D_E([0,T])$.

Our interest in this book is primarily in convergence for measure-valued stochastic processes in which cases we are mainly interested in complete (i.e., every Cauchy sequence in a given space converges in that space) and separable metric spaces (i.e., a topological space containing a countable everywhere-dense set). Under the appropriate metric, d (defined below), the space $D_E([0,T])$ is a separable metric space if E is separable, and $(D_E([0,T]), d)$ complete if (E,r) is a complete metric space (this is a theorem). In particular, letting Z be the collection of Lipschitz continuous, strictly increasing functions $\zeta : [0,\infty) \mapsto [0,\infty)$ such that

$$\gamma(\zeta) = \sup_{0\le t<s} \left|\log \frac{\zeta(s) - \zeta(t)}{s-t}\right| < \infty,$$

and $q = r \wedge 1 = \min\{r, 1\}$, set, for $x, y \in D_E([0,T])$

$$d(x,y) = \inf_{\zeta\in Z} \left[\gamma(\zeta) \vee \int_0^T e^{-u} \left(\sup_{t\ge 0} q\left(x_{t\wedge u}, \gamma(\zeta_t \wedge u)\right)\right) du\right].$$

An important result in this direction is that if $X^n, X \in D_E([0,T])$, then $\lim_{n\to\infty} d(X^n, X) = 0$ if and only if there exists a sequence $\{\zeta^n\}$ in Z such that $\lim_{n\to\infty} \gamma(\zeta^n) = 0$ and $\lim_{n\to\infty} \sup_{0\le t\le T} r(X_t^n, X_{\zeta_t^n}) = 0$.

We say that the sequence $\{X^n\}$ is relatively compact if the sequence of the corresponding probability measures $\mathbb{P}^n$ is compact. The following two results are very useful in practice and are used routinely in Chapters 19 and 20.

Theorem A.24 (Theorem 7.2, Chapter 3 of **[EK86]**). *Assume that (E,r) is a complete and separable metric space and consider the family of stochastic processes $\{X^n\}$ with sample paths in $D_E([0,T])$. Then $\{X^n\}$ is relatively compact if and only if the following hold.*

- *For every rational number* $t \in [0,T]$ *and* $\epsilon > 0$, *there exists a compact set* $K(t,\epsilon) \subset E$ *such that*

$$\liminf_{n\to\infty} \mathbb{P}\left[X_t^n \in K(t,\epsilon)\right] \geq 1-\epsilon. \tag{A.1}$$

- *For every* $0 < T' \leq T$ *and* $\epsilon > 0$, *there exists a* $\delta > 0$ *so that*

$$\limsup_{n\to\infty} \mathbb{P}\left[w(X^n,\delta,T') \geq \epsilon\right] \leq \epsilon,$$

where $w(x,\delta,T')$ *is the modulus of continuity of* x *in the interval* $[0,T']$ *defined as*

$$w(x,\delta,T') = \inf_{\text{partition}\{t_i\}\subset[0,T'],\min(t_i-t_{i-1})>\delta} \max_i \sup_{s,t\in[t_{i-1},t_i)} r(x_s,x_t).$$

Remark A.25 (Remark 7.3 in Chapter 3 of [**EK86**]). Let $\{X^n\}$ be a relatively compact sequence. Then for every $T < \infty$ and $\epsilon > 0$, there exists a compact set $K(T,\epsilon) \subset E$ such that

$$\liminf_{n\to\infty} \mathbb{P}\left[X_t^n \in K(T,\epsilon) \text{ for } 0 \leq t \leq T\right] \geq 1-\epsilon. \tag{A.2}$$

Theorem A.26 (Theorem 8.6, Chapter 3 of [**EK86**]). *Assume that* (E,r) *is a complete and separable metric space and consider the family of stochastic processes* $\{X^n\}$ *with sample paths in* $D_E([0,T])$. *Assume that condition* (A.1) *holds. Then the following are equivalent.*

(1) $\{X^n\}$ *is relatively compact.*

(2) *For each* $0 < T' \leq T$, *there exists* $\beta > 0$ *and a family* $\kappa^n(\delta)$ *with* $0 < \delta < 1$ *of non-negative random variables such that the inequality holds*

$$\mathbb{E}\left[q^\beta(X_{t+u}^n, X_t^n)|\mathcal{F}_t^n\right] q^\beta(X_t^n, X_{t-v}^n) \leq \mathbb{E}\left[\kappa^n(\delta)|\mathcal{F}_t^n\right],$$

for $0 \leq t \leq T'$, $0 \leq u \leq \delta$, $0 \leq v \leq \delta \wedge t$, $\mathcal{F}_t^n$ *is the* σ*-algebra generated by* X^n *up to time* t, *and in addition*

$$\lim_{\delta\to 0} \limsup_{n\to\infty} \mathbb{E}\left[\kappa^n(\delta)\right] = 0$$

and

$$\lim_{\delta\to 0} \limsup_{n\to\infty} \mathbb{E}\left[q^\beta(X_\delta^n, X_0^n)\right] = 0.$$

(3) *For each* $0 < T' \leq T$, *there exists* $\beta > 0$ *such that*

$$\lim_{\delta\downarrow 0} \limsup_{n\to\infty} \left[\sup_{\tau\in S^n(T')} \sup_{0\leq u\leq\delta} \mathbb{E}\left(\sup_{0<v\leq\delta\wedge\tau} q^\beta(X_{\tau+u}^n, X_\tau^n) q^\beta(X_\tau^n, X_{\tau-v}^n)\right)\right] = 0,$$

where $S^n(T')$ *is the collection of all discrete* $\{\mathcal{F}_t^n\}$*-stopping times bounded by* T', *and in addition*

$$\lim_{\delta\to 0} \limsup_{n\to\infty} \mathbb{E}\left[q^\beta(X_\delta^n, X_0^n)\right] = 0.$$

holds.

Theorem A.27 (Theorem 9.1 in Chapter 3 of [**EK86**]). *Assume that* (E, r) *is a complete and separable metric space, and consider the family of stochastic processes* $\{X^n\}$ *with sample paths in* $D_E([0, T])$. *Assume that condition* (A.2) *holds. In the topology of uniform convergence on compacts, consider a dense subset G of* $C_b(E)$. *Then* $\{X^n\}$ *is relatively compact if and only if for each* $g \in G$, $\{g(X^n)\}$ *is relatively compact in* $D_{\mathbb{R}}([0, T])$.

Note that in all of the above we can take T to be ∞, in which case the relevant space is $D_E([0, \infty)])$. The topic of characterization and convergence of stochastic processes is very rich and very well developed in probability theory. An excellent classic manuscript covering many of the basic and more advanced related results is [**EK86**], to which the interested reader is referred to for further details and proofs.

A.5. Some Limiting Results and Concentration Bounds

Two of the most classical results in probability and statistics are the law of large numbers and central limit theorem. We present them below.

Theorem A.28 (Law of large numbers). *Let* $\{X^n\}$ *be a sequence of independent and identically distributed random variables with* $\mathbb{E}X^n = \mu < \infty$. *Define* $Y^N = \frac{1}{N}\sum_{n=1}^{N} X^n$. *Then, we have that*

(1) $Y^N \overset{a.s.}{\to} \mu$.

(2) *If* $\mathbb{E}|X^n|^p < \infty$, *then* $Y^N \overset{\mathcal{L}^p}{\to} \mu$.

Theorem A.29 (Central limit theorem). *Let* $\{X^n\}$ *be a sequence of independent and identically distributed random variables with* $\mathbb{E}|X^n|^2 < \infty$ *such that* $\mathbb{E}X^n = \mu$ *and* $\operatorname{Var} X^n = \sigma^2$. *Define* $S^N = \frac{\sum_{n=1}^{N} X^n - N\mu}{\sqrt{N}\sigma}$. *Then, we have that* $S^N \overset{w}{\to} N(0, 1)$.

Another useful result is that of Chernoff-type concentration bounds. We present a special case of interest to us below. Bounds of this type are often called Chernoff-Hoeffding bounds.

Lemma A.30 (Chernoff bound for Bernoulli random variables). *Let* $\{X^n\}$ *be a sequence of independent and identically Bernoulli distributed random variables with* $\mathbb{P}(X^n = 1) = p$ *and* $\mathbb{P}(X^n = 0) = 1 - p$. *Define* $Y^N = \frac{1}{N}\sum_{n=1}^{N} X^n$. *Then for any* $\delta > 0$, *we have the estimate*

$$\mathbb{P}\left(|Y^N - p| > \delta\right) \leq 2e^{-\frac{N\delta^2}{2p(1-p)}}.$$

Note that the Chernoff bound we just saw says something useful. It says that if we want the average Y^N to be within a ball of radius δ from its expectation p with probability at least $1-\epsilon$, then we would need to have

$$N \geq \frac{2p(1-p)}{\delta^2} \log \frac{2}{\epsilon}$$

samples in our disposal.

If we do not have information about p, then using the property that $p(1-p) \leq 1/4$ (true since $p \in (0,1)$), we obtain the cruder (but perhaps more useful) bounds

$$\mathbb{P}\left(|Y^N - p| > \delta\right) \leq 2e^{-2N\delta^2}$$

and

$$N \geq \frac{1}{2\delta^2} \log \frac{2}{\epsilon}.$$

To state the next convergence result, we need to introduce the notion of quadratic variation.

Definition A.31. Let X_t be a square integrable martingale with respect to $\mathcal{F}_t$, i.e., $\mathbb{E}X_t^2 < \infty$ for all $t \in \mathcal{T}$. Let $\{t_n^N\}$ be a partition of $[0,t]$. Then, we define the quadratic variation of X_t to be

$$\langle X \rangle_t = \lim_{N\to\infty} \sum_{n=1}^{N} \left(X_{t_{n+1}^N} - X_{t_n^N}\right)^2, \text{ in } \mathcal{L}^1.$$

Theorem A.32 (Martingale central limit theorem). *Let $\{M_t\}$ be a sequence of right continuous, square integrable martingales on a probability space $(\Omega, \mathcal{F}, \mathbb{P})$ with respect to a filtration $\mathcal{F}_t$. Let $\langle M \rangle_t$ be its quadratic variation. Assume that*

- *$M_0 = 0$ almost surely.*
- *M_t has stationary increments.*
- *There is a constant $\sigma > 0$ such that*

$$\lim_{t\to\infty} \mathbb{E}\left[\left|\frac{\langle M \rangle_t}{t} - \sigma\right|\right] = 0.$$

Then, we have that

(1) $\frac{1}{\sqrt{t}} M_t \overset{w}{\to} N(0,\sigma)$ *as $t \to \infty$.*

(2) $\epsilon M_{t/\epsilon^2} \overset{w}{\to} \sqrt{\sigma} W_t$, *as $\epsilon \downarrow 0$ where W_t is a Brownian motion.*

A.6. Itô Stochastic Integral

In this section we shall introduce the Itô stochastic integral and go over some of its main properties. We refer the interested reader to excellent textbooks such as [**KS98**, **Pro05**, **RW00a**, **RW00b**] for a complete and rigorous treatment of this subject.

Definition A.33. Let $f(t) : [a,b] \mapsto \mathbb{R}$ and $\{t_n\}_{n=0}^N$ be a partition of $[a,b]$, i.e., $a = t_0 < t_1 < \cdots < t_{N-1} < t_N = b$. Then,

- The variation of $f(t)$ on $[a,b]$ is defined to be

$$\text{Variation}(f(t)) = \sup_{\{t_n\}_{n=0}^N} \sum_{n=0}^{N-1} |f(t_{n+1}) - f(t_n)|.$$

- $f(t)$ is of bounded variation if $\text{Variation}(f(t)) < \infty$.

At this point we mention that the standard Riemann-Stieltjes integral of the form $\int_a^b g(t)df(t)$ is well-defined only in the classical sense if $f(t)$ is of bounded variation.

Then the question arises. Given that the Wiener process W_t is not of bounded variation, how then does one make sense of the integral $\int_a^b g(t)dW_t$?

For this purpose, we first state the following definition.

Definition A.34. For real numbers $a < b$, we will say that $g : [a,b] \times \Omega \mapsto \mathbb{R}$ belongs in the class $\mathcal{L}[a,b]$ if the following hold.

- g is $\mathcal{B}([a,b]) \times \mathcal{F}$-measurable.
- $g(t,\cdot)$ is $\mathcal{F}_t$-measurable for every t.
- $\int_a^b \mathbb{E}[g^2(t,\cdot)]dt < \infty$.

Then, if $g \in \mathcal{L}[0,t]$, the Itô integral is defined to be the $\mathcal{L}^2$ limit of the Riemann sum

$$I(t) \overset{\mathcal{L}^2}{=} \lim_{N\to\infty} \sum_{n=1}^{N-1} g(t_{n-1})(W_{t_n} - W_{t_{n-1}}),$$

where $\{t_n\}_{n=1}^N$ is a partition of $[0,t]$. We shall write $I(t) = \int_0^t g(s)dW_s$, and $I(t)$ is called stochastic integral.

In contrast to the Riemann-Stieljes integral, where the point at which we evaluate the function g in the approximation above does not matter, the situation is different when it comes to stochastic integrals. In particular, if we write $I(t) = \int_0^t g(s)dW_s \approx \sum_{n-1}^{N-1} b_n(W_{t_n} - W_{t_{n-1}})$, then the following hold:

- If $b_n = g(t_{n-1})$, then we get the Itô stochastic integral that we defined above.

- If $b_n = \frac{1}{2}[g(t_{n-1}) + g(t_n)]$, then we get the so-called Stratonovich integral which is not the same as the Itô stochastic integral defined above.

Let us now close this section by collecting here some of the properties of the stochastic integral $I(t) = \int_0^t g(s)dW_s$.

(1) $\mathbb{E}\left[\int_0^t g(s)dW_s\right] = 0$.

(2) $\mathbb{E}\left(\int_0^t g(s)dW_s\right)^2 = \mathbb{E}\int_0^t g^2(s)ds$ (Itô isometry).

(3) $\int_0^t (af(s) + bg(s))\,dW_s = a\int_0^t f(s)dW_s + b\int_0^t g(s)dW_s$.

(4) $\int_a^b f(s)dW_s = \int_a^c f(s)dW_s + \int_c^b f(s)dW_s$.

(5) $\int_a^b f(s)dW_s$ is $\mathcal{F}_b$-measurable.

(6) If $\lim_{n\to\infty}\mathbb{E}\int_a^b (f_n(t) - f(t))^2dt = 0$, then $\lim_{n\to}\int_a^b f_n(t)dW_t = \int_a^b f(t)dW_t$ in the $\mathcal{L}^2$ sense.

(7) $\mathbb{E}\left[\int_0^t g(s)dW_s|\mathcal{F}_\rho\right] = \int_0^\rho g(s)dW_s$ for every $\rho \leq t$.

(8) $\mathbb{E}\left[\int_\rho^t g(s)dW_s|\mathcal{F}_\rho\right] = 0$.

(9) $\mathbb{E}\left[\left(\int_\rho^t g(s)dW_s\right)^2|\mathcal{F}_\rho\right] = \mathbb{E}\left[\int_\rho^t g^2(s)ds|\mathcal{F}_\rho\right]$.

A.7. Very Basics of Itô Stochastic Calculus

We refer the interested reader to excellent textbooks such as [**KS98**, **Pro05**, **RW00a**, **RW00b**] for a comprehensive treatment of stochastic calculus. Here we review the immediate concepts of interest used in this book.

Let $(\Omega, \mathcal{F}, \mathbb{P})$ be a probability space equipped with a filtration $\{\mathcal{F}_t\}_{t\geq 0}$. Let W_t be an $\mathcal{F}_t$ Wiener process defined on $(\Omega, \mathcal{F}, \mathbb{P})$ on d-dimensions.

Definition A.35. An Itô stochastic process on $[0, T]$ with values in $\mathbb{R}^d$ is a continuous stochastic process $\{X_t, t \geq 0\}$ such that for every $t \geq 0$,

$$X_t = X_0 + \int_0^t b(s)ds + \int_0^t \sigma(s)dW_s, \quad \mathbb{P} \text{ almost everywhere},$$

with X_0 being $\mathcal{F}_0$-measurable, $b(s), \sigma(s)$ being $\mathcal{F}_s$-measurable such that

$$\mathbb{P}\left[\sum_{i=1}^d \int_0^t |b^i(s)|ds + \sum_{i=1}^d \sum_{j=1}^m \int_0^t |\sigma^{i,j}(s)|^2 ds < \infty\right] = 1$$

and for all $t \in [0, T]$.

Sometimes, we often write the differential form $\dot{X}_t = b(t) + \sigma(t)\dot{W}_t$, but we always mean the integral form stated above.

The celebrated Itô formula is nothing else but chain rule involving stochastic integrals. In particular, let X_t be an Itô stochastic process as defined above, $f \in \mathcal{C}^2([0,\infty) \times \mathbb{R}])$. Then, we have that

$$\begin{aligned} f(t,X_t) &= f(0,X_0) \\ &+ \int_0^t \Bigg[\frac{\partial f}{\partial t}(s,X_s) + \sum_{i=1}^m b_i(s) \frac{\partial f}{\partial x_i}(s,X_s) \\ &+ \frac{1}{2} \sum_{j=1}^d \sum_{i,k=1}^m \sigma_{i,j}(s)\sigma_{k,j}(s) \frac{\partial^2 f}{\partial x_i \partial x_k}(s,X_s) \Bigg] ds \\ &+ \int_0^t \sum_{j=1}^d \sum_{i=1}^m \sigma_{i,j}(s) \frac{\partial f}{\partial x_i}(s,X_s) dW_s^j. \end{aligned}$$

Notice that if we define

$$\mathcal{L}^s f(x) = \frac{\partial f}{\partial t}(s,x) + \sum_{i=1}^m b_i(s) \frac{\partial f}{\partial x_i}(s,x) + \frac{1}{2} \sum_{j=1}^d \sum_{i,k=1}^m \sigma_{i,j}(s)\sigma_{k,j}(s) \frac{\partial^2 f}{\partial x_i \partial x_k}(s,x),$$

then we can write

$$f(t,X_t) = f(0,X_0) + \int_0^t \mathcal{L}^s f(X_s) ds + \int_0^t \sum_{j=1}^d \sum_{i=1}^m \sigma_{i,j}(s) \frac{\partial f}{\partial x_i}(s,X_s) dW_s^j.$$

In particular, with $f \in C^2(\mathbb{R}^d)$ the operator $\mathcal{L}^s f(x)$ is called the infinitesimal generator of the process X_t and is defined to be

$$\lim_{t \to 0} \frac{\mathbb{E} f(t,X_t) - f(0,x)}{t} = \mathcal{L}^0 f(x).$$

To see this is indeed true, let $f \in C_b^2(\mathbb{R}^d)$, apply the Itô formula, and take expectation (the mean of the stochastic integral will be zero) to obtain

$$\frac{1}{t} \left[\mathbb{E} f(t,X_t) - f(0,x) \right] = \mathbb{E} \left[\frac{1}{t} \int_0^t \mathcal{L}^s f(X_s) ds \right].$$

Due to the fact now that X_t is continuous, which implies that $s \mapsto \mathcal{L}^s f(X_s)$ is continuous and $\mathcal{L} f(\cdot)$ is bounded, we indeed obtain that

$$\lim_{t \to 0} \frac{\mathbb{E} f(t,X_t) - f(0,x)}{t} = \mathcal{L}^0 f(x).$$

A useful corollary of the general Itô formula is the so-called product formula. In particular, consider two Itô stochastic processes,

$$X_t = X_0 + \int_0^t b_1(s)ds + \int_0^t \sigma_1(s)dW_s,$$

$$Y_t = Y_0 + \int_0^t b_2(s)ds + \int_0^t \sigma_2(s)dW_s.$$

Then, by applying the Itô formula to $f(x, y) = xy$, we have that

$$X_tY_t = X_0Y_0 + \int_0^t [b_1(s)Y_s + b_2(s)X_s + \sigma_1(s)\sigma_2(s)]\, ds + \int_0^t [\sigma_1(s)Y_s + \sigma_2(s)X_s]\, dW_s.$$

We conclude this section, by presenting the notion of strong solution to a stochastic differential equation.

Definition A.36. Let b, σ be Borel-measurable functions, and let $\xi = (\xi_1, \ldots, \xi_d)$ be $\mathcal{F}_0$-measurable. X_t is called a strong solution to

$$X_t = \xi + \int_0^t b(s, X_s)ds + \int_0^t \sigma(s, X_s)dW_s, \tag{A.3}$$

if the following hold:

(1) $t \mapsto X_t$ is continuous.

(2) X_t is $\mathcal{F}_t$-measurable.

(3) $\mathbb{P}(X_0 = \xi) = 1$.

(4) $\mathbb{P}\left[\sum_{i=1}^d \int_0^t |b^i(s)|ds + \sum_{i=1}^d \sum_{j=1}^m \int_0^t |\sigma^{i,j}(s)|^2 ds < \infty\right] = 1$, for all $t \in [0, T]$.

(5) X_t satisfies the differential equation (A.3) $\mathbb{P}$ almost surely.

Then, we have the following theorem.

Theorem A.37. *Let b, σ be such that*

- $|b(t, x) - b(t, y)| + |\sigma(t, x) - \sigma(t, y)| \le K|x - y|$,
- $|b(t, x)|^2 + |\sigma(t, x)|^2 \le L(1 + |x|^2)$,

where $0 < K, L < \infty$. Let $\mathbb{E}\xi^2 < \infty$. Then, the equation (A.3) *has a unique strong solution such that*

$$\mathbb{E}(|X_t|^2) \le C(1 + \mathbb{E}|\xi|^2)e^{Ct}$$

with $0 < C < \infty$ and $t \in [0, T]$.

Appendix B

Background Material in Analysis

B.1. Basic Inequalities Used in the Book

We present here some classical inequalities that are used in this book.

- Cauchy-Schwarz inequality: for any $a_k, b_k \in \mathbb{R}$

$$\left(\sum_{k=1}^{n} a_k b_k\right)^2 \leq \left(\sum_{k=1}^{n} a_k^2\right)\left(\sum_{k=1}^{n} b_k^2\right).$$

  In particular if $b_k = 1$ for all k, then we obtain the special case

$$\left(\sum_{k=1}^{n} a_k\right)^2 \leq n\left(\sum_{k=1}^{n} a_k^2\right).$$

- Hölder inequality: for any $a_k, b_k \in \mathbb{R}$ and $p, q > 1$ such that $\frac{1}{p} + \frac{1}{q} = 1$,

$$\sum_{k=1}^{n} |a_k b_k| \leq \left(\sum_{k=1}^{n} |a_k|^p\right)^{1/p} \left(\sum_{k=1}^{n} |b_k|^q\right)^{1/q}.$$

- Young's inequality with $\epsilon > 0$,

$$|ab| \leq \frac{\epsilon}{2} a^2 + \frac{1}{2\epsilon} b^2.$$

- Jensen inequality: for any convex function φ and the sequence $\{a_k\}$ such that $\sum_{k\geq 1}^{n} a_k = 1$, we have

$$\varphi\left(\sum_{k=1}^{n} a_k x_k\right) \leq \sum_{k=1}^{n} a_k \varphi(x_k).$$

- Basic Grönwall inequality: for a non-negative function $\gamma(t)$, if the inequality

$$f(t) \leq g(t) + \int_{\alpha}^{t} \gamma(s) f(s) ds \quad \text{for all } t \in [\alpha, T]$$

holds, then we have

$$f(t) \leq g(t) + \int_{\alpha}^{t} g(s)\gamma(s) e^{\int_s^t \gamma(r)dr} ds \quad \text{for } t \in [\alpha, T].$$

If, in addition g is a non-decreasing function, then we have

$$f(t) \leq g(t) e^{\int_\alpha^t \gamma(r)dr} \quad \text{for } t \in [\alpha, T].$$

B.2. Basic Background in Analysis

We review some minimal background in analysis that will be useful for proofs of mainly the universal approximation theorem results of Chapter 16. For a more expanded discussion on real and functional analysis, the interested reader is referred to classical manuscripts such as [**Bre11, Lax02, RF10**].

Theorem B.1 (Riesz representation theorem). *Let $\phi : H \mapsto \mathbb{R}$ be a bounded linear functional on a Hilbert space H endowed with the inner product $(\cdot, \cdot)$. Then, there is a unique element $y_\phi \in H$ such that $\phi(x) = (x, y_\phi)$ for all $x \in H$. In addition $\|\phi\| = \|y_\phi\|$.*

An important example is the case where $H = L^2(K)$, the space of square integrable functions, say on $K \subset \mathbb{R}$. Consider for example the case of $K = [0, 1]$. Then, if $\phi : L^2([0, 1]) \mapsto \mathbb{R}$ is a bounded linear functional, then there is a unique $h \in L^2([0, 1])$ such that $\phi(g) = \int_0^2 g(x)h(x)dx$ for all $g \in L^2([0, 1])$.

For $p \neq 2$, the space $L^p([0, 1])$ is not a Hilbert space, but a similar result still holds. In particular if $\phi \in L^p([0, 1]) \mapsto \mathbb{R}$ is a bounded linear functional, then there exists a unique $h \in L^q([0, 1])$ such that $\phi(g) = \int_0^1 g(x)h(x)dx$ where $\frac{1}{p} + \frac{1}{q} = 1$.

For the next result, we denote by $C(K)$ the space of continuous functions on K.

Theorem B.2. *Let K be a compact set such that $K \subset \mathbb{R}^d$, and let ϕ be a bounded linear functional on $C(K)$. Then, there is a unique finite signed Borel measure μ on K such that*

$$\phi(g) = \int_K g(x)\mu(dx) \quad \text{for every } g \in C(K), \text{ and } \|\phi\| = |\mu|(K).$$

Theorem B.3. *Let K be a compact set such that $K \subset \mathbb{R}^d$, and let ϕ be a positive linear functional on $C(K)$. Then, there is a unique finite Borel measure μ on K such that*

$$\phi(g) = \int_K g(x)\mu(dx) \quad \text{for every } g \in C(K).$$

Definition B.4. A subspace U of X is dense in X with respect to the norm $\|\cdot\|$ if for all $x \in X$, there exists a sequence $\{u_k\} \in U$ such that $u_k \to x$ as $k \to \infty$.

Conversely, a subspace U of X is not dense in X if there exists $x_0 \in X$ such that no elements $u \in U$ are close to x_0.

Lemma B.5. *Let X be a normed, linear space, and let $U \subset X$ be a linear, non-dense subset of X. Then there exists a bounded, linear functional F on X, such that $F \neq 0$ on X (i.e., there is at least one point $x' \in X$ such that $F(x') \neq 0$) and $F(u) = 0$ for every $u \in U$.*

For the next result we define $I_d = [0,1]^d$ to be the hypercube in d dimensions, and we let $\mathcal{M}(I_d)$ be the space of finite signed measures on I_d.

Lemma B.6. *Consider $U \subset C(I_d)$ to be a linear, non-dense subset of $C(I_d)$. Then, we have that there exists a measure $\mu \in \mathcal{M}(I_d)$ with the property that $\int_{I_d} g(x)\mu(dx) = 0$ for all $g \in U$.*

Proof of Lemma B.6. Let us consider Lemma B.5 with $X = C(I_d)$. Then, there exists a bounded linear functional $F : C(I_d) \mapsto \mathbb{R}$ with the properties that $F \neq 0$ on $C(I_d)$ and $F(u) = 0$ for all $u \in U$. By Theorem B.3, there exists a $\mu \in \mathcal{M}(I_d)$ such that

$$F(f) = \int_{I_d} f(x)\mu(dx), \text{ for all } f \in C(I_d).$$

Thus, for any $g \in U$, we get that $F(g) = \int_{I_d} g(x)\mu(dx) = 0$, concluding the proof of the lemma. □

Remark B.7. Note that $F \neq 0$ in the previous lemma actually implies that $\mu \neq 0$.

We conclude this section with the important Stone-Weierstrass theorem that gives conditions on when a given set A is a dense subset of another set. This then means that such a set A can be used for approximation purposes. Before we present the theorem, we introduce the definition of algebra that separates points.

Definition B.8. A subset $A \subset C(K)$ is called an algebra, if the following hold:

(1) If $f, g \in A$, then $f + g \in A$.

(2) if $f \in A$ and $c \in \mathbb{R}$, then $cf \in A$.

(3) If $f, g \in A$, then $fg \in A$.

In addition, we say that an algebra A separates points in K if for all distinct $x, y \in K$, there exists $f \in A$ such that $f(x) \neq f(y)$.

As an example of a set being an algebra, consider the set of polynomial functions on an interval, say $[a, b]$:

$$A = \left\{ f(x) = \sum_{n=1}^{N} c_n x^k, x \in [a, b], c_n \in \mathbb{R}, N = 1, 2, \ldots \right\}.$$

Then, one can check that A is an algebra on $C([a, b])$ and in addition one can see that it separates points in $[a, b]$.

Theorem B.9 (Stone-Weierstrass theorem). *Consider K to be a compact set $K \subset \mathbb{R}^d$. Let A be an algebra of continuous real-valued functions on K. Assume that A contains the constant functions and that it separates points in K. Then, we have that A is a dense subset of $C(K)$.*

Bibliography

[ABMM18] R. Arora, A. Basu, P. Mianjy, and A. Mukherjee, *Understanding deep neural networks with rectified linear units*, ICLR (2018), 1–17.

[ACB17] Martin Arjovsky, Soumith Chintala, and Léon Bottou, *Wasserstein generative adversarial networks*, Proceedings of the 34th International Conference on Machine Learning (Doina Precup and Yee Whye Teh, eds.), Proceedings of Machine Learning Research, vol. 70, PMLR, 06–11 Aug 2017, pp. 214–223.

[ADBB17] K. Arulkumaran, M.P. Deisenroth, M. Brundage, and A.A. Bharath, *A brief survey of deep reinforcement learning*, arXiv:1708.05866 (2017).

[ADH+19] Sanjeev Arora, Simon Du, Wei Hu, Zhiyuan Li, and Ruosong Wang, *Fine-grained analysis of optimization and generalization for overparameterized two-layer neural networks*, 36th International Conference on Machine Learning, ICML 2019, 01 2019, pp. 477–502.

[Agr15] Alan Agresti, *Foundations of linear and generalized linear models*, Wiley Series in Probability and Statistics, John Wiley & Sons, Inc., Hoboken, NJ, 2015. MR3308143

[AGS08] Luigi Ambrosio, Nicola Gigli, and Giuseppe Savaré, *Gradient flows in metric spaces and in the space of probability measures*, 2nd ed., Lectures in Mathematics ETH Zürich, Birkhäuser Verlag, Basel, 2008. MR2401600

[ALM23] Andrea Agazzi, Jianfeng Lu, and Sayan Mukherjee, *Global optimality of elman-type RNNs in the mean-field regime*, Proceedings of the 40th International Conference on Machine Learning (Andreas Krause, Emma Brunskill, Kyunghyun Cho, Barbara Engelhardt, Sivan Sabato, and Jonathan Scarlett, eds.), Proceedings of Machine Learning Research, vol. 202, PMLR, 23–29 Jul 2023, pp. 196–227.

[AOY19] D. Araujo, R. I. Oliveira, and D. Yukimura, *A mean-field limit for certain deep neural networks.*

[Bac24] F. Bach, *Learning theory from first principles*, MIT Press, 2024.

[Bar94] A. Barron, *Approximation and estimation bounds for artificial neural networks*, Machine Learning **14** (1994), no. 1, 115–133.

[BB24] Christopher M. Bishop and Hugh Bishop, *Deep learning—foundations and concepts*, Springer, Cham, 2024, DOI 10.1007/978-3-031-45468-4. MR4719738

[BCN18] Léon Bottou, Frank E. Curtis, and Jorge Nocedal, *Optimization methods for large-scale machine learning*, SIAM Rev. **60** (2018), no. 2, 223–311, DOI 10.1137/16M1080173. MR3797719

[BD19] Dimitris Bertsimas and Jack Dunn, *Machine learning under a modern optimization lens*, first ed., Dynamic Ideas, LLC, Belmont, Massachusetts, 2019.

[BDK+22] Jeremiah Birrell, Paul Dupuis, Markos A. Katsoulakis, Yannis Pantazis, and Luc Rey-Bellet, *(f, Γ)-divergences: interpolating between f-divergences and integral probability metrics*, J. Mach. Learn. Res. **23** (2022), Paper No. [39], 70. MR4420764

[BEJ19] Christian Beck, Weinan E, and Arnulf Jentzen, *Machine learning approximation algorithms for high-dimensional fully nonlinear partial differential equations and second-order backward stochastic differential equations*, J. Nonlinear Sci. **29** (2019), no. 4, 1563–1619, DOI 10.1007/s00332-018-9525-3. MR3993178

[Ber03] P. Dimitris Bertsekas, *Convex analysis and optimization*, Athena Scientific, 2003.

[BHLM19] Peter L. Bartlett, Nick Harvey, Christopher Liaw, and Abbas Mehrabian, *Nearly-tight VC-dimension and pseudodimension bounds for piecewise linear neural networks*, J. Mach. Learn. Res. **20** (2019), Paper No. 63, 17. MR3960917

[BHMM19] Mikhail Belkin, Daniel Hsu, Siyuan Ma, and Soumik Mandal, *Reconciling modern machine-learning practice and the classical bias-variance trade-off*, Proc. Natl. Acad. Sci. USA **116** (2019), no. 32, 15849–15854, DOI 10.1073/pnas.1903070116. MR3997901

[Bil95] Patrick Billingsley, *Probability and measure*, 3rd ed., Wiley Series in Probability and Mathematical Statistics, John Wiley & Sons, Inc., New York, 1995. A Wiley-Interscience Publication. MR1324786

[Bil99] Patrick Billingsley, *Convergence of probability measures*, Wiley Series in Probability and Mathematical Statistics, 1999.

[Bis06] Christopher M. Bishop, *Pattern recognition and machine learning*, Information Science and Statistics, Springer, New York, 2006, DOI 10.1007/978-0-387-45528-0. MR2247587

[BKH16] Lei Jimmy Ba, Jamie Ryan Kiros, and Geoffrey E. Hinton, *Layer normalization*, CoRR **abs/1607.06450** (2016).

[BMR+20] Tom Brown, Benjamin Mann, Nick Ryder, Melanie Subbiah, Jared D Kaplan, Prafulla Dhariwal, Arvind Neelakantan, Pranav Shyam, Girish Sastry, Amanda Askell, Sandhini Agarwal, Ariel Herbert-Voss, Gretchen Krueger, Tom Henighan, Rewon Child, Aditya Ramesh, Daniel Ziegler, Jeffrey Wu, Clemens Winter, Chris Hesse, Mark Chen, Eric Sigler, Mateusz Litwin, Scott Gray, Benjamin Chess, Jack Clark, Christopher Berner, Sam McCandlish, Alec Radford, Ilya Sutskever, and Dario Amodei, *Language models are few-shot learners*, Advances in Neural Information Processing Systems (H. Larochelle, M. Ranzato, R. Hadsell, M.F. Balcan, and H. Lin, eds.), vol. 33, Curran Associates, Inc., 2020, pp. 1877–1901.

[BMR21] Peter L. Bartlett, Andrea Montanari, and Alexander Rakhlin, *Deep learning: a statistical viewpoint*, Acta Numer. **30** (2021), 87–201, DOI 10.1017/S0962492921000027. MR4295218

[Bot12] Léon Bottou, *Stochastic gradient descent tricks*, pp. 421–436, Springer Berlin Heidelberg, Berlin, Heidelberg, 2012.

[BPM90] Albert Benveniste, Michel Métivier, and Pierre Priouret, *Adaptive algorithms and stochastic approximations*, Applications of Mathematics (New York), vol. 22, Springer-Verlag, Berlin, 1990. Translated from the French by Stephen S. Wilson, DOI 10.1007/978-3-642-75894-2. MR1082341

[Bre11] Haim Brezis, *Functional analysis, Sobolev spaces and partial differential equations*, Universitext, Springer, New York, 2011. MR2759829

[BT00] Dimitri P. Bertsekas and John N. Tsitsiklis, *Gradient convergence in gradient methods with errors*, SIAM J. Optim. **10** (2000), no. 3, 627–642, DOI 10.1137/S1052623497331063. MR1741189

[Cal20] Ovidiu Calin, *Deep learning architectures—a mathematical approach*, Springer Series in the Data Sciences, Springer, Cham, 2020, DOI 10.1007/978-3-030-36721-3. MR4240268

[CB18] L. Chizat and F. Bach, *On the global convergence of gradient descent for over-parameterized models using optimal transport*, Advances in Neural Information Processing Systems (NeurIPS) (2018), 3040–3050.

[Cha22] Pratik Chaudhari, *Ese 546: Principles of deep learning*, Lecture notes available at: https://pratikac.github.io/, 2022.

[CHLSS23] S. Chun-Hei Lam, J. Sirignano, and K. Spiliopoulos, *Kernel limit of recurrent neural networks trained on ergodic data sequences*, arXiv:2308.14555, 2023.

[CHM+15] Anna Choromanska, Mikael Henaff, Michael Mathieu, G/'erard Ben Arous, and Yann LeCun, *The loss surfaces of multilayer networks*, Journal of Machine Learning Research **38** (2015), 192–204.

[CMV03] José A. Carrillo, Robert J. McCann, and Cédric Villani, *Kinetic equilibration rates for granular media and related equations: entropy dissipation and mass transportation estimates*, Rev. Mat. Iberoamericana **19** (2003), no. 3, 971–1018, DOI 10.4171/RMI/376. MR2053570

[CS17] Pratik Chaudhari and Stefano Soatto, *Stochastic gradient descent performs variational inference, converges to limit cycles for deep networks*, ICLR 2018 (2017).

[CSV+19] Sarath Chandar, Chinnadhurai Sankar, Eugene Vorontsov, Samira Ebrahimi Kahou, and Y. Bengio, *Towards non-saturating recurrent units for modelling long-term dependencies*, Proceedings of the AAAI Conference on Artificial Intelligence, vol. 33, 2019, pp. 3280–3287.

[Cyb89] G. Cybenko, *Approximation by superpositions of a sigmoidal function*, Math. Control Signals Systems **2** (1989), no. 4, 303–314, DOI 10.1007/BF02551274. MR1015670

[CYLW19] Q. Cai, Z. Yang, J. D. Lee, and Z. Wang, *Neural temporal-difference learning converges to global optima*, Advances in Neural Information Processing Systems 32 (NeurIPS 2019) (2019).

[DCLT19] Jacob Devlin, Ming-Wei Chang, Kenton Lee, and Kristina Toutanova, *Bert: Pre-training of deep bidirectional transformers for language understanding*, North American Chapter of the Association for Computational Linguistics, 2019.

[DCM+12] Jeffrey Dean, Greg S. Corrado, Rajat Monga, Kai Chen, Matthieu Devin, Quoc V. Le, Mark Z. Mao, Marc'Aurelio Ranzato, Andrew Senior, Paul Tucker, Ke Yang, and Andrew Y. Ng, *Large scale distributed deep networks*, NIPS (2012).

[DHS11] John Duchi, Elad Hazan, and Yoram Singer, *Adaptive subgradient methods for online learning and stochastic optimization*, J. Mach. Learn. Res. **12** (2011), 2121–2159. MR2825422

[DLL+19] S. Du, J. Lee, H. Li, L. Wang, and X. Zhai. G, *Gradient descent finds global minima of deep neural networks*, Proceedings of the 36th International Conference on Machine Learning, Long Beach, California, PMLR 97 (2019).

[DMBM17] David M. Blei, Alp Kucukelbir, and Jon D. McAuliffe, *Variational inference: a review for statisticians*, J. Amer. Statist. Assoc. **112** (2017), no. 518, 859–877, DOI 10.1080/01621459.2017.1285773. MR3671776

[DS98] Norman R. Draper and Harry Smith, *Applied regression analysis*, 3rd ed., Wiley Series in Probability and Statistics: Texts and References Section, John Wiley & Sons, Inc., New York, 1998. With 1 IBM-PC floppy disk (3.5 inch; DD); A Wiley-Interscience Publication, DOI 10.1002/9781118625590. MR1614335

[E17] Weinan E, *A proposal on machine learning via dynamical systems*, Commun. Math. Stat. **5** (2017), no. 1, 1–11, DOI 10.1007/s40304-017-0103-z. MR3627592

[ea19] Paszke et al., *Pytorch: An imperative style, high-performance deep learning library*, NeurIPS (2019).

[EK86] Stewart N. Ethier and Thomas G. Kurtz, *Markov processes: Characterization and convergence*, Wiley Series in Probability and Mathematical Statistics: Probability and Mathematical Statistics, John Wiley & Sons, Inc., New York, 1986, DOI 10.1002/9780470316658. MR838085

[GAA+17] Ishaan Gulrajani, Faruk Ahmed, Martin Arjovsky, Vincent Dumoulin, and Aaron Courville, *Improved training of Wasserstein GANs*, Proceedings of the 31st International Conference on Neural Information Processing Systems (Red Hook, NY, USA), NIPS'17, Curran Associates Inc., 2017, pp. 5769–5779.

[GAG+17] Jonas Gehring, Michael Auli, David Grangier, Denis Yarats, and Yann Dauphin, *Convolutional sequence to sequence learning*, arXiv:1705.03122, 2017.

[GB10] X. Glorot and Y. Bengio, *Understanding the difficulty of training deep feedforward neural networks*, Proceedings of Machine Learning Research 9 (13th International Conference on Artificial Intelligence and Statistics), ML Research Press, 2010, pp. 249–256.

[GBC16] Ian Goodfellow, Yoshua Bengio, and Aaron Courville, *Deep learning*, Adaptive Computation and Machine Learning, MIT Press, Cambridge, MA, 2016. MR3617773

[Gil08] Mike B. Giles, *Collected matrix derivative results for forward and reverse mode algorithmic differentiation*, Advances in automatic differentiation, Lect. Notes Comput. Sci. Eng., vol. 64, Springer, Berlin, 2008, pp. 35–44, DOI 10.1007/978-3-540-68942-3_4. MR2531677

[GK20] L. Graesser and W. L. Keng, *Foundations of deep reinforcement learning: Theory and practice in Python*, Addison-Wesley data and analytics series, Addison-Wesley, 2020.

[GMMM20] Behrooz Ghorbani, Song Mei, Theodor Misiakiewicz, and Andrea Montanari, *When do neural networks outperform kernel methods?*, J. Stat. Mech. Theory Exp. **12** (2021), Paper No. 124009, 110, DOI 10.1088/1742-5468/ac3a81. MR4412837

[GPAM+14] Ian Goodfellow, Jean Pouget-Abadie, Mehdi Mirza, Bing Xu, David Warde-Farley, Sherjil Ozair, Aaron Courville, and Yoshua Bengio, *Generative adversarial nets*, International Conference on Neural Information Processing Systems (NIPS 2014), 2014, pp. 2672–2680.

[GW08] Andreas Griewank and Andrea Walther, *Evaluating derivatives: Principles and techniques of algorithmic differentiation*, 2nd ed., Society for Industrial and Applied Mathematics (SIAM), Philadelphia, PA, 2008, DOI 10.1137/1.9780898717761. MR2454953

[GWFM+13] Ian Goodfellow, David Warde-Farley, Mehdi Mirza, Aaron Courville, and Yoshua Bengio, *Maxout networks*, Proceedings of the 30th International Conference on Machine Learning, PMLR **28** (2013), 1319–1327.

[Han19] B. Hanin, *Universal function approximation by deep neural nets with bounded width and ReLU activations*, Mathematics **7** (2019), 1–17.

[HBK+20] Yang Hu, Alka Bishnoi, Rachneet Kaur, Richard Sowers, and Manuel E Hernandez, *Exploration of machine learning to identify community dwelling older adults with balance dysfunction using short duration accelerometer data*, 2020 42nd Annual International Conference of the IEEE Engineering in Medicine & Biology Society (EMBC), IEEE, 2020, pp. 812–815.

[HGS16] H. Van Hasselt, A. Guez, and D. Silver, *Deep reinforcement learning with double q-learning*, In Thirtieth AAAI conference on artificial intelligence (2016).

[HJ20] Jeffrey Humpherys and Tyler J. Jarvis, *Foundations of applied mathematics. Vol. 2—Algorithms, approximation, optimization*, Society for Industrial and Applied Mathematics, Philadelphia, PA, 2020. MR4077176

[HK70] Arthur E. Hoerl and Robert W. Kennard, *Ridge regression: Biased estimation for nonorthogonal problems*, Technometrics **12** (1970), no. 1, 55–67.

[HLLL19] Wenqing Hu, Chris Junchi Li, Lei Li, and Jian-Guo Liu, *On the diffusion approximation of nonconvex stochastic gradient descent*, Ann. Math. Sci. Appl. **4** (2019), no. 1, 3–32, DOI 10.4310/AMSA.2019.v4.n1.a1. MR3921998

[Hor91] K. Hornik, *Approximation capabilities of multilayer feedforward networks*, Neural Networks **4** (1991), no. 2, 251–257.

[HS15] M. Hausknecht and P. Stone, *Deep recurrent q-learning for partially observable mdps*, AAAI 2015 Fall Symposium (2015).

[HSW89] K. Hornik, M. Stinchcombe, and H. White, *Multilayer feedforward networks are universal approximators*, Neural Networks **2** (1989), no. 5, 359–366.

[HSW90] K. Hornik, M. Stinchcombe, and H. White, *Universal approximation of an unknown mapping and its derivatives using multilayer feedforward networks*, Neural Networks **5** (1990), no. 3, 551–560.

[HTF10] Trevor Hastie, Robert Tibshirani, and Jerome Friedman, *The elements of statistical learning: Data mining, inference, and prediction*, 2nd ed., Springer Series in Statistics, Springer, New York, 2009, DOI 10.1007/978-0-387-84858-7. MR2722294

[HVU+19] Cheng-Zhi Anna Huang, Ashish Vaswani, Jakob Uszkoreit, Ian Simon, Curtis Hawthorne, Noam Shazeer, Andrew M. Dai, Matthew D. Hoffman, Monica Dinculescu, and Douglas Eck, *Music transformer: Generating music with long-term structure.*, ICLR (Poster), OpenReview.net, 2019.

[HZRS15] K. He, X. Zhang, S. Ren, and J. Sun, *Delving deep into rectifiers: Surpassing human-level performance on imagenet classification*, 2015 IEEE International Conference on Computer Vision (ICCV) (Los Alamitos, CA, USA), IEEE Computer Society, Dec 2015, pp. 1026–1034.

[HZRS16] Kaiming He, Xiangyu Zhang, Shaoqing Ren, and Jian Sun, *Deep residual learning for image recognition*, 2016 IEEE Conference on Computer Vision and Pattern Recognition (CVPR), 2016, pp. 770–778.

[IS15a] Sergey Ioffe and Christian Szegedy, *Batch normalization: Accelerating deep network training by reducing internal covariate shift*, CoRR arXiv:1502.03167, 2015.

[IS15b] Sergey Ioffe and Christian Szegedy, *Batch normalization: Accelerating deep network training by reducing internal covariate shift*, Proceedings of the 32nd International Conference on Machine Learning, ICML 2015, Lille, France, 6-11 July 2015 (Francis R. Bach and David M. Blei, eds.), JMLR Workshop and Conference Proceedings, vol. 37, JMLR.org, 2015, pp. 448–456.

[Ito96] Yoshifusa Ito, *Nonlinearity creates linear independence*, Adv. Comput. Math. **5** (1996), no. 2-3, 189–203, DOI 10.1007/BF02124743. MR1399380

[Jak86] Adam Jakubowski, *On the Skorokhod topology* (English, with French summary), Ann. Inst. H. Poincaré Probab. Statist. **22** (1986), no. 3, 263–285. MR871083

[JGH18] A. Jacot, F. Gabriel, and C. Hongler, *Neural tangent kernel: Convergence and generalization in neural networks*, Advances in Neural Information Processing Systems 31 (S. Bengio, H. Wallach, H. Larochelle, K. Grauman, N. Cesa-Bianchi, and R. Garnett, eds.), Curran Associates, Inc., 2018.

[JGR19] Daniel Jakubovitz, Raja Giryes, and Miguel R. D. Rodrigues, *Generalization error in deep learning*, Compressed sensing and its applications, Appl. Numer. Harmon. Anal., Birkhäuser/Springer, Cham, 2019, pp. 153–193, DOI 10.1007/978-3-319-73074-5_5. MR4182531

[JKO98] Richard Jordan, David Kinderlehrer, and Felix Otto, *The variational formulation of the Fokker-Planck equation*, SIAM J. Math. Anal. **29** (1998), no. 1, 1–17, DOI 10.1137/S0036141096303359. MR1617171

[KB15] Diederik P. Kingma and Jimmy Ba, *Adam: A method for stochastic optimization*, Proceedings of the 3rd International Conference on Learning Representations (ICLR) (2015).

[KCM+20] Rachneet Kaur, Zizhang Chen, Robert Motl, Manuel Enrique Hernandez, and Richard Sowers, *Predicting multiple sclerosis from gait dynamics using an instrumented treadmill—a machine learning approach*, IEEE Transactions on Biomedical Engineering (2020).

[KH91] C. Kuan and K. Hornik, *Convergence of learning algorithms with constant learning rates*, IEEE Transactions on Neural Networks **2** (1991), no. 5, 484–489.

[KH09] A. Krizhevsky and G. Hinton, *Learning multiple layers of features from tiny images*, Tech. Report 1(4), University of Toronto, 2009.

[KKHS20] Rachneet Kaur, Maxim Korolkov, Manuel E Hernandez, and Richard Sowers, *Automatic identification of brain independent components in electroencephalography data collected while standing in a virtually immersive environment—a deep learning-based approach*, 2020 42nd Annual International Conference of the IEEE Engineering in Medicine & Biology Society (EMBC), IEEE, 2020, pp. 95–98.

[KLM+23] Rachneet Kaur, Joshua Levy, Robert W. Motl, Richard Sowers, and Manuel E. Hernandez, *Deep learning for multiple sclerosis differentiation using multi-stride dynamics in gait*, IEEE Transactions on Biomedical Engineering (2023).

[KLS20] Dmitry Kobak, Jonathan Lomond, and Benoit Sanchez, *The optimal ridge penalty for real-world high-dimensional data can be zero or negative due to the implicit ridge regularization*, J. Mach. Learn. Res. **21** (2020), Paper No. 169, 16. MR4209455

[KMSH22] Rachneet Kaur, Robert W. Motl, Richard Sowers, and Manuel E. Hernandez, *A vision-based framework for predicting multiple sclerosis and Parkinson's disease gait dysfunctions—a deep learning approach*, IEEE Journal of Biomedical and Health Informatics (2022).

[KMZ+19] Rachneet Kaur, Sanjana Menon, Xiaomiao Zhang, Richard Sowers, and Manuel E. Hernandez, *Exploring characteristic features in gait patterns for predicting multiple sclerosis*, 2019 41st Annual International Conference of the IEEE Engineering in Medicine and Biology Society (EMBC), IEEE, 2019, pp. 4217–4220.

[KNJK18] Rahul Kidambi, Praneeth Netrapalli, Prateek Jain, and Sham M. Kakade, *On the insufficiency of existing momentum schemes for stochastic optimization*, In 2018 Information Theory and Applications Workshop (ITA), IEEE. **16** (2018), 1–9.

[Kol14] Vassili N. Kolokoltsov, *Nonlinear Markov processes and kinetic equations*, Cambridge Tracts in Mathematics, vol. 182, Cambridge University Press, Cambridge, 2010, DOI 10.1017/CBO9780511760303. MR2680971

[KP92] Peter E. Kloeden and Eckhard Platen, *Numerical solution of stochastic differential equations*, Applications of Mathematics (New York), vol. 23, Springer-Verlag, Berlin, 1992, DOI 10.1007/978-3-662-12616-5. MR1214374

[KP12] J. Kober and J. Peters, *In reinforcement learning*, In Reinforcement Learning. Springer **518** (2012), 579–610.

[KS98] Ioannis Karatzas and Steven E. Shreve, *Brownian motion and stochastic calculus*, 2nd ed., Graduate Texts in Mathematics, vol. 113, Springer-Verlag, New York, 1991, DOI 10.1007/978-1-4612-0949-2. MR1121940

[KSH12] Alex Krizhevsky, Ilya Sutskever, and Geoffrey E. Hinton, *Imagenet classification with deep convolutional neural networks*, Advances in Neural Information Processing Systems (F. Pereira, C. J. Burges, L. Bottou, and K. Q. Weinberger, eds.), vol. 25, Curran Associates, Inc., 2012.

[KST+20] Rachneet Kaur, Clara Schaye, Kevin Thompson, Daniel C Yee, Rachel Zilz, RS Sreenivas, and Richard B. Sowers, *Machine learning and price-based load scheduling for an optimal iot control in the smart and frugal home*, Energy and AI (2020), 100042.

[KSZ+19] Rachneet Kaur, Rongyi Sun, Liran Ziegelman, Richard Sowers, and Manuel E. Hernandez, *Using virtual reality to examine the neural and physiological anxiety-related responses to balance-demanding target-reaching leaning tasks*, 2019 IEEE-RAS 19th International Conference on Humanoid Robots (Humanoids), IEEE, 2019, pp. 1–7.

[KW52] J. Kiefer and J. Wolfowitz, *Stochastic estimation of the maximum of a regression function*, Ann. Math. Statistics **23** (1952), 462–466, DOI 10.1214/aoms/1177729392. MR50243

[KW13] Diederik P. Kingma and Max Welling, *Auto-encoding variational bayes*, CoRR arXiv:1312.6114, 2013.

[KWRL17] Justin Ker, Lipo Wang, Jai Rao, and Tchoyoson Lim, *Deep learning applications in medical image analysis*, IEEE Access **6** (2017), 9375–9389.

[KY03] Harold J. Kushner and G. George Yin, *Stochastic approximation and recursive algorithms and applications*, 2nd ed., Applications of Mathematics (New York), vol. 35, Springer-Verlag, New York, 2003. Stochastic Modelling and Applied Probability. MR1993642

[Lax02] Peter D. Lax, *Functional analysis*, Pure and Applied Mathematics: A Wiley Series of Texts, Monographs and Tracts, 2002.

[LB20] Chaoyue Liu and Mikhail Belkin, *Accelerating SGD with momentum for over-parameterized learning*, ICLR 2020 (2020), 1–25.

[LBBH98] Y. LeCun, L. Bottou, Y. Bengio, and P. Haffner, *Gradient-based learning applied to document recognition*, Proceedings of the IEEE **86** (1998), no. 11, 2278–2324.

[Lec88] Yann Lecun, *A theoretical framework for back-propagation*, Proceedings of the 1988 Connectionist Models Summer School, CMU, Pittsburg, PA (D. Touretzky, G. Hinton, and T. Sejnowski, eds.), Morgan Kaufmann, 1988, pp. 21–28 (English (US)).

[Led16] Sean Ledger, *Skorokhod's M*1 *topology for distribution-valued processes*, Electron. Commun. Probab. **21** (2016), Paper No. 34, 11, DOI 10.1214/16-ECP4754. MR3492929

[LFK+22] Sanae Lotfi, Marc Finzi, Sanyam Kapoor, Andres Potapczynski, Micah Goldblum, and Andrew Gordon Wilson, *Pac-Bayes compression bounds so tight that they can explain generalization*, NeurIPS, 2022.

[LL17] Scott M Lundberg and Su-In Lee, *A unified approach to interpreting model predictions*, Advances in Neural Information Processing Systems (I. Guyon, U. Von Luxburg, S. Bengio, H. Wallach, R. Fergus, S. Vishwanathan, and R. Garnett, eds.), vol. 30, Curran Associates, Inc., 2017.

[LLF98] Isaac Lagaris, Aristidis Likas, and Dimitrios Fotiadis, *Artificial neural networks for solving ordinary and partial differential equations*, IEEE Transactions on Neural Networks **9** (1998), 987–1000.

[LLFZ18] Omer Levy, Kenton Lee, Nicholas FitzGerald, and Luke Zettlemoyer, *Long short-term memory as a dynamically computed element-wise weighted sum*, Proceedings of the 56th Annual Meeting of the Association for Computational Linguistics (Volume 2: Short Papers) (Melbourne, Australia), Association for Computational Linguistics, July 2018, pp. 732–739.

[LLP00] Isaac Lagaris, Aristidis Likas, and Dimitrios Papageorgiou, *Neural-network methods for boundary value problems with irregular boundaries*, IEEE transactions on neural networks / a publication of the IEEE Neural Networks Council **11** (2000), 1041–9.

[LPB+16] Cesar Laurent, Gabriel Pereyra, Philemon Brakel, Ying Zhang, and Yoshua Bengio, *Batch normalized recurrent neural networks*, 2016 IEEE International Conference on Acoustics, Speech and Signal Processing (ICASSP), 2016, pp. 2657–2661.

[MBM+16] V. Mnih, A.P. Badia, M. Mirza, A. Graves, T. Lillicrap, T. Harley, D. Silver, and K. Kavukcuoglu, *Asynchronous methods for deep reinforcement learning*, International Conference on Machine Learning (2016).

[MKS+13] V. Mnih, K. Kavukcuoglu, D. Silver, A. Graves, I. Antonoglou, D. Wierstra, and M. Riedmiller, *Playing Atari with deep reinforcement learning*, arXiv:1312.5602, 2013.

[MKS+15] V. Mnih, K. Kavukcuoglu, D. Silver, A. A. Rusu, J. Veness, M. G. Bellemare, A. Graves, M. Riedmiller, A. K. Fidjeland, G. Ostrovski, et.al., *Human-level control through deep reinforcement learning*, Nature (2015), 529–533.

[MM23] Theodor Misiakiewicz and Andrea Montanari, *Six lectures on linearized neural networks*.

[MMM19] Song Mei, Theodor Misiakiewicz, and Andrea Montanari, *Mean-field theory of two-layers neural networks: dimension-free bounds and kernel limit*, Proceedings of the Thirty-Second Conference on Learning Theory (Alina Beygelzimer and Daniel Hsu, eds.), Proceedings of Machine Learning Research, vol. 99, PMLR, 25–28 Jun 2019, pp. 2388–2464.

[MMN18] Song Mei, Andrea Montanari, and Phan-Minh Nguyen, *A mean field view of the landscape of two-layer neural networks*, Proc. Natl. Acad. Sci. USA **115** (2018), no. 33, E7665–E7671, DOI 10.1073/pnas.1806579115. MR3845070

[MP43] Warren S. McCulloch and Walter Pitts, *A logical calculus of the ideas immanent in nervous activity*, Bull. Math. Biophys. **5** (1943), 115–133, DOI 10.1007/bf02478259. MR10388

[MP16] H. N. Mhaskar and T. Poggio, *Deep vs. shallow networks: an approximation theory perspective*, Anal. Appl. (Singap.) **14** (2016), no. 6, 829–848, DOI 10.1142/S0219530516400042. MR3564936

[MP17] Marvin Minsky and Seymour A. Papert, *Perceptrons: An introduction to computational geometry*, MIT Press (Reissue of the 1988 Expanded Edition), 2017.

[MPV21] Douglas C. Montgomery and Elizabeth A. Peck, *Introduction to linear regression analysis*, Wiley Series in Probability and Mathematical Statistics, John Wiley & Sons, Inc., New York, 1982. MR646067

[MSN08] Charles E. McCulloch, Shayle R. Searle, and John M. Neuhaus, *Generalized, linear, and mixed models*, 2nd ed., Wiley Series in Probability and Statistics, John Wiley & Sons, Inc., Hoboken, NJ, 2008. MR2431553

[Mur22] Kevin P. Murphy, *Probabilistic machine learning: An introduction*, MIT Press, 2022.

[MV19] Pierre Moulin and Venugopal V. Veeravalli, *Statistical inference for engineers and data scientists*, Cambridge University Press, Cambridge, 2019. MR3972165

[MZ22] Andrea Montanari and Yiqiao Zhong, *The interpolation phase transition in neural networks: memorization and generalization under lazy training*, Ann. Statist. **50** (2022), no. 5, 2816–2847, DOI 10.1214/22-aos2211. MR4500626

[Nau08] Uwe Naumann, *Optimal Jacobian accumulation is NP-complete*, Math. Program. **112** (2008), no. 2, Ser. A, 427–441, DOI 10.1007/s10107-006-0042-z. MR2361931

[Nes83] Yu. E. Nesterov, *A method for solving the convex programming problem with convergence rate* $O(1/k^2)$ (Russian), Dokl. Akad. Nauk SSSR **269** (1983), no. 3, 543–547. MR701288

[Nes04] Yurii Nesterov, *Introductory lectures on convex optimization: A basic course*, Applied Optimization, vol. 87, Kluwer Academic Publishers, Boston, MA, 2004, DOI 10.1007/978-1-4419-8853-9. MR2142598

[Nes07] Yuri Nesterov, *Gradient methods for minimizing composite objective functions*, Techincal Report, CORE, 2007.

[Ngu19] P.-M. Nguyen, *Mean field limit of the learning dynamics of multilayer neural networks*, (2019).

[NKB+20] Preetum Nakkiran, Gal Kaplun, Yamini Bansal, Tristan Yang, Boaz Barak, and Ilya Sutskever, *Deep double descent: where bigger models and more data hurt*, J. Stat. Mech. Theory Exp. **12** (2021), Paper No. 124003, 32, DOI 10.1088/1742-5468/ac3a74. MR4412831

[NP23] Phan-Minh Nguyen and Huy Tuan Pham, *A rigorous framework for the mean field limit of multilayer neural networks*, Math. Stat. Learn. **6** (2023), no. 3-4, 201–357, DOI 10.4171/msl/42. MR4656973

[OBPR16] I. Osband, C. Blundell, A. Pritzel, and B. Van Roy, *Deep exploration via bootstrapped DQN*, In Advances in Neural Information Processing Systems (2016).

[PB17] Jeffrey Pennington and Yasaman Bahri, *Geometry of neural network loss surface via random matrix theory*, Proceedings of the 34th International Conference on Machine Learning, PMLR **70** (2017), 2798–2806.

[PBJ12] John William Paisley, David M. Blei, and Michael I. Jordan, *Variational Bayesian inference with stochastic search*, International Conference on Machine Learning, 2012.

[PDGB14] Razvan Pascanu, Yann N. Dauphin, Surya Ganguli, and Yoshua Bengio, *On the saddle point problem for non-convex optimization*, (2014).

[PMB13] Razvan Pascanu, Tomas Mikolov, and Yoshua Bengio, *On the difficulty of training recurrent neural networks*, Proceedings of the 30th International Conference on International Conference on Machine Learning - Volume 28, ICML'13, JMLR.org, 2013, pp. III–1310–III–1318.

[Pol67] B. T. Poljak, *Some methods of speeding up the convergence of iterative methods* (Russian), Ž. Vyčisl. Mat i Mat. Fiz. **4** (1964), 791–803. MR169403

[Pri23] Simon J. D. Prince, *Understanding deep learning*, The MIT Press, 2023.

[Pro05] Philip E. Protter, *Stochastic integration and differential equations*, 2nd ed., Applications of Mathematics (New York), vol. 21, Springer-Verlag, Berlin, 2004. Stochastic Modelling and Applied Probability. MR2020294

[PVU+18] Niki J. Parmar, Ashish Vaswani, Jakob Uszkoreit, Lukasz Kaiser, Noam Shazeer, Alexander Ku, and Dustin Tran, *Image transformer*, International Conference on Machine Learning (ICML), 2018.

[RF10] H. L. Royden and P. M. Fitzpatrick, *Real analysis, 4th edition*, Pearson, 2010.

[RHK23] Sanjiban Sekhar Roy, Ching-Hsien Hsu, and Venkateshwara Kagita, *Deep learning applications in image analysis*, vol. 129, Springer Nature, 2023.

[RHW86] David E. Rumelhart, Geoffrey E. Hinton, and Ronald J. Williams, *Learning representations by back-propagating errors*, Nature **323** (1986), 533–536.

[RM51] Herbert Robbins and Sutton Monro, *A stochastic approximation method*, Ann. Math. Statistics **22** (1951), 400–407, DOI 10.1214/aoms/1177729586. MR42668

[RNSS18] Alec Radford, Karthik Narasimhan, Tim Salimans, and Ilya Sutskever, *Improving language understanding by generative pre-training*, 2018.

[Ros58] F. Rosenblatt, *The perceptron: A probabilistic model for information storage and organization in the brain*, Psychological Review **65** (1958), 386–408.

[RPK19] M. Raissi, P. Perdikaris, and G. E. Karniadakis, *Physics-informed neural networks: A deep learning framework for solving forward and inverse problems involving nonlinear partial differential equations*, J. Comput. Phys. **378** (2019), 686–707, DOI 10.1016/j.jcp.2018.10.045. MR3881695

[RS07] Alvin C. Rencher and G. Bruce Schaalje, *Linear models in statistics*, 2nd ed., Wiley-Interscience [John Wiley & Sons], Hoboken, NJ, 2008. MR2401650

[RSB12] Nicolas Roux, Mark Schmidt, and Francis Bach, *A stochastic gradient method with an exponential convergence rate for finite training sets*, NIPS 2012 (2012), 2672–2680.

[RVE18] G. M. Rotskoff and E. Vanden-Eijnden, *Neural networks as interacting particle systems: Asymptotic convexity of the loss landscape and universal scaling of the approximation error*, (2018).

[RW00a] L. C. G. Rogers and David Williams, *Diffusions, Markov processes, and martingales: Volume 1, Foundations*, Cambridge Mathematical Library, Cambridge University Press, Cambridge, 2000, DOI 10.1017/CBO9781107590120. MR1780932

[RW00b] L.C.G. Rogers and D. Williams, *Diffusions, Markov processes, and martingales: Volume 2, Itô calculus*, Cambridge University Press, 2000.

[RWC+19] Alec Radford, Jeff Wu, Rewon Child, David Luan, Dario Amodei, and Ilya Sutskever, *Language models are unsupervised multitask learners*, (2019).

[SB18] Richard S. Sutton and Andrew G. Barto, *Reinforcement learning: an introduction*, 2nd ed., Adaptive Computation and Machine Learning, MIT Press, Cambridge, MA, 2018. MR3889951

[SGS15] Rupesh Srivastava, Klaus Greff, and Jurgen Schmidhuber, *Highway networks*, (2015).

[SH17] Johannes Schmidt-Hieber, *Nonparametric regression using deep neural networks with ReLU activation function*, Ann. Statist. **48** (2020), no. 4, 1875–1897, DOI 10.1214/19-AOS1875. MR4134774

[SHK+14] Nitish Srivastava, Geoffrey Hinton, Alex Krizhevsky, Ilya Sutskever, and Ruslan Salakhutdinov, *Dropout: a simple way to prevent neural networks from overfitting*, J. Mach. Learn. Res. **15** (2014), 1929–1958. MR3231592

[SKZ+19] Rongyi Sun, Rachneet Kaur, Liran Ziegelman, Shuo Yang, Richard Sowers, and Manuel E. Hernandez, *Using virtual reality to examine the correlation between balance function and anxiety in stance*, 2019 IEEE International Conference on Bioinformatics and Biomedicine (BIBM), IEEE, 2019, pp. 1633–1640.

[SMS23] Justin Sirignano, Jonathan MacArt, and Konstantinos Spiliopoulos, *PDE-constrained models with neural network terms: optimization and global convergence*, J. Comput. Phys. **481** (2023), Paper No. 112016, 35, DOI 10.1016/j.jcp.2023.112016. MR4559355

[SMSM00] R. S. Sutton, D. A. McAllester, S. P. Singh, and Y. Mansour, *Policy gradient methods for reinforcement learning with function approximation*, Advances in Neural Information Processing Systems 12 (NIPS 1999) (2000).

[SP97] M. Schuster and K. Paliwal, *Bidirectional recurrent neural networks*, IEEE Transactions on Signal Processing **45** (1997), no. 10.

[SS17] Justin Sirignano and Konstantinos Spiliopoulos, *Stochastic gradient descent in continuous time*, SIAM J. Financial Math. **8** (2017), no. 1, 933–961, DOI 10.1137/17M1126825. MR3732944

[SS18] Justin Sirignano and Konstantinos Spiliopoulos, *DGM: a deep learning algorithm for solving partial differential equations*, J. Comput. Phys. **375** (2018), 1339–1364, DOI 10.1016/j.jcp.2018.08.029. MR3874585

[SS19] Justin Sirignano and Konstantinos Spiliopoulos, *Scaling limit of neural networks with the Xavier initialization and convergence to a global minimum*, arXiv:1907.04108, 2019.

[SS20a] Justin Sirignano and Konstantinos Spiliopoulos, *Mean field analysis of neural networks: a central limit theorem*, Stochastic Process. Appl. **130** (2020), no. 3, 1820–1852, DOI 10.1016/j.spa.2019.06.003. MR4058290

[SS20b] Justin Sirignano and Konstantinos Spiliopoulos, *Mean field analysis of neural networks: a law of large numbers*, SIAM J. Appl. Math. **80** (2020), no. 2, 725–752, DOI 10.1137/18M1192184. MR4074020

[SS20c] Justin Sirignano and Konstantinos Spiliopoulos, *Stochastic gradient descent in continuous time: a central limit theorem*, Stoch. Syst. **10** (2020), no. 2, 124–151, DOI 10.1287/stsy.2019.0050. MR4119247

[SS21] Justin Sirignano and Konstantinos Spiliopoulos, *Mean field analysis of deep neural networks*, Math. Oper. Res. **47** (2022), no. 1, 120–152, DOI 10.1287/moor.2020.1118. MR4403748

[SS22] Justin Sirignano and Konstantinos Spiliopoulos, *Asymptotics of reinforcement learning with neural networks*, Stoch. Syst. **12** (2022), no. 1, 2–29. MR4414343

[SSB16] Stanislau Semeniuta, Aliaksei Severyn, and Erhardt Barth, *Recurrent dropout without memory loss*, (2016).

[SSBD14] Shai Shalev-Shwartz and Shai Ben-David, *Understanding machine learning: From theory to algorithms*, Cambridge University Press, 2014.

[SSS+17] D. Silver, J. Schrittwieser, K. Simonyan, I. Antonoglou, A. Huang, A. Guez, T. Hubert, L. Baker, M. Lai, A. Bolton, et al., *Mastering the game of Go without human knowledge*, Nature **550** (2017), 354.

[STC04] J. Shawe-Taylor and N. Cristianini, *Kernel methods for pattern analysis*, Cambridge University Press, 2004.

[STIM18] Shibani Santurkar, Dimitris Tsipras, Andrew Ilyas, and Aleksander Madry, *How does batch normalization help optimization?*, 32nd Conference on Neural Information Processing Systems (NeurIPS 2018), November 2018.

[SY21] Jiahui Yu and Konstantinos Spiliopoulos, *Normalization effects on shallow neural networks and related asymptotic expansions*, Found. Data Sci. **3** (2021), no. 2, 151–200, DOI 10.3934/fods.2021013. MR4619395

[TH12] Tijmen Tieleman and Geoffrey Hinton, *Lecture 6.5-RMSProp: Divide the gradient by a running average of its recent magnitude*, COURSERA: Neural Networks for Machine Learning **4** (2012), 26–31.

[Tik63] A. N. Tikhonov, *On the solution of incorrectly put problems and the regularisation method*, Outlines Joint Sympos. Partial Differential Equations (Novosibirsk, 1963), Academy of Sciences of the USSR, Siberian Branch, Moscow, 1963, pp. 261–265. MR211218

[Tsi94] J. N. Tsitsiklis, *Asynchronous stochastic approximation and q-learning*, Machine Learning **16** (1994), 185–202.

[Vap99] Vladimir N. Vapnik, *The nature of statistical learning theory*, 2nd ed., Statistics for Engineering and Information Science, Springer-Verlag, New York, 2000, DOI 10.1007/978-1-4757-3264-1. MR1719582

[VC71] V. N. Vapnik and A. Ja. Červonenkis, *The uniform convergence of frequencies of the appearance of events to their probabilities* (Russian, with English summary), Teor. Verojatnost. i Primenen. **16** (1971), 264–279. MR288823

[Vil09] Cédric Villani, *Optimal transport*, Grundlehren der mathematischen Wissenschaften [Fundamental Principles of Mathematical Sciences], vol. 338, Springer-Verlag, Berlin, 2009. Old and new, DOI 10.1007/978-3-540-71050-9. MR2459454

[VSP+17] Ashish Vaswani, Noam Shazeer, Niki Parmar, Jakob Uszkoreit, Llion Jones, Aidan N Gomez, Ł ukasz Kaiser, and Illia Polosukhin, *Attention is all you need*, Advances in Neural Information Processing Systems (I. Guyon, U. Von Luxburg, S. Bengio, H. Wallach, R. Fergus, S. Vishwanathan, and R. Garnett, eds.), vol. 30, Curran Associates, Inc., 2017.

[Wat89] C.I.C.H. Watkins, *Learning from delayed rewards*, PhD thesis, University of Cambridge, Cambridge, UK (1989).

[WD92] C.I.C.H. Watkins and P. Dayan, *Q-learning*, Machine Learning **8** (1992), 279–292.

[WS05] Shuning Wang and Xusheng Sun, *Generalization of hinging hyperplanes*, IEEE Trans. Inform. Theory **51** (2005), no. 12, 4425–4431, DOI 10.1109/TIT.2005.859246. MR2243179

[WSH+16] Z. Wang, T. Schaul, M. Hessel, H. Hasselt, M. Lanctot, and N. Freitas, *Dueling network architectures for deep reinforcement learning*, ICML'16 Proceedings of the 33rd International Conference on International Conference on Machine Learning **48** (2016), 1995–2003.

[Yar17] Dmitry Yarotsky, *Error bounds for approximations with deep ReLU networks*, Neural Networks **94** (2017), 103–114.

[YS23] Jiahui Yu and Konstantinos Spiliopoulos, *Normalization effects on deep neural networks*, Found. Data Sci. **5** (2023), no. 3, 389–465, DOI 10.3934/fods.2023004. MR4622926

[ZSBRV21] Chiyuan Zhang, Moritz Hardt Samy Bengio, Benjamin Recht, and Oriol Vinyals, *Understanding deep learning requires rethinking generalization*, Communications of the ACM **64** (2021), 107–115.

[ZSKS17] Julian Georg Zilly, Rupesh Kumar Srivastava, Jan Koutník, and Jürgen Schmidhuber, *Recurrent highway networks*, Proceedings of the 34th International Conference on Machine Learning, Sydney, Australia, PMLR, vol. 70, 2017.

Index

Selected Published Titles in This Series

252 **Konstantinos Spiliopoulos, Richard B. Sowers, and Justin Sirignano,** Mathematical Foundations of Deep Learning Models and Algorithms, 2025
251 **Eric Carlen,** Inequalities in Matrix Algebras, 2025
250 **David Eisenbud and Joe Harris,** The Practice of Algebraic Curves, 2024
249 **David Damanik and Jake Fillman,** One-Dimensional Ergodic Schrödinger Operators, 2024
248 **J. I. Hall,** Introduction to Lie Algebras, 2024
247 **Michel L. Lapidus and Goran Radunović,** An Invitation to Fractal Geometry, 2024
246 **Ale Jan Homburg and Jürgen Knobloch,** Bifurcation Theory, 2024
245 **Bennett Chow and Yutze Chow,** Lectures on Differential Geometry, 2024
244 **John M. Lee,** Introduction to Complex Manifolds, 2024
243 **J. M. Landsberg,** Quantum Computation and Quantum Information, 2024
242 **Jayadev S. Athreya and Howard Masur,** Translation Surfaces, 2024
241 **Thorsten Theobald,** Real Algebraic Geometry and Optimization, 2024
240 **Nam Q. Le,** Analysis of Monge–Ampère Equations, 2024
239 **K. Cieliebak, Y. Eliashberg, and N. Mishachev,** Introduction to the h-Principle, Second Edition, 2024
238 **Julio González-Díaz, Ignacio García-Jurado, and M. Gloria Fiestras-Janeiro,** An Introductory Course on Mathematical Game Theory and Applications, Second Edition, 2023
237 **Michael Levitin, Dan Mangoubi, and Iosif Polterovich,** Topics in Spectral Geometry, 2023
236 **Stephanie Alexander, Vitali Kapovitch, and Anton Petrunin,** Alexandrov Geometry, 2024
235 **Bennett Chow,** Ricci Solitons in Low Dimensions, 2023
234 **Andrea Ferretti,** Homological Methods in Commutative Algebra, 2023
233 **Andrea Ferretti,** Commutative Algebra, 2023
232 **Harry Dym,** Linear Algebra in Action, Third Edition, 2023
231 **Luís Barreira and Yakov Pesin,** Introduction to Smooth Ergodic Theory, Second Edition, 2023
230 **Barbara Kaltenbacher and William Rundell,** Inverse Problems for Fractional Partial Differential Equations, 2023
229 **Giovanni Leoni,** A First Course in Fractional Sobolev Spaces, 2023
228 **Henk Bruin,** Topological and Ergodic Theory of Symbolic Dynamics, 2022
227 **William M. Goldman,** Geometric Structures on Manifolds, 2022
226 **Milivoje Lukić,** A First Course in Spectral Theory, 2022
225 **Jacob Bedrossian and Vlad Vicol,** The Mathematical Analysis of the Incompressible Euler and Navier-Stokes Equations, 2022
224 **Ben Krause,** Discrete Analogues in Harmonic Analysis, 2022
223 **Volodymyr Nekrashevych,** Groups and Topological Dynamics, 2022
222 **Michael Artin,** Algebraic Geometry, 2022
221 **David Damanik and Jake Fillman,** One-Dimensional Ergodic Schrödinger Operators, 2022
220 **Isaac Goldbring,** Ultrafilters Throughout Mathematics, 2022
219 **Michael Joswig,** Essentials of Tropical Combinatorics, 2021
218 **Riccardo Benedetti,** Lectures on Differential Topology, 2021

For a complete list of titles in this series, visit the
AMS Bookstore at **www.ams.org/bookstore/gsmseries/**.